Functional Properties of Food Components

FOOD SCIENCE AND TECHNOLOGY

A SERIES OF MONOGRAPHS

Series Editors

A complete list of the books in this series appears at the end of the volume.

Functional Properties
of Food Components

Yeshajahu Pomeranz

Grain Marketing Research Laboratory
United States Department of Agriculture
Manhattan, Kansas

1985

ACADEMIC PRESS, INC.

(Harcourt Brace Jovanovich, Publishers)

Orlando   San Diego   New York   London
Toronto   Montreal   Sydney   Tokyo

The opinions in this book are those of the author and do not necessarily reflect the views and policies of the U.S. Department of Agriculture.

ACADEMIC PRESS, INC.
Orlando, Florida 32887

United Kingdom Edition published by
ACADEMIC PRESS, INC. (LONDON) LTD.
24/28 Oval Road, London NW1 7DX

Library of Congress Cataloging in Publication Data

Pomeranz, Y. (Yeshajahu), Date
    Functional properties of food components.

    (Food science and technology)
    Bibliography: p.
    Includes index.
    1. Food--Composition. I. Title. II. Series.
TX551.P59 1984   641   83-21434
ISBN 0-12-561280-X (alk. paper)

PRINTED IN THE UNITED STATES OF AMERICA

85 86 87 88   9 8 7 6 5 4 3 2 1

Contents

Preface

Many books on food chemistry have been published in recent years. Generally, they have dealt only indirectly with the role and function of specific components in foods. A review of those roles and functions is the subject of this book.

The four questions most commonly asked by workers in the biological sciences are *What? How much? Where?* and *What function?* In the food sciences, the first question—*What?*—relates to the identity of the components in the food system under investigation. Although much useful information can be gained from a number of sciences and technologies, the fundamental understanding of a food system comes from a knowledge of the biochemical identity of its components. *How much?* questions the quantity of components identified in the food system, recognizing that one page of sound, well-substantiated quantitative data is superior to hundreds of pages of loose talk. *Where?* is asked in order to localize a given component in plant or animal tissue or in processed food.

The search for answers to these three questions is common to both basic and applied researchers—to those interested in theoretical and practical aspects of a problem, to those who call themselves pure scientists and to those who do not. The two groups are differentiated by the answer they seek to the fourth question: *What function?* Here, the theoretical researcher is interested either in the basic physiological function(s) of a component in a plant or animal or in collecting physical, chemical, or physicochemical data that characterize a component or system. The applied researcher wants to know what the components can do (or be made to do) toward producing a useful product—food. This book reviews studies in order to answer the above four questions with regard to the functional properties of food components. It covers

1. WHAT in the biochemical make-up of components makes them "tick" in the production of desirable and acceptable foods?
2. WHY do those components/entities perform the way they do and, often, why do they fail to perform as expected?
3. WHICH functions continue to be elusive and require more searching and probing?

This book is intended for the serious undergraduate who has a background in the general biochemistry of natural materials but is interested in specific information on the function of those components in foods, for the food scientist or technologist who is familiar with food formulation and production, and for any other interested reader with an appropriate background, whether managerial or scientific.

I

The Components

1

Water

I. Water Content of Foods

The water content of foods varies widely. The relative abundance of water and other major constituents in some representative foods is summarized in Table 1.1. Fluid dairy products (whole milk, nonfat milk, and buttermilk) contain 87–91% water; various dry milk powders contain about 4% water. Cheeses can be classified according to their moisture content as (a) soft (55–80% moisture) (e.g., cottage, cream, impastata, neuchatel, and ricotta), (b) semisoft (45–55% moisture) (e.g., blue, brie, camembert, and mozzarella), (c) semihard (35–45% moisture) (e.g., brick, cheddar, edam, provolone, and swiss), and (d) hard (less than 35% moisture) (e.g., dry ricotta, gjetost, mysost, parmesan, and romano). The water content of butter is about 15%, of cream 60–70%, and of ice creams and sherbet approximately 65%. Pure oils and fats contain practically no water, but processed lipid-rich materials may contain substantial amounts of water (from about 15% in margarine and mayonnaise to 40% in salad dressings).

Some fresh fruits contain more than 90% water in the edible portion. Melons contain 92–94%, citrus fruits 86–89%, and various berries 81–90% water. Most raw tree and vine fruits contain 83–87% water; the water content of ripe guavas is 81%, of ripe olives 72–75%, and of avocados 65%. After commercial drying, fruits contain up to 25% water. Fresh fruit juices and nectars contain 85–93% water; the water content is lowered in concentrated or sweetened juice products.

Cereals are characterized by low moisture contents. Whole grains designed for

TABLE 1.1

Relative Abundance of Water and Other Major Constituents in Some Foods[a]

Material	Approximate % by weight (wet basis)				Approximate molar concentration (carbohydrate and protein as monosaccharide and amino acid units, respectively)			
	Water	Carbohydrate (total)	Protein	Fat	Water	Main carbohydrate component	Protein	Fat
Milk	87.0	4.8	3.4	3.7	48.3	0.27 (sugar)	0.24	0.044
Apple	84.1	14.0	0.3	0.1	46.7	0.68 (sugar)	0.02	0.001
Cod	79.2	—	18.0	0.9	44.0	—	1.24	0.011
Potato	75.8	22.9	2.1	0.1	42.1	1.12 (starch)	0.15	0.001
Egg	73.4	0.7	11.9	12.3	40.7	0.04 (sugar)	0.84	0.15
Beef (lean)	68.3	—	19.3	10.5	37.9	—	1.36	0.12
Bacon	40.9	—	13.1	44.6	22.7	—	0.92	0.53
Bread (white)	38.3	52.7	7.8	1.4	21.3	2.82 (starch)	0.55	0.017
Cheese (cheddar)	37.0	—	25.4	34.5	20.6	—	1.79	0.41
Jam	29.8	69.0	0.6	0.1	16.6	3.83 (sugar)	0.04	0.001
Honey	23.0	76.4	0.4	0.05	12.8	4.24 (sugar)	0.03	0.001
Sultana raisins	18.3	71.7	1.7	0.1	10.2	3.59 (sugar)	0.12	0.001
Macaroni	12.4	79.2	10.7	2.0	6.9	4.40 (sugar)	0.75	0.024
Dried egg	7.0	2.6	43.4	43.3	3.9	0.14 (sugar)	3.04	0.51
Dehydrated potato	6.5	80.3	6.6	0.3	3.6	3.15 (starch)	0.46	0.004
Dried milk (skimmed)	5.0	49.2	34.5	0.3	2.7	2.73 (sugar)	2.43	0.004
Almonds	4.7	18.6	20.5	53.5	2.6	0.79 (cellulose)	1.44	0.63
Boiled sweets	2.7	87.3	—	—	1.5	4.82 (sugar)	—	—

[a]From Duckworth, 1976.

long-term storage have 10–12% water. The water content of breakfast cereals is less than 4%, of macaroni 9%, and of milled grain products (flour, grits, semolina, germ) 10–13%. Among baked cereal products, pies are rich in water (43–59%); bread and rolls are intermediately moist) (35–40% and 28%, respectively); and crackers and pretzels are relatively dry (5–8%). Ripe raw nuts generally contain 3–5% water, or less, after roasting. Fresh unroasted chestnuts contain about 53% water.

The water content of meat and fish depends primarily on the fat content; it varies to a lesser degree with the age, source, and growth season of the animal or fish. The water content ranges from 50 to 70%, but some organs may contain up to 80% water. The water content of sausage varies widely. Water in poultry meats is between 50% (geese) and 75% (chicken). Fresh eggs have about 74% and dried eggs about 5% water.

White sugar (cane or beet), hard candy, and plain chocolate contain 1% or less water. In fruit jellies, jams, marmalades, and preserves up to 35%; in honey 20%; and in various syrups 20–40% is water. Sweet potatoes contain less water (69%) than white potatoes (78%). Radishes have most (93%) and parsnips least (79%) water among the common root vegetables. Among other vegetables, a wider range is found. Green lima beans have about 67%, and raw cucumbers over 96% water. Dry legumes contain 10–12% water, and the water content of commercially dried vegetables is 7–10% (preferably less than 8%).

II. Significance of Water

Water properties are related to food engineering operations and to changes of state, such as specific heat, latent heat of fusion, heat of vaporization, thermal conductivity and diffusion, and several heat and mass transfer properties. These properties are important in freezing, concentrating, and drying foods. Deteriorative changes during heat processing of foods are accelerated at elevated temperatures and slowed down as moisture is reduced. Properties relevant to biochemical stability of foods are related to the abilities of water to act as a solvent, to interact with other food components, and to act as a medium for diffusion and reaction of those components (Labuza, 1971). Such properties include surface tension, viscosity, dipole moment, dielectric constant, among others. The presence of various compounds in a solution or colloidal suspension imparts to the system certain colligative properties, which may affect and be expressed by decreases in vapor pressure, surface tension, and freezing point; increases in viscosity or boiling point; or creation of gradients of osmotic pressure.

Water determination is one of the most important and most widely used measurements in the processing and testing of foods. Since the amount of dry matter in a food is inversely related to the amount of water it contains, water content is of direct economic concern to the processor and the consumer. Of even greater significance, however, is the effect of water on the stability and quality of foods. For example, grain that contains too much water is subject to rapid deterioration from mold growth, heating, insect damage, and sprouting. Small differences in water content may be responsible for unexpected cases of spoilage in commercially stored grain.

Table 1.2 describes the equilibrium moisture contents of common grains, seeds, and animal feed ingredients at relative humidities of 65 to 90% and the fungi likely to be encountered. Table 1.3 lists the minimal limits of moisture percentage needed for growth of major storage fungi in cereals.

The rate of browning of dehydrated vegetables and fruits, or oxygen absorption by egg powders increases with an increase in water content. Water content determination is an important factor in many food industry problems, for example, in the evaluation of a food material's biochemical balance or in analyzing processing losses. We must know the water content (and sometimes its distribution) for optimal processing of foods, that is, in the milling of cereals, mixing of dough to desired consistency, and the production of bread with the best grain, texture, and freshness retention. Water content must be known to establish the nutritive value of a food, to express results of analytical determinations on a uniform basis, and to meet compositional standards or laws.

TABLE 1.2

Equilibrium Moisture Content of Common Grains, Seeds, and Feed Ingredients at 65–90% Relative Humidity and Probable Fungi[a]

Relative humidity (%)	Starchy cereal seeds,[b] defatted soybean and cottonseed meal, alfalfa pellets, most feeds[c]	Soybeans[c]	Sunflower, safflower seeds, peanuts, copra[c]	Fungi
65–70	13.0–14.0	12.0–13.0	5.0–6.0	*Aspergillus halophilicus*
70–75	14.0–15.0	13.0–14.0	6.0–7.0	*A. restrictus, A. glaucus, Wallemia sebi*
75–80	14.5–16.0	14.0–15.0	7.0–8.0	*A. candidus, A. ochraceus,* plus the above
80–85	16.0–18.0	15.0–17.0	8.0–10.0	*A. flavus, Penicillium,* plus the above
85–90	18.0–20.0	17.0–19.0	10.0–12.0	*Penicillium,* plus the above

[a]From Christensen and Sauer, 1982.

[b]Wheat, barley, oats, rye, rice, millet, maize, sorghum, and triticale.

[c]% wet weight. The figures are approximations; in practice variations up to ±1.0% can be expected.

TABLE 1.3

Lower Limits of Moisture (%) Needed for Major Storage Fungi Growth
in Cereals[a]

Fungi	Grains		
	Cereal grains	Soybeans	Sunflower, safflower, peanuts, and copra
Aspergillus restrictus	14.0–14.5	12.0–12.5	8.5–9.0
A. glaucus	14.5–15.0	12.5–13.0	9.0–9.5
A. candidus	15.5–16.0	14.5–15.0	9.0–9.5
A. ochraceus	15.5–15.0	14.5–15.0	9.0–9.5
A. flavus	17.0–18.0	18.0–18.5	10.0–10.5
Penicillium	16.5–20.0	17.0–20.0[b]	10.0–15.0[b]

[a]From Christensen and Sauer, 1982.
[b]Species vary.

III. Forms of Water in Foods

Water may occur in foods in at least three forms. A certain amount may be present as free water in the intergranular spaces and within the pores of the material. Such water retains its usual physical properties and serves as a dispersing agent for the colloidal substances and as a solvent for the crystallizing compounds. Part of the water is absorbed on the surface of the macromolecular colloids (starches, pectins, cellulose, and proteins). This water is closely associated with the absorbing macromolecules by forces of absorption attributable to van der Waals forces or to hydrogen-bond formation. Finally, some of the water is in a bound form—in combination with various substances, that is, as water of hydration. This classification, though convenient, is quite arbitrary. Attempts to determine quantitatively the amounts of various forms of water in foods have been unsuccessful. Unlike water in some inorganic compounds, which shows a distinct discontinuous sorption isotherm, water content in foods shows a continuous spectrum of the types of water binding.

A useful approach to the study of water adsorption by solid foods is by means of an isotherm. An isotherm is a curve that describes, at a specified moisture, the equilibrium relation of the amount of water sorbed by the food components, and vapor pressure or relative humidity. Depending on whether water is given off or taken up in approaching the equilibrium between vapor pressure and moisture content, a desorption or an adsorption isotherm results. The isotherm is an S-shaped curve for wheat flour (see Fig. 1.1) (Hlynka and Robinson, 1954). In the low humidity range the isotherm is concave in relation to the humidity axis. In the intermediate range, the isotherm has a region of inflection that is basically

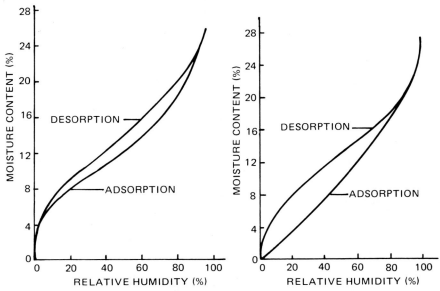

Fig. 1.1 (Left) Isotherms for wheat flour at 25°C. (Right) Isotherms for wheat at 25°C. (From Hlynka and Robinson, 1954.)

linear. In the high humidity range it is concave in relation to the moisture content axis. The lower portion of the isotherm represents the adsorption of the first layer of water vapor onto the surface of the adsorbing food. The region of inflection represents the deposition of a second layer of water molecules and the final curved portion represents the adsorption of additional layers. In the initial portion of the isotherm, the water pressure–moisture content relation is governed by the energy of binding between the water molecules and the adsorbing surface. That binding energy depends on the physical structure of the surface, its composition, and the properties of the water. In the intermediate portion, water molecules are deposited on water molecules already present in the first layer and to a small extent on the nonpolar sites. The energy is mainly that of condensation of water; the adsorption mainly depends on water vapor pressure. In the high moisture range, the vapor pressure is governed mainly by the second layer, which covers the surface. The addition of successive layers is basically the result of a capillary condensation. The water in each new layer has similar properties to water in the preceding layer. In this high moisture range, water sorption increases rapidly but the vapor pressure is affected little.

The adsorption and desorption isotherms for a given food are not always the same. In many foods the desorption isotherm is displaced to the left of the adsorption isotherm. This is known as hysteresis effect (Fig. 1.1).

Because the terms free, absorbed, and bound are relative, and because the true

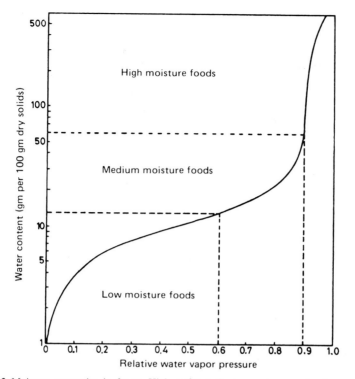

Fig. 1.2 Moisture categories in foods. High moisture foods, most or much of water usually relatively free of restrictive interaction with the solids; medium moisture foods, most, if not all, of the water physicochemically restricted by interaction with the solids; low moisture foods, most, if not all, of the water severely restricted by interaction with the solids. (From Duckworth, 1976.)

water content is not known, the conditions selected for water determination are arbitrary. The classifications just described, however, have been used by Duckworth (1976) to divide foods into three main categories according to the proportion of water they contain. The categories are low, medium, and high moisture foods. General limits for these categories are indicated in Fig. 1.2 wherein the water content–water vapor pressure relations are represented in the form of a hypothetical water sorption isotherm.

As mentioned before, many aspects of food quality can be influenced by the water content, mainly through the effect of water on the chemical and physical properties of the food. The chemical roles of water in foods are listed in Table 1.4. In the low moisture range, for example, small amoints of water in chocolate have a marked effect on the viscosity of the melted chocolate. This, presumably, is due to the increased friction between the solid particles resulting from the absorption at the crystal surfaces of water molecules.

TABLE 1.4

Chemical Roles of Water in Foods[a]

Role	Moisture range(s)	Mechanisms of effect	Quality attribute(s) affected
Solvent	All, except very low	Solution	All
Reaction medium	All, except very low	Facilitation of chemical change	All, especially in relation to keeping quality
Mobilizer of reactants	Low	Facilitation of movement of reactant molecules over hydrated surfaces	Keeping quality of dehydrated foods
Reactant	All	Hydrolysis of lipids, proteins, polysaccharides, glycosides, and so on	Especially flavor, but also texture and particularly in relation to keeping quality
Antioxidant	Low	Hydration of trace metal catalysts	
	Low	Hydrogen-bonding to hydroperoxides	
	Low	Promotion of free radical recombinations of alternative reactions	Flavor, color, texture, and nutritive value, especially on storage
	Low	Reduction of metal catalyst concentrations due to precipitation as hydroxides	
	Low	Hydrogen-bonding to functional groups of proteins and polysaccharides	Texture, ? color
Prooxidant	Medium	Reduction in viscosity increasing mobility of reactants and catalysts	
	Medium	Dissolution of precipitated catalysts	Flavor, color, texture, and nutritive value especially on storage
	Medium	Swelling of solid matrices exposing new catalytic surfaces and oxidizable groups	
	Medium and high	Forms the main source of free radicals in irradiated foods	Especially flavor and color after irradiation
Structural (intramolecular)	All	Maintenance of integrity of protein molecules	Texture and other attributes influenced by enzymes
Structural (intermolecular)	Low	Hydrogen-bonding to potential cross-linking groups of structural macromolecules	Texture (of dehydrated material on reconstitution)
	Low	Hydrogen-bonding to surface groups on solid particles, increasing friction between them	Viscosity in chocolate

TABLE 1.4 (*Continued*)

Role	Moisture range(s)	Mechanisms of effect	Quality attribute(s) affected
Structural (intermolecular)	Medium	Molecular interactions influencing lipid-binding to flour proteins	Rheological properties of doughs
	Medium and high	Influence on the conformation and interaction of gel-forming polysaccharides and proteins	Textural properties of gels
	Medium and high	Influence on the structure of emulsions through interactions with surface-active lipids	Rheological properties of emulsions

*a*From Duckworth, 1976.

In the medium moisture range the effects of water in wheat flour and dough exemplify a wide range of binding and of effects on physical, chemical, and end-use (functional) properties. The molecules of water in flour are considered to be associated with specific chemical groups in starch, proteins, gums, and other constituents. Water is especially important in hydrophobic bonding between proteins and lipids. The relations between water activity (a_w) or relative water vapor pressure and moisture content (MC) of wheat flour, as in most other natural high polymers, are characterized by a sigmoid adsorption isotherm (Fig. 1.1). Water is tightly bound as a monomolecular layer in a region of MC = 0–6.5% and a_w = 0.2, fairly strongly bound as a second monolayer in a region of MC = 6.5–14% and a_w = 0.7–0.95, and free in the true liquid phase and separate from the flour particles only in the high moisture region of MC > 23%, a_w > 0.95.

When flours are wetted and doughs are mixed, lipids extractable from flour with nonpolar solvents become bound. In fact, about one-half to two-thirds of the free lipids in wheat flour (some free nonpolar and practically all free polar components) become bound in dough or gluten formation. During dry-mixing of flour defatted with petroleum ether, for example, almost no added nonpolar flour lipids are bound, but substantial amounts of free polar flour lipids are bound in flour with an MC as low as 4.4%; the binding increases with higher moisture content.

The critical moisture level for lipid binding is between 20 and 30%; in a work-free system an approximately equal amount of lipid is bound whether a minimum (30%) or an excess of water is added. Extractable free lipids decrease steadily as flour moisture increases between 20 and 40%. Over this moisture range, binding

of polar lipids is far more extensive than binding of the neutral, mainly triglyceride (TG), lipids. While free neutral lipids fall from 55 to 40% of the total lipids,
very little free polar lipids remain free in flour of 40% moisture. It has been
concluded that while binding of the polar lipids generally may be explained on
the basis of electrostatic interaction with the charged surface of hydrated proteins, the forces involved in the binding of neutral TG are less clearly defined.
Neither the presence of polyunsaturated free lipids nor a proportion of free polar
lipids is required for neutral lipids to be bound during work-free wetting of the
flour (Daniels, 1975).

   All the preceding effects of water and lipid binding contribute to, or govern,
the rheological properties of doughs and the keeping quality of baked goods. In
gels, mobility of water may have a determining effect on processing, textural,
nutritional, keeping, and consumer acceptance attributes of foods. The next three
sections describe the role of water in bread baking, quality of snack foods, and
water-holding capacity of meat.

A. Water in Flour, Dough, and Bread

   Water comprises about 40% of standard bread dough and over 35% of baked
bread. Consequently, next to flour water is the most abundant component of
bread (Bushuk and Hlynka, 1964). Water constitutes about 14% of flour, which
is the amount in equilibrium with a relative humidity of about 70%. Flour stored
at this water level keeps well. Water in flour is hydrogen bonded to the hydroxyl
group and to oxygen in starch, and to the peptide linkage of proteins. Most amino
acid groups are capable of interacting with water. Pentosan gums are a minor
component (about 1.0 to 1.5% of wheat flour) that binds relatively large amounts
of water.

   The physical modification of flour through the process of agglomeration is an
interesting commercial development. The wetting properties of such flour (called
"instantized" or agglomerated) are superior to those of regular flour. Agglomeration of flour began in 1915 when United States Patent no. 1,555,977 was issued
to T. T. Vernon for a "Method of Treating Flour and Product Thereof." Since
the issue of the patent, many processes have been developed for the instantizing
of food products. Some of the processes involve agglomeration of particles with
a fluid to form clusters that are then spray-dried to produce a free-flowing, dust-
free product, of uniform bulk density, and of improved wettability and dispersibility. In regular flour, fine particles (about 5 μm) cannot overcome the normal
surface tension of water and do not wet uniformly. These particles form lumps
that are sticky on the outside and dry on the inside. If the fine particles are
removed or are agglomerated to large particles, the flour disperses readily in
aqueous systems.

About 46% of the water in dough is associated with the starch, 31% with the protein, and 23% with the pentosan gums. The water absorption of a wheat flour dough is governed, in practice, by the protein content and quality and by the extent to which the starch is damaged mechanically (the greater the damage, the greater the absorption). Adding water while mixing the dough assures adequate distribution of the components and proper development of the dough. Hydration of the flour particles (which have a total surface area of about 235 m^2/g) is rapid. Adequate water absorption and mixing are essential for proper dough development and the production of good bread. When water levels are too low, dough is stiff and lacks cohesion. At excessively high water levels, a batter or flour suspension results and proper dough development cannot take place.

Water absorption and dough development are influenced by various ingredients used in bread making. As the salt (sodium chloride) concentration increases, dough development time increases and water absorption decreases; presumably from the lowered water-holding capacity of the proteins. Fats lessen the hydration capacity of dough flour. In the presence of emulsifiers, however, water absorption increases, especially in cake batters.

Water plays an important role in the major changes that take place during the baking of dough: starch gelatinization, protein denaturation, yeast and enzyme inactivation, and flavor and color formation. Water content and its distribution also govern the shelf-life of bread, which is influenced by incidence of microbial damage, softness of the crumb, crispness of the crust, crumb hardening, crumbliness, and many other changes associated with overall staling and lowered consumer acceptance.

B. Water in Snack Foods

Crispness is a crucial textural characteristic of fresh dry cereal and starch-based snack food products. Loss of crispness due to adsorption of moisture is a major cause of snack food rejection by consumers. Water affects the texture of dry snack foods by plasticizing and softening the starch–protein matrix, which alters the mechanical strength of the product (Katz and Labuza, 1981). For example, the potato chip industry considers chips with more than 3% moisture unsalable; in the United Kingdom crackers are unacceptable when the moisture content exceeds 3.5%. The critical water activity (that is, the a_w at which the product becomes unacceptable) is about 0.40 for potato and corn chips. Katz and Labuza (1981) report values in the 0.35–0.50 a_w range for potato chips, popcorn, puffed corn curls, and saltines. This is the same a_w range in which amorphous to crystalline transformations take place in simple sugar systems and in which mobilization of soluble food constituents begins. Inasmuch as no major changes occur at specific a_c values for specific foods, it is postulated that the changes should not be considered as all-or-none phenomena but rather as gradual changes in intermolecular bonding that occurs when moisture content increases above the monolayer.

C. Water-Holding Capacity of Meat

Water-holding capacity is of great importance for the quality of meat and meat products, particularly sausage and canned ham. Weight losses during storage, cooking, freezing, or thawing that are related to water binding have direct economic significance and govern meat quality changes.

The cross-striated muscles of meat animals contain about 75% water (Hamm, 1975). Water in muscle is bound primarily by the myofibrillar proteins. Only 4–5% of the total water is tightly bound on the surface of the protein molecules as hydration water. This water is little influenced by changes in the structure and charges of the muscle proteins, including those that take place under rigor mortis or during cooking. On the other hand, changes in water-holding capacity that occur during storage and processing of meat depend on the degree of immobilization of the nontightly bound water (about 95% of the total) within the microstructure of the tissue. In this category, a broad spectrum of water-holding capacity is apparent. The amount of water immobilized within the tissue depends to a great extent on the spatial molecular arrangement of the myofibrillar filament proteins, myosin and actin. To a limited extent, the immobilized water is affected by tropomyosin and by the sarcoplasmic reticulum, which promotes gelation of the extractable proteins (see Chapter 6).

Increasing electrostatic repulsion between similarly charged groups or weakening hydrogen bonds decreases the cohesion between the filament proteins of muscle. Decreased cohesion produces an enlarged network through swelling and increases water immobilization. If the process proceeds beyond a certain point, the intermolecular cohesion becomes so diminished that the network collapses and the gel is transformed into a colloidal solution of myofibrillar proteins. Increasing the electrostatic attraction or strengthening the interlinking bonds reinforces the network and some of the water is no longer immobilized and becomes free.

Raising the protein net electrical charge by adding an acid or alkali loosens the lattice structure and increases the amount of immobilized water. Water-holding capacity is minimum of pH 5.0, the isoelectric point (IP) of the main structural muscle protein, actomyosin. At the IP the net protein charge is minimum. Increasing the pH of meat by an antemortem injection of adrenalin, which lowers the muscle glycogen content, raises the water-holding capacity of meat. The addition of sodium chloride increases the water-holding capacity at pH above IP and decreases it at pH below IP.

Shortly after slaughter, a considerable reduction in water-holding capacity occurs, which has been traced to a tightening of the actin–myosin network that occurs after the breakdown of adenosine triphosphate (ATP). Later, the tightened network is relaxed somewhat as a result of proteolytic action accompanied by an increase in soluble myofibrillar proteins and by an increase in water retention.

The water-retention capacity of meat is related to tenderness, juiciness, and color and to binding of chunks in the production of meat rolls. Tightening of the myofibrillar network by heat denaturation of proteins reduces the water-holding capacity and impairs juiciness. Mincing of meat in the production of sausages destroys the sarcolemma and considerably intensifies the swelling and water-holding capacity of the myofibrillar system (Hamm, 1975). Most of the factors that increase the water-holding capacity of meat also contribute to the distribution of fat in sausages. Processing of prerigor meat makes possible the production of excellent sausages because processed meat containing ATP is high in water-holding capacity. For example, salting ground beef shortly after slaughter, and before ATP breakdown, avoids a large water-holding loss. In addition, the onset of rigor mortis and associated water-holding loss is prevented in prerigor salted meat. Freezing prerigor salted ground meat preserves for several months its high water-holding capacity. Mincing the frozen material allows the production of high quality sausages from fresh prerigor beef (warm meat).

A series of papers was presented in England in 1982 at a symposium on water in meat and meat products. Offer and Trinick (1983) put forth a unifying hypothesis for water-holding capacity in meat. They postulated that gains or losses of water in meat are due to swelling or shrinking of the myofibrils caused by expansion or shrinking of the filament lattice. Myofibrils swell rapidly to about twice their original volume in salt solutions equal to those used in meat processing. Pyrophosphate reduces the concentration of sodium chloride required for maximum swelling. The swelling force involves electrostatic repulsion but the expansion of the filament lattice requires rupture of transverse structural constraints (cross-bridges, the M-line, or the Z-line) within the lattice at a critical salt concentration. When muscle fibers are heated to 60°C, they shrink substantially, probably due to shrinkage of connective tissue surrounding the fibers. Active shrinking of myofibrils contributes significantly to water losses during cooking.

According to Nair and James (1983), the effect of freezing rate on the drip loss from frozen beef is much smaller than variations from prefrozen treatment. Jeffrey (1983) discussed principles of water-holding capacity applied to meat technology as differentiated from meat's capacity to hold its own water. The effects of adding salt, polyphosphate levels, work input, and mechanical action were reviewed. Salt alters the position of the IP and increases interfibrillar volumes. Salt acts synergistically with polyphosphate to increase frozen meat yields. Mechanical action brings water, salt, and polyphosphate into intimate contact with large meat surfaces, accelerates salt and polyphosphate effects, and forms a sticky exudate of protein that binds together meat pieces and water. According to Lillford *et al.* (1983), thermal denaturation of proteins that takes place during the cooking of meat can be assigned particular temperatures and heating rates. Meat heated for a fixed time to increasing center temperatures showed progressive shortening of relaxation times of water in the tissue. Cooking meat showed a large shift of water distribution and a substantial decrease in fiber volume from reduction in fiber cross-section. In the presence and in the absence of salt, retained water is held by changes in fiber dimension. Salt-incurred swelling of some muscles depended on postmortem conditioning. Water-holding capacity was affected by salt permeation and structural damage during fibril size reduction.

IV. Water and Keeping Quality

According to Labuza (1971), free water in foods exerts close to normal water pressure. The degree of lowering of water pressure in this free water is related to Raoult's law where

$$X_s = \frac{n_{solute}}{n_{solute} + n_{H_2O}} \qquad X_{H_2O} = 1 - X_s$$

and

$$a_{H_2O} = \frac{P}{P_0} = \gamma X_{H_2O} = \frac{\%RH}{100}$$

where X_s = mole fraction of solute, X_{H_2O} = mole fraction of water, n_{solute} = moles solute, n_{H_2O} = moles water, a_{H_2O} = water activity, P = vapor pressure of water exerted by solution at temperature (T), P_0 = vapor pressure of pure water at T, γ = activity coefficient (1 for ideal solutions), $\%RH$ = equilibrium relative humidity exerted by the solution or food.

The equilibrium relative humidity or a_w is close to 100% in foods containing about 50% (or more) moisture, on a wet basis. Preservation of foods can be accomplished either by slowing down (as in refrigeration) or by destroying microbial and enzymatic activity (as in sterilization). Alternatively, water can be removed (by dehydration) to a safe level or its activity can be lowered (by use of salts or sugars). Similarly, drying reduces but does not stop enzymatic, especially lipolytic, activity. Theoretically, to reduce a_w from 1.0–0.9 to 0.8–0.75, molality of solutes in an ideal solution must be increased from 0–6.17 to 13.9–18.5, respectively (Labuza, 1971). In practice, most compounds obey Raoult's law only to a certain degree. Equilibrium relative humidity or the a_w of a food remains close to the maximum value of 100% until the MC is lowered to well below 50%.

Bound water has been described as water in an aqueous system that is unavailable as a solvent or reactant and does not freeze. The amount of water frozen out of a food system is a function of temperature. Even at $-20°C$ about 0.3 gm unfrozen water per gm of dry nonfat solids may remain. This unfrozen water has some solvent and reaction capabilities and its existence can lead to loss of food quality in storage. For almost complete freezing of water, temperatures of $-60°C$ are required. The reason for this extremely low temperature requirement stems from tight binding of water by polar groups. Moreover, as part of the water freezes, the remaining solutes are increasingly concentrated and the freezing point is depressed. The high salt concentration in liquid water of a frozen food can exert very detrimental effects on the texture of fish, meat, and poultry. Even though reaction rates decrease exponentially with decreases in temperature, the high increase in solute concentration can raise actual reaction rates. Complex deleterious enzyme reactions can occur in frozen foods unless they are blanched prior to freezing. Similarly, enzyme reactions can occur in unblanched dehydrated foods with low a_w.

Water, through its solvent properties, can cause caking in dehydrated foods. During dehydration some food components (namely, carbohydrates) precipitate out of the solution in an amorphous glassy state in which they are highly mobile and exhibit high water absorbency. This action is important in sugar crystallization, production of milk powders, storage and hygroscopicity of soluble teas and coffees, and retention of volatiles during dehydration operations.

As a_w increases, water becomes more available as a solvent and medium for reactions; rates of enzymatic and microbiological degradation also increase. Approximate lower limits of a_w for microbial growth are 0.91 for bacteria, 0.88 for yeasts, 0.80 for molds, 0.75 for halophilic bacteria, 0.65 for xerophilic fungi, and 0.60 for osmophilic yeasts. The increased availability of water accelerates nonenzymatic "browning" reactions and loss of nutritional value. The browning rate is usually maximum in intermediate moisture foods in which a_w is about 0.7. Water activity and the growth of microorganisms in foods are summarized in Table 1.5. Minimum water activity for the growth and toxicity of some selected microorganisms is given in Table 1.6.

The stability of foods and their resistance to oxidation is a function of moisture content (Labuza, 1971). The main mechanisms of water interaction in this respect are

1. Water–hydrogen bonds are formed with peroxides produced during a free radical reaction. Water that migrates to the lipid surface ties up the peroxides and slows the rate of peroxide initiation. An antioxidant effect increases until the lipid interface is saturated with water, after which point no additional effect occurs.
2. As trace metals are hydrated, they lose their capacity to catalyze the rate of oxidative initiation. Alternatively, catalysts are rendered ineffective by formation of insoluble hydroxides.
3. As water content increases, initiation, production, and stability of free radical decreases.

TABLE 1.5

Water Activity and Growth of Microorganisms in Food[a]

Range of a_w	Microorganisms generally inhibited by lowest a_w in this range	Foods generally within this range
1.00–0.95	*Pseudomonas, Escherichia, Proteus, Shigella, Klebsiella, Bacillus, Clostridium perfringens,* some yeasts	Highly perishable (fresh) foods and canned fruits, vegetables, meat, fish, and milk; cooked sausages and breads; foods containing up to approximately 40% (w/w) sucrose or 7% sodium chloride
0.95–0.91	*Salmonella, Vibrio parahaemolyticus, Clostridium botulinum, Serratia, Lactobacillus, Pediococcus,* some molds, yeasts (*Rhodotorula, Pichia*)	Some cheeses (cheddar, swiss, muenster, provolone), cured meat (ham), some fruit juice concentrates; foods containing 55% (w/w) sucrose or 12% sodium chloride
0.91–0.87	Many yeasts (*Candida, Torulopsis, Hansenula*), *Micrococcus*	Fermented sausage (salami), sponge cakes, dry cheeses, margarine; foods containing 65% (w/w) sucrose (saturated) or 15% sodium chloride
0.87–0.80	Most molds (mycotoxigenic penicillia), *Staphylococcus aureus,* most *Saccharomyces* spp., *Debaryomyces*	Most fruit juice concentrates, sweetened condensed milk, chocolate syrup, maple and fruit syrups; flour, rice, pulses containing 15–17% moisture; fruit cake; country-style ham, fondants, high-ratio cakes
0.80–0.75	Most halophilic bacteria, mycotoxigenic aspergilli	Jam, marmalade, marzipan, glacé fruits, some marshmallows
0.75–0.65	Xerophilic molds (*Aspergillus chevalieri, A. candidus, Wallemia sebi*), *Saccharomyces bisporus*	Rolled oats containing approximately 10% moisture, grained nougats, fudge, marshmallows, jelly, molasses, raw cane sugar, some dried fruits, nuts
0.65–0.60	Osmophilic yeasts (*Saccharomyces rouxii*), few molds (*Aspergillus echinulatus, Monascus bisporus*)	Dried fruits containing 15–20% moisture; some toffees and caramels; honey
0.50	No microbial proliferation	Pasta containing approximately 12% moisture; spices containing approximately 10% moisture
0.40	No microbial proliferation	Whole egg powder containing approximately 5% moisture
0.30	No microbial proliferation	Cookies, crackers, bread crusts, etc. containing 3–5% moisture
0.20	No microbial proliferation	Whole milk powder containing 2–3% moisture; dried vegetables containing approximately 5% moisture; corn flakes containing approximately 5% moisture; fruit cake; country-style cookies, crackers

[a]From Beuchat, 1981.

TABLE 1.6

Minimal a_w for Some Microorganisms of Significance to Public Health[a]

	Minimal a_w	
Microorganism	*Growth*[b]	*Toxin production*
Bacteria		
Staphylococcus aureus	0.86	
	0.86	
		<0.90 (enterotoxin A)
		0.87 (enterotoxin A)
		0.97 (enterotoxin B)
Salmonella spp.	0.93	
	0.94–0.95	
	0.92	
Vibrio parahaemolyticus	0.94	
	0.94	
Clostridium botulinum	0.93(A)	
	0.95(A)	0.95(A)
		0.94(A)
	0.93(B)	
	0.94(B)	0.94(B)
		0.94(B)
	0.95(E)	
	0.97(E)	0.97(E)
	0.95(E)	
	0.97(E)	
Clostridium perfringens	0.93–0.95	
	0.95	
Bacillus cereus	0.95	
	0.93	
	0.95	
Molds		
Aspergillus flavus	0.78	
		0.84 (aflatoxin)
	0.80	0.83–0.87
A. parasiticus	0.82	0.87 (aflatoxin)
A. ochraceus		0.85 (ochratoxin)
	0.83	0.83–0.87
	0.77	
Penicillium cyclopium	0.81	0.87–0.90 (ochratoxin)
	0.82	
	0.83	
	0.85	
P. viridicatum	0.83	0.83–0.86 (ochratoxin)
Aspergillus ochraceus	0.81	0.88 (penicillic acid)
		0.80
	0.76	0.81

(continued)

TABLE 1.6 (*Continued*)

Minimal a_w for Some Microorganisms of Significance to Public Health[a]

	Minimal a_w	
Microorganism	Growth[b]	Toxin production
Molds		
Penicillium cyclopium	0.87	0.97 (penicillic acid)
	0.82	
P. martensii	0.83	0.99 (penicillic acid)
	0.79	
P. islandicum	0.83	
P. patulum	0.83–0.85	0.95 (patulin)
	0.81	
	0.83	
		0.85
P. expansum	0.83–0.85	0.99 (patulin)
	0.83	
	0.83	
Aspergillus clavatus	0.85	0.99 (patulin)
Byssochlamys nivea	0.84	
Trichothecium roseum	0.90	
	0.90	
Stachybotrys atra	0.94	0.94 (stachybotryn)

[a]From Beuchat, 1981.
[b](A), (B), and (E), types of enterotoxins and *Clostridium*.

As water activity increases to 0.50 the antioxidant effect is progressive. The increase in a_w is accompanied by an increase in effectiveness of water-soluble antioxidants, such as metal-chelating ethylene diamine tetraacetic acid (EDTA) and radical-reacting butylated hydroxyanisole (BHA). The enhanced effectiveness apparently results from higher mobility of the reactants in the surface water (Labuza, 1971). As the a_w reaches the intermediate moisture range, oxidation of lipids increases again because the higher mobility and solubility enhance movement of trace metals to oxidation sites where they overcome the antioxidant effects of water. The complex relationship between a_w activity and relative reaction rates of deteriorative functions of foods is summarized in Fig. 1.3. In the intermediate moisture range, several reactions may affect or govern deterioration. Their simultaneous control presents a major challenge for the food industry.

According to Pothast (1978), most studies concerning the shelf life of stored foods have been conducted using model systems. Results from model systems, however, are not always applicable to actual foods, although it is true that biochemical reactions in foods depend more on a_w than on absolute water content. Enzymatic reactions in particular depend on water activity. For chemical

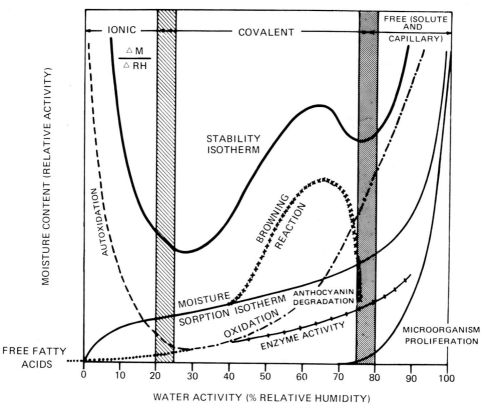

Fig. 1.3 Diagramatic representation of the influence of water activity on chemical, enzymatic, and microbiological changes and on overall stability and moisture sorption properties of food products. (From Rockland and Nishi, 1980.)

and biochemical changes in foods at low residual water levels, however, the absolute water content may be more important than a_w. Most biochemical reactions are accelerated by increasing water activity, but there are others for which the speed of reaction is greatest at low levels of relative humidity. For instance, auto-oxidative processes are particularly favored by very low a_w values (about 0.2). Nonenzymatic browning reactions and enzymatic catalyses, on the other hand, exert appreciable influences on the storage stability of dried products only in the range above 0.2 a_w.

In addition to its activity as a solvent and reactant, water can participate directly in hydrolytic cleavage (i.e., in the formation of invert sugar from sucrose under acidic conditions), the products of which can participate in nonenzymatic browning. The latter may be responsible for objectionable colors, off flavors, and reduced nutritive and functional properties of stored food.

V. Intermediate Moisture Foods

Preservation of foods at low temperatures requires a sophisticated system of refrigerated distribution and cold storage. In dehydration, the structure, texture, and overall quality (even after blanching to inactivate enzymes) are often impaired. In addition, prolonged storage of dehydrated foods may produce objectionable nonenzymatic browning and oxidative deterioration.

In intermediate moisture foods (IMF), water content is lowered to a level that prevents microbial damage. Most of the bound water is retained in the food so that the the texture of the original food is essentially preserved. The addition of humectants permits relatively high moisture levels in ready-to-eat foods. Some other tools available for engineering such foods are judicious adjustments of pH and the addition of antimicrobial agents. Typical examples of ''old time'' IMF are jams, cheeses, and certain confections. The technology of IMF has been expanded to include foods for space programs and pet foods. These programs require new production techniques such as the following:

1. Moist-infusion. Solid food pieces are soaked or cooked in a solution to produce a product of desired water activity.
2. Dry-infusion. Solid food pieces are dehydrated and then infused by soaking in a solution that contains the desired osmotic agents.
3. Blending. The components are combined by blending, cooking, or extrusion to yield a product of desired water activity.

The means to attain those objectives include (Labuza et al., 1970)

1. Reduction of water activity by dehydration or evaporation
2. Microbiological stabilization by thermal or chemical means
3. Inhibition of enzymatic activity by blanching
4. Control of physical and/or chemical deterioration by adding antioxidants, chelating agents, emulsifiers, or stabilizers
5. Enrichment by the addition of nutrients

Water activity in IMF can be reduced by adding polyhydric alcohols, sugars, and/or salt. The main antimycotic agents are propylene glycol and sorbates. Several physical and chemical treatments are available to improve organoleptic properties and consumer acceptance of stored foods.

Swift and Company has developed for the United States Department of Defense and the National Aeronautics and Space Administration several military and space rations. The products, made by the dry-infusion method of freeze-dehydrated and blended solid ingredients, include ready-to-eat, bite-size cubes of barbecued beef or chicken, roast beef or pork, chicken à la king, stew or corned beef, chili with beans, sausage, and ham. The products contain about 14–25% water and 2.5–4.5% salt; their a_w is 0.61–0.79 and their pH ranges from 4.9 to

5.9. Including about 7% glycerol, about 5% gelatin, and about 3% sorbitol in the infusion solution and about 10% fat in the finished product provides adequate physical binding, texture, and shelf life.

Because IMF offer a combination of extended shelf life, convenience, ease of nutritional engineering, and safety, the potential for such foods is large. Although widely accepted by the pet food industry and shown to be useful for specialized uses (for example, space feeding and clinical nutrition), IMF acceptance by the general public has been very slow. Consumer reception requires production of IMF with much improved organoleptic acceptability; new formulations containing humectants; better antimicrobial agents; control of storage-induced organoleptic deterioration, mainly from nonenzymatic browning; and development of economically feasible processes for large-scale production of IMF.

It is questionable whether the IMF developed thus far answer the needs of the sophisticated and critical consumer, who demands novelty, convenience, and taste satisfaction. The consumer asks for retention of the intrinsic qualities of fresh food, a reduction (not an increase) in types and quantities of additives, convenience without sacrificing quality, improved nutritional standards for fresh foods, and a demonstrated advantage of new foods over high-quality traditional foods already produced and preserved by the best available science and technology.

REFERENCES

Beuchat, L. R.(1981). Microbial stability as affected by water activity. *Cereal Foods World* **26,** 345–349.

Bushuk, W., and Hlynka, I. (1964). Water as a constituent of flour, dough, and bread. *Baker's Dig.* **38**(6), 43–46, 92.

Christensen, C. M., and Sauer, D. B. (1982). Microflora. *In* "Storage of Cereal Grains and Their Products" (C. M. Christensen, ed.), 3rd ed., Chapter 7, pp. 219–240. Am. Assoc. Cereal Chem., St. Paul, Minnesota.

Daniels, N. W. R.(1975). Some effects of water in wheat flour doughs. *In* "Water Relations of Foods" (R. B. Duckworth, ed.), pp. 573–586. Academic Press, New York.

Duckworth, R. B. (1976). The roles of water in foods. Factors influenced by water in foods. *Chem. Ind. (London)*, pp. 1039–1042.

Hamm, R. (1975). Water-holding capacity of meat. *In* "Meat" (D. A. A. Cole and R. A. Lawrie, eds., pp. 321–338. Butterworth, London.

Hlynka, I., and Robinson, A. D. (1954). Moisture and its measurement. *In* "Storage of Cereal Grains and Their Products" (J. A. Anderson and A. W. Alcock, eds.), 1st ed., Chapter 1, pp. 1–45. Am. Assoc. Cereal Chem., St. Paul, Minnesota.

Jeffrey, A. B. (1983). Principles of water binding applied to meat technology. *J. Sci. Food Agric.* **34,** 1020–1021.

Katz, E. E., and Labuza, T. P. (1981). Effect of water activity on the sensory crispness and mechanical deformation of snack food products. *J. Food Sci.* **46,** 403–409.

Labuza, T. P. (1971). Properties of water as related to the keeping quality of foods. *Proc. SOS/70, Int. Congr. Food Sci. Technol. 3rd, 1970* pp. 618–635.

Labuza, T. P., Tannenbaum, S. R., and Karel, M. (1970). Water content and stability of low-moisture and intermediate-moisture foods. *Food Technol.* **24**, 543, 544, 546, 548, 550.

Lillford, P. J., Regenstein, J. M., and Wilding, P. (1983). Process effects of structure and water binding in meat. *J. Sci. Food Agric.* **34**, 1021–1022.

Nair, C., and James, S. J. (1983). The effects of freezing rate on the drip loss from frozen beef. *J. Sci. Food Agric.* **34**, 1020.

Offer, G., and Trinick, J. (1983). A unifying hypothesis for the mechanism of changes in the water-holding capacity of meat. *J. Sci. Food Agric.* **34**, 1018–1019.

Pothast, K. (1978). Influence of water activity on enzymic activity in biological systems. *In* "Dry Biological Systems" (J. H. Crowe and J. S. Clegg, eds.), pp. 323–342. Academic Press, New York.

Rockland, L. B., and Nishi, S. K. (1980). Influence of water activity on food product quality and stability. *Food Technol.* **34**(4), 42, 44–46, 48–51, 59.

2

Carbohydrates: Starch*

I. Starch Composition, Occurrence, and Uses

After water, carbohydrates are the most abundant and widely distributed food component. Carbohydrates include, first, monosaccharides (polyhydroxy aldehydes or ketones), among which are 5-carbon compounds, such as xylose or arabinose, and 6-carbon compounds, such as glucose and fructose. The second group includes oligosaccharides, in which a hydroxyl group of one monosaccharide has condensed with the reducing group of another monosaccharide. If

---

*This chapter is based on a condensed, updated, and modified version of unpublished lectures presented by the late T. J. Schoch. The material in those lectures was published in part by Schoch in 1952 and 1961, by Schoch and Elder in 1955, and by Kite *et al.* in 1963 (reproduced by permission of Corn Products Co., Argo, Illinois).

two sugar units are joined in this manner, a disaccharide results; a linear array of three to eight monosaccharides joined by glycosidic linkages gives oligosaccharides. The third group of carbohydrates includes polysaccharides, which may be separated into two broad groups, the so-called structural polysaccharides (i.e., cellulose, hemicellulose, lignin), which constitute or are part of rigid, mechanical structures in plants, and nutrient polysaccharides (i.e., starch, glycogen), which are metabolic reserves in plants and animals.

The most important of the known monosaccharides is D-glucose, found in the blood of animals, the sap of plants, and many fruit juices. It also forms the structural unit of the most common polysaccharides. It is produced in green plants, though its conversion into starch, cellulose, and other polysaccharides may prevent its detection. Fructose is found in fruit juices and honey. An abundant source of both glucose and fructose is the disaccharide sucrose, which can be hydrolyzed to yield one mole of each of them. Another disaccharide is the milk sugar, lactose; it constitutes about 5% of cow's milk and about 6% of human milk, and on hydrolysis yields glucose and galactose. Maltose, another disaccharide in which one molecule of glucose is joined through an α-1,4-glycosidic linkage to a second molecule of glucose. This disaccharide is formed abundantly by amylolytic breakdown of polysaccharides during malting or digestion in the animal body. Another disaccharide is cellobiose, a degradation product of cellulose resembling maltose except that the two glucose units are joined through a β-glycosidic linkage. The most important freely occurring trisaccharide is the sugar beet raffinose, in which galactose is linked to a sucrose unit.

In addition to their nutritional and metabolic function, carbohydrates are used as natural sweeteners, raw materials for various fermentation products including alcoholic beverages, and the main ingredient of cereals. Carbohydrates influence the rheological properties of most foods of plant origin. The involvement of carbohydrates in the "browning" reaction may either improve or impair consumer acceptance and the nutritional value of many foods.

In food composition tables, the carbohydrate content is usually given as total carbohydrates by difference, that is, the percentages of water, protein, fat, and ash subtracted from 100. Another widely used expression is nitrogen-free extract, calculated as components other than water, nitrogenous compounds, crude fiber, crude fat, and minerals.

Fruits are a rich source of mono- and disaccharides. Dates contain up to 58% sucrose and dried figs contain a mixture of 31% fructose and 42% glucose. The sucrose content of most other fruits and fruit juices is low, though some varieties of melons, peaches, pineapples, and tangerines contain 6–9% sucrose, and mangos 12% sucrose. Reducing sugars (primarily a mixture of fructose and glucose) are the main soluble carbohydrate in most fruits, and account for 70% of seedless raisins. Partly ripe bananas are relatively rich in starch (9%), uncooked prunes in cellulose and hemicellulose, and citrus fruits in pectins. Vegetables contain

TABLE 2.1

Physical and Chemical Properties of Common Starches

Starch	Granule size (µm) Range	Average	Amylose (%)	Swelling power (at 95°C)	Solubility at 95°C (%)	Gelatinization range (°C)	Source	Taste	General description of granules
Barley	2–35	20	22	—	—	59–64	Cereal	Low	Round, eliptical
Corn									
regular	5–25	15	26	24	25	62–72	Cereal	Low	Round, polygonal
waxy	5–25	15	~1	64	23	63–72	Cereal	Low	Round, oval indentations
high amylose	—	15	up to 80	6	12	85–87	Cereal	Low	Round
Potato	15–100	33	24	1000	82	56–69	Tuber	Slight	Egg-like, oyster indentations
Rice	3–8	5	17	19	—	61–78	Cereal	Low	Polygonal clusters
Rye	2–35	—	23	—	—	57–70	Cereal	Low	Eliptical, lenticular
Sago	20–60	—	27	97	—	60–72	Pith	Low	Egg-like, some truncate forms
Sorghum	5–25	15	26	22	22	68–75	Cereal	Low	Round, polygonal
Tapioca (cassava)	5–35	20	17	71	48	52–64	Root	Fruity	Round–oval, truncated on side
Wheat	2–35	—	25	21	41	62–75	Cereal	Low	Round, eliptical
Oats	—	25	27	—	—	—	Cereal	Low	Round

substantially less glucose and fructose than fruits, and the only significant vegetable sources of sucrose are sugar beets and cane sugar. Fresh corn, white potatoes, and sweet potatoes each contain about 15% or more starch. Practically the only carbohydrate present in unsweetened milk and milk products is lactose. Nuts are generally a poor source of monosaccharides and disaccharides; chestnuts contain up to 33% starch. The main component of cereals and cereal products is starch. In milled products, the starch content increases with the degree of refinement; it is about 70% in white flour compared with about 60% in whole grain. The increase in starch is accompanied by a parallel decrease in cellulose, hemicellulose, and pentosan gums. Spices contain reducing sugars; 9–38% in cloves and black pepper, respectively, and pepper has as much as 34% starch. Commercial white sugar contains 99.5% (or more) sucrose; corn sugar about 87.5% glucose; honey about 75% reducing sugars. Most syrups and sweets have various amounts of sucrose (up to about 65%), reducing sugars (up to 40%), and dextrins (up to 35%).

This chapter deals with the functional properties of starch. (Chapters 3 and 4 review the functional properties of other carbohydrates.) Physical and chemical properties of starches are given in Table 2.1; properties of starch components in Table 2.2; and characteristics of gelatinized and cooled starch suspensions in Table 2.3.

The primary use of starch is to provide the principal source of carbohydrates in the diet. Of the world's foodstuffs, starch from many origins is of major importance. Starch's secondary usefulness includes a wide variety of industrial applications, where various chemically and physically modified starches perform a multitude of functions, from sizing paper and textiles to wet-refining mineral ores. Most of these industrial uses depend on the peculiar behavior of starch as a

TABLE 2.2

Properties of Starch Components

Property	Amylose	Amylopectin
General structure	Essentially linear	Branched
Average chain length	$\sim 10^3$	20 to 25
Degree of polymerization	$\sim 10^3$	10^4 to 10^5
Iodine complex	blue (~ 650 nm)	purple (~ 550 nm)
Iodine affinity	19–20%	1%
Blue value	1.4	0.05
Stability of aqueous solution	Retrogrades readily	Stable
Conversion to maltose (%)		
with β-amylase	100	55–60
with limit dextrinase and β-amylases	100	100

TABLE 2.3

Characteristics of Gelatinized and Cooled Starch Dispersions

Starch	Hot-paste viscosity	Texture	Clarity	Stability to retrograde	Freeze–thaw stability	Resistance to shear
Corn	Medium	Short-stiff	Opaque	Low	Low	Medium
Waxy maize	Medium-high	Soft-cohesive	Clear	High	Medium	Low
Potato	Very high	Long-cohesive	Clear	Medium	Low	Low
Rice	Medium-low	Short-stiff	Slightly opaque	Low	Low	Medium
Sago	Medium-high	Long-cohesive	Clear-translucent	Medium	Low	Low
Tapioca	High	Long-cohesive	Clear-translucent	Medium	Low	Low
Wheat	Medium-low	Soft-short	Slightly opaque	Low	Low	Medium

high-polymeric colloid. For reviews of the uses of starches see Banks and Greenwood (1975), Bautlecht (1953), Hahn (1969), Kerr (1950), Kite *et al.* (1963), Knight (1965, 1969), Petersen (1975), Radley (1954, 1976), Whistler (1971), Whistler and Paschall (1965, 1967), and Wurzburg and Szymanski (1970).

II. Forms of Starch

Starch occurs in most green-leafed plants in various locations: in the seed (cereal grains), in the root and tuber (tapioca and potato), in the stem-pith (sago), and in the fruit (banana). All of these forms are used as food. The function of starch in plants is as a storage deposit of food for energy during dormancy or germination, until green leaves can photosynthesize sugar. Under the microscope, starch is seen to consist of minute spherules or granules. This form was first described by van Leeuwenhoek in 1719. A pound of corn starch contains 750 billion granules. Two types of granules occur in starch from ordinary dent corn: (a) angular granules from the high-protein horny portion of the endosperm, and (b) round granules from the low-protein floury endosperm. As the corn kernel field-dries, those granules in the gluten matrix are squeezed together to form pressure facets. Cuzco floury corn (Peru) contains floury endosperms only, hence only round granules. In contrast, flint corn has mostly horny endosperms, and the granules are predominantly angular. Granules of each starch species have characteristic size, shape, and markings (Fig. 2.1).

Potato starch granules are large, oval, 15–100 μm in diameter, with pronounced oystershell-like striations. Corn starch is medium-sized, polygonal or round, and 10–25 μm in diameter. Rice starch is small, polygonal, and 3–8 μm

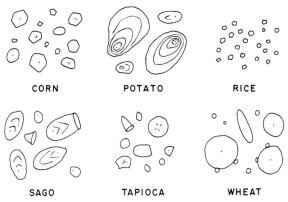

CORN   POTATO   RICE

SAGO   TAPIOCA   WHEAT

Fig. 2.1 Microscopic appearance of various granular starches. (From T. J. Schoch, Corn Products Co., Inc., Argo, Illinois.)

in diameter. Extremes in size of starch granules are seen in taro or dasheen root (Polynesia), which is only 1 μm in diameter, and calculated at 122 trillion granules per pound; in contrast, the *root of Phaius grandifolius* (Australia) is 80–100 μm in diameter. Little is known about differences in granule size of various starches in relation to botanical growth and function. Small-granule starches are relatively rare. They have potential commercial value as dusting starches, for example in cosmetics, candy dusting, rubber-tire mold releasing agents, and so on.

In some cases, pronounced striations around the *hilum* or botanical point of origin of the granule are seen. These are believed to be growth rings, and are best seen in "fruit-of-potato" strach. Photomicrographs of barley, corn, rice, and rye starches (under regular and polarized light) at the same magnification are compared in Fig. 2.2. Originally, it was thought that the granule grew by intussusception or by stuffing successive portions of carbohydrate through the central hilum (like an onion). It is now known that growth is by apposition, or deposition of successive layers on outside (like a pearl). Proof was obtained by carbon-14 labeling, followed by sectioning the granule and locating the activity. Under polarized light, the starch granule shows a strong interference cross through the hilum, suggesting an orderly arrangement of the starch substance in a spherocrystalline form. X-ray spectra of granular starches prove this orderly crystallinity extends down to the molecular level. Also, some radial organization is present, since the granule shows radial fissures and even fractures into pielike segments when crushed under a microscope cover glass, suggesting radial lines of crystal cleavage.

III. Swelling of Granular Starches

Starch granules are insoluble in cold water. They swell slightly (in a manner similar to water-logged wood) but shrink back to their original size and consistency on drying. When heated in a water suspension to progressively higher temperatures, very little happens until a certain critical temperature is reached. At that point the granule begins to swell, simultaneously losing polarization crosses. This is termed gelatinization. Not all granules in a starch sample gelatinize and lose polarization crosses at the same point; rather this occurs over a range of temperature, depending on species of starch.

Gelatinization temperature range is best determined with a Kofler hot stage on a polarizing microscope. In this method, a droplet of an aqueous starch suspension is placed on a microscope slide, and heated at 1.5°C/min on the hot stage. Percentage of granules losing polarization crosses (as estimated visually) is plotted against temperature to obtain gelatinization temperature curves (Fig. 2.3).

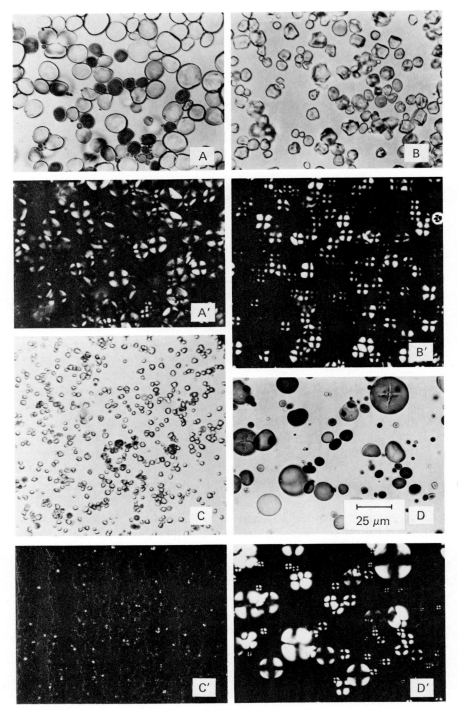

Fig. 2.2

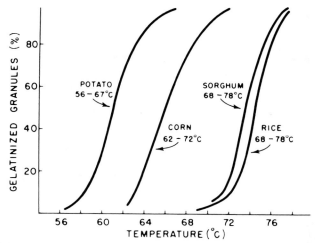

Fig. 2.3 Gelatinization temperature curves of various unmodified starches. (From T. J. Schoch, Corn Products Co., Inc., Argo, Illinois.)

Thus, expressed as initiation and completion of loss of polarization crosses, potato starch gelatinizes at 56–67°C, corn starch at 62–72°C, and rice and sorghum starches at 68–78°C. Wide variations in gelatinization temperatures of different types of rice starch occur. The sample in Fig. 2.3 is a commercial rice starch with a relatively high gelatinization temperature. Other varieties of rice starch gelatinize at a substantially lower temperature. After initial gelatinization, starch granules continue to swell as temperature is raised, until their outlines become vague when viewed with a regular microscope, but are still visible with a phase-contrast microscope. Eventually these swollen granules become so large that they imbibe all the free water and begin to crowd one another, producing the thick-bodied consistency of a cooked starch paste. A cooked starch paste is not a solution; it owes its character to persistence of the undissolved swollen granules as elastic gel particles. Only by pressure cooking at 120–150°C can one truly dissolve a swollen granule.

According to Olkku and Rha (1978), gelatinization of starch granules involves the following events:

1. Hydration and swelling to several times original size
2. Loss of birefringence
3. Increase in clarity
4. Marked, rapid increase in consistency and attainment of peak

Fig. 2.2 Photomicrographs of starch granules. (A, A′) Barley. (B, B′) Corn. (C, C′) Rice. (D, D′) Rye. (A–D) Picture taken in regular light. (A′–D′) Picture taken in polarized light.

5. Dissolution of linear molecules and diffusion from ruptured granules
6. Retrogradation of mixture to a paste-like mass or gel

Wheat starch gelatinizes in three stages:

1. Limited swelling takes place between 60 and 70°C. This involves disruption of weakly bound or readily accessible amorphous sites.
2. Subsequent rapid swelling at 80–90°C involves disruption of more strongly bound or less accessible sites.
3. Upon continuation of heating, the swollen granules fragment.

IV. Factors that Govern–Affect Starch Gelatinization

Characteristics of gelatinized and cooled dispersions are listed in Table 2.3. Starch gelatinization is affected by many factors. The most obvious is starch concentration. Swollen starch particles are vulnerable to different degrees of shear. The small amount of protein in commercial starches (about 0.2%) has a limited effect on starch gelatinization. A much greater effect is exerted by the relatively high levels of wheat flour protein. Protein and starch interact due to attraction of their opposite charges and during gelatinization form complexes. The interaction is low at alkaline pH levels, at which both starch and protein have negative charges, and high at acid pH, at which proteins bear a positive charge. Wheat proteins lose their association capacity as a result of heat denaturation. Protein forms complexes with starch molecules on the granule surface, thus preventing escape of exudate from the granule and interfering with a decrease in viscosity. Interaction of lipids with/and/or starch–proteins also has significant effects on gelatinization temperature and peak viscosity.

Wheat flour contains 2–3% pentosan gums, one-fourth to one-third of which are water soluble. The pentosans have a very high water-binding capacity and thereby exert a large effect on availability of water for starch gelatinization. The soluble pentosans yield very viscous solutions. The insoluble pentosans hydrate and bind large amounts of water. The water-soluble pentosans absorb 9.2 times and the water-insoluble pentosans absorb 4.8 times their weight in water in a gluten–water system, 6.9 times in a starch–water system, and 6.5 times in a reconstituted gluten–prime starch dough (Jelaca and Hlynka, 1971). The absorption curve of water for pentosans depends on the rate of shear in mixing.

Pasting properties of defatted starch reimpregnated with polar lipids show reduced viscosity at both steps of pasting by the CMC method (Medcalf *et al.*, 1968). In this method, starch pasting curves are obtained with the use of the carboxymethyl cellulose (CMC)–amylograph technique. The temperature is raised from 25 to 95°C and then held at 95°C for 15 min. Two values obtained from those curves are used to characterize pasting characteristics. The viscosity at 76°C is the first step at which the curve levels off. The viscosity at 95°C is the second step, at which time maximum viscosity is recorded. Nonpolar lipids increase maximum height of the second step. The initial step in pasting is related to hydration of the amorphous, more accessible regions of the starch granule. Polar lipids form a complex with starch and thus retard hydration and swelling and lower peak viscosity. During pasting, consistency increases until water penetrates crystalline or micellar regions of the starch granule (Leach et al., 1959). Once micellar regions are hydrated, the granules lose their integrity. Because polar lipids interfere with the release of exudate from the starch granules and also destroy the micellar arrangement, the consistency of the gelatinized starch decreases. The primary effect of nonpolar lipids is in preventing hydration of the micellar region.

Surface-active agents reduce viscosity of heated suspensions after the first peak. The granules continue to swell, but consistency continues to be low as little exudate is released to the solution. Adding salt increases peak viscosity. The salt seems to enhance granule integrity before fragmentation. Sugar inhibits swelling of starch and retards its gelatinization, partly by binding water and making it unavailable. High sugar concentrations inhibit swelling and raise the temperature at which birefringence disappears.

Traditionally, the increase in viscosity that takes place when starch suspensions are heated was attributed to the swelling that accompanies water inhibition. Studies undertaken by Miller et al. (1973), however, have shown that increase in viscosity is due mainly to the formation of an exudate network in the fluid around the starch granules. If the amount of water is not a limiting factor, the exudate is released in such large amounts that viscosity is chiefly a function of temperature up to about 90°C.

The behavior of starch during pasting, and especially during cooling, is determined to a large extent by the affinity of hydroxyl groups. During cooling, starch molecules aggregate and crystallize out of the solution. The linear amylose chains in particular are hydrogen bonded; they form aggregates of low solubility and in high starch concentrations they form gels. This is the phenomenon of retrogradation, which primarily involves amylose (see pages 39 and 40).

A. Chemical Structure and Physical Behavior of Starch Fractions

Starch is a natural high polymer, built up through successive condensation of glucose units by enzymatic processes in the plant (Fig. 2.4). This gives a long linear chain of about 500 to 2000 glucose units. This polysaccharide is present in

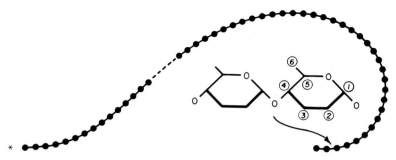

Fig. 2.4 Structure of linear starch fraction. (●) Glucose unit. (*) Aldehydic terminus of the chain. Chain length is 500–2000 glucose units. Interglucose bonding is α-1,4 as shown in the inset, with carbon atoms in the glucose unit numbered in conventional fashion. (From T. J. Schoch, Corn Products Co., Inc., Argo, Illinois.)

most starches and is termed the linear fraction or amylose. There is also a second enzyme mechanism in the plant that (according to one theory) breaks down some of these long linear chains into short lengths of 25 glucose units and recombines these fragments into a multiple branched or tree-like structure (Fig. 2.5). This is the branched fraction or amylopectin, which coexists with the linear fraction in most common starches. It contains hundreds of short linear branches, with an average branch length of 25 glucose units; hence, molecular weight is at least several million (see also Table. 2.2).

Common starches (corn, wheat, tapioca, potato) contain both linear and branches fractions. The ratio is fairly constant in any one species: 17% in tapioca, 20–25% in potato, 26% in corn. The Northern Regional Laboratory of the U.S. Department of Agriculture (Peoria, Illinois) investigated some 50–60 varieties of corn from all over the world and found no great variation from this content of the linear fraction in the starch. Some evidence shows that rice starch may vary widely in content of the linear fraction, especially between *Japonica* and *Indica* varieties. Certain varieties of cereals have grains with starch composed entirely of branched molecules, that is, the "waxy" cereals, including corn, rice, barley, and sorghum; thus far, they have not been found in wheat or root starches. Waxy corn (more properly, waxy maize) was not originally known in the Western Hemisphere, which was the birthplace of corn. American corn was exported to China and Burma over a century ago, where it apparently mutated to a waxy variety. The commercial strain now used in the Western Hemisphere was obtained from China.

Waxy starches have unique properties that make them preferred for various uses. Waxy maize and waxy sorghum starches are widely marketed and waxy rice flour and barley starch are available in limited commercial quantities. At the opposite end of the scale, three known starches are predominantly linear, from

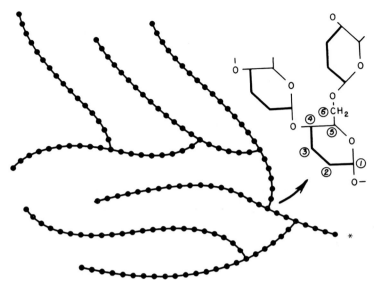

Fig. 2.5 Structure of a small section of the branched starch molecule. Branching is effected through an α-1,6 linkage, as shown in the inset. (From T. J. Schoch, Corn Products Co., Inc., Argo, Illinois.)

wrinkled-seeded garden peas, from so-called high-amylose corn, and from barley. High-linear pea starch was originally discovered by the Western Regional Laboratory of The United States Department of Agriculture (Berkeley, California). The content of its linear fraction is as high as 75%. In corn, this genetic characteristic is induced or enhanced by the genes *du* (dull), *su* (supersugary), and *ae* (amylose extender). Commercial high-amylose corn starches containing 55 and 70% of the linear fraction are marketed, and experimental varieties approaching 80% are known.

Starches can be fractionated by pressure cooking in water to completely dissolve the granule, then cooling in contact with a polar agent such as butyl or amyl alcohol, thymol, or the nitroparaffins. A complex of the linear fraction with these agents precipitates as microscopic needle clusters or as spherocrystals, which can be isolated by supercentrifuging. To explain this complex formation, it is assumed that molecules of the linear fraction are quite flexible, and can twist and coil in solution in a manner similar to overcooked spaghetti. This ability is due to the presence of α-1,4-glucosidic bonds between glucose units; in contrast, the β-1,4-bonds of cellulose produce a straight and rigid chain (Fig. 2.6). Principally from X-ray evidence, we presume that the flexible molecules of the linear fraction coil into a helical clathrate, with alcohol or other precipitant in lengthwise position at the center of the helix.

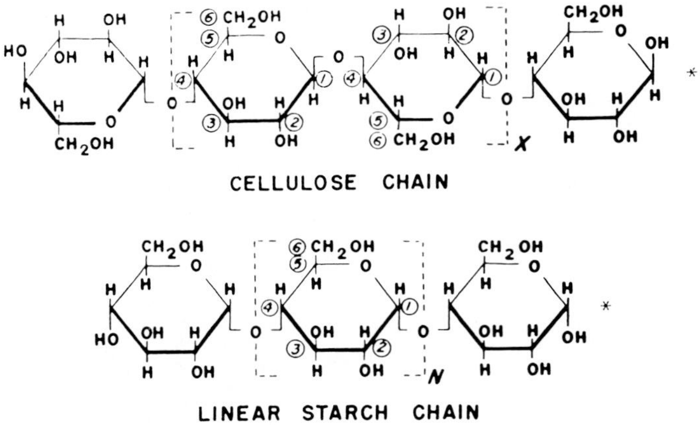

CELLULOSE   CHAIN

LINEAR   STARCH   CHAIN

Fig. 2.6 Polymeric chain structures of cellulose and amylose. (*) Aldehyde terminus of the chain. (From T. J. Schoch, Corn Products Co., Inc., Argo, Illinois.)

Fractionation of starch with amyl alcohol is excellent in the laboratory, but impractical on a large scale. Both corn and potato starches are fractionated commercially by other means, and linear and branched fractions are marketed. The processes used are

1. Starch is pressure cooked at 140–150°C in a magnesium sulfate solution to dissolve the total starch substance; then cooled to 90–100°C to precipitate the linear fraction. Subsequent cooling to room temperature precipitates the branched fraction. The process is essentially a selective salting-out or desolvation by the sulfate ion. Products are of good purity, though perhaps slightly degraded by high temperature.

2. Starch is pressure cooked in a water medium by steam injection, then cooled slowly over 24 to 36 hr to precipitate the linear fraction as well-formed spherocrystals. The process is a modernized version of the one reported by Beijerinck in 1915. The fractions are not as pure as those obtained by the sulfate precipitation method.

The linear fraction also forms insoluble complexes with other polar agents such as fatty acids and monoglycerides. The linear fraction displays an intense blue color with iodine, while the pure branched fraction shows only a weak violet-red color. This color difference is attributed to helical conformation of linear molecules, with iodine (I_2) molecules in the center of the helix. A peculiar electrical resonance of this system gives rise to a blue color. The linear molecules never assume a helical structure except in the presence of iodine or other clathrating agents (Fig. 2.7). If long linear chains are progressively fragmented into shorter chains, iodine color changes from blue to purple to red (Table 2.4). Thus

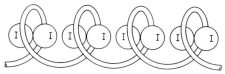

Fig. 2.7 Helical structure of the linear fraction with iodine. One I_2 molecule is enclosed therein for each turn of the helix. Similar helical conformations are also obtained with butyl and amyl alcohols, higher fatty acids, and monoglycerides. (From T. J. Schoch, Corn Products Co., Inc., Argo, Illinois.)

the branched fraction shows a red color because the effective linearity of its branches is 25 glucose units. The color obtained with iodine is used to differentiate microscopically between normal and waxy starches. Normal corn starch granules stain very dark blue or purplish-black; waxy granules stain light-red.

Another property of dissolved linear starch molecules is a tendency to associate closely by reason of attractive forces or hydrogen-bonding between hydroxyl groups on adjacent chains. This phenomenon, called retrogradation, can occur in two different ways (Fig. 2.8). If a dilute starch solution stands for a long time (e.g., an iodometric starch indicator left on a laboratory shelf), linear molecules slowly line up and "zipper" together in parallel bundles, resulting in particles too large to remain in solution. The system slowly becomes opaque and deposits a sediment. If a fairly concentrated starch system is cooled (e.g., a cooked 7% corn starch paste), it sets almost immediately into a rigid gel. A plausible reason is that the linear molecules do not have enough time to associate into insoluble particles, but must "zipper" together wherever they can. This yields the elastic network of a gel. Areas of associated linear chains or portions of chains are truly crystalline, and are called micelles or (more rarely) crystallites. The branched fraction does not ordinarily associate or retrograde on standing, because the outer branches are not sufficiently long (only 25 glucose units). Therefore, waxy starch pastes remain fluid and noncongealing.

TABLE 2.4

Relation of Linear Chain Length to Iodine Color

Chain length (glucose units)	Number of helix turns	Color
12	2	None
12–15	2	Brown
20–30	3–5	Red
35–40	6–7	Purple
>45	9	Blue

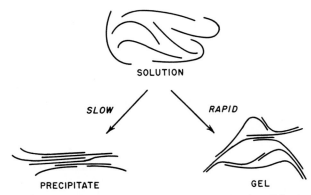

Fig. 2.8 Mechanisms of retrogradation of the linear starch fraction. Dilute solutions slowly deposit an insoluble precipitate (left). Concentrated solutions rapidly set up to a rigid gel (right). (From T. J. Schoch, Corn Products Co., Inc., Argo, Illinois.)

B. Structure of the Starch Granule

According to accepted theory, the starch granule is a spherocrystal consisting of concentric layers or growth rings. In each layer, intermingled and intertangled linear and branched molecules are laid down in radial fashion (Fig. 2.9A). Wherever possible, adjacent linear molecules and the outer linear branches of the branched fraction associate in a parallel fashion to create radially oriented micelles. A long linear chain may pass through several micelles, or the outer fringes of a single branched molecule may participate in a number of micelles. Between these organized micellar areas are looser and more amorphous regions. This pattern of locally crystalline areas giving rise to radially oriented micelles explains both polarization crosses and X-ray spectra.

When the granule is crushed, it fractures along radial lines of weakness in

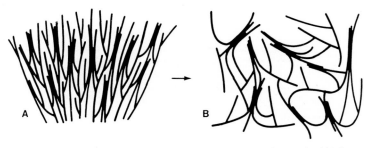

Fig. 2.9 Micellar structure and mechanism of swelling of the starch granule. (A) Segment of an unswollen granule, with associated micelles represented as thickened sections. (B) Segment of a swollen granule, showing disoriented but still persistent micelles. (From T. J. Schoch, Corn Products Co., Inc., Argo, Illinois.)

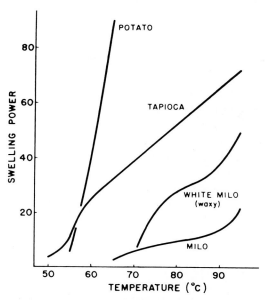

Fig. 2.10 Swelling patterns of unmodified starches. (From T. J. Schoch, Corn Products Co., Inc., Argo, Illinois.)

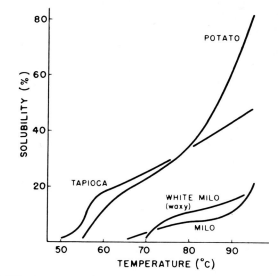

Fig. 2.11 Solubilization patterns of unmodified starches. (From T. J. Schoch, Corn Products Co., Inc., Argo, Illinois.)

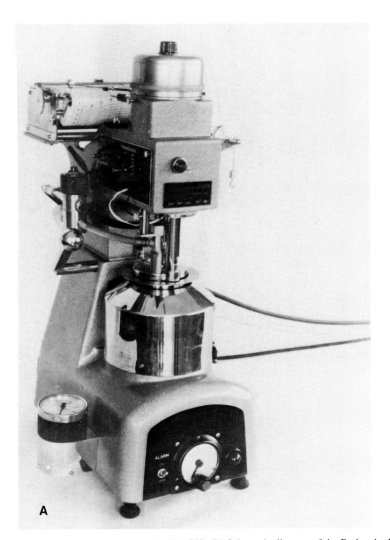

A

Fig. 2.12 (A) Brabender VISCO/amylo/GRAPH. (B) Schematic diagram of the Brabender VIS-CO/amylo/GRAPH. The container bowl (1) and the stirrer (2) are made of stainless steel. Stirrer is connected to highly sensitive interchangeable measuring spring cartridge (3). Bowl is rotated at uniform speed (4) and stirrer deflects, depending on viscosity of the test material. Resistance encountered is transmitted to the spring system and continuously recorded by the recording mechanism (5). Container is heated by a radiant source (6). Temperature inside the container is controlled by a thermoregulator (7) extending into the material. Cooling is effected by a cooling probe (not shown) immersed in the material. A solenoid valve controls the supply of water from the tap. Thermoregulator is set to desired temperature. When increasing or decreasing temperatures during a test, a synchronous motor (not shown) drives the thermoregulator up or down by the gear train (8). Pilot light (9) indicates heating. Main switch (10) serves to turn instrument on and off. Entire system is programmed by a presetting timer (11) which monitors test run, automatically shuts off system, and triggers alarm buzzer advising of test completion. (From C. W. Brabender Co., Duisburg, West Germany.)

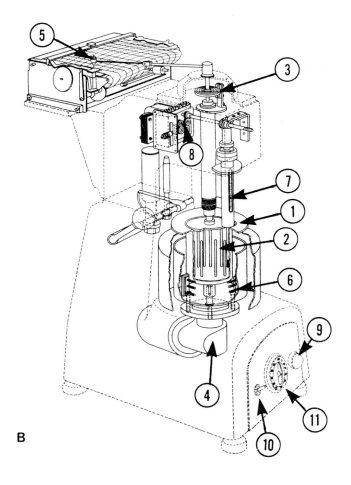

B

amorphous areas, although this also means mechanical breakage of occasional primary-valence glucosidic bonds. Prior to gelatinization, water merely seeps in and saturates the open intermicellar areas in the granule. When the gelatinization temperature is reached, the loose bonding in these amorphous intermicellar regions is progressively weakened, and the granule starts to swell tangentially. The optical polarization cross and X-ray spectrum are lost as soon as the radial orientation of micelles is disturbed. Swelling continues as temperature is raised, further relaxing the granule network and permitting more water to enter. Throughout swelling, most of the micelles remain intact, and, therefore, eventually arrive at an enormously swollen network state, elastic in nature because it is still being held together by persistent micelles (Fig. 2.9B).

Some of the shorter linear molecules are released from the entangling network and leach out into the surrounding aqueous substrate. Branched molecules are apparently much too completely enmeshed to become soluble. If this paste is violently agitated (as in a food blender), opposing forces may shear swollen granules and tear apart the network, severing primary glucosidic bonds by force. Similarly, starch paste pumped through the jet of a spray-dryer or through the orifice of a homogenizer may undergo rather extensive hydrolytic cleavage of molecules, due to the enormous shearing action.

C. Granule Swelling and Paste Viscosity

Each species of starch swells differently, reflecting varietal differences in the molecular organization within the granule. For example, corn starch gelatinizes at a relatively high temperature of 62–72°C, and thereafter swells rather slowly. In contrast, potato starch gelatinizes at a lower temperature and swells much more rapidly.

One can follow the pattern of swelling by pasting a weighed amount of starch in excess water at an appropriate temperature, then centrifuging to sediment the swollen granules, pouring off the aqueous supernate, and weighing the precipitated swollen granules. Swelling power is calculated as weight of swollen granules per gram of dry starch. One can obtain a complete swelling pattern of a starch by determining its swelling power at 5°C intervals over the temperature range from initial gelatinization to 100°C (Fig. 2.10). Tapioca and particularly potato starches show evidence of weak internal organization; they swell freely and enormously. In contrast, milo (sorghum) starch exhibits a restricted swelling in two stages, indicative of two sets of bonding forces within the granule: a weaker set relaxing just above the gelatinization temperature, and a stronger set persisting until 90–95°C is reached. Corn starch is similar to sorghum. Waxy sorghum (milo) has intermediate swelling between the root starches and the common cereal starches.

During determination of swelling power, one can make a dry-weight assay on an aliquot of the supernatant solution after centrifuging to obtain a solubilization pattern of the starch (Fig. 2.11). The extensive solubilization of potato and tapioca, the restricted two-stage solubilization of milo (sorghum), and the intermediate position of waxy sorghum are all similar to their respective swelling patterns.

A practical consequence of granule swelling is paste viscosity. When granules swell sufficiently to imbibe all free water and crowd neighboring granules, they produce a thick consistency or viscosity. Exudate as described by Miller *et al.* (1973) is also important. The best way to measure standard viscosity is with the Brabender amylograph, which automatically charts changes in consistency throughout a controlled heating and cooling cycle (Fig. 2.12).

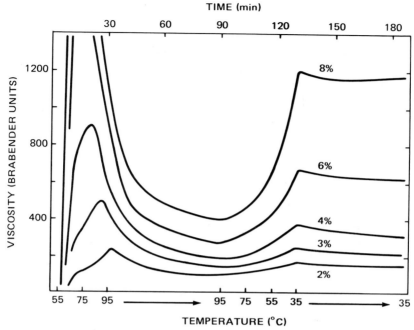

Fig. 2.13 Brabender viscosity—potato starch. (From T. J. Schoch, Corn Products Co., Inc., Argo, Illinois.)

In the amylograph, one can heat a continuously stirred starch suspension at a rate of 1.5°C/min, hold it at any desired temperature, and cool it at a controlled constant rate. To obtain an example of a free-swelling starch, Brabender viscosities were determined for potato starch at concentrations of 2–8% (Fig. 2.13). The pastes were cooked to 95°C, held for an hour, then cooled to 35°C, and held another hour. Outstanding features of these viscosity curves are

1. The high peak viscosity obtained, in accord with the free-swelling character of potato starch.
2. The great breakdown of viscosity during cooking, reflecting the fragility of the enormously swollen granules. This starch had a swelling power of 440 at 95°C, that is, 1 gm could potentially occupy 440 ml. Because there simply is no room for this degree of swelling with the starch concentrations employed here, the fragile swollen granules are torn apart by the continuous stirring, and the viscosity drops sharply.
3. The "setback" or increase in viscosity on cooling is rather minor. Note that potato starch is not prone to retrogradation.

Corn starch, with a swelling power of only 34 at 95°C, gives a markedly different set of Brabender curves (Fig. 2.14):

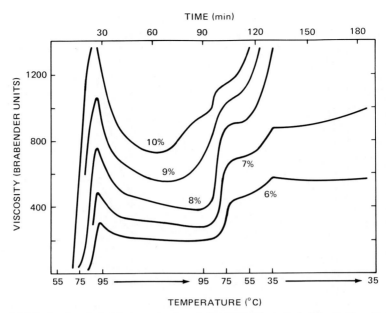

Fig. 2.14 Brabender viscosity—corn starch pasted at 95°C. (From T. J. Schoch, Corn Products Co., Inc., Argo, Illinois.)

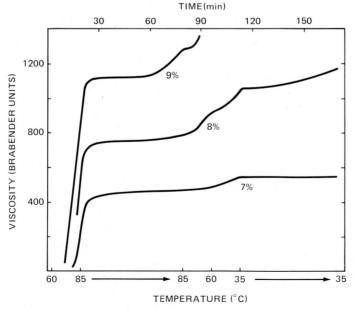

Fig. 2.15 Brabender viscosity—corn starch pasted at 85°C. (From T. J. Schoch, Corn Products Co., Inc., Argo, Illinois.)

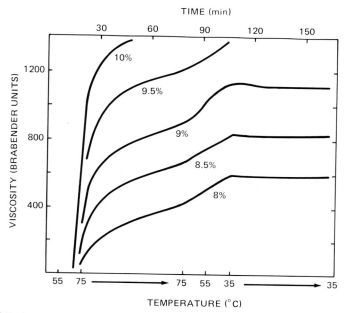

Fig. 2.16 Brabender viscosity—corn starch pasted at 75°C. (From T. J. Schoch, Corn Products Co., Inc., Argo, Illinois.)

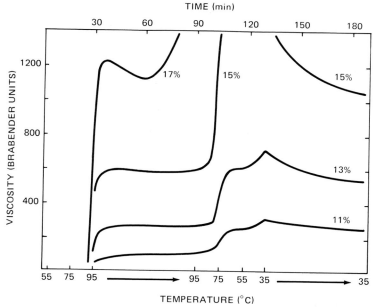

Fig. 2.17 Brabender viscosity, "Amylomaize I" corn starch at 95°C. (From T. J. Schoch, Corn Products Co., Inc., Argo, Illinois.)

TIME (min)

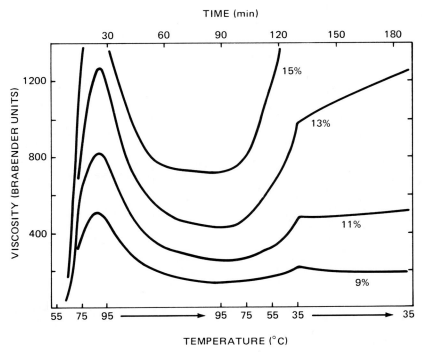

Fig. 2.18 Brabender viscosity, "Amylomaize I" in NaOH at 95°C. (From T. J. Schoch, Corn Products Co., Inc., Argo, Illinois.)

1. The peak viscosity is much lower, in accord with its lower swelling power.
2. Being less fragile, the extent of starch breakdown during cooking is much less.
3. There is a great setback on cooling, due to retrogradation of the linear molecules to a gel network.

Interdependence between granule swelling and the Brabender curves can be illustrated by analysis of the viscosities of corn starch at lower cooking temperatures (Fig. 2.15 and 2.16). Obviously, the type of Brabender curve is not specific for and characteristic of any particular starch, but instead depends on the extent of swelling of the granules under the conditions of cooking employed.

As a further illustration, the high content (55%) of the linear fraction in high-amylose corn starch rigidifies the micellar network and greatly restricts granule swelling (swelling power = 8 at 95°C). This means high concentrations of this starch must be used to obtain a Brabender curve (Fig. 2.17). Because the granules are so strongly bonded internally, the paste shows little or no breakdown during cooking at 95°C in water. But if the same starch is pasted in 0.1 *N* NaOH at 95°C, its swelling power increases to 83 (Fig. 2.18). Here the Brabender pattern for 0.1 N NaOH shows a high peak viscosity and extensive breakdown on cooking, approaching that of potato starch in water.

V. Food Applications

A. Functions of Starch in Foods

Starches are used in foods for six reasons:

1. Thickening agents (sauces, cream soups, pie fillings)
2. Colloidal stabilizers (salad dressings)
3. Moisture retention (cake toppings)
4. Gel-forming agents (gum confections)
5. Binders (wafers, ice cream cones)
6. Coating and glazing agents (nut meats, candies)

In our culture, the main purposes of starch are aesthetic rather than nutritional: to make foods more pleasing to the eye, to impart better texture and "mouth-feel," to prevent separation of ingredients, and to provide a carrier for delicate flavors.

1. Starch Thickeners

Thickeners represent a major part of the total starch sales to the food industry. Starch imparts to foods a thick-bodied consistency. Traditional food starches have been the thick-cooking unmodified cereal starches: corn, wheat, and rice. A disadvantage of these cereal starches is their tendency to congeal into rigid gels, due to retrogradation of linear molecules. This was quite acceptable in earlier days for such foods as the old-fashioned gel-type of corn starch pudding, usually chocolate or vanilla flavored and cast in a decorative mold. But present tastes in desserts requires heavy-bodied consistency with little or no gel structure.

In food starches, the most important and widely used development has been cross-bonding of granular starches by chemical means. This is particularly applied to waxy starches, where absence of the linear fraction eliminates congelation. These cross-bonded starches have a variety of desirable properties in food use and bring out the different qualities desired in food thickeners. Because waxy starches swell excessively, they exhibit an unstable Brabender viscosity. We can introduce a few chemical cross-linkages between adjacent molecules within the unswollen starch granule, and thereby supplement the micellar network and greatly stabilize the granule against excessive swelling and consequent deterioration of viscosity. This is accomplished simply by suspending the ungelatinized granular starch in dilute caustic alkali, and adding trace amounts (less than 1 part

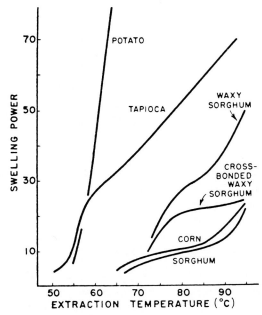

Fig. 2.19 Swelling patterns of granular starches. (From T. J. Schoch, Corn Products Co., Inc., Argo, Illinois.)

per thousand) of such agents as sodium trimetaphosphate or epichlorohydrin. These agents introduce ester or ether cross-linkages comparable to occasional "spot-welds" within the granule network. Such cross-bonded starches show greatly reduced swelling power, depending on the degree of cross-linking introduced. In Fig. 2.19 we have cross-bonded waxy sorghum starch with trimetaphosphate, thereby reducing the swelling power to half the original value. As anticipated, solubilization of the starch is likewise greatly reduced by cross-bonding.

Indeed, reduction in solubility constitutes the most reliable and sensitive analytical test for cross-bonding in starches. This stabilization of granule swelling is likewise reflected in the Brabender viscosity (see Fig. 2.20). The pasting peak is eliminated, and there is no reduction in viscosity during cooking. On a weight basis, this cross-bonded product is a much more efficient producer of viscosity than any of the unmodified thick-boiling starches, requiring only 27.5 gm compared with 40 gm needed for the other three starches.

The Brabender amylograph can be used to show the progressive change in paste viscosity as increasing amounts of cross-bonding are introduced into a waxy starch (Fig. 2.21). Alkaline suspensions of waxy sorghum starch were treated with various amounts of sodium trimetaphosphate, ranging from 0.002 to

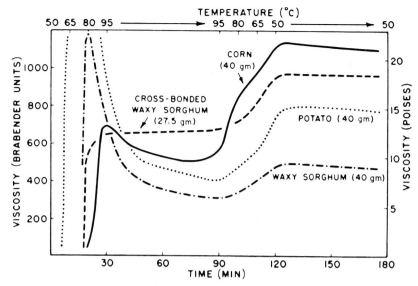

Fig. 2.20 Brabender viscosities of various starches. Concentration of starch is given in grams (dry basis) per 500 ml of paste. (From T. J. Schoch, Corn Products Co., Inc., Argo, Illinois.)

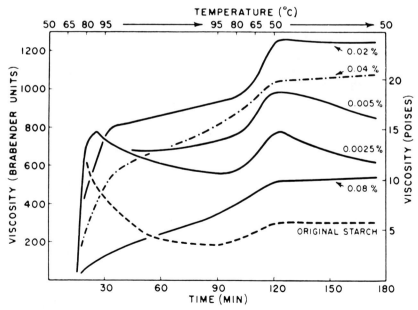

Fig. 2.21 Effect of the degree of cross-bonding on the Brabender viscosity of waxy sorghum starch. Degree of cross-bonding expressed as percentage of metaphosphate on dry-starch basis. Starch concentration is 30 gm/500 ml. (From T. J. Schoch, Corn Products Co., Inc., Argo, Illinois.)

0.08%, based on dry starch. After a suitable reaction period, the mixture was neutralized, and the starch filtered, washed, and dried. Brabender curves show that maximum viscosity and viscosity stability are obtained with only 2 parts of added metaphosphate per 10,000 parts starch. We do not know how much of this trace amount of metaphosphate actually reacted to form cross-linkages, since some of the reagent probably did not react and was, therefore, washed out. This can be determined with a radio-tagged metaphosphate, since the amount of phosphorus naturally present in the base starch is substantially higher than that introduced by cross-linkage. At higher levels of metaphosphate, viscosity is rapidly reduced, and the product becomes much less efficient as a thickening agent. By introducing a still larger extent of cross-linkage, products are obtained that will not swell even when heated in water to 120°C; such starches have been used as dusting powder for surgical gloves and to resist steam sterilization. Unlike talc, any residue of this starch left in a surgical wound is slowly broken down and assimilated by body enzymes. Cross-bonding exemplifies the extreme sensitivity of high-polymeric substances to slight chemical modifications. Two parts or less of phosphate per 10,000 parts of starch probably corresponds to the introduction of only several cross-links per giant macromolecule of starch. Yet, this amount is sufficient to alter completely the behavior of the starch granule.

Not only high viscosity but, even more important, a certain kind of viscosity is usually desired. For example, potato, tapioca, and unmodified waxy starches may give high viscosity, but of a "long" cohesive nature. The paste will stretch out and snap back when spooned, in an action called "viscoelasticity." This occurs because granules swell excessively on cooking and become overly deformable and stretchable. Viscoelasticity can be avoided by lightly cross-bonding the granules to reduce swelling during cooking. Virtually all manufacturers of food starches now market specialty products of this type, derived from tapioca, waxy maize, and waxy sorghum.

The following simple method demonstrates viscoelasticity of starch pastes (Fig. 2.22). Cooked 5% pastes of various starches were poured into shallow trays, and a tracer line was ruled on surface of the paste with a methyl-violet pencil. A spatula was then drawn edgewise through the paste at right angles to the tracer line, and the width of distortion of this line was determined.

1. Viscoelastic starches such as potato and waxy sorghum exhibit widest distortion, due to extensibility of greatly swollen granules.
2. "Short" pastes, such as cross-bonded starches, give least distortion, since granules do not overswell to become extensible.

In the second connection, monoglyceride is advocated for addition to such products as dehydrated mashed potatoes to impart a "short" fluffy consistency to the reconstituted material, without cohesiveness or excessive breakdown on whipping. Advertisements for monoglyceride additives are rather clever: "You

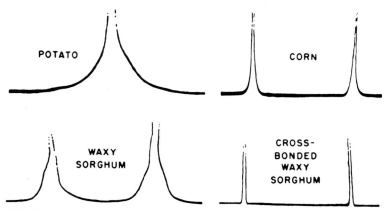

Fig. 2.22 Viscoelasticity of 5% starch pastes, as indicated by width of distortion of pigmented line on surface of paste. (From T. J. Schoch, Corn Products Co., Inc., Argo, Illinois.)

don't want wallpaper paste, you want mashed potatoes.'' In explanation of this action, such fatty adjuncts as monoglycerides and higher fatty acids form helical clathrates with the linear fraction of the starch. This can also occur when segments of linear chains pass through the intermicellar regions of the starch granule (Fig. 2.23). In this case, the granule is drawn more tightly together, and swelling is restricted by complexing potato starch with either stearic acid or monoglyceride (Myverol brand) (Fig. 2.24). The solubilization is markedly repressed. Finally, the effect of these fatty adjuncts on paste viscosity is shown by complexing potato starch with various amounts of stearic acid, and then determining Brabender amylograph curves for the products (Fig. 2.25). Viscosity is greatly stabilized by incorporation of only 1% stearic acid. No such effect is observed with waxy starches, which lack a linear fraction. These considerations must be kept in mind whenever starches are used in conjunction with fatty adjuncts. Quite

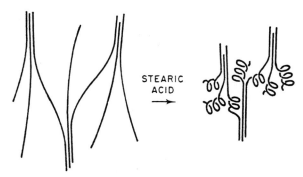

Fig. 2.23 Restriction of swelling of micellar network by stearic acid. (From T. J. Schoch, Corn Products Co., Inc., Argo, Illinois.)

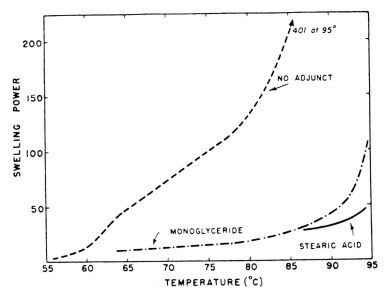

Fig. 2.24 Effect of stearic acid and monoglyceride on swelling of potato starch. (From T. J. Schoch, Corn Products Co., Inc., Argo, Illinois.)

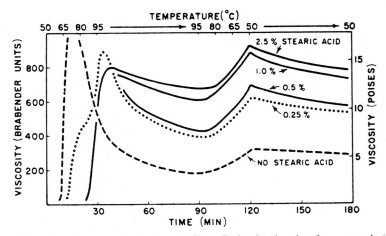

Fig. 2.25 Effect of various amounts of stearic acid on Brabender viscosity of potato starch. (From T. J. Schoch, Corn Products Co., Inc., Argo, Illinois.)

TABLE 2.5

Effect of Autoclaving on Brookfield Viscosity and Solubility of Starches

Starch	Viscosity poises[a]		% Soluble starch	
	Initial	Final	Initial	Final
Potato	54	2	39	94
Waxy sorghum	56	14	21	74
Corn	44	10[b]	24	51
Cross-bonded waxy sorghum	56	50	5	8

[a]Cold-paste viscosity after 2 h standing.
[b]Includes increase due to setback.

different consistencies may be obtained if the starch is cooked in the presence of the fatty material (with consequent repression of granule swelling), of if the adjunct is added after the starch is fully swollen by cooking.

Resistance of Starch Thickeners to Heat, Shear, and Acidity. Another important consideration is stability of starch thickeners in the face of high temperature, as employed in sterilization of canned goods by autoclaving or retorting. Resistance to thinning during heat sterilization is readily evaluated by measuring the decrease in Brookfield viscosity on autoclaving 5% starch pastes (Table 2.5) As a result of heat sterilization

1. Fragile, high-swelling potato and waxy sorghum starches are drastically thinned and solubilized.
2. Corn starch is substantially solubilized, though subsequent congelation of its linear fraction gives an apparent high viscosity.
3. Viscosity and solubility of cross-bonded waxy starch remain virtually unchanged.

During processing in a food plant, the starch thickener must not thin excessively under conditions of agitation and shear, as may be encountered in pumping, mixing, and homogenization. Action of shear on paste parallels the effect of autoclaving—both involve energy expenditure that may fragment and solubilize swollen granules. Mechanical breakdown is evaluated by stirring cooked 5% starch pastes at various speeds for 20 min, and measuring change in Brookfield viscosity (Fig. 2.26). As a result of intensive agitation and shear

1. The highly swollen granules of potato, tapioca, and waxy starches are drastically thinned at all rates of shear.
2. Viscosity of corn starch survives low shear, but decreases sharply at higher shear rates.
3. Only cross-bonded starches can resist this sort of maltreatment.

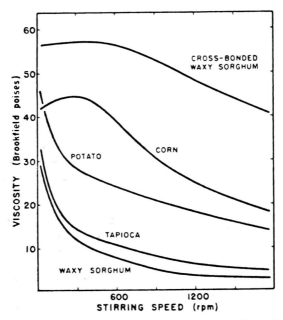

Fig. 2.26 Effect of shear on Brookfield viscosity of starch pastes. (From T. J. Schoch, Corn Products Co., Inc., Argo, Illinois.)

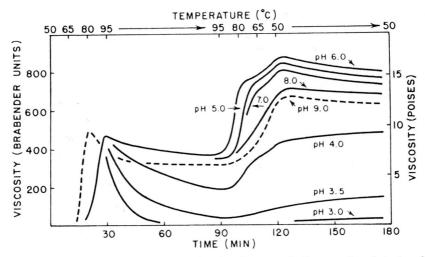

Fig. 2.27 Effect of pH on the Brabender viscosity of corn starch. Concentration of starch = 35 gm/500 ml of paste. (From T. J. Schoch, Corn Products Co., Inc., Argo, Illinois.)

It is important to stress that only intact swollen granules can give paste viscosity, and not fragmented granules or solubilized starch substance. Likewise, of major importance is resistance to thinning under acid conditions of many foods, for example, when used as thickener in fruit pies with pH as low as 3.0–3.5. As an experiment, canned cherry pie fillings were thickened with two waxy sorghum starches cross-bonded to different levels. After the fillings were pressure cooked, the pie shells were filled and baked. The acidities were identical, but the consistency of one cherry pie filling was thinned excessively by the canning and subsequent baking. The other filling (containing starch sufficiently cross-bonded to resist acidity) retained its viscosity and did not leak from the pie shell. Type and content of any starch modification must be carefully adjusted to fit conditions of use. To illustrate, Brabender viscosities were determined for corn starch at various pH levels (Fig. 2.27).

Maximum final viscosity is obtained at pH 6.0, with considerable thinning below pH 4.5 Obviously, such a starch cannot be used satisfactorily in acid media. To compare acid stabilities of other starches, the final 50°C viscosities at the end of each Brabender test have been plotted against pH (Fig. 2.28). If relative heights of these curves, which depend only on starch concentration, are disregarded and if only viscosity change with pH is considered, Fig. 2.28 shows that

1. Corn starch exhibits maximum stability at pH 6, with considerable thinning at pH 4.5 and below.
2. Potato and waxy sorghum are highly acid-sensitive, and thin rapidly below pH 5.
3. Cross-bonded waxy sorghum is most stable, withstanding a pH as low as 3.5.

By slightly increasing the degree of cross-bonding (that is, from 0.02 to 0.04% trimetaphosphate), we can obtain starches that develop viscosity only when cooked in acid medium (Fig. 2.29). Apparently, starch granules are so stabilized by cross-bonding that swelling and viscosity can be developed only after mild hydrolysis to loosen the internal granule network and permit swelling. This explains the unusual circumstance of increasing the apparent viscosity of high-polymeric molecules by hydrolysis.

2. Emulsifying Properties

Starch is not a high-efficiency emulsifying agent, primarily because it lacks the polyelectrolyte character necessary for emulsion stability. To measure average droplet size of oil-in-water emulsions, thus permitting evaluation of emulsion stability, one dissolves a red dye in the oil, which is then emulsified into the starch paste. The smaller the size of the dispersed oil droplets, the lighter the

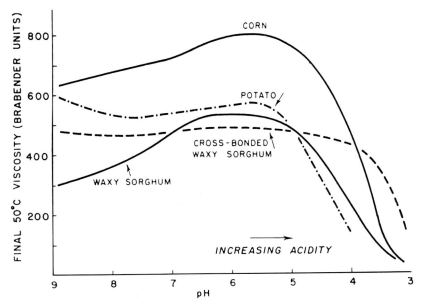

Fig. 2.28 Effect of pH on the final 50°C Brabender viscosities of starches. Relative heights of the curves are not significant. (From T. J. Schoch, Corn Products Co., Inc., Argo, Illinois.)

color of the emulsion. Average droplet size can be calculated by measuring reflectance of blue light in a reflecting spectrophotometer.

When this technique is applied to oil emulsions in pastes of various modified starches, we find that only strongly charged sulfate ester and cationic starch ether are effective in producing and maintaining a finely divided emulsion. This is evident both from the smaller initial droplet size and from the absence of any coalescence of droplets on storage. Neither starch is food-acceptable, however.

In general, the food use of starch in emulsifying fats and oils depends merely on the mechanical separation of oil droplets by swollen granules. A typical example is salad dressing, in which cross-bonded starches are used.

3. Paste Clarity

Next to viscosity, the most important aesthetic quality of a starch paste is its visual appearance with respect to transparency or opacity. In some instances, as with starches used to thicken salad dressings, opacity is desirable to enhance a light bright color. In other cases, as with fruit pie fillings, maximum transparency is desired. A pie thickened with corn starch has a dull unappetizing appearance, because of retrogradation of its linear fraction to an insoluble opaque

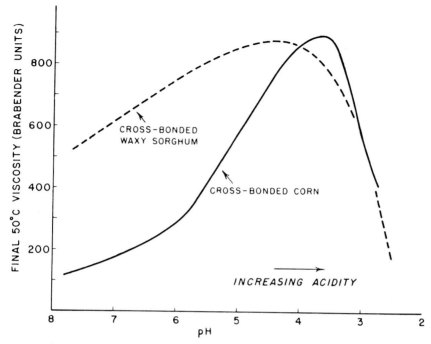

Fig. 2.29 Effect of pH on the final 50°C Brabender viscosities of highly cross-bonded corn and waxy sorghum starch. (From T. J. Schoch, Corn Produets Co., Inc., Argo, Illinois.)

state. A pie thickened with cross-bonded waxy starch has a bright transparent appearance, due to absence of retrogradation.

To evaluate the opacity of starches, one can measure light transmission through 1-cm cells of the paste. In Fig. 2.30 the percent of light transmission is plotted against paste concentration. Certain of these results seem quite predictable: the high clarity of potato starch, the somewhat lower value for waxy sorghum, and the high opacity of corn starch. However, cross-bonded waxy sorghum appears more opaque than corn starch, which totally disagrees with the previously described results obtained using cherry pies thickened with the two starches. The inconsistency is explicable because a consumer does not judge clarity by amount of light transmitted through the paste, but rather by the amount of light reflected from the paste. For this reason, it is necessary to measure paste reflectance with a reflecting spectrophotometer. The starch is cooked over a range of concentrations and stored for 1 day at 10°C to permit any retrogradation to occur. The paste is then placed in a black capsule and its light reflectance measured. The system is calibrated on magnesium oxide as 100% (Fig. 2.31).

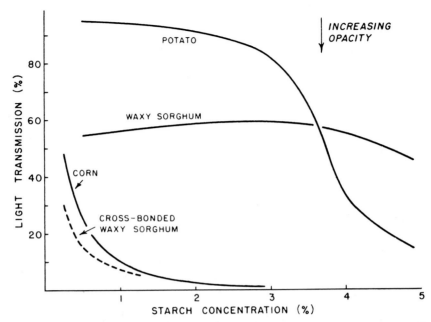

Fig. 2.30 Light transmission versus concentration of various starches. Pastes aged 1 day at 10°C. (From T. J. Schoch, Corn Products Co., Inc., Argo, Illinois.)

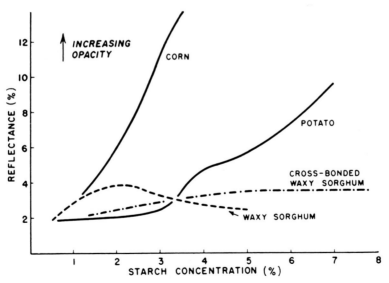

Fig. 2.31 Paste reflectance versus paste concentration of various starches. (From T. J. Schoch, Corn Products Co., Inc., Argo, Illinois.)

Results obtained by this technique are reasonable and in close agreement with visual appraisals. They show that

1. Corn starch is by far the most opaque, in accordance with its strong retrogradation tendencies.
2. Potato starch is much clearer, but develops opacity at higher concentrations.
3. Waxy sorghum and its cross-bonded modification are most clear and show little change with increasing concentration.

4. Storage Stability of Food Starches

Storage stability of starch-thickened foods is of primary importance to the manufacturer, especially if products must be frozen or stored under refrigeration. Conditions during cold storage are optimum for promoting retrogradation. Such changes can be readily determined by measurement of paste reflectance during storage (Fig. 2.32). During cold storage

1. Corn starch retrogrades and becomes increasingly opaque much more rapidly at 4°C than at room temperature.
2. Potato starch develops opacity primarily during the first week of cold storage; retrogradation thereafter is very slow.
3. The waxy starches (both unmodified and cross-bonded) show little change over 4 weeks' cold storage.

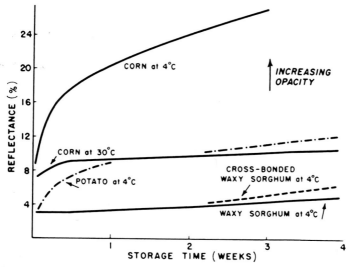

Fig. 2.32 Change in paste reflectance of 3% starch paste on storage at pH 6.0. (From T. J. Schoch, Corn Products Co., Inc., Argo, Illinois.)

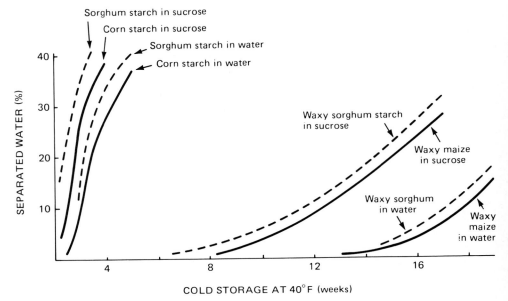

Fig. 2.33 Cold storage stability of various starches (5% pastes in water and in 37° Brix sucrose solution, stored at 40°C, and centrifuged—pH = 6.0). (From T. J. Schoch, Corn Products Co., Inc., Argo, Illinois.)

As an extreme case, poor storability may result in syneresis of a watery fluid from the food, which is literally pushed out of the swollen granule by the gradual association of starch molecules. When this happens, the consumer assumes that the food has spoiled through microbiological activity.

One can evaluate cold storage stability by holding the starch paste at 4°C (40°F) for 4 to 6 months and periodically measuring the amount of water that separates when the paste is centrifuged (Fig. 2.33).

1. Aqueous pastes of waxy maize and waxy sorghum begin to show water separation after 14 weeks' storage.
2. The same starches in sugar syrup show separation after only 8 weeks. The sugar probably sequesters water and thus interferes with hydration of the starch.
3. Normal sorghum and corn starches (which have a ratio of amylose to amylopectin of about 1:3) will not withstand even 1 week of cold storage, whether pasted in water or in sugar syrup.

This procedure is much too slow for routine testing or research, however. A more vigorous and rapid test subjects a 5% starch paste to repeated cycles of freezing and thawing and then determines the water separated on centrifuging the

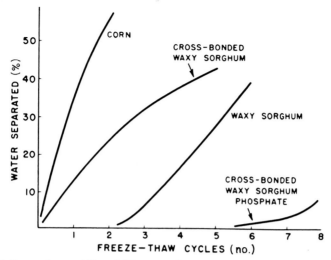

Fig. 2.34 Freeze–thaw stability of 5% pastes of various starches, as indicated by syneresis of water after repeated cycles of freezing overnight at −7°C, followed by thawing at 30°C. A trace of freshly precipitated silver iodide was added to the pastes to act as nuclei for ice crystals, and thus prevent supercooling. (From T. J. Schoch, Corn Products Co., Inc., Argo, Illinois.)

thawed paste. One such freeze–thaw cycle is equivalent to 2–3 weeks' cold storage (Fig. 2.34).

1. Corn starch retrogrades very rapidly under these conditions and does not survive even one freeze–thaw cycle.
2. Waxy sorghum is good with regard to water separation, but much too viscoelastic for most food uses.
3. When waxy sorghum is cross-bonded to reduce cohesiveness, freeze–thaw stability is impaired.

This problem of water separation is solved by introducing ionic phosphate groups into the cross-bonded waxy starch. Mutual repulsion between negatively charged phosphate sites dilates starch molecules and, thus, maintains the high water-holding capacity of the paste. By increasing the amount of ionic phosphate groups, one can increase freeze–thaw stability to any desired degree. Indeed, starches have survived 25 freeze–thaw cycles with no apparent water separation, a survival rate far beyond practical need. The high freeze–thaw stability of waxy rice starch is especially interesting. (See Fig. 2.35.) In water at pH 6, this starch commences to show syneresis only after 20 freeze–thaw cycles. Its stability is reduced in a sucrose or acid medium, but its water-holding capacity is still adequate for normal use under those conditions. No explanation exists for this phenomenon. It is not related to the small granule size, because freeze–thaw instability of normal rice starch is the same as that of corn starch.

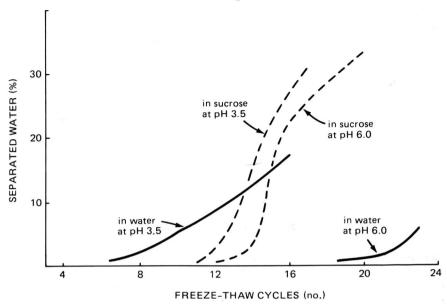

Fig. 2.35 Freeze–thaw stability of waxy rice starch (5% pastes in water and in 37 Brix sucrose solution, frozen at −8°C, thawed, and centrifuged). (From T. J. Schoch, Corn Products Co., Inc., Argo, Illinois.)

VI. Bread Making: Role of Starch

The unique gluten-forming properties of proteins are known to contribute to the excellent baking quality of wheat flour in the production of white bread. The gluten forms a film containing starch granules and gas bubbles. Starch participates in this process (Khoo *et al.*, 1975) as the granules take on various shapes to accommodate vacuoles and to produce the desired crumb grain and texture. During dough mixing, fully hydrated and developed gluten proteins form a web of fibrils with numerous microscopic vacuoles. After fermentation, this protein lattice structure contains larger air cells. Many of the small air cells enmesh minute starch granules within them. The veil-like protein coating on the surface of the starch granules, due mainly to increase in the size of the air cells, stretches and rolls up into fibrils. These fibrils, after kneading and proofing, aggregate and form longer and larger fibrils. A bread crumb is characterized by thin walls and large gas cells. The strands are swollen and fused, and the veil-like protein and starch form a cohesive mass.

Bechtel *et al.* (1978) found by light and transmission electron microscopy that protein strands provided a matrix network in a mixed wheat flour dough, and that matrix formation required adequate mixing. Fermentation of a complete dough produced gas vacuoles. After oven spring, protein strands were thin and had small vacuoles. Starch granules in the bottom center of a load varied in degree of gelatinization after oven spring; starch started to gelatinize from the interior of the granule and appeared fibrous. In the baked bread, most of the starch was gelatinized into fibrous strands interwoven with thin protein strands. Fretzdorff and co-workers (1981) examined by freeze–fracture microscopy the structures of isolated flour components of mixed doughs containing several combinations of ingredients, of fermented doughs, and of bread crumbs. Protein development was studied from a protein network in a flour–water dough to a sheet-like protein in a complete dough. Fermenting the complete dough altered the sheet-like protein to a fine network. Protein–starch interactions in dough were intensified by fermentation. In the bread crumbs, protein and starch were tightly connected.

Chabot and colleagues (1979) reported that air cells in bread were about 20 μm thick. Starch granules were embedded in a matrix but in most cases were disguised by the protein covering them. Small vacuoles were covered in the protein layer covering the granules. The protein covering may be so complete that it may prevent iodine staining of starch granules. During baking, starch granules gelatinize and are flexible enough to "fit" around the air cells. The granules remain intact and are identifiable, partly because limited water is available during gelatinization. Under conditions of limited water availability, a strong bond between starch and proteins may be formed. Additional studies on the structure of white bread crumb were summarized by Pomeranz (1976).

Modifications in starch during baking are of particular interest in relation to the role of starch in wheat bread and rye bread. According to Kulp and Lorenz (1981), the functionality of wheat starch as a bread ingredient remains elusive. The integrity of the wheat starch granule is essential for its optimal performance both in bread and cake systems. Mechanical, chemical, or biochemical disruptions of the native granule organization have adverse effects on the bread-forming properties of wheat starch. The effect of excessive mechanical starch damage is further aggravated by abnormally high levels of cereal α-amylase from malt or sprouted grain.

Yasunaga and colleagues (1968) analyzed the gelatinization of starch in white bread crumbs by pasting a slurry of the crumb in the Brabender amylograph. Degree of gelatinization depended mainly on the available moisture but also on the temperature during baking. Starch in the outer layers of the crumb was gelatinized more than starch in the center. It was concluded that starch was only partly gelatinized during baking. Higher baking absorption, higher baking temperature, and longer baking time each produced more extensive gelatinization. Derby *et al.* (1975) stated that under baking conditions, changes in starch depend upon temperature and amount of available water. If the temperature is maintained at approximately 100°C, changes in starch depend primarily on the amount of available water. The availability of water is determined by the formula

TABLE 2.6

Stages of Granular Dispersion of Starch in Baked Goods[a]

Swollen	Scottish short bread	Biscuits	Cakes	Bread	Wafers
↓					
Gelatinized					
↓					
Disrupted					
↓					
Dispersed					
↓					
Enzymatically degraded					

[a]From Greenwood, 1976.

or recipe used and by the presence of ingredients or components such as proteins, pentosan gums, or sugars, which compete with starch for water. The amount of moisture available for gelatinization is also affected by the degree of protection against water absorption, provided by fat to the starch particles. Starch in bread was at an intermediate stage of gelatinization, in which birefringence was retained in a relatively small number of granules. Stages of granular dispersion of starch in baked goods are given in Table 2.6.

Lineback and Wongsrikasem (1980) measured starch gelatinization in baked products by microscopic and enzymatic methods. Starch isolated from white bread had collapsed granules, but folding and deformation were not complete, indicating that pasting was incomplete before water availability became a limiting factor. The starch had lost all birefringence and was 96% susceptible to the action of glucoamylase. Their results must be interpreted with caution, because in the study the isolated starch was only 54% of the total starch in the bread.

Vassileva and co-workers (1981), using scanning electron microscopy (SEM), examined changes in starch structure in the crumb and crust of bread. Starch modification increased during baking as crumb temperatures increased from 93–94°C to 97–98°C. Maintaining the crumb temperature at 97–98°C caused relatively small changes in starch appearance but improved bread quality. Light microscopic studies of starch isolated from bread crumbs heated during baking at 93–94°C and 100°C showed little birefringence. The results were used to explain why crumbs of bread heated for a short time during baking at 97–98°C are sticky and "ball" during chewing. It is known that there is a transfer of water from denatured gluten to gelatinized starch during baking. That transfer may be suboptimal if the gelatinization is very limited. Starch granules located 200–300 μm beneath the outer crust surface were more gelatinized than granules in the outer crust layer. This phenomenon may be attributed to higher water availability for gelatinization beneath the outer crust surface.

Changes in structure of white bread resulting from addition of single cell protein (SCP) were studied by Evans and colleagues (1977). When flour and water were mixed into a well-developed dough, the starch granules were enveloped by a continuous sheet of gluten protein. As a result of protein–starch interaction during dough development, a smooth, veil-like network stretched over the starch granules. That network was disrupted after adding 6% SCP. Supplementation with SCP thickened the gluten sheets, masked the contours of starch granules, and increased the rupture of gluten. There was no translucency to the poorly formed gluten sheets; the gluten mass was opaque, rough-textured, thick, and lacked cohesiveness. Adding sodium stearoyl-2-lactylate (SSL) to dough improved gluten sheeting, and adding SSL to the SCP dough partially counteracted the adverse effects of SCP on the gluten layer.

A. Starch in Wheat and Rye Breads

Information on the structure of rye breads is limited. Wassermann and Dorfner (1974) compared by SEM structures of white, mixed, and rye bread crumbs. It was calculated that the surfaces of starch and gluten structures in baked wheat bread are 1000 and 2000 dm^2/g, respectively. It was postulated that there are no basic differences between structures of wheat and rye breads; in both, crumb cell walls are permeated by small micropores. In *Knäckebrot* (crisp bread), however, a mass of gelatinized starch–protein forms a continuous wall free of micropores.

According to Drews and Seibel (1974) and Stephan (1982), water-imbibing and swelling substances (proteins and pentosans), as well as amounts and properties of starch, are of major significance in rye bread making. Wheat starch, which gelatinizes at higher temperatures (above 70°C) than rye starch (above 50°C), is less susceptible to enzymatic attack. Consequently, wheat products can be baked in relatively neutral media and may require light acidification only if produced from extensively sprouted grain. In white wheat bread the significance of starch is much less than in rye bread. More attention is directed to the contribution of gluten quality and quantity. Gluten can exert its unique viscoelastic properties more distinctly in the relatively neutral wheat dough than in the more acidic rye dough. In addition, wheat breads are baked from predominantly white flours, which contain relatively small amounts of pentosan gums, which are concentrated in the outer kernel layers. Those pentosans absorb water and swell, but they produce only with difficulty porous and well-leavened baked products; pentosans may actually exert a strain on the whole bread system.

In wheat bread, the well-leavened gluten system is reinforced during baking by the gelatinized starch. The protein components cannot realize their full potential in the acidic, highly viscous rye dough. Consequently, the degree of leavening of rye bread is reduced. Starch is a major structural component rather than a reinforcing contributor in rye bread. The role played by the starch is affected by the pH of the system, dough composition and formulation, enzymatic activities, and extent of starch degradation (mechanical or enzymatic).

Use of a coarse meal, rather than a flour, is especially important in the production of bread from extensively sprouted rye. In the meal, surface area, starch damage, and effective enzymatic activities are reduced, dough acidification is enhanced, and excessively high, stress-causing levels of water are eliminated. In some parts of Central Europe much ripe grain has some degree of sprout damage. Starch damage in rye bread making, unlike in wheat bread making, is highly undesirable. Similarly, the tendency to use finely milled flour in wheat bread production to accelerate manufacturing processes and produce more uniform products increases water absorption and causes starch damage. Use of finely milled flour is even more of a disadvantage in rye bread processing. In addition to these reasons, practice has shown that some deficiencies in rye bread production, usually caused by raw material of somewhat inferior quality, can be mitigated by the use of coarse meals rather than flours.

Pomeranz (1984a) reported that in white wheat flour doughs, structure was based primarily on formation of a protein matrix that became finer and more uniform as dough mixing and/or fermentation proceeded. "Stringing" of small starch granules contributed to dough structure; large starch granules seemed to contribute little to dough structure in white wheat flour systems. In the bread, protein-gelatinized starch (mainly large granules) interaction was of major importance. Much of the starch was gelatinized, but in some (especially the small starch granules) gelatinization appeared very limited. The mixed wheat and rye flour system resembled to some extent the white wheat flour system. In addition, some unknown interaction appeared to involve gumlike substances and modifications of gluten and starch by organic acids. In rye meal systems, gum materials are one of the major contributors to dough structure. They facilitated "stringing" of small starch granules and their adherence to large granules. The bran-aleurone-rich chunks were incorporated by the gum-like materials in the acidified system into the dough. It was concluded that whereas in white wheat or mixed wheat and rye systems, bran-aleurone particles are part of the problem, in the rye meal system they are part of the solution in that they provide a coherent and continuous dough or bread structure. The major quantitative and qualitative contributor to rye meal bread structure was gelatinized starch.

In a subsequent study, structures of rye flour doughs, crumb and crust of wheat, and rye and wheat, rye flour, and rye and meal breads were examined by SEM (Pomeranz et al., 1984b). In the rye and wheat flour dough, in addition to a protein matrix gum-like materials apparently contributed to the structure. In the rye meal dough, structure was based on contributions of several components, including starch aggregates, large meal chunks, aleurone-pericarp particles, and probably gum-like materials.

The crumb of wheat flour bread had a fine network of protein strands and sheets enmeshing and covering, to widely differing degrees, expanding and interacting starch granules. The crumb of the rye and wheat flour bread is crumbly and fragile.

Starch modification–expansion–interaction was responsible for crumb formation and structure to a larger degree in rye than in wheat bread. The modification was less in protected areas inside vacuoles compared with areas outside vacuoles. Fracture of freeze-dried crumb showed the interaction among starch granules themselves and a matrix or gum-like membrane and formation of vacuoles. Extent of starch modification in the crust seems related to water availability. Crumb and crust texture differed in number, size, and distribution of small–intermediate vacuoles. They were greater and more uniform in wheat bread than in rye bread and in rye flour bread than in rye-meal-based bread.

B. Bread Staling

In bread staling changes in texture and moistness of the crumb are primarily, but not entirely, starch reactions. While the branched starch fraction does not retrograde under ordinary circumstances, it can slowly undergo a weak association when the amount of water is limited. Under these conditions, the outer branches apparently associate, and the branched molecules collapse. This is in part the cause of stale bread.

In conventional bread a sequence of starch reactions occurs during baking, cooling, storage, and reheating of the loaf (Fig. 2.36).

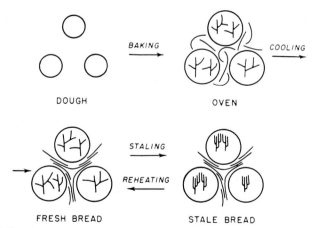

Fig. 2.36 Roles of starch fractions during baking and aging of normal (nonsoftened) bread. Linear molecules diffuse from the swollen granules and set up a permanent gel network between granules. Staling is attributed to association of branched molecules within the swollen granules. (From T. J. Schoch, Corn Products Co., Inc., Argo, Illinois.)

1. The granules of wheat starch first gelatinize during baking.
2. Some of the linear fraction leaches out of these swollen granules and becomes concentrated in the small amount of interstitial water between the granules.
3. On cooling the loaf, these intergranular linear molecules set into a rigid gel. This is normal fresh bread, which has an elastic structure. If compressed and then released, it regains its original form.
4. On standing, particularly at refrigeratored temperatures, the branched molecules within the swollen granules undergo a slow association to rigidify the swollen granules. This hardens the crumb structure, causing stale bread.
5. This association of branched molecules is relatively weak, and is readily reversed merely by heating to 50°C. Heating restores the elastic gel structure of the original fresh bread. Mild heating cannot solubilize the intergranular network of retrograded linear molecules.

When we compare the reactions that occur in bread softened by the addition of monoglyceride (Fig. 2.37), we note that

1. During baking, the granules swell, but the linear fraction is immobilized within the granule by insoluble complex formation with monoglyceride.
2. No gel network is developed between the swollen granules, therefore the bread is oversoft.
3. Transportation from the bakery oven to the ultimate consumer may require 24–36 hr. During this time, the branched molecules within the swollen granules associate sufficiently to produce a firm crumb structure comparable to freshly baked bread. Further association of these branched molecules continues, eventually resulting in the hardened crumb structure of stale bread.
4. If this stale bread is reheated, dissociation of branched molecules restores the swollen granules to their original soft state. However, there is no permanent gel network to maintain the crumb structure, and hence the bread is oversoft.

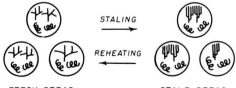

FRESH BREAD   STALE BREAD

Fig. 2.37 Roles of the starch fractions during aging of softened bread. Monoglyceride forms an insoluble helical complex with linear molecules, which are thereby immobilized within the swollen granules. (From T. J. Schoch, Corn Products Co., Inc., Argo, Illinois.)

It is speculated that starch is involved in the disappearance of flavor when bread stales. Fresh bread owes its yeasty and buttery aroma and taste to a variety of flavor components. These "disappear" during staling and a flavor of rancid fat gradually develops. When canned bread that was 2 years old was tested, it had dry, pulpy consistency, rancid odor and flavor, and dark color (sugar–protein "browning"). The resealed can was heated 15 min in boiling water: soft tender gel structure of fresh bread and a yeasty butter-like aroma and taste resulted. Except for color, the bread was restored to an oven-fresh condition. One may theorize that the linear starch fraction complexes with flavor ingredients to produce insoluble helical clathrates imperceptible to taste. Reheating dissociates the clathrate and liberates flavor. Little is known about the complex mechanochemical reactions of the starch fractions with other food ingredients, particularly flavor constituents, proteins, and fats.

VII. Modified Starches for Industrial Applications

In most food applications, starch is used as a thickening agent and hence ultimately consumed in the form of cooked paste. Industrial usage of starch is quite different; starches are usually cooked and then dried into a film or coating. Such coatings are employed as sizing for paper or textiles, or as adhesive film between two surfaces. Another important difference is the great diversity of modified or chemically altered starches for industrial use. This section reviews industrial modified starches in light of their potential use in food-related products. Comparative viscosity ranges of starches are shown in Fig. 2.38. Modified starches in the food industry are listed in Table 2.7.

A. Physically and Chemically Modified Starches for Industrial Use

Various physical types and chemical modifications of starch marketed for industrial use are summarized in the following sections.

1. "Pearl" and Powdered Starches

The physical form of a starch is of considerable importance to the user. "Pearl" or small-lump starch is usually preferred, because it is nondusting, easy to move with bulk-handling equipment, and dispersed quickly in water. Powdered starches are needed for dry-dusting, that is, as a dusting and dry-release agent for rubber-forming molds; as a cosmetic powder; and in printing to

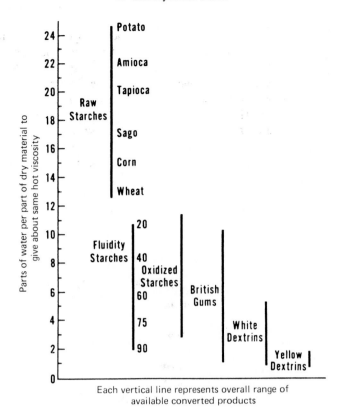

Each vertical line represents overall range of
available converted products

Fig. 2.38 Comparative viscosity ranges of different types of starches. (Courtesy of the Corn Refiners Association, Washington, D.C.)

TABLE 2.7

Modified Starches in the Food Industry[a]

Modified starch	Major properties	Examples of uses
Pregelatinized	Cold water soluble	Cake mixes
		Instant desserts
Cross-linked	Delayed thickening	Pie fillings, soups, baby foods,
	Stable to wide pH range and high temperature	sauces, and gravies
Cross-linked	Clarity, delayed thickening	Frozen foods, canned foods
etherified	Freeze–thaw stability	
esterified	Stable to pH range and high temperature	Pie fillings
Oxidized	High clarity, thin-boiling	Jellies, lemon curd
Acid-thinned	Gel strength	Jellies, gums, pastilles
Dextrins	High solubility	Toffees, glazes

[a]From Selby, 1977.

prevent ink "off-set" on the backs of paper sheets. The primary objection to powdered starch is its explosion hazard. Ideally, all starches should disperse to a monogranular form in cold water. Any persistent lumps are due to slight gelatinization during improper drying; such grit particles cook into microlumps, which are troublesome in industrial coating or sizing.

2. Pregelatinized Starches

Pregelatinized starches represent the simplest type of modification, achieved by precooking the starch in water, then drying the paste in a spray-dryer or on heated rolls. These starches are designed as a convenience item for users who do not have facilities to cook starch. Example uses are as a home laundry starch (as sizing agent), as a binder for wet sand cores in metal-casting foundries (as adhesive), and as a bodying agent for muds used in oil well drilling. Less viscosity or adhesiveness is evident in pregelatinized starch compared with freshly cooked paste; it achieves an 80% efficiency, at best.

3. Acid-Modified Thin-Cooking Starches

Acid-modified thin-cooking starch is prepared by treating a warm aqueous starch suspension with dilute mineral acid. Hydrolysis of a few bonds within the ungelatinized granule weakens micellar network, so that the granule actually dissolves on subsequent cooking. Brabender viscosity curves of typical acid-thinned corn starch show interesting characteristics (Fig. 2.39).

1. Very high starch concentrations are employed—approximately five times that used for unmodified corn starch.

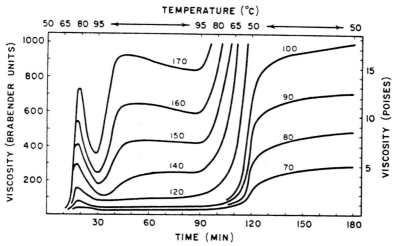

Fig. 2.39 Brabender curves of 80-fluidity acid-thinned corn starch. Concentration shown in grams per 500 ml. (From T. J. Schoch, Corn Products Co., Inc., Argo, Illinois.)

2. Low pasting peak is immediately followed by breakdown in viscosity. Acid-weakened granules begin to swell, then fragment and dissolve.
3. Extreme and unusual retrogradation takes place during cooking at 95°C; this probably represents congelation of short linear chain fragments, induced by close proximity due to high concentration.
4. Some evidence of viscosity decrease is evident during later stages of cooking; this probably represents mechanical breakdown of gel structure by agitation.
5. Extreme retrogradation occurs during cooling, with formation of a strong gel structure.

Acid-thinned starch products are used where soluble starch sizing is required, for example, for textile warp threads to increase tensile strength and prevent "fuzz" during weaving. Its greater solubility and lower viscosity forms a strong protective film around threads. Also, because of high congelation during cooling, acid-thinned starches are used to impart gel qualities to such confections as gum drops.

4. Enzyme-Thinned Starches

In textile or paper mills cooked starch pastes frequently are treated with liquefying α-amylases to decrease the molecular size and thereby thin and solubilize the starch (Figs. 2.40 and 2.41). The reaction must be conducted on the pasted starch, because enzymes have little effect on ungelatinized granular starches. These products are used primarily because of low cost: a few cents worth of enzyme can thin 100 lb of starch.

5. White Dextrins

White dextrins represent a continuation of the acid-modified starches, in other words, starches produced by higher degrees of simple glucosidic hydrolysis.

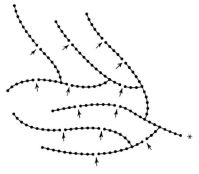

Fig. 2.40 Action of α-amylase on amylopectin. (*) Aldehydic terminus of the chain. (From T. J. Schoch, Corn Products Co., Inc., Argo, Illinois.)

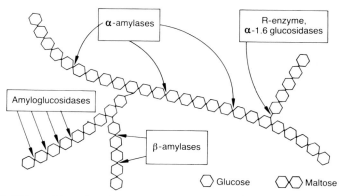

Fig. 2.41 Effects of the action of different types of amylases on amylopectin. (Courtesy of Röhm, G.m.b.H., Darmstadt, West Germany.)

These starches exhibit appreciable solubilities in cold water and hence cannot be produced in water suspensions as are regular acid-modified starches, since yield loss of solubles on filtration would be prohibitive. For this reason white dextrins are manufactured by heating dry starch with traces of mineral acid in an agitated heating vessel. These products display low viscosity and usually are completely soluble in hot water. White dextrins are used principally as paste adhesives.

6. Yellow Dextrins

Yellow dextrins are made by continued heating of dry acidified starch to higher temperature than that used for white dextrins. When moisture content drops below 3%, hydrolytic cleavage is no longer supported (Fig. 2.42). Instead, fragmented molecules recombine to form bushy, tightly branched structures. Recombination of fragments is completely random (1–2, 1–3, 1–4, 1–5, and 1–6 glucosidic linkages have been identified). There is some evidence that final yellow dextrin products can have higher molecular weight than original starch, but spatial dimensions of very bushy molecules are so small and compact that

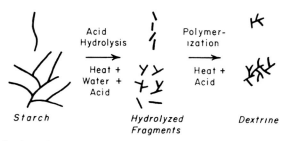

Fig. 2.42 Mechanism of hydrolysis and repolymerization of the starch fractions during heat dextrinzation. (From T. J. Schoch, Corn Products Co., Inc., Argo, Illinois.)

these starch products have very low viscosity. Because branch length must be extremely short, yellow dextrins show little or no retrogradation, even in highly concentrated solutions. They are used as a remoistening adhesive on postage stamps, labels, and envelope flaps.

7. Oxidized Starches

Oxidized starches are usually manufactured by treating an aqueous suspension of granular starch with alkaline hypochlorite. Oxidation warps the linear molecules (and likewise the branched molecules) by the introduction of carboxyl and carbonyl groups, so that the products show little or no retrogradation. Simultaneously, oxidative scission of the molecules occurs, creating low viscosity. In highly modified starches of this type, swelling power has no significance, but the solubilization pattern is of primary concern (Fig. 2.43). The oxidized corn starch becomes completely soluble in water above 70°C, indicating a weak micellar network and molecular association within the granule. In comparison, an acid-modified corn starch (designated as 60-fluidity) is still substantially associated and insoluble at 90°C. Due to warpage of linear chains, oxidized starches are

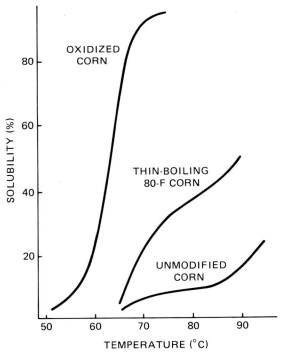

Fig. 2.43 Solubilization patterns of oxidized and acid-modified starches. (From T. J. Schoch, Corn Products Co., Inc., Argo, Illinois.)

least subject to retrogradation and give the clearest solutions and most transparent coatings. They are used for sizing high-quality writing and printing papers. Also, oxidized starches are strongly anionic due to formation of carboxyl groups; like most polyelectrolytes, they have effective protective colloid action toward dispersed materials.

8. Starch Ethers and Esters

The most important starch ethers and esters are hydroxyethyl ether and acetate ester, prepared by treating an alkaline starch suspension with ethylene oxide or acetic anhydride. While both fractions are reacted, the major benefit is from derivatization of the linear fraction, by introducing groups along the chain, which interfere with side-by-side association or retrogradation. With hydroxyethyl starch, the usual level of derivatization is 3–10 substituent groups per 100 glucose units (in high-polymer language, a "degree of substitution" or "D.S." of 0.03–0.10). Products of much higher D.S. (0.1–0.6) are available. No fragmentation of molecules occurs so these products are of high viscosity; thinning by acid-modification to any desired fluidity may be used as supplementary treatment. Starch ethers and esters yield strong, clear films and find particular application as finishing sizing for cotton textiles and for clay coating of printing papers.

Paste transparency (as measured by light transmission) is not as high as in waxy starches (Fig. 2.44). Only by reaching to higher degrees of derivatization (e.g., 0.20 D.S.) can retrogradation be completely suppressed to produce clear solutions. In medical settings, hydroxyethyl starch is used as a blood plasma extender with certain advantages over dextrans. The function of the extender is to maintain viscosity and osmotic pressure of the blood. Hydroxyethyl groups slow down action by starch-hydrolyzing enzymes in the blood, thus maintaining the desired effect of a colloidal extender of high molecular weight.

9. Cationic Starch Ethers

Cationic starch ethers are a type of industrial starch derivative in which the starch is given a positive ionic charge by introduction of quaternary ammonium groups. These unusual positively charged polyelectrolytes are used for specific flocculation, dispersion, and adsorption reactions. Some examples of their use are addition to the pulp beater in a paper mill, flocculation of titanium dioxide pigment, and facilitation of adsorption on negatively charged paper fibers.

10. Cross-Bonded Starches

The general type of cross-bonded starches was discussed in the previous section on food starches. Granular starches are treated to introduce cross-link-

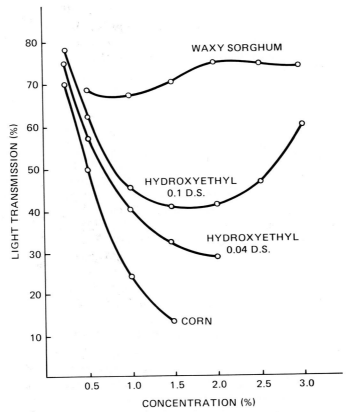

Fig. 2.44 Paste transparency of thick-boiling starches. (From T. J. Schoch, Corn Products Co., Inc., Argo, Illinois.)

ages between molecules in the micellar network, thus stabilizing the granule mechanically against excessive swelling and breakdown of viscosity by heat or shear. Industrial types of granular starches are similar, but they are cross-bonded with nonfood reagents, for example, formaldehyde. An example of industrial use is in paste adhesives containing caustic soda.

11. Dialdehyde Starches

Oxidation of granular starch with periodate to introduce two aldehyde groups per glucose unit produces a dialdehyde starch. This polyaldehyde is primarily employed as a copolymerizing agent with proteins and polyhydroxylated substances, for example, in the curing of leather or the manufacture of paper with high wet strength.

12. Linear Starch Fractions

Linear fractions (amylose) from potato and corn starches are commercially available. A high-amylose corn starch is marketed with a linear fraction content as high as 80%. Considerable research is being conducted to find uses for these materials. Because linear polymers generally form films and fibers of high mechanical strength, a possible use is as edible transparent coatings of foods, for example, casings for sausages and frankfurters, or envelopes for cooking vegetables in boiling water. It is hoped that more unique and specific uses will be found for linear starch material. Concentrated sodium hydroxide solutions in the range of 4 to 9 N are transformed into rigid thermally reversible gels with as little as 0.1% of the linear fraction. The question still remains—what uses can be found for concentrated caustic soda in gel form? Interestingly, no such gel is formed with either potassium or lithium hydroxide.

B. Dye-Staining of Ionic Starches

The staining of ionic starches with dyes is helpful in identifying ionic starch derivatives and in evaluating uniformity of their derivatization. For example, potato starch is one of the few native starches bearing an ionic charge, due to the presence of natural ionized phosphate ester groups. Potato starch stains with any cationic dye. If a mixture of corn and potato starch is treated with safranine, only the anionic potato starch is stained. Similarly, if a mixture of corn and cationic corn starches is stained with light-green SF, the exact composition can be determined by counting the proportion of stained cationic granules in a blood-counting cell. Finally, a sample of hypochlorite-oxidized corn starch can be stained in a nonuniform fashion with methylene blue, proving that the granules had not been uniformly oxidized. There is no other method known by which blends of this type can be positively identified, or the nonuniform character of a modified starch established.

VIII. Industrial Starch Applications

The paper industry is a major user of starch. Starch is employed in the formation and surface-sizing of paper, as adhesives in corrugated and laminated board, box, and bag fabrication, and in gummed paper products. Other major users are the food-processing industry and the textile (weaving, dye-printing, finishing) industry. The chief reasons for industrial use of starch are its low and stable cost and its versatility. Typical applications are

1. Dry binder (medicinal tablets)
2. Dry-dusting agent (off-set printing, rubber molds, surgical gloves)
3. Thickening agent (dye-printing of textiles, oil well drilling muds)
4. Gel-former (electrical "dry" cells)
5. Colloidal suspending agent (clay coating of paper, calcimines)
6. Sizing and coating (textile finishing, paper sizing, laundry starch)
7. Adhesives (gummed paper products, plywood, foundry core binder)
8. Oxygen barrier (encapsulation of vitamins and flavoring materials, protective coating)
9. Explosives (fireworks, nitro-starch)

Starch Films and Coatings

Adhesive films and coatings represent the largest industrial use of starch. This application exemplifies the various functions performed by modified starches. Unfortunately, broad general information on the structure and behavior of starch coatings cannot be readily obtained by the study of sized paper or textiles, due to interference by the cellulosic surface itself. The best approach is to study free starch films, cast as a wet coating on glass or silicone-coated metal plates, then dried under carefully controlled conditions, and detached as self-supporting films or flakes.

When wet coatings of various modified starches are dried on glass, we obtain immediate visual evidence of the importance of molecular size in creating a strong coherent film. Products such as white dextrin and enzyme-converted starch produce loose, flaky films, because the starch molecules have been excessively fragmented to the detriment of internal strength.

The clearest and strongest films are obtained with oxidized and etherified starches, in which retrogradation is prevented (by disturbing the linearity of the linear fraction) without excessive degradation of molecular size and strength. Strips of acid-modified starch still have sufficient molecular size to yield a strong continuous film, but retrogradation causes considerable opacity. In general, clear films are obtained from clear solutions, when clarity is measured in both cases in light transmission. In other words, any insoluble swollen starch granules or retrograded particles in an original solution persist, even in the final dried film, and cause opacity. Film opacity is caused by light scattering from an uneven dried surface of film due to insoluble protruding particles. Any insoluble particles totally embedded within the film have the same index of refraction as the surrounding starchy matrix, and hence do not scatter light to cause opacity. This point is proven by casting and drying a film of unmodified corn starch paste on glass, producing a smooth, glossy surface next to the glass and a pebbled upper surface. Light transmission of such a film is low. However, if two such films are

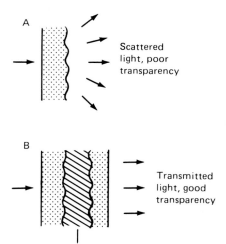

Fig. 2.45 Transparency in single (A) and double (B) starch films. (From T. J. Schoch, Corn Products Co., Inc., Argo, Illinois.)

sandwiched together, with pebbled sides on the interface, and if the space between them is flooded with clove oil (which has same index of refraction as starch), the composite sandwich has a very high light transmission (Fig. 2.45). All starch films would be crystal clear if both sides were polished perfectly smooth and flat, irrespective of the presence of undissolved swollen granules or retrograded particles within the film.

A significant application of starch is clay coating of papers for glossy magazine stock, in which sharp reproduction of color and black-and-white illustrations must be obtained on high-speed printing presses. Clay is used as white pigment and opacifier colloidally dispersed in a matrix of a dried-down starch film on the surface of the paper. The starch in a clay coating must perform a number of functions:

1. Act as a dispersing and deflocculating agent for the clay particles
2. Provide adhesion between clay particles and paper surface
3. Yield a film that is internally strong and coherent
4. Yield a film that is transparent, to enhance optical brightness of the opacifier and pigment
5. Have sufficient flexibility to fold without cracking
6. Show as much water resistance as possible, to avoid water spotting and wet "rub-off" of printed stock
7. Absorb the proper amount of ink, while preventing excessive penetration, spreading, and "wicking" of the ink impression

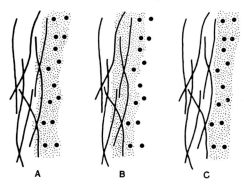

Fig. 2.46 Schematic representation of the penetration of clay coating into fibers of a paper sheet. Black circles represent clay particles; shaded area represents starch adhesive. (A), (B), and (C) indicate proper, excessive, and insufficient penetration, respectively. (From T. J. Schoch, Corn Products Co., Inc., Argo, Illinois.)

For clay coating of acid-modified paper, modified thin-boiling starches may be cooked in water to produce a concentrated solution, or unmodified starch may be pasted and converted to the proper fluidity with enzymes. Clay is then added, and the heavy slurry spread evenly on the paper, which undergoes a rapid drying operation. Viscosity of the coating must be adjusted within close limits for application to paper surface, so that the coating locks into the surface fibers (Fig. 2.46A). If starch adhesive is too thin, it may "wick" into the paper, leaving the clay particles high and dry (Fig. 2.46B). If it is too viscous, the coating may not adhere to paper fibers (Fig. 2.46C).

Ideally, starch molecules in clay coating should be as large as possible to impart film strength and continuity necessary for a proper coating. The general practice is to thin the coating starch excessively (for example, by hydrolysis) to reduce the amount of water as much as possible. Water introduced in the starch adhesive must be dried off the finished paper, which is not easily accomplished on a high-speed machine operating at 15 mph or faster. Although coating may be correctly applied to the paper, its internal strength may be low if starch molecules are excessively fragmented for the sake of fluidity. The following undesirable effects may occur:

1. Weak adhesion of coating to clay and to paper results. Dry abrasion resistance is poor, and the hard impact of printing may lift flecks of coating from the paper surface.
2. Coating exhibits brittleness and poor foldability.
3. Coating is highly sensitive to water spotting and to damp "rub-off."

A simple technique is available to study the extent of associative bonding within a free starch film and the strength of this bonding. A cooked starch paste

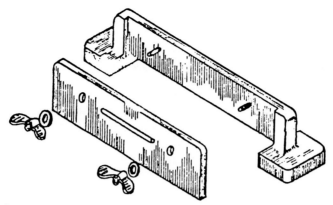

Fig. 2.47 Starch film applicator. (From T. J. Schoch, Corn Products Co., Inc., Argo, Illinois.)

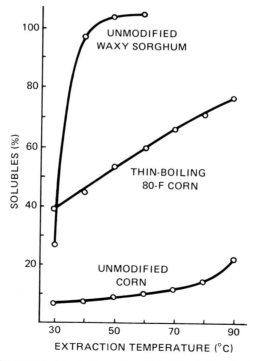

Fig. 2.48 Solubility patterns of starch films. (From T. J. Schoch, Corn Products Co., Inc., Argo, Illinois.)

or solution is coated onto a silicone-treated stainless steel plate with an adjustable applicator (Fig. 2.47). The applicator is readjusted with thickness gauges to allow a wet film thickness that will dry down to 1 mm. The film is carefully dried under controlled conditions of humidity and temperature, then ground to pass through a 50-mesh sieve (50 openings per linear inch). The water-solubility of this film is determined over a range of temperatures, in much the same fashion as the testing for the swelling power and solubility of granular starches. Solubility is then plotted as a function of extraction temperature (Fig. 2.48). For example, internal bonding within a film of unmodified corn starch is extensive because the film is less than 10% soluble at room temperature. This internal bonding is also strong; solubility is not greatly increased by extracting at 90°C. Obviously, this strength represents associative bonding of linear molecules within the film. In contrast, unmodified waxy sorghum starch is quite extensively bonded, but the strength of this bonding is weak, being completely dissociated at 45–50°C. The thin-boiling acid-modified corn starch is intermediate with both the extent and strength of its internal bonding moderate.

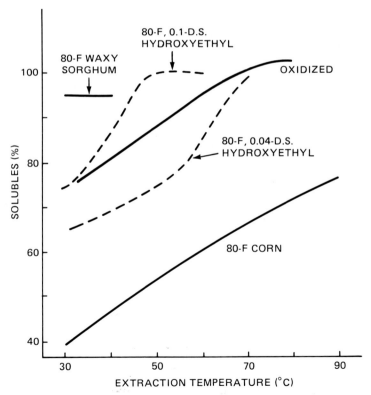

Fig. 2.49 Film solubilities of various thin-boiling starches. (From T. J. Schoch, Corn Products Co., Inc., Argo, Illinois.)

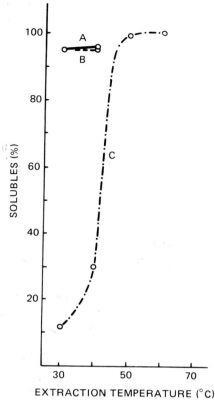

Fig. 2.50 Solubilities of films of acid-modified waxy sorghum starch dried at (A) 30°, (B) 80°, and (C) 4°C. (From T. J. Schoch, Corn Products Co., Inc., Argo, Illinois.)

If we compare films from a variety of thin-boiling starches, all of approximately the same molecular size, but representing different industrial modifications (Fig. 2.49), we find the following effects:

1. The acid-modified 80-fluidity corn starch depicted in Fig. 2.49 is the same as in Fig. 2.48, showing moderate strength and moderate associative bonding.
2. Films of oxidized corn starch show much less internal association, in accord with the oxidative warping of the linear fraction.
3. Hydroxyethyl corn starches show progressive dissociation as the level of hydroxyethyl groups is increased from 4 to 10 per 100 glucose units.
4. Acid-modified waxy starch is completely dissociated. Because of its high cold water solubility, such a product is a preferred remoistening adhesive for such uses as labels and gummed paper tape.

The conditions under which a starch film is dried have an important effect on the associative bonding within the film. This fact is exemplified by the solubility patterns of 80-fluidity acid-modified waxy sorghum starch dried at 80, 30, and 4°C (designated as 80-fluidity white milo starch, as shown in Fig. 2.50). The films dried at 80°C and at 30°C are completely dissociated, and completely soluble in water at room temperature. The film slowly dried at 4°C is extensively associated, however. The strength of its associative bonding is weak, and the film becomes completely soluble in water at 45–50°C. The effect follows exactly the same principle as that described for the staling of bread: the slow associative bonding of branched starch molecules at low temperature is completely reversed merely by heating to 50°C in the presence of water. This "staling" reaction has only recently been recognized. Whenever branched starch molecules are held in concentrated solution or exposed to high humidity for extended periods of time, particularly at low temperature, the normally dilated branches tend to fold and to associate with one another. The reaction parallels retrogradation of the linear fraction; however, since the branch length is so short (only 25 glucose units), this associative bonding is much weaker than retrogradation and is readily reversed by heating to 50°C. Staling of branched starch systems is actually quite common. Various examples may be cited:

1. Tests at a paper mill showed that both wet-adhesive and dry-adhesive strength of starch coatings were substantially improved by slow drying of the coated paper at room temperature. Unfortunately, this is not practical in present high-speed paper processing.

2. When roll-dried acid-modified corn starch (80-fluidity) was held at 93% relative humidity for 1 and 2 weeks, its film solubility dropped substantially (Fig. 2.51). The high humidity introduces enough moisture into the pregelatinized starch to permit some free motion of the molecules. Short linear segments slowly associate, and the rewetting quality of the pregelatinized starch slowly deteriorates.

3. Prolonged and improper warehousing of pregelatinized corn starch for use in foundry core binders deteriorated the so-called green adhesive strength of the binder when admixed with wet sand. In this case, the wet sand core may sag and deform before it is dried.

4. When a garment of light-blue percale is sized with any common home laundry starch, then dried in a heated dryer, the common practice is to sprinkle the dried garment with water before ironing. This may cause extensive water spotting, as the droplets of water dissolve the dissociated and soluble starch sizing. But if the starched garment is dried on an outdoor clothesline instead of in a heated dryer, the starch sizing "stales" to an insoluble condition and will not water spot when sprinkled with water before ironing.

Two major starch problems attract considerable attention among research staffs of paper companies. The first concerns the intentional insolubilization of

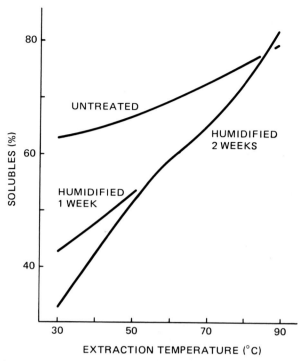

Fig. 2.51 Effect of 93% relative humidity on solubility of roll-dried, acid-modified corn starch. (From T. J. Schoch, Corn Products Co., Inc., Argo, Illinois.)

starch coatings and adhesives after they are applied. Ideally, it should be possible to apply a starch coating or adhesive to the paper in a completely dissociated and soluble state and then to introduce some chemical curing *in situ* to produce an insoluble network. This would provide a starch with maximum wet and dry strength, which could be used as coating or as an adhesive for laminated or corrugated cardboard. Although this has not been accomplished in a successful commercial sense, an attempt has been made by incorporation of aldehydic resins that copolymerize and cross-bond with the starch coating or adhesive after application. For example, the addition of 5 to 10% of urea–formaldehyde (U–F) just before casting the film markedly reduces the solubility of acid-modified corn starch. (See Fig. 2.52.) Such coatings show improved water resistance, but they are not actually waterproof. Many of these copolymerizing agents are much more effective on the acid side of the pH level, that is, below pH 4.0. Paper deteriorates slowly at this acidity, however, and many printing inks will not cure properly under these conditions.

A second problem concerns the reduction in weight of coated papers. A substantial part of the weight of a magazine is the clay pigment used as opacifier.

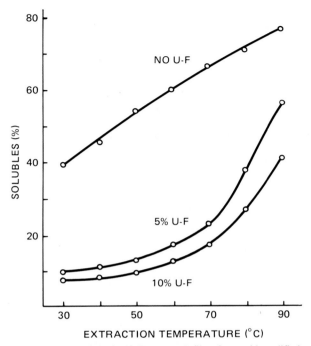

Fig. 2.52 Effect of urea–formaldehyde (U–F) on starch films from acid-modified corn starch (80-fluidity). (From T. J. Schoch, Corn Products Co., Inc., Argo, Illinois.)

With the rapid rise in postal rates, subscription magazine mailing charges have become an important consideration. Efforts are being made to "foam coat" printed paper with an emulsion of air in starch or other adhesive matrix. If the air bubbles could be reduced to several tenths of a micron in diameter, such a foam coating might have optical opacity and brightness equal to clay—and an air opacifier costs nothing to ship!

REFERENCES

Banks, W., and Greenwood, C. T. (1975). "Starch and its Components." Wiley, New York.
Bautlecht, C. A.(1953). "Starch." Van Nostrand–Reinhold, Princeton, New Jersey.
Bechtel, D. B., Pomeranz, Y., and de Francisco, A. (1978). Breadmaking studied by light and transmission electron microscopy. *Cereal Chem.* **55,** 392–401.
Chabot, J. F., Hood, L. F., and Liboff, M. (1979). Effect of scanning electron microscopy preparation methods on the ultrastructure of white bread. *Cereal Chem.* **56,** 462–464.
Derby, R. I., Miller, B. S., Miller, B. F., and Trimbo, H. B. (1975). Visual observation of wheat starch gelatinization in limited water systems. *Cereal Chem.* **52,** 702–713.

Drews, E., and Seibel, W. (1974). Erkentnisse über die Roggenmehlbackfähigkeit und ihre Auswirkung auf die Diagrammführung. *Getreide, Mehl Brot* **28**, 307–315.

Evans, L. G., Volpe, T., and Zabik, M. E. (1977). Ultrastructure of bread dough with yeast single cell protein and/or emulsifiers. *J. Food Sci.* **42**, 70–74.

Fretzdorff, B., Bechtel, D. B., and Pomeranz, Y. (1982). Freeze-fracture ultrastructure of wheat flour ingredients, dough, and bread. *Cereal Chem.* **59**, 113–120.

Greenwood, C. T. (1976). Starch. *Adv. Cereal Sci. Technol.* **1**, 119–157.

Hahn, R. R. (1969). Tailoring starches for the baking industry. *Baker's Dig.* **43**(4), 48–53, 64.

Jelaca, S. L., and Hlynka, I. (1971). Water binding capacity of wheat flour crude pentosans and their relation to mixing characteristics of dough. *Cereal Chem.* **48**, 211–222.

Kerr, R. W. (1950). "Chemistry and Industry of Starch." Academic Press, New York.

Khoo, U., Christianson, D. D., and Inglett, G. E. (1975). Scanning and transmission microscopy of dough and bread. *Baker's Dig.* **49**(4), 24–26.

Kite, F. E., Maywald, E. C., and Schoch, T. J. (1963). Functional properties of food starches. *Staerke* **15**, 131–138.

Knight, J. W. (1965). "Chemistry of Wheat Starch and Gluten." Leonard Hill, London.

Knight, J. W. (1969). "The Starch Industry." Pergamon, Oxford.

Kulp, K., and Lorenz, K. (1981). Starch functionality in white pan breads—new developments. *Baker's Dig.* **55**(5), 24, 25, 27, 28, 36.

Leach, H. W., McCowen, L. D., and Schoch, T. J. (1959). Structure of the starch granule. I. Swelling and solubility patterns of various starches. *Cereal Chem.* **36**, 534–544.

Lineback, D. R., and Wongsrikasem, F. (1980). Gelatinization of starch in baked products. *J. Food Sci.* **145**, 71–74.

Medcalf, D. G., Youngs, V. L., and Gilles, K. A. (1968). Wheat starches. II. Effect of polar and nonpolar lipid fractions on pasting characteristics. *Cereal Chem.* **45**, 88–95.

Miller, B. S., Derby, R. I., and Trimbo, H. B. (1973). A pictorial explanation of the increase in viscosity of a heated wheat starch water suspension. *Cereal Chem.* **50**, 271–280.

Olkku, J., and Rha, C. (1978). Gelatinisation of starch and wheat flour starch. *Food Chem.* **3**, 293–317.

Petersen, N. B. (1975). "Edible Starches and Starch-Derived Syrups." Noyes Data Corp., Park Ridge, New Jersey.

Pomeranz, Y. (1976). Scanning electron microscopy in food science and technology. *Adv. Food Res.* **22**, 205–307.

Pomeranz, Y., Meyer, D., and Seibel, W. (1984a). Wheat, wheat-rye, and rye dough and bread studied by scanning electron microscopy. *Cereal Chem.* **61**, 53–59.

Pomeranz, Y., Meyer, D., and Seibel, W. (1984b). Structures of wheat, wheat-rye, and rye breads. *Getreide, Mehl Brot* **38**, 138–146.

Radley, J. A., ed. (1954). "Starch and Its Derivatives." Wiley, New York.

Radley, J. A., ed. (1976). "Starch Production Technology." Appl. Sci. Publ., Ltd., London.

Schoch, T. J. (1952). Fundamental developments in starch for paper coating. *Tappi* **35**(7), 1–8.

Schoch, T. J. (1961). Starches and amylases. *Proc. Am. Soc. Brew. Chem.*, pp. 83–92.

Schoch, T. J., and Elder, A. L. (1955). Starches in the food industry. *Adv. Chem. Ser.* **12**, 31–34.

Selby, K. (1977). The role of cereal-based products. *Chem. Ind. (London)* pp. 494–498.

Stephan, H. (1982). Merkmale verschiedener Sauerteigführung und Brotqualität. *Getreide, Mehl Brot* **36**, 16–18.

Vassileva, R., Seibel, W., and Meyer, D. (1981). Backparameter und Brotqualität. IV. Strukturveränderungen der Stärke in der Brotkrumme beim Backen. *Getreide, Mehl Brot* **35**, 303–305.

Wasserman, L., and Dorfner, H. H. (1974). Raster-Elektron-Mikroskopie von Gebäcken. *Getreide, Mehl Brot* **28**, 324–328.

Whistler, R. L. (1971). Alsberg-Schoch Award Lecture: Starch and polysaccharide derivatives in the food and nonfood industries. *Cereal Sci. Today* **16,** 54–59, 73.

Whistler, R. L., and Paschall, E. F., eds. (1965). "Starch: Chemistry and Technology," Vol. 1. Academic Press, New York.

Whistler, R. L., and Paschall, E. F., eds. (1967). "Starch: Chemistry and Technology," Vol. 2. Academic Press, New York.

Wurzburg, O. B., and Szymanski, C. D. (1970). Modified starches for the food industry. *Agric. Food Chem.* **18,** 997–1001.

Yasunaga, T., Bushuk, W., and Irvine, G. N. (1968). Gelatinization of starch during baking. *Cereal Chem.* **45,** 269–279.

3

Carbohydrates: Structural Polysaccharides, Pectins, and Gums

I. Introduction

According to Sanderson (1981) and Glicksman (1962, 1963, 1969, 1976), polysaccharides in foods include (a) starches (raw, pregelatinized, and modified), (b) cellulose and cellulose derivatives, (c) seaweed extracts (alginates, carrageenans, agar, furcellaran), (d) plant exudates or gums (arabic, karaya, and tragacanth), (e) seed gums (locust bean, guar), (f) plant extracts (pectins), and (g) microbial gums (xanthan). Starches were described in Chapter 2; this chapter describes the remaining types of polysaccharides. Polysaccharides encompass three functional properties (Zimmermann, 1979). First, they serve functions in the plant itself as sources of reserve nutrients, structural entities, water-binding moieties, and components that govern osmotic pressure. Second, they fulfill functions in foods as moieties that govern–control–influence the shape, texture, water binding, and overall sensory properties. Finally, polysaccharides have functions in nutrition as sources of nutrients and dietary fiber (see Kraus, 1975; Spiller and Amen, 1975a,b; Sharma, 1981). Functional properties in foods affect (Glicksman, 1969) the following:

1. Water-binding capacity
2. Rheological properties
3. Capacity to form films or gels
4. Capacity to bind flavor compounds

5. Osmotic pressure
6. Hygroscopicity
7. Chemical reactivity
8. Sweetening and taste enhancement
9. Resorption

These properties have numerous technological and application capabilities that include thickening, emulsification, stabilization of emulsions and foams, flow properties, texture, softness retention, browning, fermentability, control of microbial and enzymatic modifications, and stabilization of taste and flavor.

The basic components of polysaccharides in foods are glucose, galactose, and mannose, the hexuronic acids, and the pentoses arabinose, rhamnose, and xylose. While a few polysaccharides are made from simple, single sugars, most are heterogeneous. For example, guar gum is composed of D-mannose and D-galactose, and xanthan gum is composed of D-glucose, D-mannose, and D-glucuronic acid. Properties of the polysaccharides are governed by molecular weight, configuration, and the properties of individual components–groups, that is, reducing carbonyl groups, free hydroxyl groups, carboxyl groups in hexuronic acids, sulfate ester groups in carrageenan or methyl ester groups in galacturonic acid of pectins. In addition, functional groups can be modified through introduction of phosphate esters in starches (see Chapter 2), carboxymethyl groups in carboxymethylcellulose, or methoxyl groups in methylcellulose (Zimmermann, 1979). These groups influence solubility, the effects of salt and pH, and the capacity to interact with polyelectrolytes (i.e., carrageenan with proteins). The effects depend on degree of substitution, localization, and distribution of substitution groups. The linkages in starch are 1,4 and 1,6. The linkages in pectins are 1,4 (see Fig. 3.1) and in other polysaccharides mainly 1,4 and 1,3.

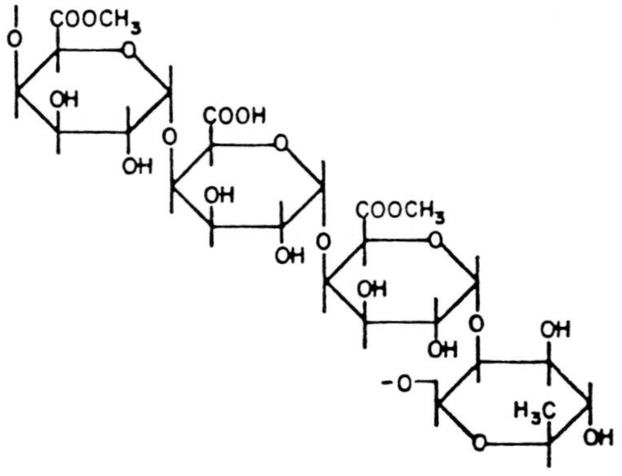

Fig. 3.1 Segment of pectin molecule containing three galacturonic acid units and one rhamnose unit. (From Baker, 1981; copyright by the American Chemical Society.)

Following is a list of the compositional characteristics of carbohydrates that affect their functional properties (Zimmermann, 1979):

1. *Molecular components*
   Hexoses, pentoses, uronic acids
   Configuration
   Ring configuration
2. *Functional groups*
   Type: carboxyl, sulfate ester, phosphate ester, hydroxyl, methyl ester
   Localization in the molecule
   Degree of substitution
3. *Structure*
   Glycosidic linkage
   Type of structure
   Degree of branching
   Degree of polymerization
   Distribution of basic units
4. *Conformation*
   Helical, band, net, association, micelle

Functional properties of polysaccharides are affected by noncovalent bonds, in particular hydrogen bonds (as they relate to interaction with water and polar compounds and to associations), ionic linkages, and complexing of lipids. The α-helical structure of amylose discussed in Chapter 2 is depicted in Fig. 3.2. Formation of a three-dimensional network of a gel and of the egg-box model of calcium pectate are shown in Fig. 3.3.

Stable conformations require large numbers of noncovalent bonds (hydrogen, dipolar, ionic). Associations require long segments with regular sequences. According to Morris (1973), polysaccharides are highly hydrophilic giant molecules

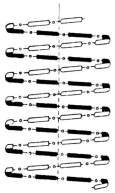

Fig. 3.2 Helical structure of amylose. (From Zimmermann, 1973.)

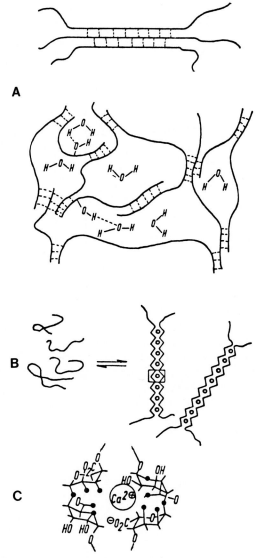

Fig. 3.3 Conformation of polysaccharides. (A) Top, micelle formation; bottom, three-dimensional gel network. (B) Association of polygalacturonic acid sequences through chelation of Ca^{++} ions according to the egg-box model. Left, random association; right, organized association (bands). (C) Chelation formula. ●, Oxygen atoms bind Ca ions coordinatively. (From Zimmermann, 1979.)

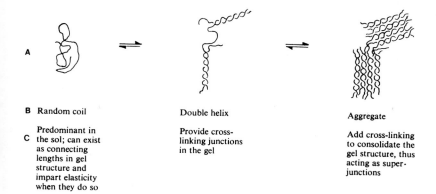

A

B Random coil   Double helix   Aggregate

C Predominant in   Provide cross-   Add cross-linking
the sol; can exist   linking junctions   to consolidate the
as connecting   in the gel   gel structure, thus
lengths in gel   acting as super-
structure and   junctions
impart elasticity
when they do so

D   Sol ⇌ Incipient gel ⇌ Clear elastic gel ⇌ Stiff gel ⇌ Turbid rigid gel ⇌ Phase separation; syneresed gel

Fig. 3.4 States of polysaccharide molecules and their role in gel properties. (A) Pictorial view of molecular states. (B) Name. (C) Significance for gel structure. (D) Corresponding range of gel properties. (From Rees, 1972.)

that can affect radically the physical properties of up to 100 times their own weight of water. They are, therefore, used widely in the food industry as thickeners, texture modifiers, and emulsion stabilizers. Their water-binding capacities can be shown dramatically in gel formation. For example, 1% calcium alginate can hold 99% water in a gel that has a firm and crisp structure. Maintenance of shape in such gels arises from the cooperative association of long, structurally regular regions of chains in ordered conformations. Two types of linkages are involved. In the uronic acid polymers, alginate and pectin, the junction zones are microcrystalline regions, in which several chains align with divalent counterions sandwiched in between them. They have a geometric regularity of microscopic crystals. In carrageenan, structurally regular junctions are formed by chains that intertwine into double helices. These helices aggregate by cooperative interactions (Fig. 3.4). In each case, the junctions are terminated by structural irregularities. Those irregularities destroy the geometric repetitiveness and prevent any complete aggregation that might lead to precipitation of the polysaccharide. Even minute changes in the number of such breaks in junctions may alter radically the gel structure. In some cases a geometric fitting of structurally regular regions of two polysaccharides takes place. This results in the formation of strong gels, as in locust bean gum and certain bacterial polysaccharides, neither of which gels on its own. Such interactions have many practical implications in governing the texture of natural foods, in the use of polysaccharides as food additives, and in the interactions of polysaccharides and other food components (usually proteins) (Morris, 1973).

Rheological properties of polysaccharides are affected by (Zimmermann,

1979) molecular size and conformation, orientation of molecule, association of molecule, water binding and swelling, concentration, particle size (distribution, shape, and specific surface as they affect solubility and dispersibility), degree of dispersion, and viscosity of the dispersing medium. Flow properties of gums useful to the food industry are described by Krumel and Sarker (1975).

Viscosity of hydrocolloidal polysaccharides increases during solubilization until maximum swelling and water binding takes place; at the stage of molecular dispersion, viscosity decreases somewhat. Hydrated, stretched polysaccharides occupy approximately spherical spaces in solution. Contact–interaction among the spheres increases viscosity in dilute solutions. The concentration effect is not linear (Zimmermann, 1979). Excess water facilitates unfolding and high mobility. High concentrations result in large increases in viscosity through binding of water and interaction of particles. Linear polymers have higher viscosities than branched ones, at comparable concentrations and relative molecular weights. Highly branched hydrated polysaccharides have lower effective volumes than linear polymers of equal molecular weight. Salt inhibits hydration and thereby reduces viscosity. Acids and bases have little effect on viscosity of non-ionic colloids in the 3.0–10.0 pH range.

II. Cellulose and Other Cell Wall Components

Cell wall components contain many undefined polymers varying in size and in composition. Part of the cellulose is soluble in alkali. The insoluble fraction is called α-cellulose. The fraction in the alkaline extract precipitated upon acidification is β-cellulose, and the fraction remaining in solution is γ-cellulose, which contains mainly hemicelluloses.

Hemicelluloses comprise a mixture of alkali-soluble polysaccharides, the amount and composition of which vary with the extraction procedure. These include polymers of mannose, galactose, xylose, and arabinose, as well as polymers of uronic acids and of methyl- or acetyl-substituted monoses. Cellulose and hemicelluloses are among the most abundant and widely distributed carbohydrates in plant materials. They are the main cell wall constituents of fruits, vegetables, and cereals. Their content increases with maturity; they are highly concentrated in pericarp tissues of cereal grains and present in low concentration in the starchy endosperm (see Sterling, 1963). *Lignin* is found in trace amounts in high quality fruits and vegetables, but is present in larger amounts in the bran of cereal grains and in some spices. *Pentosans* are polymers of pentoses and methyl pentoses. Pentosans of the wheat endosperm belong to the arabinoxylan series of polysaccharides (D'Appolonia and Kim, 1976). Some are present in the free form while others are conjugated with proteins. Wheat flour contains about 3% total pentosans, of which less than one-third is water soluble.

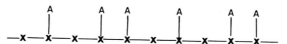

Fig. 3.5 Structure of highly branched arabinoxylan. A, L-arabinofuranose; X, D-xylopyranose. (From D'Appolonia and Kim, 1976.)

The basic structure of water-soluble pentosans from wheat flour is given in Fig. 3.5. The main characteristics of the repeating unit are a straight chain of anhydro-D-xylopyranose residues linked β-1,4, to which are attached anhydro-L-arabinofuranose residues at the 2 or 3 positions of the D-xylose units. Wheat flour pentosans contribute significantly to dough consistenty. They have two unique properties that distinguish them from the common hemicelluloses of the plant cell wall. First, they are readily dispersible in water and form highly viscous solutions. Second, they form gels upon addition of oxidizing agents. A hypothetical structure of a glycoprotein from wheat flour that causes gelation is given in Fig. 3.6. The proposed mechanism of oxidative gelation is illustrated in Fig. 3.7. In addition to arabinoxylans, an arabinogalactan has been isolated (see Fig. 3.8). The role of pentosans in brewing is described by Steiner (1968).

The most important cellulose derivative for food applications is carboxymethylcellulose (CMC). CMC is available in various forms that differ in molecular weight and degree of substitution. The basic function of CMC is to bind water or impart viscosity to the aqueous phase, stabilize the other ingredients, or

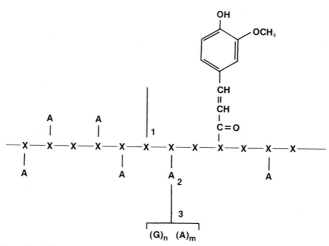

Fig. 3.6 Hypothetical structure of glycoprotein 2 from wheat flour as proposed by Neukom *et al.* (1967). Vertical line at 1 indicates polypeptide chain; X, β-D-xylopyranose units; A, α-L-arabinofuranose units; G, galactose units; 1, 2, 3, possible linkages between carbohydrates and protein. Linkage at 3 was established by Fincher *et al.* (1974).

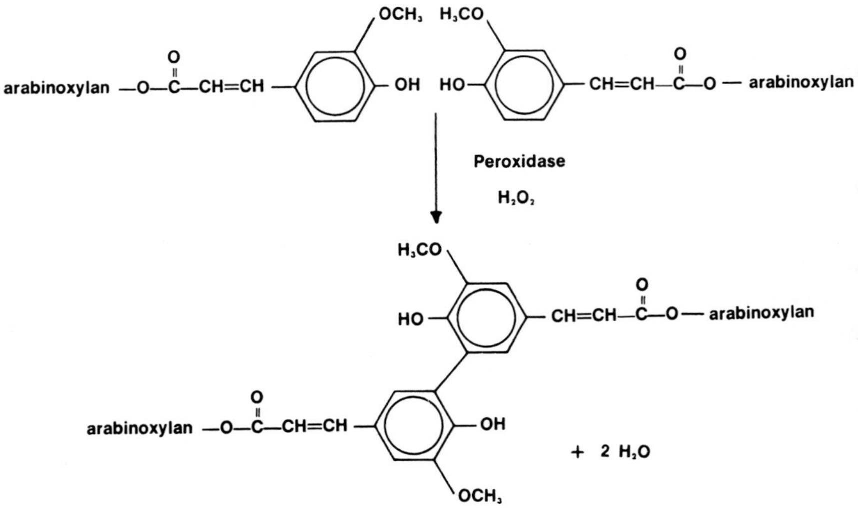

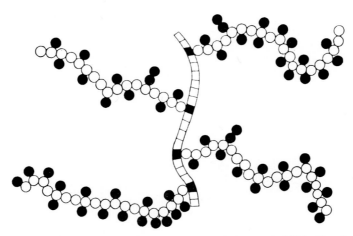

Fig. 3.7 Proposed mechanism for oxidative gelation by Geissmann and Neukom (1973). Oxidative phenolic coupling of ferulic acid residues forms diferulic acid cross-linkages.

Fig. 3.8 Proposed model of arabinogalactan-peptide by Fincher *et al.* (1974). ○, galactose; ●, arabinose; ■, hydroxyproline; □, other amino acids. Arabinose is present in the α-L-arabinofuranosyl configuration. The D-galactopyranose residues are linked to hydroxyproline by glycosidic linkages in the β-anomeric configuration.

TABLE 3.1

Uses of Cellulose Hydrocolloids, Seasonings, Flavor Enhancers, and Gelatinized Wheat Starch[a]

Food	Product	Use level (%)	Functions
Extruded foods	Klucel[b]	5–20	Thermoplastic binder for extrusion
	Cellulose gum[c]	0.2–0.4	Binder
Fruits and vegetables			
Cereals			
Canned fruits and vegetables	Cellulose gum	0.1–0.5	Improves texture; thickener
Instant cereals	Cellulose gum	0.025–0.5	Improves texture; fast rehydration
Low-calorie products			
Foods and beverages (low calorie)	Cellulose gum	0.05–0.3	Improves mouth-feel and body; suspending aid
Syrups (low calorie)	Cellulose gum	0.1–0.4	Thickener; improves mouth-feel
Spreads (low calorie)	Cellulose gum	0.1–0.3	Provides body
Pet foods			
Canned pet foods	Cellulose gum	0.2–0.5	Thickener; processing aid
Dry pet foods (gravy type)	Cellulose gum	0.3–0.5	Instant gravy thickener; extrusion aid
Semimoist pet foods	Cellulose gum	0.2–0.4	Improves moisture retention; extrusion aid; improves appearance
Semimoist pet foods	HVP[d]	0.5	Flavor enhancer
Protein foods			
Batters for deep-fat frying	Cellulose gum	0.1–0.4	Binder; fat and oil barrier; viscosity control and "pickup"
Meats	HVP	0.5–5.0	Seasoning and flavor enhancer
Meat dishes	HVP	0.5–5.0	Meat tenderizer; flavoring agent
Seafood	HVP	0.5–5.0	Flavor enhancer
Seasonings, soups, and sauces			
Bouillon	HVP	5–50	Flavor ingredient
Gravies	HVP	0.5–2.0	Flavor ingredient
	Cellulose gum	0.1–0.3	Thickener
Sauces	HVP	0.5–2.0	Flavoring agent
	Cellulose gum	0.1–0.5	Thickener
Seasonings	HVP	1–50	Flavor ingredient
Soups	HVP	0.5–5.0	Flavoring agent
	Cellulose gum	0.2–0.4	Processing aid; thickener
Snack and specialty foods			
Dips	HVP	0.25–0.75	Flavor ingredient
	Cellulose gum	0.1–0.3	Thickener; stabilizer
	Klucel	0.1–0.4	Thickener; stabilizer

(continued)

TABLE 3.1 (*Continued*)

Food	Product	Use level (%)	Functions
Eggs (frozen or dried)	Cellulose gum	0.25	Stabilizer; viscosity control
Jams and jellies	Cellulose gum	0.1–0.4	Thickener; stabilizer
Nuts	Klucel	0.1–0.4	Protective coating; oil barrier
	Cellulose gum	0.1–0.3	Binder for salt, spices, etc.
Snack items	HVP	0.1–1.0	Flavor ingredient
	Klucel	5–20	Thermoplastic binder for extrusion
Table syrup	Cellulose gum	0.1–0.15	Thickener; stabilizer

[a]From Hercules, Inc., Wilmington, Delaware.
[b]Klucel is hydroxypropylcellulose.
[c]Cellulose gum is Na-carboxymethylcellulose.
[c]HVP is hydrolyzed vegetable protein.

prevent syneresis (Sanderson, 1981). Stabilization is necessary to control the formation of ice crystals in ice cream, for example. In addition, CMC gives a product pleasing texture, body, and melting characteristics. CMC is often used in combination with gelatin, pectin, or locust bean gum. Because of its high water-binding capacity CMC can be used as a bulking agent in dietetic foods. CMC can prevent precipitation of soy or milk proteins at pH values close to the IP of those proteins. A disadvantage of CMC is that at low pH it loses viscosity and has a tendency to precipitate.

Other ethers of cellulose are also permitted in foods. Methylcellulose and hydroxypropylmethylcellulose gel upon heating. In this respect they are similar to some proteins. The gels are oil-repellent and so reduce oil absorption. Inclusion of these ethers in batters and coatings inhibits oil absorption and improves adhesion of the batter or coating to a food product. These cellulose derivatives also can function as emulsifiers; they exhibit stability because they contain no negatively charged groups whose ionization is suppressed at low pH levels, which could result in precipitation or gelation of the polysaccharide chains. Hydroxypropylcellulose and methylethylcellulose precipitate, rather than gel, in hot water. Uses of cellulose hydrocolloids in various foods are listed in Tables 3.1 and 3.2.

Microcrystalline cellulose (MCC) is produced by controlled acid hydrolysis of cellulose fibers. It is manufactured in two basic forms: fine powder and colloid MCC (Thomas, 1982). The principal applications of the powder MCC are food tablet formation, anticaking agents and flow aids, and nonnutritive sources of food solids and dietary fiber in dietetic foods. Some of the uses of colloidal MCC include stabilization of emulsions and foams, stabilization at high temperatures, nonnutritive fillers and thickeners, aqueous stabilizers and gelling agents, texture

TABLE 3.2

Uses of Cellulose Hydrocolloids, Seasonings, Flavor Enhancers, and Gelatinized Wheat Starch[a]

Food	Product	Use level (%)	Functions
Baked goods			
Baked goods	Cellulose gum[b]	0.1–0.4	Improves moisture retention
Cakes	Cellulose gum	0.1–0.3	Better moisture retention and keeping qualities
Cakes (dry mixes)	Cellulose gum	0.1–0.3	Controls batter viscosity; improves cake quality
Doughnuts	Cellulose gum	0.1–0.25	Improves moisture retention, keeping qualities, and oil "holdout"
Icings	Cellulose gum	0.05–0.5	Improves moisture retention and texture; controls sugar crystal size
Meringues	Cellulose gum	0.1–0.5	Tenderizer; stabilizer; inhibits "weeping"
Pie fillings	Cellulose gum	0.1–0.5	Thickener; inhibits syneresis in presence of starch
Beverages			
Beverages	Cellulose gum	0.025–0.3	Improves mouth-feel and body
Fruit drinks	Cellulose gum	0.1–0.3	Improves mouth-feel and body; pulp suspension
Juices	Cellulose gum	0.025–0.3	Improves mouth-feel and body; suspending aid
Milk drinks	Cellulose gum	0.1–0.5	Thickener
Orange-flavored milk beverages	Cellulose gum	0.2–0.3	Stabilizer casein; thickener
Orange juice (frozen concentrate or dry mix)	Cellulose gum	0.1–0.5	Improves mouth-feel and body; suspending aid
Protein beverages	Cellulose gum	0.01–0.1	Thickener; protein stabilizer; suspending aid
Syrups			
fountain	Cellulose gum	0.1–0.5	Stabilizer; thickener
slush	Cellulose gum	0.1–0.3	Controls ice crystal size; improves mouth-feel
Soya beverages	Cellulose gum	0.1–0.3	Thickener; protein stabilizer
Desserts			
Confections	Cellulose gum	0.1	Increases moisture retention
	Cellulose gum	0.2–0.4	Controls sugar crystal size
	Klucel[c]	0.1	Edible coatings and glazes
Frostings	Cellulose gum	0.1–0.3	Controls texture and sugar crystal size

(*continued*)

TABLE 3.2 (*Continued*)

Food	Product	Use level (%)	Functions
Frozen desserts	Cellulose gum	0.1–0.3	Improves mouth-feel, body, and texture; controls ice crystal size
Gelatin desserts	Cellulose gum	0.1–0.3	Improves texture; inhibits syneresis
Ice cream	Cellulose gum	0.15–0.4	Improves mouth-feel, body, and texture; controls ice crystals size
Mousses	Klucel	0.25	Whipping aid; stabilizer
	Cellulose gum	0.1	Stabilizer
Products thickened with starch	Cellulose gum	0.1–0.3	Inhibits syneresis; improves thickening and texture
Puddings	Starch	12–26	Ingredient
	Cellulose gum	0.2–0.4	Improves texture; inhibits syneresis; permits use of less starch
Toppings	Klucel	0.2–0.5	Stabilizer; whipping aid
	Cellulose gum	0.1	Stabilizer

[a]From Hercules, Inc., Wilmington, Delaware.
[b]Cellulose gum is Na-carboxymethylcellulose.
[c]Klucel is hydroxypropylcellulose.

modifiers, suspending agents, and control of ice crystal formation in frozen desserts. Table 3.3 shows the typical properties of commercial types of MCC.

III. Seaweed Extracts

Seaweed extracts include the natural gums (agar, carrageenan, furcellaran, and alginates) and the modified (semisynthetic) gum propylene glycol alginate (Kelcoloid) (Whistler and Smart, 1953; Lawrence, 1973, 1976; Guisley, 1968; Glicksman, 1969; Sanderson, 1981). Agar is used to a very limited extent in foods. Its main usefulness is resistance to heat. Agar, furcellaran, and carrageenan are extracts of red seaweeds (Rhodophyceae). They are all galactose polymers characterized by thermoreversible gelation. Agar has a particularly low setting and high melting temperature. Idealized formulas of covalent structures of polysaccharides having helix cross-linkages in their frameworks [i.e., agarose (agar), furcellaran (Danish agar), κ-carrageenan, and ι-carrageenan] are depicted in Fig. 3.9. Other structural features do, of course, exist and are important for gel properties. Carrageenans are the most important red seaweed polysaccharides used by the food industry. They contain three fractions—lambda (λ), iota (ι),

TABLE 3.3

Typical Properties of Commercial MCC Types[a]

Form	% MCC	Equipment required to activate	Use levels in foods (%)	Average particle size (μm)	Set-up viscosity (centipose)	Primary use
Powdered (spray-dried)	100	Dry blend	0.5–5	50	—	Tableting; oil carrier, flow aid, dietetic filler
	100	Dry blend	5–15	100	—	Tableting
	100	Dry blend	0.5–1.5	20	—	Opacifier, dietetic filler
				% colloidal (<0.2 μm)		
Colloidal						
bulk-dried	91	Homogenizer	0.5–3.0	30	(2.1%) 750–1300	Whipped toppings
	89	Homogenizer	0.3–0.8	70	(1.2%) 800–1450	Heat-stable emulsions, Frozen desserts, Emulsions
spray-dried	88	High speed mixer	0.3–1.0	70	(1.2%) 900–1600	General food stabilizer, Thixotropic gels
Coprocessed Whey	85	High speed mixer	0.2–2.5	70	(1.2%) 50–300	Pourable systems
	22	Dry blend or slow mixing in H₂O	2–4	15	(4.0%) 500–2000	Dry-blended foods

[a]Courtesy of FMC Corporation, Philadelphia, Pennsylvania.

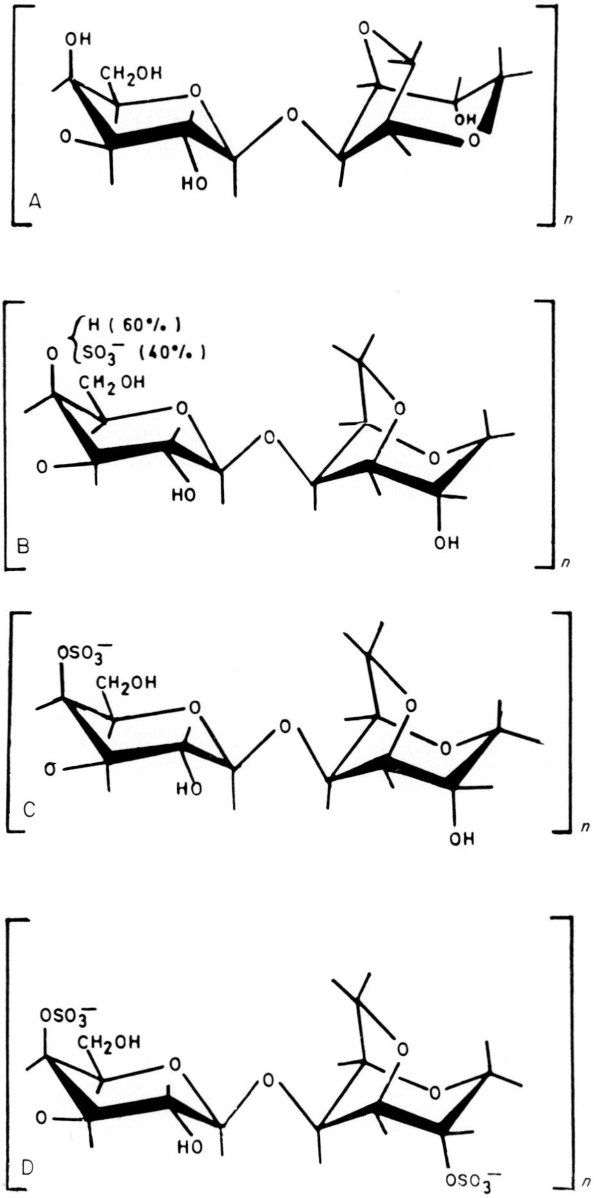

Fig. 3.9 Covalent structures (idealized formulas) of polysaccharides having helix cross-linkages in their gel frameworks. (A) Agarose (agar). (B) Danish agar. (C) κ-carrageenans. (D) ι-carrageenans. (From Rees, 1972.)

and kappa (κ)—which differ in sulfate ester and 3,6-anhydrogalactose content. The λ-fraction requires potassium ions for gelation and yields brittle gels, which are prone to syneresis. The ι-carrageenan gels are formed in the presence of calcium ions and are more elastic. The κ-carrageenan is nongelling. Commercial products contain a mixture of the three fractions. Solubility is related to the level of hydrophilic sulfate ester groups in the carrageenan molecule. Carrageenan has the ability to stabilize milk proteins in milk products ("milk reactivity"). It produces water gels important to the production of low-calorie jams and jellies. Carrageenans are also used in combination with other hydrocolloids as thickeners and stabilizers. Furcellaran (also known as Danish agar) has properties similar to those of carrageenan. It is used in Europe in the preparation of jams, marmalades, and jellies.

Alginates are extracted from brown seaweeds (Phaeophyceae). They are linear copolymers of D-mannuronic acids (M-blocks) and L-glucuronic acids (G-blocks) and contain regions composed entirely of one of the acids (Fig. 3.10). Solubility properties of alginates depend on the M-blocks/G-blocks ratio and the

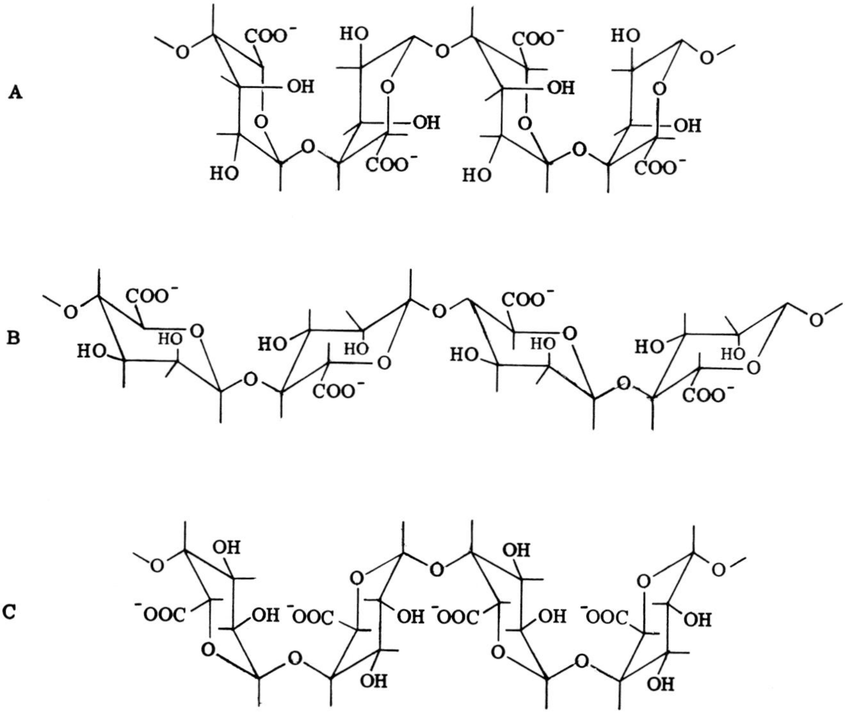

Fig. 3.10 Block shapes in alginates and pectins. (A) α-1,4-linked L-glucuronic acid units in alginates. (B) β-1,4-linked D-mannuronic acid units in alginates. (C) α-1,4-linked D-galacturonic acid units in pectins. (Copyright by the Institute of Food Technologists.)

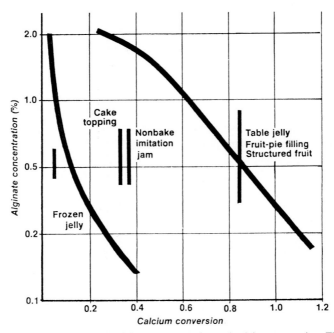

Fig. 3.11 Type of gel varies with alginate concentration and calcium conversion. Three types of gelation are shown here. (Left) Thickening occurs, as with frozen jelly. (Right) Table jelly and structured fruits in area of firm gels or suspensions. (Center) Thixotropic mixtures, such as toppings. These areas can be varied by careful formulation. (From Hannigan, 1983.)

block structure. Gels prepared with alginates that are rich in glucuronic acid are stronger, more brittle, and less elastic than those prepared with alginates rich in mannuronic acid. Alginate gels are formed under cold conditions by calcium-induced intermolecular associations involving the G-block regions, which have a geometry suited to the cooperative binding of calcium ions. The role of alginate–calcium reaction in production of new dessert gels is described in Fig. 3.11 (Hannigan, 1983). Alginates are often used in the production of difficult-to-manufacture foods, for example, frozen desserts with soft unfrozen fruit pieces or pie fillings in which the fruit pieces should retain their shape during heat processing. Because of their heat stability, alginate gels are often used in preference to some thermoreversible gelling systems described previously. Alginates can be used as gelling agents in ice cream or as stabilizers in combination with phosphates, which sequester the calcium in milk and allow the alginate to dissolve. Alginates have excellent thickening properties and stabilizing capabilities, which are desirable in sauces and gravies, milk drinks, bakery fillings such as whipped cream, icings, and cooked puddings. At low pH levels, alginates gel or precipitate; therefore they cannot be added as stabilizers to such products as fruit juices, water ices, French dressings, or salad dressings. In these foods, one may use propylene glycol alginate in which the carboxylic acid groups have been esterified, in part, to prevent gelation and precipitation. Propylene glycol alginate acts both as emulsifier and emulsion stabilizer. It imparts, at low concentrations, stability to beer foams (Sanderson, 1981). Typical food applications of alginates are listed in Table 3.4.

TABLE 3.4

Typical Food Applications of Alginates[a]

Use	Product	Amount used (%)	How used	Purpose
Bakery jelly	Sodium alginate	0.1 of jelly	Cook with pectin	Retards syneresis; soft, easily spread gel
Pie filling	Sodium alginate	0.3–0.5 of total filling	Dissolve in part of water, mix with cooked starch, add sugar, fruit	Soft, smooth gel body; improved clarity, gloss, full flavor, prevents separation, cracking
Ice cream	Milk-soluble sodium alginate	0.15–0.4	Water dispersion, dry-mix with sugar	Maintains smooth texture during storage, smooth meltdown without serum drainage, improved whipping
Confectionary gels	Sodium alginate	0.1–0.7	Dissolve in water, add sugar, cook, add acid, etc.	Makes clear candy gel for tender-to-chewy eating qualities; retains moisture, retards syrup separation
French dressing	Propylene glycol alginate	0.5	Dissolve in water, add vinegar, oil, etc.	Uniform emulsion with long shelf life
Meat sauce	Propylene glycol alginate	0.5–1.0	Dissolve in water, add salt, condiment, etc.	Smooth, heavy, uniform body, prevents separation
Flavor emulsion	Propylene glycol alginate	1.0–3.0	Dissolve in water	Stable emulsion
Dessert gels	Sodium alginate	0.8	Dry-mix with sugar, citric acid, etc. Dissolve in hot water	Clear, firm gel; quick set
Milk puddings	Milk-soluble sodium alginate	0.8	Dry-mix with sugar, cocoa, etc. Dissolve in cold milk	Quick, easy mix with cold milk to form smooth, firm gel in 1 hr

(continued)

TABLE 3.4 *(Continued)*

Use	Product	Amount used (%)	How used	Purpose
Syrups, toppings, purees	Sodium alginate, propylene glycol alginate	0.1–0.5	Dissolve in water, add to syrup	Heavy, smooth body without excessive sweetness; no masking of flavor, prevents separation
Frozen fruit	Propylene glycol alginate	2.0	Dry-mix with sugar, sift into fruit, mix	Uniform, heavy body prevents syrup separation; smooth texture when frozen; bright color, glossy appearance, full flavor
Dietetic foods	Sodium alginate, propylene glycol alginate	0.2–2.0	Dry-mix or in water, add protein, etc.	Thickener; smooth, heavy body
Dry mixes	Sodium alginates, propylene glycol alginates	0.05–1.0	Dry blend	Easy mixing; improved texture and shelf life of finished convenience foods
Icing	Gel-forming algin	3–8 oz/ 100 lb sugar	Dissolve in water	Quick set and drying; nonsticking, retards melting; tender, soft body

[a]From Gibsen and Rothe, 1955.

IV. Plant Exudates

Plant exudates include the gums arabic, karaya, and tragacanth (Whistler and BeMiller, 1973; Smith and Montgomery, 1959; Sanderson, 1981). These gums are complex heteropolysaccharides obtained from shrubs and trees found mainly in Africa and Asia. Secreted or exuded in response to injury of plant tissue, on exposure to air, the gums form hard, glossy flakes, which vary in color from white to dark-brown, depending on the level of impurities. These gum exudates are relatively expensive and vary widely in their quality. The most important is gum arabic, an excellent emulsifier used in the production of oil-in-water flavor emulsions for soft drinks and beverages. This gum is used in the confectionery industry in the preparation of candies. Here gum arabic functions as an emulsifier, texturing agent, or inhibitor of sugar crystallization. It is also effective in the production of spray-dried or encapsulated flavors. Gum arabic is unique among water-soluble polysaccharides in that solutions up to 50% can be obtained. Good stabilization is obtained at those high concentrations. This gum is often used in combination with other polysaccharides, since arabic alone provides little viscosity at low concentrations. Karaya gum, on the other hand, has high solution viscosity and can be used as stabilizer in such products as imitation whipped cream. Gum tragacanth retains its viscosity at low pH levels and can be used to stabilize dressings containing vinegar. Sources, characteristics, and applications of water-soluble gums are listed in Table 3.5. Table 3.6 lists properties of water-soluble gums. Typical functions of gums in specific food products are given in Table 3.7.

V. Seed Gums

Seed gums from guar and locust bean (or carob) are polysaccharides in which the mannan is rendered soluble by the presence of single unit galactose side chains, which makes them galactomannans (Glicksman, 1969; Whistler and BeMiller, 1973; Sanderson, 1981). Guar gum with a mannose to galactose ratio of about 2:1 is more highly substituted than locust bean gum, which has a mannose to galactose ratio of 4:1 and is more water soluble. Guar gum can be dissolved in cold water, but almost all grades of locust bean gum require hot water for dissolution. Guar and locust bean gums are highly viscous in low concentrations and are useful in thickening, stabilization, and water-binding applications. They are not charged, are not affected adversely by low pH, and are effective in acidic products. These polysaccharides can stabilize such foods as ice cream, fruit drinks, and salad dressings. Seed gums impart a smooth texture and reduce syneresis in processed cheeses; they can be used to control syneresis in a variety of gels. Guar gum is an efficient water binder in comminuted meat products, canned meats, pet foods, and icings. Guar gums and locust bean gums are compatible with most other polysaccharides. When combined with carrageenans and xanthan gums, a synergistic interaction takes place. This interaction is especially notable in combination with locust bean gum, in which it is caused by intermolecular associations that involve unsubstituted or ''smooth'' regions in the locust bean gum molecule. Such interactions depend on the relative proportions and overall concentration of the respective polysaccharides and can result in the formation of a permanent gel network. In guar gum, the degree of substitution is higher and there are fewer ''smooth'' regions. Consequently, interaction is less pronounced; this gives rise to a synergistic increase in viscosity rather than to gelation.

TABLE 3.5

Sources, Characteristics, and Applications of Water-Soluble Gums[a]

Product	Source	pH	Characteristics	Applications
Gum arabic	*Acacia* tree (regions of Africa)	4.5–5.5	Low viscosity; Newtonian flow; cold water soluble; excellent emulsifier	Flavor emulsions, clouding agents, filler in dietetic formulas, candy and snack coatings, icings
Gum tragacanth	*Astralgus* shrub (Middle East)	5.1–5.9	Creamy texture, acid stable, bi-functional emulsifier	Heavy-bodied flavor emulsions, low-calorie salad dressings, relishes, sauces, sandwich spreads
Locust bean	*Ceratonia siliquia* plant (Mediterranean)	~5.5	Partial hydration in cold water, full hydration when heated; viscosity builder, smooth texture, milk reactive, synergistic with carrageenans	Thickener in sauces, gravies, aids in "whey-off" prevention in ice cream, buttermilk, cream cheese, cottage cheese; prevents "watering off" in dietetic jams and jellies, pie fillings and frozen foods; bodying agent in beverages
Gum guar	Endosperm of seed from *Cyanopsis tetragonolobus* plant (India and Pakistan)	5.5–6.1	Cold water soluble, high viscosity, milk reactive	Prevents ice crystallization in ice cream, provides body in other dairy applications; thickener for instant soups, sauces, gravies; bodying agent in beverages; suspension aid in batters; moisture retention in cakes, donuts, frozen foods

		pH	Properties	Uses
Gum agar-agar	Red algae *Geldium* or *Gracalaria* seaweed (Far East and Mediterranean)	~7.0	Extremely strong, rigid gels, resistant to heat, pH	Icing stabilizer, pie fillings, meringues, piping gels
Furcellaran	*Furcellaria fastigiata* seaweed	5.5–9.0	Strong but flexible clear gels	Dietetic jams, jellies, dessert gels, syrup thickener, flans
Gum karaya	*Sterculia urens* tree (India)	4.3–5.5	Hydrates rapidly in cold water	Paper, pharmaceutical preparations
Gum ghatti	*Anogeissus latifolia* tree (India and Ceylon)	~4.5	Excellent emulsifier and film former	Pharmaceutical and industrial emulsions
Tamarino	*Tamarindus indica* (India)	6.0–7.0	High water absorption, stable viscosity	Textiles, paper
Carrageenan (Irish moss extract)	*Chondrus crispus* (Nova Scotia, Mediterranean)	~7.0	Protein reactive	Dairy products
Alginates	*Laminaria* seaweed (offshore waters of U.S., England, Norway, France, Japan, and Canada)	7.5	Thickener, suspending and gelling agent	Dietetic and regular salad dressings, puddings, pie fillings, ice cream, sherbet and icings

[a]Courtesy of TIC Gums Inc., New York.

TABLE 3.6

Unique Properties of Water-Soluble Gums[a]

	Gum arabic	Gum tragacanth	Locust bean	Gum guar	Agar	Gum ghatti	Gum karaya	Colloid 600	Carrageenan	Alginates
Complete solubility in cold water	+									
High cold water viscosity		+		+			+			+
High viscosity after cooking		+	+	+				+	+	+
High gel strength after cooking					+			+	+	
Excellent suspending agent in acid media		+		+						
Excellent general purpose suspending agent		+		+		+				
Good emulsifying properties	+					+		+	+	
High swelling properties in cold water				+			+			
Good film former	+					+				+
High milk reactivity			+					+	+	+
Good water absorber		+	+	+			+			

[a]Courtesy of Tragacanth Importing Corporation, New York.

TABLE 3.7

Typical Functions of Gums in Food Products[a]

Function	Example
Adhesive	Bakery glaze
Binding agent	Sausages
Calorie control agent	Dietetic foods
Crystallization inhibitor	Ice cream, sugar syrups
Clarifying agent	Beer, wine
Cloud agent	Fruit juice
Coating agent	Confectionery
Emulsifier	Salad dressing
Encapsulating agent	Powdered flavors
Film former	Sausage casings, protective coatings
Flocculating agent	Wine
Foam stabilizer	Whipped toppings, beer
Gelling agent	Puddings, desserts, aspics
Molding	Gum drops, jelly candies
Protective colloid	Flavor emulsifiers
Stabilizer	Beer, mayonnaise
Suspending agent	Chocolate milk
Swelling agent	Processed meats
Thickening	Jams, pie fillings, sauces
Water binding (prevents syneresis)	Cheese, frozen foods
Whipping agent	Toppings, icings

[a]From Glicksman, 1976.

VI. Plant Extracts

The important plant extracts, pectins, occur naturally in fruits and vegetables. Pectins are important as gelling and thickening agents. The pectins occur, bound to calcium mainly in the middle lamella, in the growing tissues of many higher plants, and bound to cellulose in the primary cell membrane. Pectins are generally obtained commercially by processing the parenchymatic tissues of plant materials. During fruit ripening, the cell pectins are solubilized enzymatically and are present in fruit juices. They serve as the cementing agents of the cell and regulate the water content (see also Deuel and Stutz, 1958; Hirst and Jones, 1946; Kertesz, 1951; Joslyn, 1962; Pilnik and Zwicker, 1970; Rombouts and Pilnik, 1972; Blanshard and Mitchell, 1979.)

Chemically, pectins are polymers of galacturonic acid connected by α-1,4-linkages to long chains. They can be divided and defined as

1. *Pectic substances* Material comprising all polygalacturonic acid-containing materials
2. *Protopectins* Water-insoluble materials, in bound form, yielding pectins upon hydrolysis
3. *Pectin* Partly esterified polygalacturonic acids; generally methyl esters, in some (rapeseeds and beets) also esterified with acetic acid. They can be divided into low methoxy and high methoxy pectins, depending on whether they contain less or more than 7% methoxy esterified galacturonic acids. A completely (100%) esterified polygalacturonide theoretically has 16.3% methoxyl groups.
4. *Pectinic acids* All carboxyl groups are in the free form and are water insoluble. Salts of pectinic acids are water soluble.

Pectic substances from plant materials often contain various arabans and galactans. The isolated substances are a heterogeneous mixture that varies in molecular weight, degree of esterification, and araban and galactan content. The solubility of pectins increases with an increase in the degree of esterification and with a decrease in molecular weight. The less soluble a pectin is in water, the easier it can be precipitated by adding an electrolyte. Pectins with an esterification grade of up to 20% are precipitated by sodium chloride solutions, with a 50% grade by calcium chloride solutions, and with a 70% grade by aluminum chloride or copper chloride solutions. Completely esterified pectins are not precipitated by electrolytes. The pectins can also be precipitated with organic solvents, such as acetone, methyl, ethyl, or propyl alcohol. The concentration of alcohol that is required increases with the degree of esterification.

The usefulness of pectins stems mainly from their capacity to form stable gels or films, and increase the viscosity of acidified, sugar-containing solutions. Completely esterified pectins can be gelled without acid or electrolyte addition. Pectins with a high esterification grade gel rapidly at a relatively low acidity. With a decrease in esterification, the rate of gel formation decreases and the required optimum pH level decreases. The gelling capacity is greatly enhanced by the presence of calcium salts, which also increase the stability and decrease the dependence on pH and sugar concentration. Viscosity increases, at a given esterification grade, with an increase in molecular weight; decreased esterification lowers viscosity. The effects on viscosity depend also on the total concentration and the electrolytes present. The stability of gels is affected by numerous factors—including temperature, the presence of acids or alkali, enzymes, and mechanical treatment. The pectin content and its quality (determined as jelly grade, that is, the pounds of sugar that 1 lb of pectin would set to a gel under standardized conditions) are important in the manufacture of jellies, jams, and preserves. The food applications of pectins are listed in Table 3.8.

TABLE 3.8

Food Applications of Pectins[a]

Type	Typical uses	Approximate amount in final product (%)	Important factors
Regular pectin (jelly grade 150)	Fruit jellies, jams, preserves, also bakers' jellies	0.1–0.8	Soluble solids for jellies 65%, jams 60%; pH 2.8–3.2
	Confectionery jelly pieces	0.85–1.25	Soluble solids 80–82%; buffer salt needed; pH (50–50 mixture jelly, water) 3.4–3.7; equal weights glucose and sucrose
	Home jelly making mixtures	Usually 3 oz to 6–8 glasses	Mixtures of dextrose, pectin, fruit acid (jelly grade about 10)
	Thickeners for low-calorie fruit syrups and beverages	Few tenths	Care needed in preparing solution when sugar is not included as spacing agent
	Flavor emulsions, salad dressings	2–3 of water phase	Best for oil contents 15–20% and higher
	Cream whipping aids, baker's glazes, malted milk thickeners	—	These and other specialty uses require specific instructions
Low methoxyl pectin (pectin L.M.)	Salad and dressing gels		
	a. Imitation flavor and color (for home consumption)	0.8–1.5	$Ca(H_2PO_4)_2 \cdot 2H_2O$, 8–16% weight of pectin; sometimes sodium citrate and fruit acid are added
	b. Fruit and vegetable juices (canned gels)	1.0–1.8	$CaCl_2 \cdot 2H_2O$, 6–14% weight of pectin; pH 3.6–3.8; pectin in solution before adding calcium salt
	Milk gels and puddings	0.8–1.5	No calcium salt needed
	Low-calorie jam-like fruit gels for dietetic use	0.8–1.5	Need proper balance of pectin and calcium; small amounts of sorbitol and glycerol desirable
	Frozen strawberries	0.1–0.15 berry weight	Most effective when in sugar syrup added to packed berries
	Fruit and berry gels for use in ice cream	0.8–1.5	40–50% sugar; fruit acid, sometimes calcium salt

[a]From Joseph, 1953.

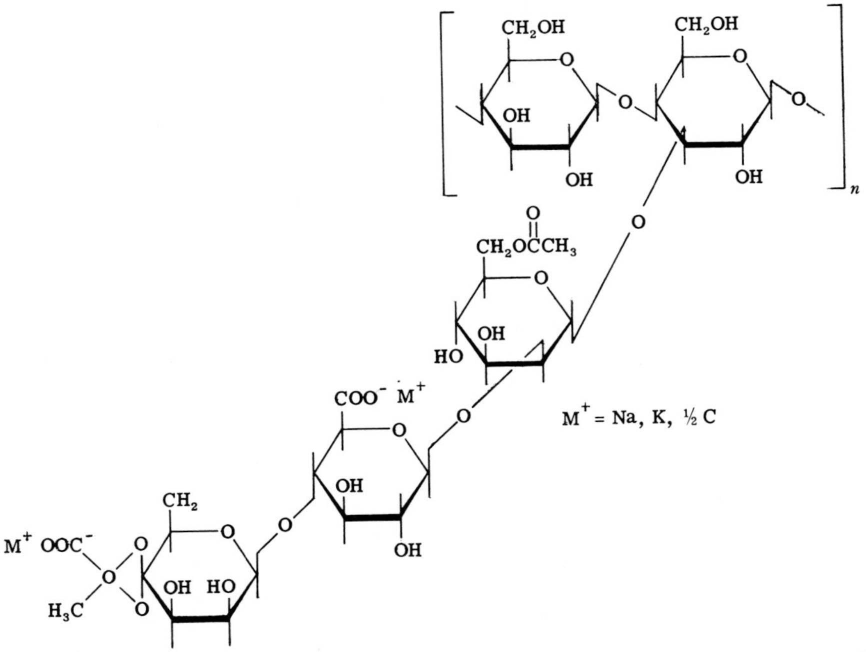

Fig. 3.12 Primary structure of xanthan gum. (From Sanderson, 1981; copyright by the Institute of Food Technologists.)

VII. Microbial Gums

Xanthan is the only microbal gum permitted for use in foods (Sanderson, 1981). This gum has a cellulose backbone, which is made water soluble by the presence of short side chains attached to every second glucose molecule in the main chain (see Fig. 3.12) Consequently, the molecule exists in solution as a rigid rod stabilized by noncovalent interactions between the backbone and the side chains, as shown in Fig. 3.13. This conformation makes possible the following properties.

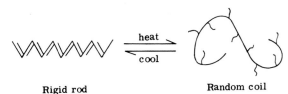

Rigid rod   Random coil

Fig. 3.13 Molecular structure of xanthan gum in solution. (From Sanderson, 1981; copyright by the Institute of Food Technologists.)

TABLE 3.9

Hydrocolloid Gelling Systems[a]

Hydrocolloid	Solubility Hot	Solubility Cold	Affected by electrolytes	Effect of heat	Gelling mechanism Thermal	Gelling mechanism Chemical	Special conditions	Type of texture	Appearance	Applications
Gelatin	X		No	Melts at room temperature	X			Tender, elastic	Clear	Desserts, confections, canned meat
Agar	X		No	Can withstand autoclaving	X			Firm, brittle	Clear	Canned meat, confections, bakery icings
κ-Carrageenan	X		No	Does not melt at ambient temperatures	X		Requires potassium cation for gelling	Brittle	Clear	Desserts, canned meats and pet foods, flan puddings
κ-Carrageenan and locust bean gum	X		No		X		Requires potassium for gelation	Elastic	Cloudy	Desserts
ι-Carrageenan	X		No		X		Requires calcium for gelation	Tender, elastic	Clear	Aseptic canned desserts
Furcellaran	X		No		X		Requires potassium for gelation	Brittle water gel, tender milk gel	Clear	Flan, puddings
Sodium alginate		X	Yes	Nonmelting and irreversible		X	Reacts with calcium to gel	Brittle	Clear	Dessert gels, milk puddings
Pectin	X		No		X		Requires sugar and acid to gel	Spreadable	Clear	Jams and jellies
Low methoxyl pectin		X	Yes			X	Reacts with calcium to gel	Brittle	Clear	Canned fruit and milk desserts
Gum arabic		X	No		X			Soft, chewy	Clear	Confections
Starches	X	X	Yes	Retrograde on storage	X			Spreadable to soft rigid texture	Cloudy, opaque	Puddings, desserts
Xanthan gum and locust bean gum	X		No		X		Both components required for gelation	Elastic, rubbery	Cloudy	

[a]From Blanshard and Mitchell, 1979.

1. High viscosity at rest or low shear, even at low concentrations
2. Low yield value
3. High pseudoplasticity; the solutions are highly shear-thinning
4. Increased viscosity in the presence of salt
5. Viscosity stability at elevated temperatures and over a wide pH range in the presence of salt
6. Synergistic interaction with guar gum or locust bean gum. Guar gum creates increased viscosity, locust bean gum produces thermoreversible gels.

High viscosity at low shear and yield value both facilitate formulation of stable suspensions, emulsions, and foams. Dressings flow readily when poured from the bottle and then recover their initial viscosity on the salad. Use of xanthan permits the desired "cling" or adhesion to the foodstuff. Xanthan gum is readily dispersible in water and high viscosity is obtained rapidly in both cold and hot systems. It can be used in a range of dry-mix products, such as milkshakes, bakery fillings, sauces, beverages, and desserts. Variations in texture can be attained through inclusion of locust bean or guar gum. Generally, inclusion of polysaccharides impairs the flavor of foods; xanthan gum solutions thin under shear in the mouth and facilitate flavor release. In starch-thickened foods, flavor release is markedly improved by replacing part of the starch with xanthan gum. The pseudoplasticity of xanthan gum also facilitates pumping and filling operations. In some canned foods, the gum is used as a partial replacement for starch to enhance heat penetration and thus reduce heat damage during processing. The inclusion of small amounts of xanthan gum in starch-thickened frozen foods improves their freeze–thaw stability (Sanderson, 1981). Properties of hydrocolloid gelling systems are summarized in Table 3.9.

REFERENCES

Baker, R. A. (1981). The role of pectin in citrus quality and nutrition. *ACS Symp. Ser.* **143**, 110–128.

Blanshard, J. M. V., and Mitchell, J. R., eds. (1979). "Polysaccharides in Foods." Butterworth, London.

D'Appolonia, B. L., and Kim, S. K. (1976). Recent developments on wheat flour pentosans. *Baker's Dig.* **50**(3), 45–49, 53, 54.

Deuel, H., and Stutz, E. (1958). Pectic substances and pectic enzymes. *Adv. Enzymol.* **20**, 341–382.

Fincher, G. B., Sawyer, W. H., and Stone, B. A. (1974). Chemical and physical properties of an arabinoxylan–peptide from wheat endosperm. *Biochem. J.* **139**, 535–545.

Geissmann, T., and Neukom, H. (1973). On the composition of the water-soluble wheat flour pentosans and their oxidative gelation. *Lebensm.-Wiss. Technol.* **6**, 59.

Gibsen, K. F., and Rothe, L. B. (1955). Algin, versatile food improver. *Food Eng.* **27**(10), 87–89.

Glicksman, M. (1962). Utilization of natural polysaccharide gums in the food industry. *Adv. Food Res.* **11**, 109–200.

Glicksman, M. (1963). Utilization of synthetic gums in the food industry. *Adv. Food Res.* **12,** 283–366.

Glicksman, M. (1969). "Gum Technology in the Food Industry." Academic Press, New York.

Glicksman, M. (1976). Hydrocolloid utilization in fabricated foods. *Cereal Foods World* **21,** 17–23, 26.

Glicksman, M. (1979). Gelling hydrocolloids in food product applications. *In* "Polysaccharides in Foods" (J. M. V. Blanshard and J. R. Mithcell, eds.), pp. 185–204. Butterworth, London.

Guisley, K. B. (1968). Seaweed colloids. *Kirk-Othmer Encycl. Chem. Technol., 2nd ed.* **17,** 763.

Hannigan, K. (1983). Algin–calcium reaction produces new dessert gels. *Food Eng.* **55**(9), 66–67.

Hirst, E. L., and Jones, J. K. N. (1946). The chemistry of pectic materials. *Adv. Carbohydr. Chem.* **2,** 235–251.

Joslyn, M. A. (1962). The chemistry of protopectin: A critical review of historical data and recent developments. *Adv. Food Res.* **11,** 1–107.

Joseph, G. H. (1953). Better pectins. *Food Eng.* **25**(6), 71–73, 114.

Kertesz, Z. I. (1951). "The Pectic Substances." Wiley (Interscience), New York.

Kraus, B. (1975). "The Guide to Fiber in Foods." New American Library, New York.

Krumel, K. L., and Sarkar, N. (1975). Flow properties of gums useful to the food industry. *Food Technol.* **29**(4), 36–44.

Lawrence, A. A. (1973). "Edible Gums and Related Substances," Food Technol. Rev. No. 9. Noyes Data Corp., Park Ridge, New Jersey.

Lawrence, A. A. (1976). "Natural Gums for Edible Purposes." Noyes Data Corp., Park Ridge, New Jersey.

Morris, E. R. (1973). Polysaccharide conformation as a basis of food structure. *In* "Molecular Structure and Function of Food Carbohydrate" (G. G. Birch and L. F. Green, eds.), pp. 125–132. Wiley, New York.

Neukom, H., Providoli, H., Gremli, H., and Hui, P. A. (1967). Recent investigations on wheat flour pentosans. *Cereal Chem.* **44,** 238–244.

Pilnik, W., and Zwicker, P. (1970). Pectins. *Gordian* **70,** 202–204, 252–257, 302–305, 343–346.

Rees, D. A. (1972). Polysaccharide gels. A molecular view. *Chem. Ind. (London)*, pp. 630–636.

Rombouts, F. M., and Pilnik, W. (1972). Research on pectin depolymerases in the sixties—a literature review. *CRC Crit. Rev. Food Technol.* **3,** 1–87.

Sanderson, G. R. (1981). Polysaccharides in foods. *Food Technol.* **35**(7), 50–57, 83.

Sharma, S. C. (1981). Gums and hydrocolloids in oil–water emulsions. *Food Technol.* **35**(1), 59–67.

Smith, F., and Montgomery, R. (1959). "The Chemistry of Plant Gums and Mucilages and Some Related Polysaccharides," Am. Chem. Soc. Monogr. Ser. No. 141. Van Nostrand-Reinhold, Princeton, New Jersey.

Spiller, G. A., and Amen, R. J., eds. (1975a). "Fiber in Human Nutrition." Plenum, New York.

Spiller, G. A., and Amen, R. J. (1975b). Dietary fiber in human nutrition. *CRC Crit. Rev. Food Sci. Nutr.* **5,** 39.

Steiner, K. (1968). The role of pentosans in brewing. *Brewer's Dig.* **43**(3), 70–73, 76, 77.

Sterling, C. (1963). Texture and cell-wall polysaccharides in foods. *In* "Recent Advances in Food Science" (J. M. Leitch and D. N. Rhodes, eds.), pp. 259–281. Butterworth, London.

Thomas, W. R. (1982). The practical application of microcrystalline cellulose in foods. *Prog. Food Nutr. Sci.* **6,** 341–351.

Whistler, R. L., and BeMiller, J. N., eds. (1973). "Industrial Gums." Academic Press, New York.

Whistler, R. L., and Smart, C. L. (1953). "Polysaccharide Chemistry." Academic Press, New York.

Zimmermann, R. (1979). Die funktionellen Eigenschaften der Kohlenhydrate. *Lebensmittelindustrie* **26**(2), 57–62.

4

Corn Sweeteners and Wheat Carbohydrates

In some parts of the world, potato, tapioca, sweet potatoes, rice, and, more recently, wheat are the main sources of starch; in the United States corn is the basic raw material. The main by-products of industrial corn processing are corn oil and animal feeds. The growth of the corn sweetener industry is predicated on the abundant supply of raw materials, the diversity of products, and the sustained development of required science, engineering, and technology (Anderson and Watson, 1982; Inglett, 1974; Birch, 1971; Johnson, 1976). The long history of corn sweeteners extends over the period from 1923, when crystalline commercial dextrose was developed; through the development of enzymatic methods for the manufacture of "tailored" syrups of specific compositions and end-use properties; to the most recent capabilities of the industry to process enzyme catalyzed isomerization of glucose (dextrose) to fructose (levulose). For general reviews of corn sweetener development, see Birch (1971), Birch and Green (1973), Birch and Shallenberger (1977), Eisenberg (1955), Katz (1972), Schultz *et al.* (1969), and Shallenberger and Birch (1975).

I. Corn Sweeteners

A. Production of Corn Syrup

The manufacture of corn sweeteners is a multistep continuous process. The wet milling process of corn is depicted diagrammatically in Fig. 4.1. Acid conversion and enzyme conversion processes are described in Figs. 4.2 and 4.3, respectively. To convert the starch granules in the slurry to corn syrup, the granules must be gelatinized and the starch depolymerized in a conversion process that is halted as soon as the desired composition is reached. Two or more interrelated processes may be involved. For instance, an acid primary conversion may be followed by an enzymatic conversion. The products of the conversion may be used in the production of isomerized corn syrups. Starch conversion products are classified by their dextrose equivalent (DE). This is a measure of the reducing sugar content calculated as anhydrous dextrose and expressed as a percentage of total dry substance. DE is a useful parameter in classifying corn syrup, but it does not provide full information about actual composition.

While the most common methods used in the production of corn syrup are the acid and acid–enzyme processes, some syrups are produced by multiple enzyme processes. In the acid conversion process, a starch slurry of about 35 to 40% dry matter is acidified with hydrochloric acid to pH of about 2.0 and pumped to a converter. In the converter, the steam pressure is adjusted to about 30 lb/in.^{2} and the starch is gelatinized and depolymerized to a predetermined level. The process is terminated by adjusting the pH to 4–5 with an alkali. The liquor is clarified by filtration and/or centrifugation and concentrated by evaporation until it contains about 60% dry matter. The syrup is further clarified and decolorized by treatment with powdered and/or granular carbon, refined by ion exchange to remove soluble minerals and proteins and to deodorize and decolorize, and further concentrated in large vacuum pans or continuous evaporators.

In the acid–enzyme process, the liquor containing a partially converted product is treated with an appropriate enzyme or combination of enzymes to complete the conversion. For example, in the production of a 42-DE high maltose syrup, the acid conversion is continued until dextrose production is negligible. At this point, β-amylase (a maltose-producing enzyme) is added and the conversion is allowed to continue. The enzyme is deactivated and the purification and concentration are continued as in the acid process.

In enzyme–enzyme processes, the starch granules are cooked and preliminary

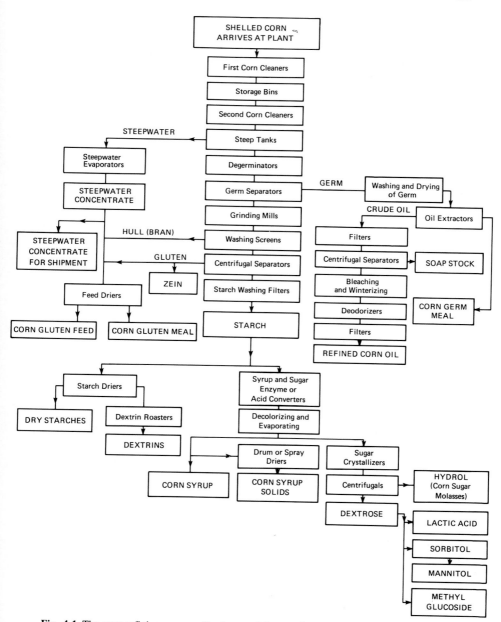

Fig. 4.1 The corn refining process. Products and intermediate points between processes are in capital letters; equipment and processes are in capital and lowercase letters. (Reprinted with permission of the Corn Refiners Association.)

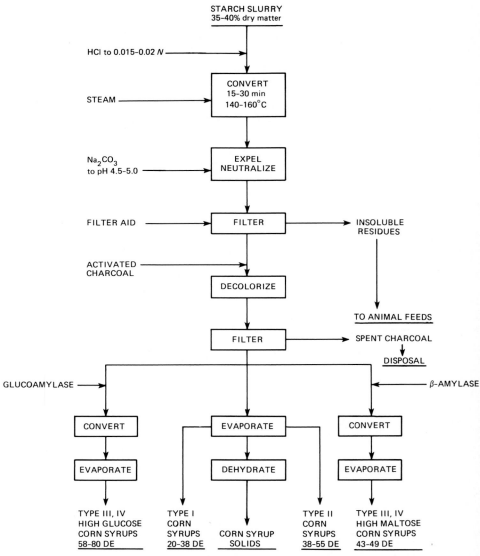

Fig. 4.2 Acid-converted corn syrup process. (Reprinted with permission of the Corn Refiners Association.)

starch depolymerization is accomplished by starch-liquefying α-amylase. The final depolymerization is achieved by a single enzyme or a combination of enzymes. Combinations of enzymes make possible the production of syrups with specific composition and/or properties, such as high maltose syrup or highly fermentable syrup.

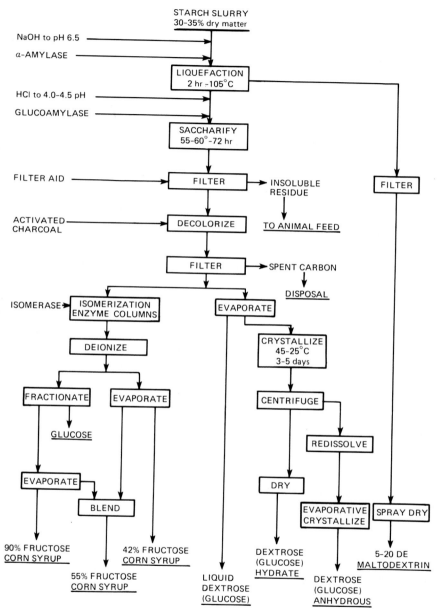

Fig. 4.3 Enzyme-converted corn sweetener process. (Reprinted with permission of the Corn Refiners Association.)

B. Types of Corn Sweeteners

There are five types of corn sweeteners.

1. *Corn syrup (glucose syrup)* is the purified concentrated aqueous solution of saccharides obtained from edible starch. It has a DE of 20 or more.

2. *Dried corn syrup (dried glucose syrup)* is corn syrup from which the water has been partially removed. Refined corn syrups are spray- or vacuum drum-dried to achieve a low moisture content, and to form granular, crystalline, or powdery amorphous products. These products are mildly sweet and moderately hygroscopic. Because of their hygroscopicity they are packed in multiwall moistureproof paper bags. Such dried syrups are comparable in chemical composition to liquid corn syrups, except for their lower moisture contents.

3. *Maltodextrin* is a purified concentrated aqueous solution of saccharides or the dried product derived from that solution obtained from starch. It has a DE of less than 20. Maltodextrins are produced in the same manner as corn syrups except that the conversion process is stopped at an early stage to keep the DE below 20. Both acid and enzyme processes can be used. Maltodextrins are usually dried into white free-flowing powders and are packed in multiwall bags. In some cases, however, moderately concentrated solutions of maltodextrins are sold.

4. *Dextrose monohydrate* is purified and crystallized D-glucose containing one molecule of water of crystallization per molecule of D-glucose. For the manufacture of dextrose, complete depolymerization of the starch substrate and recovery of the product by crystallization are required. The starch slurry is gelatinized as in the manufacture of corn syrup and is partially converted by acid or α-amylase. A purified glucoamylase enzyme, free of transglucosylase activity, is added to the intermediate substrate. When the dextrose conversion is complete, the enzyme is deactivated and the dextrose liquor is filtered to remove residual suspended materials. The liquor is purified and decolorized with granular or powdered carbon. The liquor is concentrated to about 75% solids, and cooled and pumped into crystallizers. The temperature is slowly lowered to about 25°C. Crystallization is induced by seed crystals left in the crystallizer from the previous batch. Dextrose monohydrate crystallizes from the mother liquor. It is separated by centrifugation and washed in the centrifuges with a spray of water. The wet crystals are dried in warm air to about 8.5% moisture. The mother liquor is reconverted, again refined, concentrated, and crystallized to produce a second batch of dextrose hydrate.

5. *Dextrose anhydrous* is purified and crystallized D-glucose without water of crystallization. Anhydrous dextrose is obtained by redissolving dextrose hydrate and refining the solution to a highly purified and clear filtrate. The solution is evaporated to a high solids content and anhydrous α-D-glucose is precipitated by crystallizing at an elevated temperature. The anhydrous crystals are separated by

centrifugation, washed with a warm water spray, and dried. Anhydrous dextrose also can be made by direct crystallization from a high DE liquor.

Dextrose is the major dry corn sweetener. An increase in the production of corn syrup solids (dried corn syrup), however, broadens the practical range of corn syrup and makes the sweetener available for users equipped to handle only dry ingredients and for applications requiring dry sweeteners. In addition, malto-dextrins find increasing use in products that require soluble carbohydrates with relatively low sweetness.

C. Syrup Characteristics and Properties

The characteristics and functional properties of corn syrups vary according to their composition. Corn syrups are classified into four types on the basis of DE.

Type I	20 DE up to 38 DE
Type II	38 DE up to 55 DE
Type III	55 DE up to 73 DE
Type IV	73 DE and above

To evaluate adequately a syrup, one must know its actual carbohydrate composition. Results of detailed analyses are given in Table 4.1 (from Nesetril, 1967). In general, type I syrups and maltodextrins have a relatively small concentration of low molecular weight sugars such as dextrose, maltose, and maltotriose. Type III and IV syrups have relatively small concentrations of oligosaccharides above maltoheptose. Compositions of type II syrups depend on the process and the extent of conversion (Figs. 4.4 and 4.5).

TABLE 4.1

Examples of Carbohydrate Composition of Commercially Available Corn Syrups[a]

Type of conversion	Dextrose equivalent	Polymer							
		Mono-	Di-	Tri-	Tetra-	Penta-	Hexa-	Hepta-	Higher
Acid	30	10.4	9.3	8.6	8.2	7.2	6.0	5.0	45.1
Acid	42	18.5	13.9	11.6	9.9	8.4	6.6	5.7	25.4
Acid–enzyme[b]	43	5.5	46.2	12.3	3.2	1.8	1.5	—	29.5[c]
Acid	54	29.7	17.8	13.2	9.6	7.3	5.3	4.3	12.8
Acid	60	36.2	19.5	13.2	8.7	6.3	4.4	3.2	8.5
Acid–enzyme[b]	63	38.8	28.1	13.7	4.1	4.5	2.6	—	8.2[c]
Acid–enzyme[b]	71	43.7	36.7	3.7	3.2	0.8	4.3	—	7.6[c]

[a]From Nesetril, 1967.

[b]The carbohydrate composition of acid–enzyme syrups varies as a result of different processes used. The values given here are to be considered only as examples of ranges available commercially.

[c]Includes heptasaccharides.

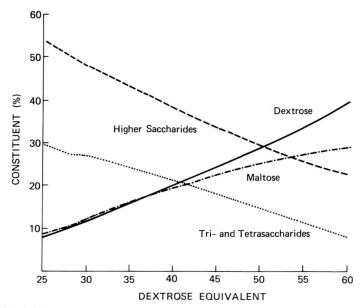

Fig. 4.4 Acid hydrolyzed corn syrup analysis to determine composition of syrup. (Reprinted with permission of the Corn Refiners Association.)

It is frequently possible to describe the composition of a syrup in terms of DE and one or more of the saccharide fractions. For example, 43-DE high maltose type syrup is made by an acid–enzyme process, in which a partially converted corn syrup is treated with β-amylase to produce a syrup high in maltose and low in dextrose. A regular type II 42-DE acid converted syrup contains about 20% dextrose and 14% maltose; a 43-DE high maltose type II syrup contains about 8% dextrose and 40% maltose.

Corn syrups are also characterized according to solids content. Most commercial corn syrups are sold on a Baumé (Bé) basis, which is a measure of the dry matter content and specific gravity. In contrast, the high fructose corn syrups are sold on the dry matter basis. Because all corn syrups (except high fructose syrups and high dextrose syrups) are very viscous at room temperature (RT), the Baumé determination is made at 60°C (140°F) and an arbitrary correction of 1.00 Bé is added to the observed reading. This designates a value called "Commercial Baumé." Most corn syrups are available in the range of 41 to 45 Bé, corresponding to a dry matter content of ∼77–85%. High fructose and high dextrose syrups are available at about 71% dry substance content.

Corn syrups and dextrose are sweet, but otherwise essentially tasteless. Many free reducing sugars exhibit the phenomenon of bitter-sweetness that is often associated with β-anomers. Reversion reactions that occur during acid hydrolysis

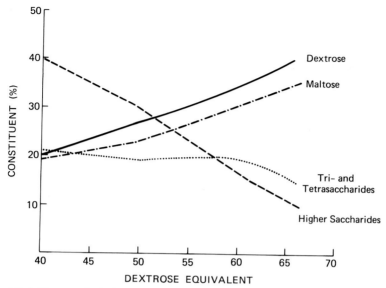

Fig. 4.5 Acid-enzyme hydrolyzed corn syrup analysis. (Reprinted with the permission of the Corn Refiners Association.)

of starch may produce disaccharides with β-linkages and levoglucosan (1,6-anhydro-β-D-glucopyranose). These sugars are present, however, only in trace amounts and their effects on taste are not significant. The sweetening power of corn sweeteners depends on percentages of sweetener solids and the *combination* of sweeteners. For example, a 2% water solution of dextrose is about two-thirds as sweet as a 2% sucrose solution. As the concentration is increased, the difference decreases. The relative sweetness of sugars is given in Table 4.2. When corn syrup or dextrose is used in combination with sucrose, the sweetness is usually greater than expected. For example, at 45% solids, a mixture of 25% 42-DE corn syrup and 75% sucrose is as sweet as a sucrose solution of 45% solids. Consequently, when corn sweeteners are used with sucrose in products high in total sweetener concentration, there is no apparent loss of sweetness. In less sweet products, or when only one sweetener is used, a decrease in sweetness may be expected if sucrose is replaced by some corn sweeteners. Sweetness is also influenced by temperature and the presence of nonsugar substances. It is difficult, therefore, to state accurately the relative sweetness of syrups and sugars for use in the food industries. Each product should be considered individually.

Corn syrup, corn syrup solids, and dextrose are compatible with other sweeteners and with food flavors. They are widely used with other sweeteners, mainly sucrose. In canned fruits, for example, a liquid packing medium containing a blend of several corn sweeteners gives optimum sweetness combined with supe-

TABLE 4.2

Relative Sweetness of Sugars[a]

Sugar	Index
Sucrose	100
Dextrose	70–80
Levulose	140
70-DE, corn syrup	70–75
High conversion corn syrup	65
Regular conversion corn syrup	50
Maltose	30–50
Lactose	20
High fructose corn syrup	
90%	120–160
55%	>100
42%	100
Invert sugar	>100
Sorbitol	50
Xylitol	100

[a]From Nesetril, 1967, Piekarz, 1968, and Young, 1981.

rior gloss and mouth-feel. In ice cream, body and texture are improved through the use of corn sweeteners; the product is smoother, has a better "melt-down", and is more resistant to "heat shock." In sherbets and ices, corn syrup sweeteners help eliminate crystallization and promote smoothness.

Physical characteristics of sweeteners are summarized in Table 4.3. These and other physical properties of the sweeteners are important to food and beverage products. Practically all corn sweeteners are readily soluble in water; aqueous solutions containing 70–80% dissolved solids can be obtained at RT. Dextrose has a negative heat of solution, that is, it cools a mixture in which it is dissolved. Anhydrous dextrose melts at 146°C (295°F) and dextrose hydrate melts at (82°C (180°F). Dried corn syrups or commercial corn syrup solids are granular, crystalline, or amorphous powders which, when heated gradually, soften or dissolve in their own trace moisture. Corn sweeteners are hygroscopic; the degree of hygroscopicity increases as DE increases. Corn syrups and dextrose are employed as moisture conditioners and stabilizers. The higher saccharides impart to corn syrup cohesive and adhesive properties. They also exhibit some attributes of vegetable gums, and contribute a chewy texture to confections and to chewing gum. Corn sweeteners, particularly the corn syrups, control crystallization of sucrose. This property is advantageous in confections, ice cream, frozen desserts, jams, jellies, and preserves. Corn syrups and sugars contribute to the attractiveness of foods. They are used in combination with sucrose to obtain a

TABLE 4.3

Physical Characteristics of Sweeteners[a,b]

Sweetener	Solids	Fermentable solids (%)	Water (%)	Viscosity centipoise (38°C)	Storage temp (°C)	% dry-basis composition		
						Dextrose	Levulose	Maltose
Dry sugar	100.0	100.0	—	—	—	—	—	—
Dry dextrose	91.5	91.5	8.5	—	—	100	—	—
Corn syrup (70 DE)	82.5	82.0	17.5	5200	35	50	—	29.0
30/70 blend[c]	77.9	86.6	22.1	1550	30	39	10.1	18.4
Liquid sugar	67.0	67.0	33.0	90	RT	—	—	—
Liquid dextrose	70.0	70.0	30.0	80	55	100	—	—

[a]From Piekarz, 1968.
[b]Based on 100 lb of product.
[c]30% liquid sugar–70% 70-DE corn syrup.

glossy appearance of fruits canned with syrups containing a corn sweetener. The sweeteners also improve the sheen and clarity of hard candies, jams, and jellies.

Dextrose and fructose have a relatively high osmotic pressure. This enhances their effectiveness in inhibiting microbial spoilage. Corn syrup of 55 DE has about the same average molecular weight as sucrose or lactose and hence about the same osmotic properties as those sugars. Corn syrups of lower DE have higher molecular weights and correspondingly lower osmotic pressures.

The effect of corn syrups and sugars on the freezing point of a solution is of significance in the manufacture of ice cream and frozen desserts. Lowering of the freezing point of a solution is inversely proportional to the molecular weight of the dissolved solids. In general, type I and type II syrups depress the freezing point somewhat less than an equal weight of sucrose. Type III corn syrups have about the same effect as sucrose on the freezing point. Dextrose and type IV corn syrups contain substantial proportions of monosaccharides and lower the freezing point to a greater extent. Corn sweeteners are used by the ice cream industry to control sweetness and improve body and texture. Viscosity of corn syrups depends on density, DE, and temperature. Viscosity decreases as DE and temperature rise and increases with rises in density.

Fermentability is another important property of corn syrups and sugars, particularly in the baking and brewing industries. Sugars with lower molecular weight, mainly the mono- and disaccharides, glucose, fructose, maltose, and sucrose, are readily fermentable by yeasts. The total fermentability of corn syrups is roughly proportional to their content of mono-, di-, and trisaccharides—the higher the DE, the higher the fermentability.

Dextrose and fructose combine with nitrogenous compounds at elevated temperatures to produce brown coloration resulting from what is called the Maillard reaction. This reaction makes corn sweeteners useful in the manufacture of caramel food colorings, promotes a golden-brown crust in baking, and produces desirable color and caramel flavor in other food products and confections.

The reducing sugars in corn sweeteners inhibit oxidative reactions in foods. This is useful in maintaining the bright-red color in tomato catsup and strawberry preserves and in retaining the characteristic color of cured meats. The chemical reducing action of syrups and sugars is measured and expressed as DE. A scale of relative reducing power of various sugars (i.e., DE of D-glucose polymers) is shown in the following tabulation:

	Theoretical	Observed
Monosaccharide	100.0	100.0
Disaccharide	52.6	58.0
Trisaccharide	35.7	39.5
Tetrasaccharide	27.0	29.8
Pentasaccharide	21.7	24.2
Hexasaccharide	18.2	20.8

In summary, as the DE of the main corn syrups increases from about 20 to 75, functional properties and characteristics of the corn syrups change in the following manner:

Bodying characteristics	Decrease
Browning reactions	Increase
Cohesiveness	Decreases
Fermentability	Increases
Flavor enhancement	Increases
Flavor transfer medium activity	Increases
Foam stabilization	Decreases
Freezing point depression	Increases
Humectancy	Variable
Hygroscopicity	Increases
Nutritive solids	Variable
Osmotic pressure	Increases
Prevention of coarse ice crystal formation in freezing	Decreases
Prevention of sugar crystallization	Decreases
Sheen production	Variable
Sweetness	Increases
Viscosity	Decreases

To maintain their typical characteristics, corn sweeteners are usually kept slightly on the acid side. They are available with a pH range of 3.5–5.5.

D. Isomerized Corn Syrups

Isomerized corn syrups have become a sizable part of the total production of syrups by the corn wet milling industry. Commercial isomerized corn syrups are clear, sweet, bland, low viscosity sweeteners high in dextrose (glucose) and levulose (fructose). High levulose corn syrups are functionally equivalent to liquid invert sugar in most foods and beverages and can be substituted for it with little or no change in formulation, processing, or final product.

The high levulose syrups (also called high fructose corn syrups) are prepared by the enzymatic action of dextrose isomerase on dextrose, which converts a portion of the dextrose to levulose. The substrate can be either dextrose or a high conversion corn syrup composed mostly of dextrose. The levulose content in the syrup may be 50% or more, depending upon the substrate, method of preparation, and so on.

The high levulose corn syrups presently on the market are composed mainly of the simple sugars dextrose and levulose. The composition of a 42% levulose syrup is similar to that of commercial invert sugar. (see Tables 4.4 and 4.5, and Junk and Pancoast, 1973, and Wardrip, 1971). The high levulose corn syrups are low in viscosity and are easy to ship, store, and blend.

According to Fruin and Scallet (1975), attributes of high levulose corn syrup of particular interest to food manufacturers are their ability to:

1. Retain moisture and/or prevent drying out
2. Control crystallization
3. Produce an osmotic pressure that is higher than for sucrose or medium invert sugar and thereby help control microbiological growth or help in penetration of cell membranes
4. Provide a ready yeast-fermentable substrate
5. Blend easily with sweeteners, acids, and flavorings
6. Provide a controllable substrate for browning and Maillard reactions
7. Impart a degree of sweetness that is essentially the same as in invert liquid sugar

High levulose corn syrups have many applications in foods and beverages. They are both convenient and economical in carbonated and still beverages. They are equivalent to invert sugar in providing humectancy and sweetness, and in controlling crystal size in some candy. In some candy products, however, there may be a problem of excessive moisture "pick-up." Large amounts of high levulose corn syrups may unduly darken, especially at high temperatures, and may soften some candies by inhibiting sugar crystallization. In bread, sucrose or invert sugar can be replaced on a pound-per-pound basis by high levulose corn syrups. Similarly, blends of high DE corn syrups with sucrose or with high

TABLE 4.4

Properties of Commercial Corn Syrups[a]

Typical analysis	Acid conversion–DE level			Acid–enzyme			Enzyme–enzyme	
	Low	Regular	Intermediate	High maltose	Regular	High DE	Glucose syrup	High fructose
Commercial Baumé	43°	43°	43°	43°	43°	43°	—	—
Solids %	80	80.3	81	80.3	82	82.2	71	71
Moisture %	20	19.7	19	19.7	18	17.8	29	29
Dry basis								
Dextrose equivalent	37	42	52	42	62	69	96	(95)
Ash (sulfated) %	0.4	0.4	0.4	0.4	0.4	0.4	0.03	0.03
Carbohydrate composition								
Monosaccharides %								
Dextrose %	15	19	28	6	39	50	93	52
Fructose %	0	0	0	0	0	0	0	42
Disaccharides %	12	14	17	45	28	27	4	3
Trisaccharides %	11	12	13	15	14	8	⎫	⎫
Tetrasaccharides %	10	10	10	2	4	5	⎬ 3	⎬ 3
Pentasaccharides %	8	8	8	1	5	3	⎭	⎭
Hexasaccharides %	6	6	6	1	2	2		
Higher saccharides %	38	31	18	30	8	5		
Viscosity centipoises								
24°C	150,000	56,000	31,500	56,000	22,000	—	—	—
37.7°C	30,000	14,500	8,500	14,500	6,000	—	—	—
44°C	8,000	4,900	2,900	4,900	2,050	—	—	—

[a]From Anderson and Watson, 1982.

TABLE 4.5

Carbohydrate Profile of Various Nutritive Sweeteners[a]

Sweetener	DP_1[b]		DP_2		DP_3	DP_4
	Fructose (%)	Dextrose (%)	Sucrose (%)	Maltose (%)	Triose (%)	and higher (%)
Sucrose			100	—	—	—
Medium invert	25	27	46	2[c]	—	—
Total invert	45	48	3	4[c]	—	—
Dextrose	—	100	—	—	—	—
42 DE, corn syrup	—	20	—	14	12	54
42 DE, high maltose	—	8	—	40	15	37
62 DE, corn syrup	—	39	—	31	7	23
42% high fructose corn syrup	42	52	—	6[b]	—	—
55% high fructose corn syrup	55	40	—	5[b]	—	—
90% high fructose corn syrup	90	9	—	1[b]	—	—

[a]From Young, 1981.
[b]DP = degree of polymerization.
[c]DP_2 and higher saccharides.

levulose syrups are functionally equivalent on a solid basis in bread, sweet goods dough, Danish pie fillings, soft or moist cookies, bakers' jellies, and soft fillings. The high levulose syrups must be used sparingly in icings that depend on sucrose crystal development for body and consistency, and in baking of light-colored and dense cakes.

According to Fruin and Scallet (1975), high levulose corn syrups can replace up to 100% sucrose in bread, pie fillings, jellies, and fillings, up to 70% in soft cookies, 25–75% in boiled and marshmallow icings, 20–50% in dark-colored cakes, up to 30% in chiffon or angel food cake, up to 20% in hard cookies or white icings, and 10–15% in flat icings. Corn sweeteners in combination with sucrose or invert sugars are available commercially. In using the combinations, one must consider the moisture content and percentage of fermentables in each blend component.

II. Wheat Carbohydrates

Carbohydrates are the most abundant components of wheat and wheat products. Great progress has been made in understanding the role of cereal carbohydrates in milling of cereal grains, in panary fermentation and bread making, in increasing storage and shelf life potentialities, and in processing whole wheat and milled wheat into food products (Shellenberger et al., 1966).

A. Wheat Starch

Starch is the major constituent of the wheat kernel, of which it constitutes 54–72% of the dry weight (the amount depends on both variety and growing conditions). In wheat, as in other plants, starch occurs in granules. There are a few small spherical starch granules in the germ, but only the endosperm is of interest to wheat processors.

Starch granules from wheat endosperm vary considerably in size. In commercial as well as in laboratory-separated samples, the granules range from <2 to 35–40 μm in diameter and the occurrence of some wheat starch granules up to 50 μm in diameter has been reported. The smallest granules are spherical and the largest are lenticular; both shapes can be identified in granules ranging 10–15 μm in diameter. The proportion of large to small granules varies from sample to sample.

The refractive index of wheat starch granules hs been reported to be 1.5245. Both the hilum and the lamellae are indistinct when an ungelatinized wheat starch granule is viewed microscopically, and birefringence is weak. The density of wheat starch has been reported to be 1.53; density of air-dried granules 1.485.

It has been well known since the early 1800s the common starches are composed of at least two types of molecules (see Chapter 2). A study of 89 wheats grown in the United States and 61 grown in other countries showed an overall range of 17–29% amylose in wheat starch; the starches from the U.S. wheats contained 20–27% amylose (Deatherage *et al.*, 1955). These values agree with the more limited data of other workers.

The glucose residues in amylose (the linear fraction) are joined together by α-1,4-glucosidic bonds. In amylopectin, some α-1,6-glucosidic bonds occur, causing the molecule to have a branched or bushy structure (see Chapter 2). Both fractions, like other high polymers, are composed of molecules within a rather broad range of molecular weights. Molecular weight determinations, therefore, give only average values. (Moreover, values obtained by different methods will seldom, if ever, closely agree because calculations are based on different characteristics of the molecules). The molecular size of wheat amylose has been reported to be 860 glucose units on the basis of osmotic pressure measurement, and 540 glucose units by periodate oxidation; these values correspond to molecular weights of 140,000 and 87,000, respectively (Potter and Hassid, 1948). Those data were interpreted to indicate that wheat amylose is slightly branched. The molecular weight of the amylopectin fraction has been reported to be 4 million. There is some evidence that amylose is the fraction first formed in the wheat plant and that amylopectin may be made from it, but that formation of amylopectin from amylose is slight after the latter has been incorporated into granules. This may explain the comparatively high ratio of amylopectin to amylose in wheat starch granules during early stages of maturation of the wheat kernel. In any case, there appears to be no segregation of amylose within the granule.

When the starch granule is immersed in water or exposed to a humid atmosphere, it readily takes up moisture. The hydration is a process of simple adsorption that causes the granule to increase about 10% in diameter or about 33% in volume. The sorption is not confined to the surface of the granule. Undamaged granules retain their birefringence under those conditions. Granules that have been damaged by shearing lose birefringence and swell more; this is true of even a damaged portion of a granule.

The phenomenon exhibited by a granule subjected to shear and placed in cold water is apparently the same as the reaction a granule undergoes on heating in water, namely, gelatinization. When wheat starch is heated in excess water, birefringence is first lost around the hilum and then progressively outward as heating continues, until none is left. Some wheat starch granules begin to lose birefringence at 58°C., whereas others in the same sample may retain all or much of their birefringence at 70°C. All birefringence of all granules in a wheat starch sample will be lost at somewhere between 70 and 78°C. On the whole, granule size is not clearly related to the gelatinization temperature of wheat starch, but the last granules to lose all birefringence are usually very small. These changes, which are accompanied by a change in X-ray spectrum, were designated by Katz (1928) as "the first order of gelatinization." They represent the condition attained by wheat starch during bread baking.

As the pasting process progresses as the temperature is increased, the large, lenticular granules swell and assume a saddle shape, while the usually smaller spherical granules swell more uniformly. A second change in X-ray pattern to an amorphous condition occurs when the starch suspension is heated from 80°C to boiling; this is "the second order of gelatinization" described by Katz and Rientsma (1930). Other workers have confirmed these two steps in gelatinization and swelling of wheat starch.

After loss of birefringence, the granules are susceptible to rapid attack by amylases. Microscopic observation of the suspension shows that some starchy material is leached from the granule. Under ordinary conditions of pasting, however, amylose goes into colloidal solution both inside and outside the granule sac, which is formed by the amylopectin; it exists as an artifact during swelling of the granule by tangential expansion.

Amylose does not remain long in solution when the temperature is lowered, however. It then retrogrades, or associates, into mostly crystalline particles. Wheat amylose has been reported to be 87% retrograded after 24 hr at 25°C (Whistler and Johnson, 1948). In this way, the amylose again becomes relatively resistant to attack by enzymes. A similar change is undergone more slowly by the amylopectin.

B. Minor Carbohydrates

Cellulose, hemicelluloses, pentosan gums, and sugars, although minor constituents of wheat are nevertheless important carbohydrate components. Cellulose is the major constituent of most of the cell walls in the kernel and plays a significant role in kernel structure. Hemicelluloses and pentosans constitute a small fraction of the wheat kernel; the former are constituents of some cell walls, while the latter exist in the mucilagenous portion of flour. About 2.5% of the total carbohydrate in wheat endosperm has been reported to be hemicelluloses, which appear to be chiefly constituents of the cell walls of the starchy endosperm. The so-called wheat gums may be associated with the inner sides of the cell walls or may lie chiefly in the starchy endosperm cell contents adjacent to the cell walls. That the pentosan gums are at least loosely associated with the cell walls is indicated by their occurrence with the latter in the "tailings" fraction from wheat flour. The sugar residues of the water-extracted gums are similar in that they consist of D-glucose and L-arabinose in a 2:1 ratio. Sucrose, levosin, glucose, fructose, maltose, raffinose, and glucofructosan have been reported to be present in small amounts in the starchy endosperm. Sucrose and raffinose have been identified as the chief sugars in wheat germ, in which only traces of glucose are found.

C. Composition of Wheat and Milled Products

A simplified summary of the approximate chemical composition of wheat and its milled products is given in Table 4.6, in which results are expressed on a 14% moisture basis. The concentrations of pentosans and ash are highest in the aleurone-rich bran; lipids, proteins and sugars are highest in the germ; and starch is concentrated in the starchy endosperm. In the more refined white flour, which contains particles originating primarily from the starchy endosperm, the con-

TABLE 4.6

Chemical Composition of Wheat and Milled Products[a]

Product	Wheat (%)	Protein (%)	Fat (%)	Ash (%)	Starch (%)	Pentosans (%)	Sugars (%)
Wheat	100.0	15.0	2.0	2.0	53.0	5.0	2.5
White wheat flour	65.0	14.0	1.0	0.5	67.0	1.5	1.0
Dark flour	18.5	17.0	3.0	3.0	45.0	4.5	4.5
Bran	16.0	16.5	4.5	6.5	12.0	18.0	5.5
Germ	0.5	31.0	12.5	4.5	10.0	4.0	16.5

[a]Expressed on 14% moisture basis.

centration of protein, lipids, minerals, and nonstarchy carbohydrates is much lower than in the dark flour, which is a mixture of starchy endosperm, aleurone, cell layer bran, and germ.

Wheat generally is considered to be a raw product that must be processed by milling and baking before it can be used for human consumption. The basic purpose in milling is to break open the firm outer coat of the kernel, called bran, so the endosperm and germ can be removed. Germ, which is higher than endosperm in fat, is removed to improve storing and baking qualities. For white flour production the endosperm is separated as completely as possible from the bran and germ, and ground into flour. The bran and germ are utilized in feeds for livestock and poultry and in other manufactured by-products. If, however, dark flour or whole wheat flour is desired, then portions or all of bran and germ remain with the endosperm in the milling process.

D. Function of Wheat Flour Carbohydrates in Bread Making

According to Neukom *et al.* (1962) carbohydrates are important in all stages of bread making (Table 4.7). First, let us consider the relatively minor (in quantity) carbohydrate components in wheat and wheat flour and their role in bread making. The most important among these are substances known as cereal gums (see also Chapter 3). Extracted with water from wheat flour, they yield viscous solutions and seem to play quite an important role in bread making. Cereal gums are largely polysaccharidic in nature, but they also contain a certain amount of protein. The polysaccharides of cereal gums are composed of glucosans and araboxylans. Water-soluble gums constitute about 20 to 25% of the total pentosan content of wheat flour and about 40% of rye flour. High levels of gums in rye flour are responsible for the high water-binding capacity, the stickiness, and

TABLE 4.7

Function of Wheat Flour Carbohydrates in Bread Making[a]

Stage	Function	Carbohydrate
Dough mixing	Water binding	Starch, soluble pentosans
Fermentation	Substrate for yeast	Mono- and disaccharides, dextrins, fructosans
Baking	Water binding	Gelatinized starch
	Crumb texture	Gelatinized starch
	Crust formation (color, flavor)	Gelatinized starch
Storage	Aging	Retrograded starch
	Freshness retention	Soluble pentosans

[a]From Neukom *et al.*, 1962.

the impaired rheological properties of doughs made from rye flour. In wheat flour, water-soluble pentosans affect hydration, dough development, mixing characteristics, and the viscosity and oxidation requirements of a flour. Bread making potentialities of wheat flour are generally improved by adding oxidizing agents in trace amounts, that is, 10–50 ppm potassium bromate. The oxidation requirement seems related to the presence of thiol-containing amino acids in wheat proteins. It is known, however, that minute amounts of oxidizing agents cause gelation of a dilute pentosan solution in an aqueous wheat flour extract. No full explanation can be given for this unusual reaction, but it seems of special interest in view of the great importance of oxidizing agents on dough and baking properties (see also Chapter 3).

The pentosan-rich gummy fractions affect rheological properties of macaroni doughs. Excessive amounts of the water-insoluble or water-soluble pentosans reduce the spread of cookies. Proper amounts of water-soluble pentosans are important in producing cakes with desirable structure and texture. Initially it was believed that pentosans increased the volume of bread, but later studies with purified preparations did not confirm the early findings. It is well established, however, on the basis of reconstitution studies that, in the absence of pentosans, doughs are soft, slack, and moist. Addition of the pentosan fraction restores to the doughs the normal "boldness," stickiness, and dryness. Poor baking characteristics of doughs rich in so-called tailing fractions are associated with high levels of pentosans in the fractions.

Views on the role of starch in baking range from those held by workers who claim that differences in loaf volume and crumb structure depend primarily on the starch content and its properties, to those of workers who report that the only role of starch in bread making is that of a filler, to dilute the protein that governs bread making potentialities (see also Chapter 2). It has been shown that differences in starch, particularly those associated with different classes of wheat, may affect loaf volume significantly. It is clear from reconstitution studies that very impaired bread is obtained if wheat starch is replaced by starches isolated from other cereal grains, but some uncertainty exists regarding the differences in performance of starches from flours from different hard wheats. The importance of starch in bread making was demonstrated by Rotsch (1954). Bread of satisfactory quality could be baked from a dough in which gluten was replaced with other gel-forming materials. No substitute could be found, however, for starch in the dough formula. These findings modify the tendency to confine bread making potentialities to protein quality or quantity.

The effect of wheat milling on starch has been studied extensively. When flour is ground, some of the starch granules rupture and so-called damaged starch is formed. A slight increase in the amount of damaged starch increases water absorption of the dough and availability of starch to amylase attack during fermentation. Excessive amounts of ruptured starch are, however, undesirable as they impair gas retention of the dough.

Starch is equally important in cake making (Myhre, 1976). Cake baking is a complex process in which the essential ingredients (flour, sugar, water, eggs, shortening, leavening agent, and salt) are mixed to produce a fluid batter emulsion into which air is incorporated. At early stages of baking, gelation increases batter viscosity. With a rise in temperature starch imbibes increasing amounts of water and gradually a rigid porous cake structure is formed. The rate of gelation is affected by the presence of certain wetting materials (which reduce the temperature of starch gelation) and of polar lipids (which inhibit starch gelation). In extremes of either case, cake quality is impaired.

III. Sugars in Bread Making

Sugars have several roles in bread making. Besides providing sweetness, they furnish a fermentable substrate from which yeast can produce carbon dioxide and alcohol, impart color to bread crust or toast, and modify the texture and the appearance of the baked product (Pomeranz and Finney, 1975). According to Nesetril (1967), sugars provide a source of fermentable substrates required to maintain adequate yeast activity and leavening action until an internal loaf temperature of about 60°C (140°F) is reached in baking. Residual sugars that remain after fermentation has been stopped by temperatures above 60°C serve several functions:

1. Fast crust color formation due to the caramelization and the interreactions of reducing sugars (the Maillard or browning reaction). This allows lower baking temperatures and faster baking time with more moisture remaining in the loaf.
2. The development of volatile acids and aldehydes, responsible for enhanced flavor and aroma.
3. Texture, grain, and crumb become smoother, softer, and whiter. This is probably related to the action of sugars in delaying the starch gelatinization and protein denaturation. Sugars are known tenderizers.
4. Extended shelf life through moisture retention due to the hygroscopic nature of certain sugars. Sucrose and dextrose hydrate are the least hygroscopic; levulose, honey, invert sugar, and high conversion corn syrups are highly hygroscopic.
5. The addition of sugars increases yield of dough.

Sugars in bread dough arise from three sources: (a) those originally in the flour, (b) those produced from oligosaccharides or polysaccharides by the action of flour or yeast enzymes, and enzymatic supplements, and (c) those added as dough ingredients. The amount of fermentable sugars added to wheat flour in the production of white bread has increased from about 2% (on flour basis) in the 1920s to about 8% in the 1970s.

A. Sugars in Flour and Dough

The four most important sugars in bread making are the disaccharides, sucrose and maltose, and the monosaccharides, glucose and fructose. In addition, lower oligomers of the glucofructosans in wheat flour support yeast fermentation. Spring wheat flour contains 0.02–0.08% fructose, 0.01–0.09% glucose, 0.19–0.26% sucrose, 0.07–0.10% maltose, and 1.26–1.31% oligosaccharides; the total amount of sugars is 1.55–1.84%. However, the maltose content of sponge and "straight" dough immediately after mixing is 10–15 times that of the flour from which it was made, when expressed on the same moisture basis. The large and rapid increase in the maltose content of doughs during mixing results from action of β-amylase on the mechanically damaged, susceptible starch in the flour.

According to Piekarz (1968), in the presence of a mixture of fermentable sugars, bakers' yeast preferentially ferments dextrose. Maltose is fermented only when dextrose has been used up or reduced to a minimal level. In commercial bread production, there is an adequate quantity of dextrose and neither maltose nor levulose is required to complete the proof and leaven the bread. Dextrose is relatively heat stable and imparts a "foxy" red color to the crust. Levulose caramelizes readily and produces a dark-brown crust.

B. Sugars Added in Bread Making

Beet or cane sugar is the traditional and most common sweetener used in bread making, but the use of sweeteners prepared by hydrolyzing corn starch with acids, enzymes, or their combination has greatly increased. The commercial products vary widely in composition, dextrose equivalent, and amounts of fermentable sugars. They range from regular corn syrup, with a DE of only 42 and a high concentration of higher saccharides (about 30% saccharides above hexasaccharides, as percentage of total carbohydrates), to liquefied (about 70% crystalline pure) dextrose, in which the carbohydrates have DE of more than 99.5% and are more than 99.5% fermentable. A corn syrup in which 42% of the total carbohydrate is fructose and 50% is glucose is available commercially. The carbohydrates in this syrup, which is as sweet as sucrose in most applications, are 95% fermentable (see pages 133–135).

Generally, corn syrups with a low DE and a high content of higher saccharides are undesirable in bread making. They provide only small amounts of fermentable sugars, their sweetening power is low, and they may cause excessive stickiness in bread. Some of the syrups with a high DE, and some of the enzymatic supplements, however, can produce bread equal in every respect to that produced from high sucrose formulation.

C. Sugars Produced by α-Amylases

In dough, β-amylase cannot attack raw, mechanically undamaged starch and it hydrolyzes damaged starch relatively slowly, but α-amylase vigorously attacks damaged starch and also attacks starch that is not sufficiently damaged to be susceptible to β-amylase. Sound wheat flour contains an abundance of the saccharifying enzyme, β-amylase, and only minute amounts of the dextrinizing enzyme, α-amylase. An adequate level of α-amylase activity is essential in doughs baked without added sugar to maintain a rate of gas production sufficient to leaven the dough and to produce desirable taste and crust color. α-Amylase activity, however, does more than insure an adequate level of fermentable sugars. Amylolytic activity affects dough consistency, modification of starch during oven baking, and the quality of the baked bread.

D. α-Amylase Sources

α-Amylases from different sources vary widely in their thermostability, starch-liquefying action, and effects on dough consistency, loaf volume, bread crumb characteristics, and crumb compressibility.

Certain strains of the mold *Aspergillus oryzae* produce an amylase of the α-type that is more thermolabile than the cereal α-amylases. The thermolability of fungal amylases offers an advantage over the bacterial amylases, which impair bread quality. Bacterial amylases have shown little promise as diastatic supplements because they produce a sticky and gummy crumb, generally attributed to their high thermostability.

Rubenthaler *et al.* (1965) compared the effects on loaf volume and bread characteristics of α-amylases from cereal, fungal, and bacterial sources. Table 4.8 shows the effects of various levels of cereal, fungal, and bacterial α-amylases on loaf volume and bread crumb with a no-sugar formula. Loaf volume increases were usually accompanied by greater crust browning. Expressed on an equal basis of dextrinogenic activity, the increase in loaf volume was greatest for bacterial amylase, smallest for fungal amylase, and intermediate for cereal amylase. The large increase in loaf volume for bacterial amylase, however, was accompanied by open and impaired crumb grain, a result of the marked thermostability of bacterial amylases. The heat-stable bacterial amylase continued to hydrolyze starch throughout practically all of the baking time, so that starch is markedly degraded. With cereal amylase, loaf volume was lower than with bacterial amylase, but crumb grain was much improved, particularly at high enzyme levels. Consequently, cereal amylases of intermediate heat stability yielded the best loaf volume and crumb grain, although optimum levels were not reached. The heat-labile fungal amylase improved bread much less than an equivalent level of dextrinogenic activity from cereal amylase. These results

TABLE 4.8

Effects of α-Amylase on Loaf Volume and Crumb Grain of Bread Baked without Sugar[a,b]

Enzyme level (SKB units)[c]	Wheat malt α-amylase		Fungal α-amylase		Bacterial α-amylase	
	Loaf volume (cm³)	Bread crumb grain	Loaf volume (cm³)	Bread crumb grain	Loaf volume (cm³)	Bread crumb grain
0	390	U	390	U	390	U
5	475	Q	465	Q–U	635	Q–U
10	570	Q	495	Q–U	765	Q–U
20	645	Q	515	Q–U	878	Q–U, open
40	780	Q–S	580	Q	900	Q–U, open
80	858	Q–S	635	Q–S	940	U, very open

[a]From Rubenthaler et al., 1965.

[b]S, satisfactory; Q, questionable; U, unsatisfactory.

[c]Per 100 gm flour.

were confirmed by baking experiments that showed the effects of two levels (40 and 80 Sandstedt–Kneen–Blish (SKB) units/100 gm flour) and each of three cereal, five fungal, and two bacterial α-amylases. Bacterial amylases gave the best loaf volumes and the poorest crumb grains; cereal amylases gave the best bread.

TABLE 4.9

Effect of Adding 10 SKB Units of α-Amylase at Given Sugar Levels on Loaf Volume and Bread Crumb Grain[a,b]

Sugar (%)	No enzyme		Wheat malt α-amylase		Fungal α-amylase		Bacterial α-amylase	
	Loaf volume (cm³)[c]	Bread crumb grain	Loaf volume (cm³)	Bread crumb grain	Loaf volume (cm³)	Bread crumb grain	Loaf volume (cm³)	Bread crumb grain
0	362	U	515	U	442	U	763	Q–U
1	400	U	790	Q–S	695	Q–U	875	Q–U
2	603	Q	860	Q–S	807	Q–S	855	Q–U
3	785	Q–S	850	Q–S	800	Q–S	830	Q–U
4	825	S	835	S	823	S	888	Q–U
5	865	S	890	S	858	S	930	Q–U
6	900	S	930	S	910	S	975	Q–U

[a]From Rubenthaler et al., 1965.

[b]S, satisfactory; Q, questionable; U, unsatisfactory.

[c]Per 100 gm flour.

Subsequent baking tests showed the effects on bread quality of adding a mixture of α-amylases from three sources (total of 40 SKB units/100 gm flour). As few as 5 SKB units of bacterial α-amylase per 100 gm flour had a considerable detrimental effect on bread crumb texture and bread crumb grain. Cereal α-amylase improved bread quality substantially. The best results were obtained with a mixture containing a high level of cereal amylase and low levels of fungal and bacterial enzymes.

The effects of amylases from various sources (Table 4.9) depended on the level of sugar in the dough formula. Although the overall effect of the various amylases was similar to that observed in the no-sugar formula, the responses obtained by adding cereal and fungal α-amylases were best in bread baked with 1 and 2% sugar. Adding amylases to formulas containing more than 2% sugar increased loaf volume by only relatively small and constant amounts for the three α-amylases at the 3–6% sugar levels.

For a given α-amylase, there was no loaf volume or crumb grain response as sucrose was increased from 2 to 3%. Similarly, for cereal and fungal α-amylases, there were no material responses when sucrose was increased from 3 to 4%. For all three α-amylases, however, loaf volume increased somewhat as sucrose was increased from 4 to 5% and from 5 to 6%.

Bread baked with 2% sugar, high levels of cereal α-amylase, and low levels of bacterial enzyme (Table 4.10) had a loaf volume equal to and bread crumb grain only slightly inferior to that of bread baked with 6% sugar.

TABLE 4.10

Effects of Adding Combinations
of α-Amylase and 2% Sugar
on Loaf Volume and Bread Crumb Grain[a,b]

Sugar (%)	SKB units of enzyme from		Loaf volume (cm³)[c]	Bread crumb grain
	Cereals	Bacteria		
0	0	0	420	U
2	0	0	603	Q
4	0	0	825	S
6	0	0	890	S
2	38	2	885	Q–S
2	39	1	875	Q–S
0	38	2	790	Q–S

[a]From Rubenthaler et al., 1965.
[b]S, satisfactory; Q, questionable; U, unsatisfactory.
[c]Per 100 gm flour.

E. Amyloglucosidase in Bread Making

Fungal enzymes of the amyloglucosidase type, which hydrolyze starch or dextrins specifically and almost completely to glucose, are of great interest in several food industries. Enzymes of this type include those with activities designated "maltase," "glucogenic enzyme," "glucamylase," and "limit dextrinase." Several commercial plants use the glucamylase process to produce crystalline dextrose of high purity.

Pomeranz et al. (1964) studied the effects of fungal α-amylase and fungal amyloglucosidase, alone and in combination, on bread quality. The effects depended on the source of the enzymes and on levels of sugar in the bread formula. A constant level of cereal or fungal enzymatic supplements in doughs containing various levels of sugar (Table 4.11) was most effective in low sugar formulas and least effective in sugar-rich doughs. In studies of various levels of amyloglucosidase and fungal α-amylase in sugar-free doughs, the loaf volume increase at the highest level of fungal α-amylase was smaller than that for amyloglucosidase. Amyloglucosidase (100 mg), in a formula containing 2.0 gm sugar and 4.0 gm nonfat milk solids (Table 4.12), produced a loaf volume nearly comparable to that of loaves baked from a dough containing 6.0 gm sugar and 0.25 gm malt.

Increasing amounts of fungal α-amylase or amyloglucosidase in doughs containing various sugar levels (Fig. 4.6) produced corresponding increases in loaf volume. Volume response reached a plateau at high levels of enzyme supplementation, but at all levels (compared on an equal α-amylase basis), amyloglucosi-

TABLE 4.11

Effects of Dough Formulation on Loaf Volume[a]

Sugar (gm)	Malt (gm)	Fungal α-amylase (SKB units)	Amyloglucosidase B (mg)	Loaf volume (cm^3)[b]
0	0	0	0	390
2	0	0	0	680
6	0	0	0	815
0	0.25	0	0	525
2	0.25	0	0	775
6	0.25	0	0	875
0	0	40	0	450
2	0	40	0	775
6	0	40	0	845
0	0	0	25	485
2	0	0	25	750
6	0	0	25	850

[a]From Pomeranz et al., 1964.
[b]Per 100 gm flour.

TABLE 4.12

Effects of Adding Amyloglucosidase to Bread Formula
Containing 2 gm Sugar and 4 gm Nonfat Milk Solids[a]

Sugar (gm)	Malt (gm)	Amyloglucosidase A (mg)	Loaf volume (cm³)[b]
6	0.25	0	960
2	0	0	700
2	0	5	760
2	0	25	875
2	0	50	900
2	0	100	925
6	0	0	875

[a]From Pomeranz et al., 1964.
[b]Per 100 gm flour.

dase increased loaf volume more than fungal α-amylase. The improving effect of amyloglucosidase depended on the level of sugar in the dough formula (Table 4.13). High levels of amyloglucosidase were needed in low sugar breads to produce loaves comparable to those containing high sugar and low enzyme levels.

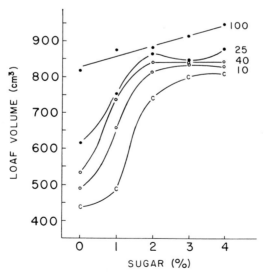

Fig. 4.6 Effects on loaf volume of adding indicated levels of fungal α-amylase or amyloglucosidase to doughs containing various levels of sugar. (C) Control. (●) Glucosidase (mg). (○) Fungal α-amylase (SKB). (From Pomeranz et al., 1964.)

TABLE 4.13

Effects of Adding Amyloglucosidase
on Loaf Volume of Bread Made from Doughs
Containing Sugar[a]

Amyloglucosidase (mg)	Loaf volume (cm³) with sugar levels (gm/100 gm flour) of				
	1	2	3	4	6
0	445	685	783	823	898
5	—	—	—	825	898
10	—	—	—	863	903
20	—	—	850	873	905
50	820	845	880	—	—
100	885	880	—	—	—

[a]Condensed from Pomeranz *et al.*, 1964.

Amyloglucosidase derived from *Aspergillus niger* gave better results than that from *Rhizopus delemar*. Despite the high gas production power of the *R. delemar* preparation, its effect on loaf volume was small, probably because the enzyme from *R. delemar* was inactivated rapidly above 55°C, whereas the enzyme from *A. niger* was relatively stable up to 70°C. Thermostabilities of α-amylases from different sources are compared in Table 4.14. The availability of more thermostable glucosidases that are compatible with bread making conditions would greatly increase their contribution to the production of fermentable sugars.

TABLE 4.14

Thermostability of Amylases[a]

Temperature (°C)	% Enzyme activity retention		
	Fungal	Malt	Bacterial
65	100	100	100
70	52	100	100
75	3	58	100
80	1	25	92
85	—	1	58
90	—	—	22
95	—	—	8

[a]From Shellenberger *et al.*, 1966.

IV. Cereal Malts in Bread Making

The most common source of amylase supplements for baked goods has been malt. When malted, barley, wheat, and rye produce relatively large amounts of α-amylase and β-amylase, but oats, sorghum, corn, and rice product only α-amylase. Because the combination of the two enzymes is more effective in hydrolyzing starch to fermentable sugars than α-amylase alone, malts from barley, wheat, and rye are more efficient for most uses than malts from other grains. Finney *et al.* (1972) evaluated the role of cereal malts in bread making to determine optimum levels of malt required to product acceptable bread baked with 0 to 6% added sugar, and to compare the effects of cereal malts (including the recently developed triticale—a wheat–rye cross) in bread making.

Unmalted grains did not contribute to the loaf volume of bread baked without added sugar (average volume of 442 cm³ of bread baked from 100 gm flour without added sugar). The loaf volume increase, effected by adding 12.3 SKB units of α-amylase, exceeded 400 cm³ for several malts and almost equaled the volume of unsupplemented bread (Table 4.15). Doubling the amount of α-amylase (24.6 SKB units/100 gm flour) generally increased loaf volume by about an additional 100 cm³. Further increases in malt had no significant additional effect. Responses to barley, wheat, and rye malts, added on an equivalent α-amylase basis, were similar. The response to a 60° Lintner (L) commercial, diastatic malt syrup was greater than the response to barley, wheat, or rye malt flours. The α-amylase content of the syrup was 12.3 SKB units per gram, compared with 14 to 57 α-amylase SKB units in the malt flours. Supplementa-

TABLE 4.15

Effects of Malts on Loaf Volume of Bread Baked from 100 gm Flour without Added Sugar[a]

Malt source	α-amylase (units/gm)	Loaf volume (cm³) for α-amylase levels of		
		0.0 units	12.3 units	24.6 units
Malt syrup	12.3	442	894	950
Commercial barley (Larker)	38.4	432	801	902
Barley (Dickson)	26.7	445	810	933
Barley (Piroline)	22.3	453	798	919
Barley (Hembar)	14.1	450	810	915
Wheat	28.7	438	829	925
Rye	46.8	443	820	913
Oats	22.0	426	648	758
Triticale 2	34.9	448	853	942
Triticale 3	57.4	440	867	940
Average	—	442	—	—

[a]From Finney *et al.*, 1972.

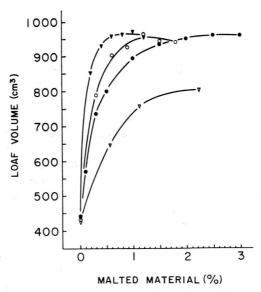

Fig. 4.7 Effects of malts from various sources on loaf volume of bread baked without added sucrose. Formulas included 100 gm flour, water as needed, 1.5 gm salt, 3 gm shortening, 2 gm yeast, 4 gm nonfat milk solids, and 3 mg potassium bromate. (▼) Triticale. (○) Larker barley. (●) 60° Lintner malt syrup. (▽) Oats. (From Finney *et al.,* 1972.)

tion was based on equivalent α-amylase SKB units, so that much more malt syrup than malt flour was usually added. The greater weight of malt syrup might have contributed more preformed fermentable sugars than malt flours.

Triticale malts are excellent supplements for two reasons: (1) on the average, they have much higher diastatic power than the other cereal malts and (2) the average soluble protein in malted triticale was 53% of the total protein (compared with 27 to 38% in the other malts). Thus, triticale could have contributed more significantly (than did other cereal malts) to fermentable sugars and low molecular weight proteins. As noted for malt syrup, the beneficial effects of triticale malts were more pronounced at the 12.3 SKB unit level than at the 24.6 SKB unit level of supplementation. Malt flour from oats was inferior to that from other grains.

The main types of malt supplements are compared in Fig. 4.7. Various amounts of the malts from triticale, commercial (Larker) barley, and 60°L malt syrup were required to attain the maximum loaf volume of 960 cm³/100 gm flour. The differences in malt requirements reflected differences in α-amylase levels of the malts. The poor performance of oat malt flour might have resulted from its low β-amylase and diastatic power (about one-tenth the values in triticale malts and about one-fourth those in barley malts). The maximum loaf volume of 960 cm³, regardless of malt supplement, was consistent with the criterion of optimum loaf volume of a given flour.

Sucrose and Barley Malt

In bread produced without malt (Fig. 4.8), loaf volume rose sharply as sucrose levels were increased from 0 to 2%, decreased somewhat at 3% sucrose, and increased significantly with further increases in sucrose. In the formulations with optimum levels of malt and various levels of sucrose (see Fig. 4.8), the malt requirement decreased as the sucrose level was increased above 4%. The optimum malt amount was higher with milk solids than without. With optimum malt, increasing sucrose amounts from 0 to 6% did not increase loaf volume. Loaves baked without added sucrose but with optimum malt were essentially equal in volume to those baked with 6% sucrose and optimum malt, and were higher than those baked with high levels of sucrose but no malt.

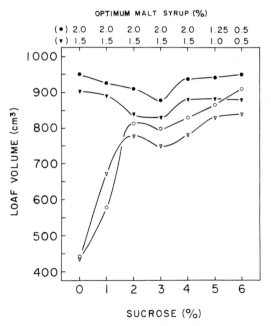

Fig. 4.8 Loaf volumes of bread baked from 100 gm flour with 0–6 gm sucrose by four formulations. Formula also included 1.5 gm salt, 3 gm shortening, 2 gm yeast, and optimum potassium bromate (3 mg with 4 gm nonfat milk solids and 1 mg without milk). (●) Milk, optimum malt. (○) Milk, no malt. (▼) No milk, optimum malt. (▽) No milk, no malt. (From Finney *et al.*, 1972.)

REFERENCES

Anderson, R. A., and Watson, C. A. (1982). The corn milling industry. *In* "Handbook of Processing and Utilization in Agriculture" (I. A. Wolf, ed.), Vol. 2, Part 1, pp. 31–61. CRC Press, Boca Raton, Florida.

Birch, G. G., ed. (1971). "Sweetness and Sweeteners." Appl. Sci. Publ., Ltd., London.

Birch, G. G., and Green, L. F., eds. (1973). "Molecular Structure and Function of Food Carbohydrates." Wiley, New York.

Birch, G. G., and Shallenberger, R. S., eds. (1977). "Developments in Food Carbohydrates," Vol. 1. Appl. Sci. Publ., Ltd., London.

Deatherage, W. L., MacMasters, M. M., and Rist, E. C. (1955). A partial survey of amylose content in starch from domestic and foreign varieties of corn, wheat and sorghum and from some other starch-bearing plants. *Trans. Am. Assoc. Cereal Chem.* **13**, 31–42.

Eisenberg, S. (1955). Use of sugars and other carbohydrates in the food industry. *Adv. Chem. Ser.* **12**, 78.

Finney, K. F., Shogren, M. D., Pomeranz, Y., and Bolte, L. C. (1972). Cereal malts in breadmaking. *Baker's Dig.* **46**(1), 36–38, 55.

Fruin, J. C., and Scallet, B. L. (1975). Isomerized corn syrup in food products. *Food Technol.* **29**(11), 40, 42, 44, 45.

Inglett, G. E. (1974). "Symposium: Sweeteners." Avi Publ. Co., Westport, Connecticut.

Johnson, J. C. (1976). "Specialized Sugars for the Food Industry." Noyes Data Corp., Park Ridge, New Jersey.

Junk, W. R., and Pancoast, H. M. (1973). "Handbook of Sugars." Avi Publ. Co., Westport, Connecticut.

Katz, J. R. (1928). The X-ray spectroscopy in starch. *In* "A Comprehensive Survey of Starch Chemistry" (R. P. Walton, ed.). Chem. Catalog Co., New York.

Katz, J. R., and Rientsma, L. M. (1930). Abhandlungen zur physikalischer Chemie der Stärke und Brotbereitung. III. Erster und zweiter Grad der Verkleisterung. *Z. Phys. Chem., Abt. A* **150**, 67–80.

Katz, M. H. (1972). Correlating physical and sensory measurements to quantify the functional properties of carbohydrates. *Food Technol.* **26**(3), 20–22.

Myhre, D. V. (1970). The function of carbohydrates in baking. *Baker's Dig.* **44**(3), 38, 39, 60.

Nesetril, D. M. (1967). Corn sweeteners: Their types and uses in baking. *Baker's Dig.* **41**(3), 28–30, 32.

Neukom, H., Kuendig, W., and Deuel, H. (1962). The soluble wheat flour pentosans. *Cereal Sci. Today* **7**, 121–123.

Piekarz, E. R. (1968). Levulose containing corn syrup. *Baker's Dig.* **42**(5), 47–69.

Pomeranz, Y., and Finney, K. F. (1975). Sugars in breadmaking. *Baker's Dig.* **49**(1), 20–22, 24, 26, 27.

Pomeranz, Y., Rubenthaler, G. L., and Finney, K. F. (1964). Use of amyloglucosidase in breadmaking. *Food Technol.* **18**(10), 138–140.

Potter, A. L., and Hassid, W. Z. (1948). Starch. II. Molecular weights of amyloses and amylopectins from starches of various plant origins. *J. Am. Chem. Soc.* **70**, 3774–37777.

Rotsch, A. (1954). Chemische und backtechnische Untersuchungen an künstlichen Teigen. *Brot Gebaeck* **8**, 129–130.

Rubenthaler, G. L., Finney, K. F., and Pomeranz, Y. (1965). Effects on loaf volume and bread characteristics of alpha-amylases from cereal, fungal, and bacterial sources. *Food Technol.* **19**(4), 239–241.

Schultz, H. W., Cain, R. F., and Wrolstad, R. R., eds. (1969). "Symposium on Foods: Carbohydrates and their Roles." Avi Publ. Co., Westport, Connecticut.

Shallenberger, R. S., and Birch, G. G. (1975). "Sugar Chemistry." Avi. Publ. Co., Westport, Connecticut.

Shellenberger, J. A., MacMasters, M. M., and Pomeranz, Y. (1966). Wheat carbohydrates—their nature and functions in baking. *Baker's Dig.* **40**(3), 32–38.

Wardrip, E. K. (1971). High fructose corn syrup. *Food Technol.* **25**(5), 47.

Whistler, R. L., and Johnson, C. (1948). Effect of acid hydrolysis on retrogradation of amylose. *Cereal Chem.* **25,** 418–424.

Young, L. S. (1981). Manufacture, use, and nutritional aspects of 90% high fructose corn sweeteners. *Conference on Formulated Foods and Their Ingredients, Agric. Food Chem., Div. Am. Chem. Soc.,* Anaheim, California, November 2–4, 1981.

5

Proteins: General

I. Introduction

The problem of providing adequate protein for an expanding world population is second only to the overall world-hunger problem. Table 5.1 summarizes the protein content of selected foodstuffs. Apart from their nutritional significance, proteins play a large part in the organoleptic properties of foods. Proteins exert a controlling influence on the texture of foods from animal sources. Protein content of wheat and flour is considered one of the best single indices of bread-making potential. The protein test, although not included generally as a grading factor in grain standards, is accepted as a marketing factor.

Proteins often occur in foods in physical or chemical combination with carbo-hydrates or lipids. The glycoproteins, glycolipids, and lipoproteins affect the rheological properties of food solutions and have technical applications as edible emulsifiers. The aging of foods is associated with chemical changes in proteins. During heating (boiling, baking, roasting), the amino acid side chains are either degraded or they interact with other food components (i.e., lysine and reducing sugars) and give typical flavors. Excessive heating, on the other hand, reduces the nutritive value of most proteins.

The primary nutritional importance of protein is as a source of amino acids. Twenty-two amino acids are generally thought to be constituents of proteins. Some amino acids are essential for physical and mental health. Of the amino acids in foods, eight are known to be essential to man; that is, they must be

TABLE 5.1

Protein Content[a] ($N \times 6.25$, in %) of Selected Foodstuffs[b]

Animal origin	Protein	Plant origin	Protein
Milk		Rice, whole	7.5–9.0
Whole, dried	22–25	Rice, polished	5.2–7.6
Skimmed, dried	34–38	Wheat, flour	9.8–13.5
Beef		Corn meal	7.0–9.4
Dried	81–90	Chick pea	22–28
Roasted	72	Soybean	33–42
Egg		Peanut	25–28
Whole, dried	35	Walnut	15–21
Whole, dried,		Potato[c]	10–13
defatted	77	Tapioca[c]	1.3
Herring[c]	81	Alfalfa[c]	18–23
	69	Chlorella[c]	23–44
		Torula yeast[c]	38–55

[a]$N \times 6.25$ (%).
[b]Unless stated otherwise, on "as is" basis.
[c]H_2O-free basis.

supplied in the diet to maintain growth and health. Table 5.2 summarizes the content of essential amino acids in some common proteins. Proteins from some sources (i.e., cereal grains) are deficient in certain amino acids (i.e., lysine). Deficient proteins must be combined with those from other sources to provide an adequate balance of the essential amino acids. Such a balance can be accomplished by a combination of wheat flour with dry skim milk or soy flour.

Chapters 5 and 6 review the functional properties of proteins. This chapter

TABLE 5.2

Essential Amino Acids in Proteins[a]

Amino acid	FAO reference	Skim milk	Soy	Beef	Egg	Fish	Yeast
Lysine	4.2	8.6	6.8	8.3	6.3	6.6	6.8
Tryptophan	1.4	1.5	1.4	1.0	1.5	1.6	0.8
Phenylalanine	2.8	5.5	5.3	3.5	5.7	4.1	4.5
Methionine	2.2	3.2	1.7	2.8	3.2	3.0	2.6
Threonine	2.8	4.7	3.9	4.5	4.9	4.8	5.0
Leucine	4.8	11.0	8.0	7.2	9.0	10.5	8.3
Isoleucine	4.2	7.5	6.0	4.7	6.2	7.7	5.5
Valine	4.2	7.0	5.3	5.1	7.0	5.3	5.9

[a]Gm/100 gm of protein.

concerns proteins in general; Chapter 6 describes proteins in specific foods. In this chapter are covered the functionality of proteins, effects of protein denaturation on functionality, and methods to modify or improve functional properties. For general reviews of protein function see Schultz and Angelmier (1964), Whitaker and Tennenbaum (1977), Milner *et al.* (1977), Lawrie (1970), Satterlee (1981), Stanley *et al.* (1981), Cherry (1981), and Pour-El (1979).

II. Functionality of Proteins

Functional properties denote characteristics that govern the behavior of proteins in foods during processing, storage, and preparation as they affect food quality and acceptance (Matil, 1971). According to Nakai and Powrie (1981), functional properties include

1. Sensory and kinesthetic properties, e.g., flavor, odor, color, and texture
2. Hydration, dispersibility, solubility, and swelling
3. Surface active properties, e.g., emulsification, foaming, and adsorption, including fat binding
4. Rheological properties including gelation and texturization
5. Other properties, e.g., adhesive, cohesive, dough making, and film and fiber making

Functionality is, in a broad sense, any property of a protein other than its nutritional value, that affects its utilization. Typical functional properties performed by proteins in food systems are listed in Table 5.3. Factors that influence the functional properties of proteins are listed in Table 5.4. Functional properties should not be viewed, however, as separate entities, either from the standpoint of individual proteins or of other food components. Aspects of protein functionality and interrelationships are summarized in Fig. 5.1.

According to Hurrell (1980), proteins are the most reactive among the major food components. Proteins can react with reducing sugars, fats, and their oxidation products, polyphenols, and many other food components (Fig. 5.2). These reactions can lead to reduction in nutritive value, browning and flavor formation, and occasionally to toxicity, yet some of these interactions are essential to functionality of proteins in foods.

The ability of proteins to bind food components (i.e., lipids, water, flavors) is important in the formulation of many foods. This binding capacity affects adhesion, film and fiber formation, and viscosity. According to Kinsella (1976), binding is influenced by pH and ionic strength, both of which affect the surface area and properties of proteins; the amounts and physical properties of most food components (fats, carbohydrates, lipids); and the mechanical, thermal, chemical,

TABLE 5.3

Typical Functional Properties Performed by Proteins in Food Systems[a]

Functional property	Mode of action	Food system
Solubility	Protein solvation, pH dependent	Beverages
Water absorption and binding	Hydrogen-bonding of HOH, entrapment of HOH (no drip)	Meats, sausages, breads, cakes
Viscosity	Thickening, HOH binding	Soups, gravies
Gelation	Protein matrix formation and setting	Meats, curds, cheese
Cohesion–adhesion	Protein acts as adhesive material	Meats, sausages, baked goods, pasta products
Elasticity	Hydrophobic bonding in gluten, disulfide links in gels (deformable)	Meats, bakery
Emulsification	Formation and stabilization of fat emulsions	Sausages, bologna, soup, cakes
Fat adsorption	Binding of free fat	Meats, sausages, donuts
Flavor binding	Adsorption, entrapment, release	Simulated meats, bakery, etc.
Foaming	Forms stable films to entrap gas	Whipped toppings, chiffon desserts, angel cakes

[a]From Kinsella and Srinivasan, 1981.

TABLE 5.4

Factors Influencing the Functional Properties of Food Proteins[a]

Intrinsic	Environmental factors	Process treatments
Composition of protein(s)	Water	Heating
	Ions	pH
Conformation of protein(s)	pH	Salts
	Temperature	Reducing/oxidizing agents
Mono- or Multi-component	Oxidizing/reducing agents	Drying
	Lipids, flavors, sugars	Physical modification
Homogeneity–heterogeneity		Chemical modification

[a]From Kinsella and Srinivasan, 1981.

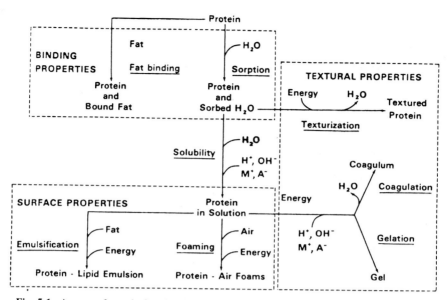

Fig. 5.1. Aspects of protein functionality and its interrelationships. (From Phillips and Beuchat, 1981; copyright of the American Chemical Society.)

and enzymatic modifications and treatments the food undergoes in production, storage, and home processing. Generally, hydrophobic proteins effectively lower surface tension and bind many lipophilic materials, such as lipids, emulsifiers, and flavor materials. The capacity of proteins to bind fat is important in the production of meat extenders and replacers, in which the absorption of fat by proteins enhances flavor retention and improves mouth-feel. The fat is absorbed

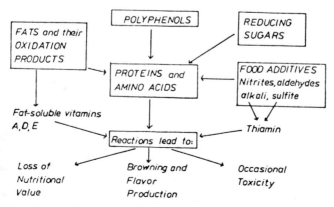

Fig. 5.2. Some important interactions of food components on processing. (From Hurrell, 1980.)

TABLE 5.5

Classification of Protein Groups

Polarity	Group	Example of amino acid
Apolar	Straight and branched aliphatic chains	Gly, Ala, Val, Leu, Ileu
Polarizable (variable dipole)	ϕ—	Phe, Tyr
Polar (permanent dipole)	R—OH	Ser, Thr
	ϕ—OH	Tyr
	—S—S	Cys, Met
	$-C\underset{NH_2}{\overset{O}{\lVert}}$	Asp-NH$_2$, Glu-NH$_2$
Charged	$-C\underset{O^-}{\overset{O}{\lVert}}$	Asp, Glu
	imidazolium	His
	$-NH_3^+$	Lys
	guanidinium	Arg

primarily through physical entrapment. Fat absorption can be increased if the protein is modified chemically to increase its bulk density.

Intrinsic functional properties represent the composite properties of the components of a food system as related to the amino acid composition, primary sequence, conformation, molecular shape and size, charge distribution, and intramolecular and intermolecular bonding. Table 5.5 classifies protein-reactive groups according to polarity. Types of bonds that govern structural forces in proteins are listed in Table 5.6. Some of these bonds are illustrated in Fig. 5.3. A high proportion of apolar residues affects interpeptide interaction, hydration,

TABLE 5.6

Forces Governing Protein Structure

Bond type	Mechanism	Energy (Kcal/mole)	Active group	Example	Role in gel matrix formation
Covalent	The atoms bound by a common electron pair	30–100	C—C, C—N, C=O, C—H, C—N—C, S—S	Bonds within amino acids, peptide bonds, disulfide bonds	Bridging, ordering
Ionic	Attraction between opposite charges	10–20	—NH$_3^+$ —COO$^-$ $\backslash$NH$^+$ // NH$_2$ —C$\backslash$NH$_2^+$	Lysine Glutamic acid Histidine Arginine	Solvent interaction, salt working
Hydrogen	Hydrogen shared between two electronegative atoms	2–10	N—H $\cdots$ O=C— —OH $\cdots$ O=C— O $\cdots$ OH C$\backslash$ $\backslash$C— HO $\cdots$ O	Amide-carbonyl Tyrosine-carbonyl Carboxyl-carbonyl	Bridging, stabilizing
Hydrophobic	Apolarity	1–3	Apolar groups	Apolar side chains	Strand thickening, strengthening, stabilizing
Electrostatic repulsion	Coulombic repulsion between particles with same charges	$\dfrac{Q_1 Q_2}{r^2}$	Polar groups	Polar groups of side chains	—
van der Waals repulsion	Repulsion of apolar groups that are close		Steric hindrance between side chain groups	All groups	—

Bond type	Functional groups involved	Disrupting solvents
Physical		
<u>Electrostatic</u> — COO⁻ ⁺NH₃ —	Carboxyl Amino Imidazole Guanido	Salt solutions High or low pH
Hydrogen bond —C=O HO— | NH	Hydroxyl Amide Phenol	Urea solutions Guanidine hydrochloride Dimethylformamide
Hydrophobic bonds 	Long aliphatic chains Aromatic	Detergents Organic solvents
Covalent		
Disulfide bonds —S—S—	Cystine	Reducing agents Sulfite Mercaptoethanol

Fig. 5.3. Types of bonds between protein chains. (From Wall and Huebner, 1981.)

solubility, and surface activity. The relationship among hydrophobicity, solubility, and charge frequency is given in Fig. 5.4. Hydrophobicity, interfacial tension, and emulsifying activity relationships are shown in Fig. 5.5. The contribution of hydrophobicity, charge frequency, and a structural parameter to functionality of proteins is summarized in Table 5.7. Hydrophobic interactions are important in the tertiary folding of proteins. They influence such properties as emulsification, foaming, and flavor binding. Charged amino acids enhance electrostatic interactions; they play a role in stabilizing globular proteins and in water binding and thereby influence hydration, solubility, gelation, and surfactancy. Sulfhydryl groups may be oxidized to form disulfide bounds; interchange reactions between thiol and disulfide groups may affect rheological properties. Formation of the β-lactoglobulin/κ-casein disulfide-linked complex minimizes heat gelation of concentrated milks and improves baking properties of milk powders.

Reactive polar groups participate in hydrogen bonding, which influences conformation of the α-helix and β-sheet structures. Acylated polar groups affect the physical properties of proteins. For example, phosphorylated caseins in milk curdle in the presence of calcium and therefore are the basis of cheese manufacture. The nature and magnitude of covalent and nonvalent bonds determine the size, shape, and surface charge of proteins. These properties can be modified by pH, temperature, or the presence of ionic groups, and can be used to change functional properties.

The formation of gels (gelation) is important in many foods. Protein gels may be visualized as three-dimensional matrixes or networks of intertwined, partially associated polypeptides in which water is entrapped. The gels have relatively high viscosity, plasticity, and elasticity (Kinsella, 1976). Examples of protein gels include gelation, coagulated egg white, soybean tofu, milk casein curd, and the myofibrillar gel formed by heating meat or fish proteins. Such gels have useful food applications because they provide a structural matrix for holding water and other food ingredients.

Gelation generally requires prior heating of the protein, which results in modification of the protein molecule, or denaturation, decribed later in this chapter. Formation of denatured protein gels requires an initial balance between attractive and repulsive forces. Gelation is a two-stage process involving an initial denaturation of native protein into unfolded polypeptides, which then may associate gradually to form the gel matrix if attractive forces and thermodynamic conditions are suitable. When the temperature coefficients of gelation are high, it

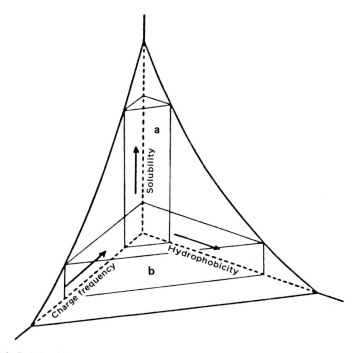

Fig. 5.4. Relationships among solubility, charge frequency, and hydrophobicity. (From Nakai and Powrie, 1981.)

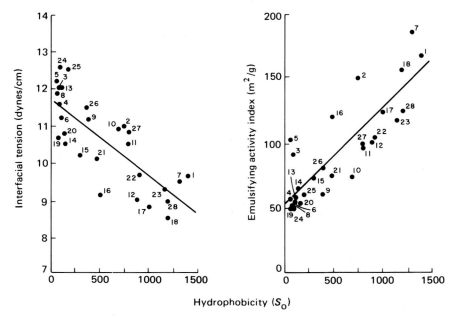

Fig. 5.5. Correlation of S_o with the interfacial tension and emulsifying activity of proteins. 1, Bovine serum albumin; 2, β-lactoglobulin; 3, trypsin; 4, ovalbumin; 5, conalbumin; 6, lysozyme; 7, κ-casein; 8–12, ovalbumin denatured by heating at 85°C for 1, 2, 3, 4, and 5 min, respectively; 13–18, denatured lysozyme by heating at 85°C for 1, 2, 3, 4, 5, and 6 min, respectively; 19–23, ovalbumin bound with 0.2, 0.3, 1.7, 5.7, and 7.9 moles of dodecylsulfate per mole protein, respectively; 24–28, ovalbumin bound with 0.3, 0.9, 3.1, 4.8, and 8.2 moles of linoleate per mole protein, respectively. (From Nakai and Powrie, 1981.)

TABLE 5.7

Contribution of Hydrophobicity, Charge Frequency,
and Structural Parameter to Functionality of Proteins[a,b]

	Hydrophobicity	Charge frequency	Structure
Solubility	−	+	−
Emulsification	+ sur	(−)	+
Foaming	+ tot	−	+
Fat binding	+ sur	(−)	−
Water holding	−	+	?
Heat coagulation	+ tot	−	+
Dough making	(+)	−	+

[a]Nakai and Powrie, 1981.

[b]+, positive contribution; −, negative contribution; sur, surface hydrophobicity; tot, total hydrophobicity; ( ), contributes to a lesser extent.

is expected that the first stage in gelation will be accomplished more quickly than the second stage, which involves network formation (Kinsella, 1976). Consequently, increasing the temperature enhances the formation of a fine and firm gel. Upon cooling, the uncoiled peptides associate to form the network. The association may involve various covalent and noncovalent interactions; disulfide bonds, hydrogen bonds, ionic attractions, hydrophobic associations, or combinations of these.

Gelation and curd formation are important in processing of milk and soy proteins; these actions essentially involve calcium coagulated protein gels. Heating milk proteins results in gelation; physical properties of the gel (such as firmness) depend on protein concentration, rate and length of heat treatment, and presence of various chemical compounds. Gelation of myofibrillar fish proteins is the basis for the manufacture of such products as fish sausage and *kamaboko* (fish paste). Gelation is also important in the manufacture of processed meats, whether processed along or in combination with vegetable extenders.

Many of the critical properties of proteins reflect interaction with water, as demonstrated by sorption, viscosity, gelation, solubility, emulsifying, surfactant, and rheological (including textural and sensory) properties. Protein solubility is related to its functionality and potential applications. Factors such as the nature of extractant, the extraction methods and conditions, and the nature, history, and concentration of proteins all affect the amounts and types of proteins extracted from a source as well as their functional properties.

Swelling, or the expansion that accompanies or results from uptake of water, is another significant factor. Imbibition of water in proteins is important in many foods, for example wheat flour dough and sausages. The limited amount of water is sufficient to affect several characteristics such as body, thickening, and viscosity, but is insufficient to dissolve and leach out food components in processing. The extent of swelling varies with the protein source, protein particle size, pH, ionic strength, and temperature. Several types of milk and soy products are some of the common protein additives used to enhance swelling and to hold water in foods. Binding of water by proteins in a such a manner that it becomes a structurally integral component of a food is important in the production of textured foods. Moreover, the availability of the water (or its activity) is important in food preservation and plays a role in the chemical or microbial deterioration of food during storage. A correlation exists between protein hydration and viscosity of a food system. The relationship is affected by the concentration of protein and such factors as pH, ionic strength, and temperature (Kinsella, 1976). Hydration of a protein is influenced by the presence of hydrophilic polysaccharides, lipids, and salts; by the pH of the food; and by the food processing history and storage conditions. In general, a positive relationship exists between the content of hydrophilic groups (hydroxyl, carboxyl, and basic moieties), fewer amides, and the water-binding capacity of proteins. Consequently, deamidation is a possible approach to increasing the water-binding capacities of proteins of vegetable origin.

Absorption of water and swelling change the hydrodynamic properties of a food system. Such properties are reflected in thickening and increased viscosity, which influence flow characteristics. Viscosity is useful in evaluating the thickening potentials in many fluids and batter-type foods. Viscosity is influenced by solubility and swelling of proteins. Highly soluble proteins (such as albumins and globulins) have low viscosities. Soluble proteins with a high, initial swelling (such as sodium caseinate) show a concentration-dependent viscosity, which reflects the content of partially solvated swollen particles. Proteins with a limited swelling capacity (such as soy sodium proteinate) have a high viscosity at relatively low concentrations. Swelling can be a useful indicator of viscosity. Some of the factors that affect protein structure and component interactions also affect swelling and viscosity of foods.

Solubility of proteins is influential in imparting a "body" to aqueous solutions and in the formation and stability of emulsions and foams. Hydrophobic interactions determine the stability of protein structure in aqueous systems and in protein–lipid interactions. Certain anions, for example, chaotropic agents, decrease water structure and polarity, thereby weakening hydrophobic interactions and increasing solubility of apolar compounds.

Most proteins have little flavor, yet they influence perceived flavor because they may contain bound "off" flavors. Some proteins modify flavor by selective binding; some produce "off" flavors (such as bitter peptides produced during hydrolysis); and some act as precursors of flavors as a result of Strecker degradation of amino acids.

Flavor, alone or in combination with mouth-feel and texture, is the most important property that governs food acceptability (Kinsella, 1976). Some protein products may modify flavor of a food either directly because of their unique composition or indirectly because of their contribution to the generation of unusual or undesirable flavors during industrial food processing, storage, or preparation in the home. For example, one of the major problems in utilization of soybean protein is its unmistakably "beany" flavor; this characteristic limits (or at least complicates) soybean use in traditional food products.

The "off" flavors of many food proteins often result from the presence of small amounts of lipids in the preparations. The lipids may undergo undesirable chemical oxidative or enzymatic (i.e., lipoxygenase) modifications. Complete extraction of the lipids (which are bound to the proteins) often requires the use of polar organic solvents; this may lower functional properties of the proteins as a result of their denaturation. Other compounds that may affect flavor of proteins are small amounts of amino acids, peptides, or nucleotides (Kinsella, 1976).

As mentioned before, proteins themselves may have no objectionable flavors but may develop such flavors as a result of thermal processing or interaction during processing with other food components. Some additives (e.g., glutamyl peptides) or treatments (e.g., with proteolytic enzymes) may mask or reduce the off flavor. Finally, proteins may affect synthetic, or engineered, foods by interacting with added flavors. The interaction may involve partial adsorption, retention, or actual chemical binding. For instance, the loss of flavor during commercial processing, storage, and home preparation are highly significant in the utilization of proteins in extended meat products. If the proteins are used alone (rather than as extenders) their capacity to absorb and hold small amounts of expensive added flavors without adverse changes in subsequent storage and use is critical.

The behavior of proteins at interfaces influences the formation of foams and emulsions. Protein-stabilized foams facilitate the production of aerated foods. Surface-active proteins control aeration and texture by allowing the uniform distribution of fine air cells throughout the protein matrix. Factors that affect the foaming properties of proteins are summarized in Table 5.8. A diagram of foaming properties is shown in Fig. 5.6. It is important that foams maintain stability when subjected to heating or in the presence of other food ingredients, additives, or pH changes. Egg white, for example, exhibits its capacity for stability as a foam at elevated temperatures. High surface viscosity is desirable from a standpoint of foam stability, but may be a problem with regard to foam formation. The best combination of properties is attained by viscoelastic proteins. Factors that affect formation and stability of protein foams include solubility and rate of diffusion to interface, and adsorption. These in turn depend on such properties as hydrophobicity, orientation and association of polypeptides,

TABLE 5.8

Factors Affecting the Foaming Properties of Proteins[a]

1. *Concentration, surface, and bulk viscosity:* solubility, diffusion rate, interaction in the disperse phase, and increase of bulk with such substances as sucrose
2. *Structure:* disordered or flexible versus globular or rigid; availability of hydrophobic and hydrophilic groups
3. *Denaturation:* ease of polypeptide unfolding
4. *Electrical double layer:* repulsion affected by counter ions such as salts in solution; availability of hydrophobic and hydrophilic groups
5. *Marongoni effect:* ability to concentrate rapidly at a stress point in the film
6. *pH:* maximum near the isoelectric point; extremes, dissociate polypeptides
7. *Temperature:* dissociate polypeptides
8. *Denaturants:* improve availability and interaction of polypeptides, e.g., thio-reducing reagents
9. *Complementary surfactants:* other proteins, polysaccharides
10. *Coagulation:* irreversible aggregations

[a]From Hermansson, 1979.

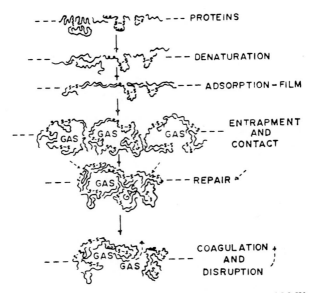

Fig. 5.6. Diagram of foaming properties of proteins. (From Cherry and McWaters, 1981.)

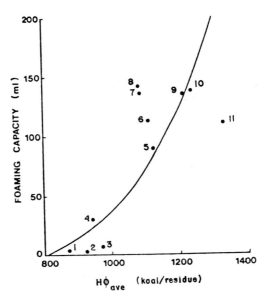

Fig. 5.7. Correlation between foaming capacity (FC) and Bigelow hydrophobicity ($H\phi_{ave}$).
$$r = .7971 \; p < .01 \; FC = -309.01 + 0.34 \; H\phi_{ave}$$
1, ribonuclease; 2, ovomucoid; 3, lysozyme; 4, trypsin; 5, blood serum albumin; 6, ovalbumin; 7, conalbumin; 8, pepsin; 9, κ-casein; 10, β-lactoglobulin; 11, β-casein. (From Nakai and Powrie, 1981.)

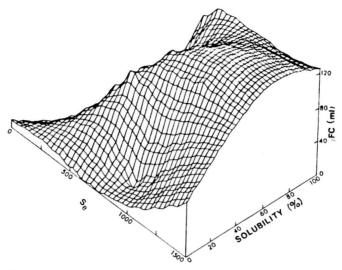

Fig. 5.8. The relationship of hydrophobicity and solubility with foaming capacity of proteins $S_e:S_o$ measured after extensive dissociation; solubility:nitrogen solubility index. (From Nakai and Powrie, 1981.)

viscoelasticity, aggregation–coagulation balance, surface charge, and hydration. (Kinsella, 1981). The correlation between foaming capacity and hydrophobicity is illustrated in Fig. 5.7. Fig. 5.8 describes the relationship of hydrophobicity and solubility with foaming capacity of proteins.

III. Denaturation of Proteins

Denaturation of proteins is significant in processing, storage, and end-use properties of foods. According to Wu and Inglett (1974), denaturation may be defined as a modification of the secondary, tertiary, or quarternary structure of the protein molecule that does not involve breaking covalent bonds. The change in protein structure is generally associated with changes in physicochemical and functional properties.

Hermansson (1979) discussed the significance of aggregation and denaturation in gel formation. Association generally pertains to changes on the molecular level as they relate to reversible monomer–dimer reactions, subunit equilibrium, and related reactions, which are characterized by noncovalent bonds. The terms "aggregation," "coagulation," and "flocculation," on the other hand, are used to describe nonspecific protein–protein interactions and/or formation of high molecular weight complexes. Aggregation is a general term to denote protein–protein interactions. Coagulation denotes random aggregation, which encompasses protein denaturation. Flocculation is a colloidal phenomenon in which the interaction between proteins is governed by the balance between van der Waals attraction and electrostatic repulsion due to the presence of an electric double layer.

The term gelation is used to describe aggregation of denatured molecules. Whereas in coagulation the aggregation is random, in gelation it involves a certain degree of order and the formation of a continuous network. The kinetics of the reaction (dissociation, swelling, denaturation, or aggregation) determines the structure and properties of a gel. In discussing denaturation and aggregation of a soy protein system and a whey protein system, as affected by heat treatment, Hermansson (1979) showed that although the two protein systems are entirely different in character, they are nevertheless capable, after heat treatment, of forming comparable gels—comparable in terms of functional properties.

The major methods of protein denaturation include application of heat, changes in pH, use of organic solvents, and addition of several organic compounds. For example, heat denaturation decreases solubility of soy proteins, and this decrease affects water absorption of soy flours. Flours in which protein solubility is intermediate (55–70%) have higher water absorption rates than flours in which protein solubility is very high (85%) or very low (10%). Generally, the higher the water-soluble protein in a soy protein concentrate, the better its emulsifying action. Similarly, whipping properties of soy products, expressed by foam expansion and foam stability, are highly and positively correlated with protein solubility. On the other hand, proteins with high solubility (indicative of low or no heat treatment) have low gelation properties. Insolubility of soy proteins requires the action of moist heat. For example, denaturation and protein insolubility are at maximum levels after heating for 2.5 hr at 127°C and 100% relative humidity. Heat causes thickening, followed by gelation, of aqueous dispersions of soy protein products in concentrations above 7% by weight. Gelling rates and gel firmness depend on the temperature, duration of heating, and protein concentration.

Heating converts the 11 S soy protein into a fast-sedimenting aggregate and a 4 S fraction. The soluble aggregate increases in size and precipitates on continued heating. Soybean 11 S protein is dissociated into subunits by exposure to temperatures above 70°C. The subunits may aggregate. The rate and extent of aggregation is enhanced by low ionic strength and is depressed at extreme acidic or alkaline pH values, as well as at high ionic strength. The trypsin inhibitor in soybeans is stable over a wide pH range at temperatures below 30°C. The protein is denatured by heating at higher temperatures. Denaturation is measured by a change in solubility and is accompanied by a loss in trypsin-inhibiting power and by an increase in susceptibility to digestion by pepsin. Mild denaturation can be reversed by cooling, but denaturation cannot be reversed after prolonged heating or after brief heating at the isoelectric point. Soybean hemagglutinin shows maximum stability toward thermal inactivation at pH 6.0–7.0. Dry heating to

100°C, steaming, or both inactivate soybean lipoxygenase. This treatment improves storability of soy products and maintains acceptable flavors.

Little decrease is observed in denaturation of gluten (a mixture of storage proteins in wheat), as measured by loss of solubility in dilute acetic acid, by heating at temperatures below 70°C. Heating at higher temperatures (e.g., 90°C) results in rapid denaturation (Pence *et al.*, 1953). See Fig. 5.9. Pence *et al.*

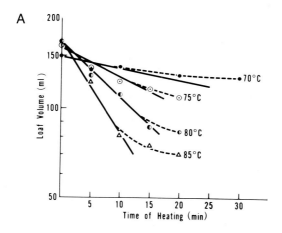

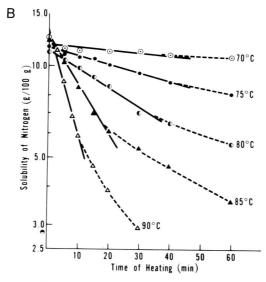

Fig. 5.9. Denaturation of wet, gum gluten by heat at various temperatures as measured by (A) decrease in loaf volume of reconstituted doughs and (B) loss of solubility in dilute acetic acid. Gluten is from commercial flour. Straight lines calculated by method of least squares. (From Pence *et al.*, 1953.)

(1953) measured the denaturation of wet gluten in heat ranging from 70 to 85°C by decreases in loaf volume of reconstituted doughs. Little loss was recorded at 70°C for 20 min. The loss was more pronounced at higher temperatures and at 85°C for 20 min much of the bread-making potential was destroyed. The relation between temperature and rate of denaturation of wet gluten by heat as measured by the baking test and by the solubility method is shown in Fig. 5.10. As shown in the figure, gluten denaturation was essentially a first-order reaction with an energy of activation of about 35,000 calories/mole by the baking test and 44,000 calories by the solubility method.

Pence *et al.* (1953) also measured the effect of moisture content on the rate of gluten denaturation by heat. The solubility curve showed that denaturation at 90°C was rapid if the moisture content was above 35%. At lower moisture levels, denaturation was much slower, and at 5% MC denaturation was negligible even at 90°C. Both solubility determination and baking tests pointed to a maximum rate of denaturation around 40% MC. Denaturation proceeded slowly at pH 4.0 and increased rapidly as pH rose to about 5.0. The rate remained constant until about pH 6.0, increased sharply thereafter, and remained fairly constant between pH 6.5 and 7.5. Variations in salt concentration had no effect on the rate of gluten denaturation by heat but influenced the damage to heat caused by low pH levels. The rates of denaturation of glutens from flours of different bread-making

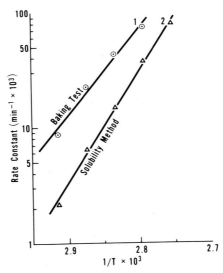

Fig. 5.10. Relationship between temperature and rate of denaturation of wet, gum gluten by heat as measured by the baking test method (curve 1) and by the solubility method (curve 2). The energy of activation (E = slope × R) equals 34,600 calories/mole by the baking test, and 43,800 by solubility. (From Pence *et al.*, 1953.)

potential varied significantly, but no correlation was seen between the rate of gluten denaturation and baking quality.

A plot of the rate of gliadin (gluten component) denaturation at 90°C (as measured by solubility) versus duration of heating, showed that an induction period of 10 min occurred before denaturation began. After this interval, denaturation attained a first-order rate about one-third as large as that of gluten under similar conditions. The fine parallelism between baking and solubility tests pointed to a high correlation between gluten denaturation and functional properties and revealed the usefulness of determining protein solubility in predicting bread-making potential of heat-damaged gluten proteins.

Dalek et al. (1970) observed the degree of denaturation of proteins in vital gluten by measuring the viscosity and optical rotation of dispersions. Rohrlich (1955) studied the effect of drying wheat on enzyme activities. Catalase activity was relatively unaffected by temperatures up to 65°C, decreased rapidly between 65 and 100°C, and was less than one-third the original value at 100°C. Protease activity fell proportionately with rises in temperature and at 100°C only about 30% of the original activity remained. The β-amylase activity decreased with rising temperature to about 40% at 100°C. Up to a temperature of about 65°C the endosperm proteins soluble in N-acetic acid increased by about 5%, and decreased under heating to higher temperatures. The safe temperature limit for drying grain is important in many countries where wheat must be harvested with a high moisture content. Finney and associates (1962) have shown that loaf volume, crumb grain, and mixing time of preripe hard red winter wheat were affected adversely when dried at high temperatures, depending on the amount of moisture in the grain. As wheat MC increased, maximum drying temperature for normal loaf volume decreased (see Fig. 5.11). Mixograms* and internal characteristics of loaves of bread baked from samples of wheat harvested at various stages of maturity and dried at various temperatures are shown in Figs. 5.12 and 5.13.

Wu and Inglett (1974) reviewed the effects of roasting peanuts on the solubility, association and dissociation, electrophoretic mobility, and antigenic properties of the peanut proteins. Protein solubility was inversely proportional to dry heat temperatures and related in a sigmoidal manner to wet heat temperature. Similarly, heat denaturation of cereal and cottonseed proteins is a function of moisture, temperature, and length of heat.

Almost all proteins are affected by hydrogen or hydroxyl ions. The 11 S soy protein in acid solution is part of a reversible association–dissociation system that contains three resolvable fractions having sedimentation coefficients of about 2, 7, and 13 S, and an unresolvable fraction having a sedimentation

---

*Mixograms are graphs recorded during mixing of doughs in a mixograph. The important parameters are time to peak (point of minimum mobility) and drop in resistance to mixing beyond the peak.

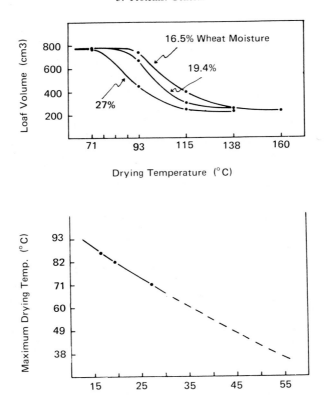

Fig. 5.11. Interrelations between loaf volume, drying temperature, and wheat moisture content. Maximum drying temperature is the highest that can be employed within a moisture level without impairing loaf volume and crumb grain. (From Finney *et al.,* 1962.)

coefficient above 13 S. The relative amounts of the fractions depend upon pH, ionic strength, and type of salt present. Low pH and low ionic strength favor dissociation into the 2 S and 7 S fractions. Soybean 11 S globulin can undergo conformational changes that involve dissociation into subunits approximately one-half and one-eighth the size of the 11 S globulin. Such changes occur at low ionic strength at alkaline pH values, and at moderate ionic strengths at acid pH values. Changes generated at acid pH values are irreversible. Conformational changes generated at alkaline pH values can be reversed by increasing the ionic strength. The results suggest that dissociation is caused by forces of electrostatic repulsion among the subunits. Dissociation is accompanied by an increase in levorotation, indicating configurational changes of the subunits which, in turn, may result under certain conditions in irreversible dissociation and aggregation.

The exposure of acid-precipitated soy protein to sodium hydroxide at high pH markedly increases its relative viscosity and rapidly shifts the sedimentation constants of the 2, 7, 11, and 15 S ultracentrifuge components to essentially 3 S. The shape and magnitude of the viscosity curve depend on the concentration of the protein and sodium hydroxide; furthermore, they vary with time.

Conversion of soy protein into coagulated fibers that can be processed into textured food products usually demands a pH in the range of 9.0 to 13.5; in practice, at least 10.5. The use of a high pH is a serious disadvantage because this tends to degrade the protein and requires complex procedures to produce a uniform extract for use in spinning the fiber. On the other hand, plant proteins can be dissolved at pH 3.0–9.0 by a 0.2 M NaCl solution. An edible protein fiber is made by extruding the aqueous protein into an aqueous medium at a temperature higher than 80°C to produce a heat-set product, rather than using the

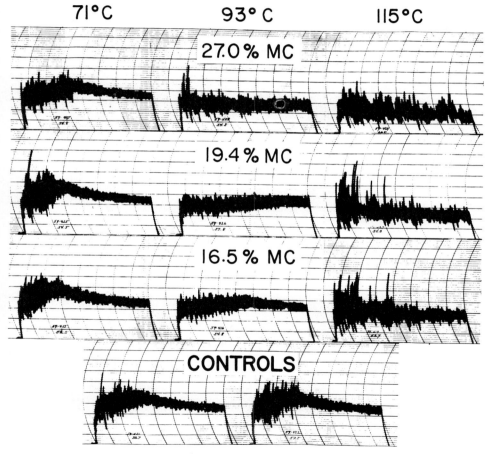

Fig. 5.12. Mixograms of wheat harvested at various stages of maturity and dried at various temperatures. MC, moisture content. (From Finney *et al.,* 1962.)

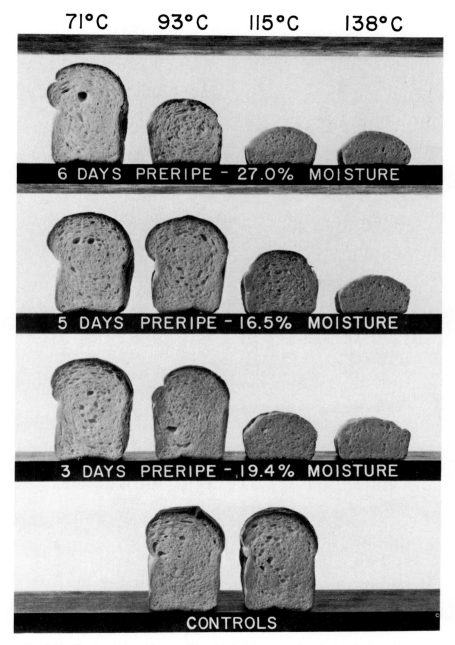

Fig. 5.13. Characteristics of loaves of bread baked from samples of wheat harvested at various stages of maturity and dried at various temperatures. (From Finney *et al.*, 1962.)

conventional method of extruding a strongly alkaline solution of soy protein into an acid–salt coagulation bath.

Methanol and ethanol extraction of soybean meal reduces the subsequent extractability with water or salt solutions of all protein components; the effect is greatest on globulin components. Hot extractions are more effective than cold extractions in reducing protein solubility. Water is generally a less effective denaturant than pure organic solvent; concentrations of 40–60% alcohol are the most effective. The 7 S component is denatured most rapidly (rendered insoluble) when wet curd is in contact with an aqueous solution containing 20% or more ethanol. The rate of denaturation of 11 S and 15 S components is slow, and the 2 S components are not denatured at all.

A systematic study was made of the denaturing ability of soybean protein by about 30 organic solvents (Fukushima, 1969). The denaturing ability of organic solvents usually depended on their hydrophobicities and on dilution by water. Highly hydrophobic solvents had little denaturing power over proteins, even at high temperatures. The denaturing power of solvents strengthened with addition of water, and that of water also rose with addition of solvents. Lower alcohols were much stronger denaturants than were the other solvents studied. The denaturing abilities of alcohols at low concentrations increased with the hydrophobicities of the alcohols; the reverse was found at high concentrations.

IV. Modification of Functional Properties

Few, if any, proteins in the native state have optimum functional properties. Hence, modification of native proteins can impart the needed functional attributes to an already available raw material and avoid the high costs of producing novel foods.

Schwenke (1978) described the parameters that affect native protein structure and functionality during isolation and processing:

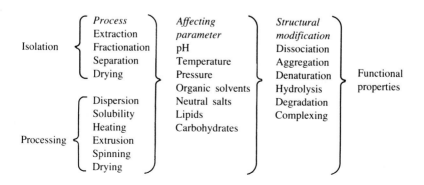

The five main types of protein modification are physical, hydrolytic (chemical and enzymatic), derivative, complexing, and fractionating. (Means and Feeney, 1971; Kinsella, 1976, 1977; Feeney and Whitaker, 1977)

A. Physical Modification

The principal techniques for physically modifying protein structure are extrusion and spinning. Other methods include formation of chewy gels, curds, or fibers, and freeze alignment. In thermoplastic extrusion, a slurry or an aqueous suspension of proteins is heated at 100–170°C under increasing pressure (1000 lb/in.) in the barrel of an extruder. As the pressure is released through the exit aperture (dies of various shapes and diameters), the protein fibers align in parallel layers. With the sudden release of pressure, a flash evaporation of moisture occurs and formation of fibrous, porous, and flaky structured protein particles results; the structured protein may be cut and shaped as desired. The process yields fibrous particles of textured protein that are chewy, with a pleasing mouthfeel.

The raw materials used to produce such extrusion products include defatted oilseed meals, full fat oilseed flours, crude protein concentrates, and cereals. The original type and condition of the raw material and the processing condition allow flexibility in the various products. The factors that can be used to control processing include MC, pH, pressure, and temperature; addition of hydrocolloid binders or emulsifiers; and configuration and size of the extrusion dies. The extruded particles may be compacted or expanded. Acidic batters yield textured products with numerous small vacuoles; these particles are tough, dense, and rehydrate slowly. Alkaline batters yield light pieces that have relatively large vacuoles and rehydrate rapidly. Porosity and texture of the final product are affected by factors that control the ease of coalescence of entrapped steam droplets, such as pressure, viscosity and texture of the protein matrix, die configuration, among others.

Texturizing protein by fiber spinning typically involves preparing a viscous alkaline dispersion above pH 10.0 with a protein content of 10–50%. This spinning dope is extruded under pressure through a wet spinning device composed of small die with 1000 to 1600 openings, each 0.002–0.006 in. in diameter. The extruded protein filaments are drawn through an acidic bath and coagulated. The coagulated fibers are picked up on rollers and stretched, and each is reduced to about 20 μm in diameter. Stretching and heating imparts tensile strength, toughness, some elasticity, and chewiness. As they leave the acid bath, the bundles of filaments or ''tows'' are washed and spun to fabricate meatlike products.

In fiber formation, various concentrations of purified protein isolates must be used. Theoretically, proteins from many sources could be used in fiber production, but in practice, the potential of fiber formation and the properties of the fibrous products depend on several protein characteristics. These characteristics include molecular size and shape; polypeptide chain length; and primary, secondary, and tertiary structure. At this time, the most common sources are soybean and wheat gluten proteins. Potential sources include most oilseed proteins.

The texture of fibers is influenced by (a) the composition, purity, and modification of the protein source; (b) operating conditions (extrusion pressure, acidity of coagulating solution, diameter of fiber, degree of stretching of the fiber and the tow, extent of heating); and (c) the nature of additives (binders, fats, emulsifiers). The parameters that affect the spinning process include dimensions of the spinnerette, draining velocity, the distance the fiber travels in the coagulation bath (spinning length), the velocity through the spinnerette, the composition and properties of the dope (including its degree of coagulation, composition, and temperature of the spinning bath), and coagulation properties affecting both the dope and the bath. To physically bind the filaments, one must use edible binders such as carbohydrates (starches, dextrins, gums, alginates, carboxymethyl cellulose) or proteins. The fibers may be coated with fats and various additives to impart desirable colors, flavors, or tastes.

Chewy gels can be prepared by heating vegetable protein dispersions to form gels of required shapes. The moderately structured gels are smooth, moist, fairly resilient, elastic, and resistant to shear or bite. In the freeze alignment technique, a 3–20% heat-coagulable protein solution is frozen in the following unidirectional manner: the solution is cooled at a rate that produces elongated ice crystals perpendicular to the cooling surface. This aligns the proteins and concentrates them in spaces between the elongated ice crystals. The integrity of the protein structure is maintained by careful removal of water (ice), usually by vacuum drying. Other methods of water removal include displacement of water with ethanol. The dry structured protein is usually stabilized by steam treatment. Factors that influence textural quality include type of protein(s), protein concentration, pH, and rate of ice crystal growth.

B. Hydrolytic Modification

The usefulness of nonfunctional preparations of food proteins can be extended by partial hydrolysis using acid, alkali, or enzymatic treatments (Kinsella, 1977). Vegetable proteins can be acid hydrolyzed in glass-lined vessels with 10–20% hydrochloric acid for about 12 hr at atmospheric pressure or 4–5 hr under higher pressures. The hydrolysate is neutralized with sodium hydroxide to pH value of 6.0. The material is then filtered to remove insoluble humin. The filtrate is treated with charcoal to remove excess color and bitter compounds. The treated liquid is refiltered and packaged as a liquid, concentrated, or spray-dried. The hydrolysates have distinct meat-like flavors and are used to enhance the taste of meat and broth used in soups and other products. Casein and corn hydrolysates have light colors and flavors and are preferred in fish, poultry, and pork preparations; soy, yeast, and gluten hydrolysates have stronger colors and flavors and are preferred in beef and mutton preparations. Hydrolysates are used in formulated foods, soups, sauces, gravies, canned meats, bakery items, and beverages.

Acid hydrolysis is rapid and inexpensive; it can be used to process crude and inexpensive raw materials. Hydrolysates contain high concentrations of salt, which can be removed by ultrafiltration. The hydrolysates may be fractionated to produce fractions that have different properties and compositions. Acid hydrolysis causes some isomerization and destruction of some amino acids (mainly tryptophan, serine, and threonine, and sulfur-containing amino acids—especially cysteine).

Alkaline treatment facilitates extraction of proteins from various foods. This method is commonly used to improve protein solubility; enhance emulsification, foaming, and fiber spinning; and inactivate enzymes and destroy toxins, enzyme inhibitors, and allergens. Alkaline treatments have been used in processing oilseeds, aquatic foods, single cell proteins, and leaf proteins. Depending upon concentration, temperature, and duration of treatment, alkali may loosen, deaggregate, depolymerize, or actually hydrolyze proteins. The extent of actual hydrolysis ranges from the limited formation of polypeptides to the release of large amounts of amino acids. Alkaline treatment causes formation of dehydroalanine from cysteine and serine by means of β-elimination-type reactions. Dehydroalanine reacts with the ε-amino group of lysine and forms lysinoalanine. This may result in cross-linking of polypeptides, rendering proteins resistant to proteolysis, and causing reduction of their nutritive value. The transformation of reactive protein side chains to lysinoalanine side chains is described in Fig. 5.14. In addition to lysinoalanine, lanthionine and ornithioalanine may be formed during the reaction of dehydroalanine with cysteine and ornithine, respectively. For this reason, alkaline treatment may reduce nutritive value of proteins by causing racemization, cross-linking, or destruction of certain amino acids (lysine, serine, threonine, arginine, isoleucine, and methionine).

Numerous studies have examined the conditions under which alkali causes the formation of undesirable reactions and compounds. Increasing alkali concentration, temperature, and duration of hydrolysis causes the progressive destruction of labile amino acids and the formation of lysinoalanine. Under normal, well-controlled processing conditions the extent of undesirable alterations is likely to be limited.

According to Friedman (1979), cross-linked amino acids have been found in hydrolysates of both alkali-treated and heat-treated proteins. It was shown that lysinoalanine (one of the cross-linked amino acids) may cause histological changes in the descending position of the proximate tubules of rat kidneys. These findings have raised concerns about the nutritional quality and safety of alkali-

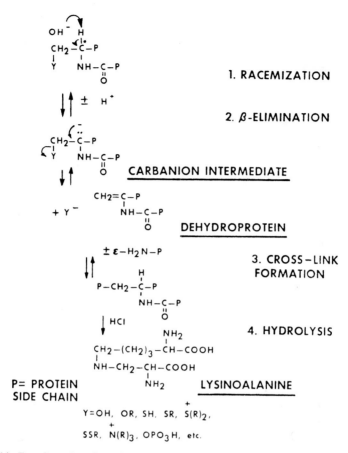

Fig. 5.14. Transformation of reactive protein side chains to lysinoalanine side chains via elimination and cross-linking formation. Hydroxide ion abstracts an acidic hydrogen atom (proton) from an α-carbon atom of an amino acid residue to form an intermediate carbanion. The carbanion, which has lost the original asymmetry of the amino acid residue, can either recombine with a proton to reform a racemized residue in the original amino acid side chain or undergo the indicated elimination to form a dehydroalanine side chain. The dehydroalanine then combines with an ε-amino group of a lysine side chain to form a cross-linked protein which on hydrolysis yields free lysinoalanine. (From Friedman, 1979.)

treated food proteins. Friedman (1979) reported that treating commercial wheat gluten under alkaline conditions at 65°C destroyed part of the serine, threonine, cysteine, lysine, arginine, and tyrosine residues. The losses were accompanied by the appearance of lysinoalanine and several unidentified ninhydrin-positive substances. It was found that acylation by acetic and succinic anhydrides effectively reduced destruction of lysine residues and inhibited formation of lysinoalanine.

Partial proteolysis may be used to solubilize proteins in oilseeds, cereal grains, and fish. Some of the treated products are finding use in many food preparations. Pepsin-treated soy proteins are used in whipping agents and as extenders in egg white in bakery and candy formulations; enzymatically hydrolyzed caseins are used in candy manufacture; proteolysis increases the amount of extractable bovine or oilseed proteins and improves the emulsifying and/or foaming properties of the hydrolyzed products. Proteolytic enzymes are relatively costly; their activity may be difficult to control (especially in batch processing) and the hydrolysates may have bitter flavors. For best results, proteolysis conditions must be carefully controlled, usually by the use of bound enzymes, especially recently available semipermeable, molecular sieve-type membranes. Proteolytic enzymes used in protein modification are listed in Table 5.9.

The plastein reaction is a special type of proteolysis (see also Chapter 11). The

TABLE 5.9

Proteolytic Enzymes in Protein Modification[a]

Food	Purpose or Action
Baked goods	Softening action in doughs; cut mixing time, increase extensibility of doughs; improvement in texture, grain and loaf volume; liberate β-amylase
Brewing	Body, flavor and nutrient development during fermentation; aid in filtration and clarification; chillproofing
Cereals	Modify proteins to increase drying rate, improve product handling characteristics; production of miso and tofu
Cheese	Casein coagulation; characteristic flavor development during aging
Chocolate–cocoa	Action on beans during fermentation
Eggs, egg products	Improve drying properties
Feeds	Waste product conversion to feeds; digestive aids, particularly for pigs
Fish	Solubilization of fish protein concentrate; recovery of oil and proteins from inedible parts
Legumes	Hydrolyzed protein products; removal of flavor; plastein formation
Meats	Tenderization; recovery of protein from bones
Milk	Coagulation in rennet puddings; preparation of soybean milk
Protein hydrolysates	Condiments such as soy sauce; bouillon; dehydrated soups; gravy powders; processed meats; special diets
Antinutrient factor removal	Specific protein inhibitors of proteolytic enzymes and amylases; phytate[a]; gossypol[a]; nucleic acid[b]
Wines	Clarification[b]
In vivo processing	Conversion of zymogens to enzymes; fibrinogen to fibrin; collagen biosynthesis; proinsulin to insulin; macromolecular assembly

[a]Whitaker, 1977.

[b]In large part caused by other than proteolytic enzymes.

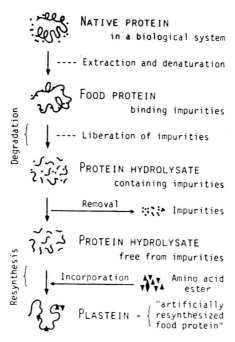

NATIVE PROTEIN
    in a biological system

  ---- Extraction and denaturation

FOOD PROTEIN
        binding impurities

Degradation {   ---- Liberation of impurities

PROTEIN HYDROLYSATE
        containing impurities

  ——Removal——→ Impurities

PROTEIN HYDROLYSATE
        free from impurities

Resynthesis { ——Incorporation—— Amino acid ester

PLASTEIN = { "artificially resynthesized food protein"

Fig. 5.15. Scheme for removing impurities from protein substrate by hydrolysis, purification, and resynthesis by means of the plastein reaction. (From Phillips and Beuchat, 1981; copyright by the American Chemical Society.)

reaction involves hydrolysis at 37°C for 30–40 hr of a 5% protein slurry with a proteolytic enzyme (that is, pepsin) at 1% of the substrate concentration. The hydrolysate is concentrated to a 20% solids content and, after the addition of chymotrypsin, incubated at 37°C for 30–70 hr. The treatment results in re-synthesis of the initial hydrolysate to form new oligopeptides or proteins. The type of final product depends on the original substrate, enzymes used, and conditions of the plastein reaction. A successful plastein synthesis usually re-quires a proteolysate of 30% total solids composed of polypeptides of average molecular weight of 1500 to 2000. A pH in the isoionic region of 4.0–6.0 assures a high concentration of non-ionized amino and carboxyl groups, which enhances peptide bond formation. The plastein reaction facilitates the removal of impurities, inactivation of antinutrients, and transesterification of proteins from various sources to create products with improved nutritional and functional at-tributes. The wide range of raw material that can be used and products that can be obtained in this manner has contributed much to the promise and acceptance of the plastein reaction.

C. Derivative Modification

Modification of proteins has been used for millennia in tanning, dyeing, and, more recently, in the production of synthetic resins. Chemical derivatization to modify the functional properties of proteins for food uses, however, has received less attention (Schwenke, 1978). The reaction used to form acylated derivatives is shown in Fig. 5.16. Acylation can be used to modify a wide range of foods to impart special properties. For example milk proteins may be modified to produce stable emulsions, emulsifiers, or special mixed beverages. Egg proteins are used in baked products and mayonnaise. Fish proteins are used in the production of emulsifiers, binding compounds, and cheeselike gels; fish hydrolysates are used in special baked goods, puddings, and ice-cream-like foods. Soy proteins are used as emulsifiers and coffee whiteners.

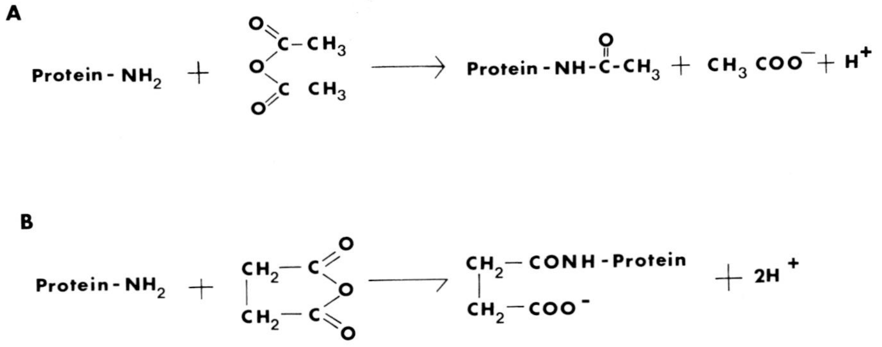

A

$$\text{Protein - NH}_2 \; + \; \begin{matrix} \text{O} \\ \diagdown \text{C}-\text{CH}_3 \\ \text{O} \\ \diagup \text{C} \;\; \text{CH}_3 \\ \text{O} \end{matrix} \; \longrightarrow \; \text{Protein-NH-}\overset{\text{O}}{\underset{}{\text{C}}}\text{-CH}_3 \; + \; \text{CH}_3\text{COO}^- + \text{H}^+$$

B

$$\text{Protein - NH}_2 \; + \; \begin{matrix} \text{CH}_2-\text{C}\diagup^{\text{O}}\diagdown_{\text{O}} \\ | \\ \text{CH}_2-\text{C}\diagdown_{\text{O}} \end{matrix} \; \longrightarrow \; \begin{matrix} \text{CH}_2-\text{CONH -Protein} \\ | \\ \text{CH}_2-\text{COO}^- \end{matrix} \; + \; 2\text{H}^+$$

Fig. 5.16. Acylation reaction of (A) acetic and (B) succinic anhydride to form acylated derivatives.

D. Complexing and Fractionating

Complexing involves interaction with lipids and carbohydrates, as well as with surface active agents. Fractionation methods are used widely in separation of milk proteins or in the washing out of vital (functional) gluten from wheat flour. Dry fractionation methods include air classification of wheat flour. Wheat flour produced by conventional roller-milling contains particles of different sizes: large endosperm chunks, small particles of free protein, free starch, and tiny chunks of protein with attached starch granules. It is possible partly to separate protein from starch by fine-grinding conventionally roller-milled flour by air classification. This process is summarized in Fig. 5.17. The schematic diagrams show how flour is fractionated, the simplified protein shift process, and the details of the fractionation method.

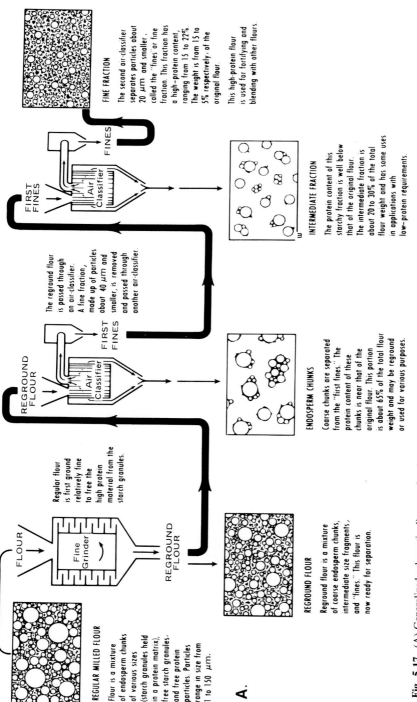

REGULAR MILLED FLOUR

Flour is a mixture of endosperm chunks of various sizes (starch granules held in a protein matrix), free starch granules, and free protein particles. Particles range in size from 1 to 150 μm.

A.

Regular flour is first ground relatively fine to free the high protein material from the starch granules.

REGROUND FLOUR

Reground flour is a mixture of coarse endosperm chunks, intermediate size fragments, and "fines." This flour is now ready for separation.

The reground flour is passed through on air-classifier. A fine fraction, made up of particles about 40 μm and smaller, is removed and passed through another air-classifier.

ENDOSPERM CHUNKS

Coarse chunks are separated from the "first fines." The protein content of these chunks is near that of the original flour. This portion is about 65% of the total flour weight and may be reground or used for various purposes.

INTERMEDIATE FRACTION

The protein content of this starchy fraction is well below that of the original flour. The intermediate fraction is about 20 to 30% of the total flour weight and has some uses in applications with low-protein requirements.

FINE FRACTION

The second air-classifier separates particles about 20 μm and smaller, called the "fines" or fine fraction. This fraction has a high-protein content, ranging from 15 to 22%. The weight is from 15 to 5% respectively, of the original flour.

This high-protein flour is used for fortifying and blending with other flours.

Fig. 5.17. (A) Generalized schematic diagram showing how wheat flour is fractionated. Flour particles are measured in μm; white areas in circles indicate starch granules; black portions indicate protein. Scale ∼ 100 ×. (B) Diagram of simplified protein shift process. (C) Simplified scheme of air classification of wheat flour. (A, Courtesy of the Wheat Flour Institute, Chicago; B and C, courtesy of Prof. A. B. Ward.)

B.

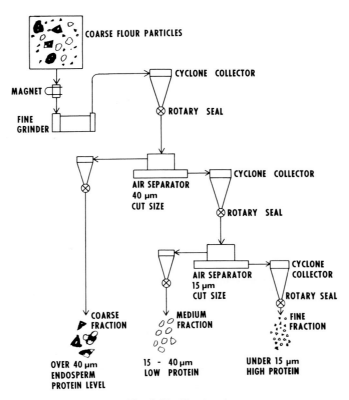

Fig. 5.17. (*Continued*)

Air classification is relatively inexpensive and its advantages are numerous: manufacture of more uniform flours from varying wheats, increase of protein content in bread flours and decrease of protein in cake and cookie flours, controlled chemical composition and particle size, and manufacture of special flours for specific uses. Protein-rich fractions (up to 30% protein) from air classification can be used to increase the bread-making quality of protein-low or functionally weak flours. In terms of amino acid balance, the air-classified protein concentrates are comparable to, or slightly poorer than, the original flours.

The subaleurone endosperm in wheat consists of a distinctive layer, one cell deep, situated adjacent to the aleurone cells on the outside and to the inner endosperm on the inside. The subaleurone and the inner endosperm differ in cell size and shape, size and abundance of starch granules, and in protein contents. The protein contents of subaleurone and inner endosperm range from 33 to 54% and 8 to 15%, respectively. A subaleurone fraction with 40.5% protein can be

C.

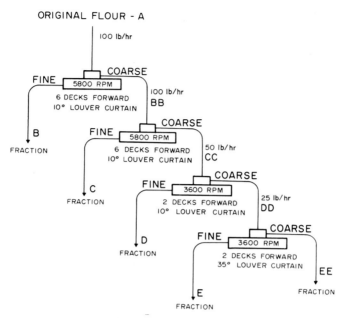

Fig. 5.17. (*Continued*)

obtained by dry-milling and air classification of a third break flour (a certain stream in conventional roller milling of wheat) with 15.1% protein.

REFERENCES

Cherry, J. P., ed. (1981). "Protein Functionality in Foods," ACS Symp. No. 147. Am. Chem. Soc., Washington, D.C.

Cherry, J. P., and McWatters, K. H. (1981). Whippability and aeration. *ACS Symp. Ser.* **147**, 149–176.

Dalek, V., Liss, W., and Kaczkowski, J. (1970). Indexes of the degree of denaturation in wheat gluten. *Bull. Acad. Pol. Sci., Ser. Sci. Biol.* **18**, 743.

Feeney, R. E., and Whitaker, J. R., eds. (1977). "Food Proteins: Improvement through Chemical and Enzymatic Modification," Adv. Chem. Ser. No. 160. Am. Chem. Soc., Washington, D.C.

Finney, K. F., Shogren, M. D., Hoseney, R. C., Bolte, L. C., and Heyne, E. G. (1962). Chemical, physical, and baking properties of preripe wheat dried at varying temperatures. *Agron. J.* **54**, 244–247.

Friedman, M. (1979). Alkali-induced lysinoalanine formation in structurally different proteins. *ACS Symp. Ser.* **92**, 225–238.

Fukushima, D. (1969). Denaturation of soybean proteins by organic solvents. *Cereal Chem.* **46**, 156.

Hermansson, A.-M. (1979). Aggregation and denaturation involved in gel formation. *ACS Symp. Ser.* **92,** 81–104.

Hurrell, R. F. (1980). Interaction of food components during processing. *In* "Food and Health: Science and Technology" (G. G. Birch and K. J. Parker, eds.), Chapter 22, pp. 369–388. Appl. Sci. Publ., Ltd., London.

Kinsella, J. E. (1976). Functional properties of proteins in foods: A survey. *Crit. Rev. Food Sci. Nutr.* **7,** 219–280.

Kinsella, J. E. (1977). Functional properties of proteins in foods. Some methods of improvement. *Chem. Ind. (London),* pp. 177–182.

Kinsella, J. E. (1981). Functional properties of proteins: possible relationships between structure and function in foams. *Food Chem.* **7,** 273–288.

Kinsella, J. E., and Srinivasan, D. (1981). Nutritional, chemical, and physical criteria affecting the use and acceptability of proteins in foods. *In* "Criteria of Food Acceptance." Foster Verlag, Switzerland.

Lawrie, R. A., ed. (1970). "Proteins as Human Food." Avi Publ. Co., Westport, Connecticut.

Matil, K. F. (1971). The functional requirements of proteins for foods. *J. Am. Oil Chem. Soc.* **48,** 477–480.

Means, G. E., and Feeney, R. E. (1971). "Chemical Modification of Proteins." Holden-Day, San Francisco, California.

Milner, M., Scrimshaw, N. E., and Wang, D. I. C., eds. (1977). "Protein Resources and Technology: Status and Research Needs." Avi Publ. Co., Westport, Connecticut.

Nakai, S., and Powrie, W. D. (1981). Modification of proteins for functional and nutritional improvements. *In* "Cereals—A Renewable Resource, Theory and Practice" (Y. Pomeranz and L. Munck, eds.), Chapter 11, pp. 217–242. Am. Assoc. Cereal Chem., St. Paul, Minnesota.

Pence, J. W., Mohammad, A., and Mecham, D. K. (1953). Heat denaturation of gluten. *Cereal Chem.* **30,** 115.

Phillips, R. D., and Beuchat, L. R. (1981). Enzyme modification of proteins. *ACS Symp. Ser.* **147,** 275–298.

Pour-El, A., ed. (1979). "Functionality and Protein Structure," ACS Symp. Ser. No. 92. Am. Chem. Soc., Washington, D.C.

Rohrlich, M. (1955). Einfluss der Trocknung des Weizens auf Fermentaktivität und Klebereiweiss. *Muehle* **92,** 320.

Satterlee, L. D. (1981). Proteins for use in foods. *Food Technol.* **35**(6), 53–70.

Schmidt, R. H. (1981). Gelation and coagulation. *ACS Symp. Ser.* **147,** 131–148.

Schultz, H. W., and Anglemier, A. F., eds. (1964). "Proteins and Their Reactions." Avi Publ. Co., Westport, Connecticut.

Schwenke, K. D. (1978). Beinflussung funktioneller Eigenschaften von Proteinen durch chemische Modifizierung. *Nahrung* **22,** 101–120.

Stanley, D. W., Murray, E. D., and Lees, D. H., eds. (1981). "Utilization of Protein Resources." Food & Nutrition Press, Inc., Westport, Connecticut.

Wall, J. S., and Huebner, F. R. (1981). Adhesion and cohesion. *ACS Symp. Ser.* **147,** 111–130.

Whitaker, J. R. (1977). Enzymatic modification of proteins applicable to foods. *Adv. Chem. Ser.* **160,** 95–155.

Whitaker, J. R., and Tannenbaum, S. R. (1977). "Food Proteins." Avi Publ. Co., Westport, Connecticut.

Wu, Y. V., and Inglett, G. E. (1974). Denaturation of plant proteins related to functionality and food applications—a review. *J. Food Sci.* **39,** 218–225.

6

Proteins: Specific Foods

The first three sections in this chapter cover proteins from animal sources: milk, eggs, and muscle (including fish). The sections on plant proteins are an update of a condensed report on concentrated seed proteins (Food Protein from Grains and Oilseeds—A Development Study), by V. D. Burrows, A. H. M. Greene, M. A. Korol, P. Melnychyn, G. G. Pearson, and I. R. Sibbald (1972). Wheat and soy proteins are covered only briefly in this section, but are discussed in greater detail in Chapters 10 and 11, which deal with specific foods. The last sections in this chapter concern single cell protein (SCP). Single-cell protein is a generic name for crude or refined proteins whose origins are bacteria, yeasts, molds, and algae. They are used as flavoring agents, emulsifiers, nutrient supplement, and extenders.

TABLE 6.1

Distribution and Characteristics of Milk Proteins[a]

Component	Approximate concentration		Genetic variants[c]	Approximate molecular weight	pI	Groups per mole		
	% of skim milk protein	gm/liter				Phosphorus	Disulfide	Sulfhydryl
Casein	78–85	(27.2)[b]			4.6			
α_s-caseins	45–55							
α_{s1}-casein		13.6	A, B, C, D	23,500	5.1	8	0	0
(α_{s0}-, α_{s2}-, α_{s3}-,								
(α_{s4}-, α_{s5}-, are								
minor components)								
β-Casein	25–35	8.2	A^a, $\underline{A}^b$, A^c, $\underline{B}$, C, D	24,000[e]	5.3	5	0	0
κ-Casein	8–15	4.1	A, B,	19,000[d]	3.7–4.2	1	0	0
γ-Casein	3–7	1.4			5.8			
$\gamma 1^e$			A^a, $\underline{A}^b$, A^c, B	20,500		1	0	0
$\gamma 2^e$			$A^{a\ or\ b}$, A^c, B	11,800		0	0	0
$\gamma 3^e$			$A^{a,b\ or\ c}$, B	11,500		0	0	0

Whey proteins								
	15–25	(6.8)						
β-Lactoglobulin	7–12	3.6	A, B, C, D	18,300	5.3	0	2	1
α-Lactalbumin	2–5	1.7	A, B	14,200	5.1	0	4	0
Immunoglobulins	1.5–2.5	0.6			4.6–6.0	0	Present and variable	
IgG1[f], IgG2				160,000	(monomers)			
IgM				900,000	(pentamer)			
IgA				400,000	(dimer)			
FSC[g]				70,000				
Serum albumin	0.7–1.3	0.4		69,000	4.7	0	0	0
Proteose-peptone[h]	2.0–4.0	0.7		4,000–40,000	3.7	0.5–2.0	17	0

[a] Compiled partially from data presented by Swaisgood (1973) and Whitney et al. (1976). From Brunner, 1977.

[b] Values for casein based on 27.2 gm/liter of total casein and 55% α_s-, 30% β-, 15% κ-, and 5% γ-caseins.

[c] Underscored variations represent those most frequently found in Western breeds.

[d] Molecular weight of carbohydrate-free species; species containing 0–5 carbohydrate moieties (~650 daltons) exist.

[e] Similar to β-casein segments represented by residue number 29–209, 106–209, and 108–209, respectively.

[f] Principal immunoglobulin in normal milk.

[g] Free secretory component.

[h] Heterogeneous mixture of glycoproteins; usually designated as components 3, 5, 8-fast, and 8-slow in ascending order of electrophoretic mobility.

I. Milk Proteins

Milk proteins can be categorized as caseins, whey proteins, and proteins associated with the lipid phase (Brunner, 1977). Distribution and characteristics of milk proteins are summarized in Table 6.1. Caseins are a family of related phosphoproteins. They constitute 2.5–3.2% of fluid milk and about 80% of milk proteins. Three main components (α_{s1}-, β-, and κ-casein) and a minor component (γ-casein) represent about 95% of the caseins. Caseins are precipitated from raw skim milk at pH 4.6 at 20°C as casein micelles with associated calcium. Casein can be modified to an insoluble paracalcium caseinate by adding rennin or related enzymes. A soluble caseinate can be obtained by raising the pH to 6.7. Caseins and sodium or potassium caseinates are selected for food product applications that require surfactant properties such as emulsification and foam stabilization. The high protein products (more than 90% protein) are highly soluble, form viscous solutions, and are resistant to heat denaturation.

Whey proteins are obtained from skim milk after separation of casein in cheese production. The two main whey proteins are β-lactoglobulin and α-lactalbumin. Additional proteins are bovine serum albumin, immunoglobulins, and proteose-peptones. In the production of cheese, of every 10 gallons of milk 9 gallons are a whey by-product. The whey contains about 6.5% solids, of which almost 5% are lactose, 1% is protein, and 0.5% are minerals. World whey production is 80 million metric tons from cheese and 3 million metric tons from casein production. Only half of the whey is utilized. Consequently, about 250,000 metric tons of high quality protein are wasted. The main outlets are for animal feed, as liquid whey, low grade whey powder, or in blended calf-milk replacers. Annual production of fluid whey in the United States is 10 million metric tons, of which only one-third is used as human food and animal feed. The remaining two-thirds of whey creates major disposal problems. Spray-dried whey contains 9–13% protein, up to 3.5% fat, 7–11% minerals, up to 75% lactose, and up to 6% acids (depending on the type of cheese from which the whey was derived) (Satterlee, 1981). Acid whey from the production of cottage cheese, cream cheese, and casein has a pH of about 4.5; sweet whey from the production of cheddar- and swiss-type cheeses has a pH of about 6.3.

The high lactose content and the relatively low protein content of whey creates problems, especially for people with lactose intolerance. Commercial methods to cleave the lactose molecule to glucose and galactose are available. Lactose in spray-dried whey has several functional properties: it enhances the flavor and browning of heated foods such as bread, cakes, and breaded meat or fish. Whey protein is soluble at a low pH levels, at which proteins from other sources coagulate and settle out. Consequently, whey proteins are well suited for fortification of acidic soft drinks. Partially delactosed whey produces foams with excellent stability, superior to egg albumin foam (Satterlee, 1981). Whey protein concentrates can be produced in an essentially undenatured form by two membrane methods: ultrafiltration and reverse osmosis. Whey protein concentrate produced by these methods has excellent emulsion capacity and whipping properties, is soluble at low pH, and is compatible as an ingredient of many diverse types of foods. The concentrates can be used in a variety of specialty products, such as infant formulations, special adult diet foods, and gravy mixes. Other potential uses are as meat extenders or in nutritional fortification; manufacture of textured vegetable proteins; improvement of extruded products (e.g., pasta); fortification of soft drinks; in combination with soy proteins as a milk substitute; as a partial replacement for eggs; or in cake formulations, ice creams, fudge, caramels, chocolate flavors, and cocoa substitutes.

The four types of whey concentrate powders, specified by the International Dairy Federation (1978) are described in Table 6.2. Ion exchange and adsorption chromatography techniques make it possible to produce powders that are essentially fat-free and contain over 90% protein. Proteins in these isolates are highly nutritious, soluble over a wide range of pH values, produce firm gels on heating at protein concentrations above 5% (optimum 10–11%), and have good fat- and water-binding properties. The functionality of whey protein concentrates can be improved by modification: selective and limited enzyme, acid, or alkaline hydrolysis and complexing with CMC. The functionality of the complex equals that of egg whites in whipping properties (foam volume and stability) and is useful in the production of cake mixes, meringues, desserts, and other foods.

Lactalbumin, a by-product of cheese whey, has an excellent nutritional value, superior to that of whole egg and whole milk proteins. It can be used in the production of whey cheese, cheese spreads, and processed cheese. It is also suitable as a meat extender and can be used in baked goods (e.g., cookies) and in some breakfast foods in which protein fortification (rather than functionality) is the most important criterion.

TABLE 6.2

Compositions (in %) of Some Ultrafiltration-Derived Spray-dried Whey Protein Concentrate Powders[a]

	35% product	50% product	60% product	80% product
Moisture	4.6	4.3	4.2	4.0
Crude protein $N \times 6.38$	36.2	52.1	63.0	81.0
True protein $N \times 6.38$	29.7	40.9	59.4	75.0
Lactose	46.5	30.9	21.1	3.5
Fat	2.1	3.7	5.6	7.2
Ash	7.8	6.4	3.9	3.1
Lactic acid	2.8	2.6	2.2	1.2

[a]From The International Dairy Federation, 1978.

TABLE 6.3

Composition of Egg White Proteins[a]

Constituents	Ovalbumin[b] (residues per 45,000 gm)	Ovotransferrin (residues per 76,000 gm)	Ovomucoid (residues per 28,000 gm)	Ovoinhibitor (residues per 49,000 gm)	Ficin inhibitor[c] (residues per 12,700 gm)	Lysozyme (residues per 14,307 gm)	Ovomucin[d] (residues per 10,000 gm)	Ovoflavo-protein (residues per 32,000 gm)	Ovomacro-globulin[e] (residues per 10,000 gm)	Avidin (residues per subunit)
Alanine	34	52	11.7	20.1	8.50	12	4.18	13.9	3.95	5
Arginine	19	33	6.3	20.5	7.18	11	2.80	5.6	2.60	8
Aspartic acid	31	79	31.9	47.3	11.3	8(13)[f]	7.49	20.0	6.50	15
Cystine/2	6[g]	22	17.5	34.7	3.35	8	4.72	15.9	1.15	2
Glutamic acid	50	69	14.9	40.7	18.0	2(3)[f]	8.60	36.4	7.82	10
Glycine	18	58	16.1	32.3	6.48	12	4.70	8.3	3.49	11
Histidine	8	13	4.3	12.9	1.25	1	1.63	9.2	1.25	1
Isoleucine	24	24	3.2	17.3	5.69	6	3.58	7.1	4.42	8
Leucine	32	48	12.2	22.4	10.0	8	5.38	14.8	6.25	7
Lysine	20	62	13.6	24.2	7.21	6	4.68	17.2	4.11	9
Methionine	15	11	1.9	3.6	2.05	2	1.50	8.2	1.43	2
Phenylalanine	20	25	5.3	6.4	3.07	3	3.15	7.0	3.45	7
Proline	16	31	7.7	17.2	2.48	2	4.19	9.7	3.65	2
Serine	36	42	12.5	26.0	10.9	10	6.14	28.8	5.05	9
Threonine	15	35	14.6	28.0	3.53	7	5.41	7.7	4.55	21

Tryptophan	3	18	0	1.0		6	1.11^h	8.8	0.5	4
Tyrosine	9	20	6.7	15.1	4.79	3	3.33	9.5	2.68	1
Valine	30	44	16.0	25.9	9.51	6	4.99	5.7	5.41	7
Sialic acid	0	0	0.3		0		$0.61(1.38)^h$	0.5	<0.01	0
Hexose	5^i		12–16.5	5–10	0		$2.84(4.57)^h$	3.09^j	2.2	5
Mannose		4	10–13.4				2.72^h	2^j		
Galactose			0.8–6.3				1.36^h	1^j		
Glucosamine	3^i	6	14.8–27.7	7–15			$2.79(3.64)^h$	4.41^j	3.21	4
Galactosamine			0	0			$0.75(0.99)^h$			0
N-Terminal	acetylglycine^i	alanine	alanine			lysine	serine and glycine	serine and glycine		alanine
C-Terminal	proline^i	phenylalanine^k	phenylalanine^k			leucine	glycine	glycine		glutamic acid

^{a}Source: Osuga and Feeney (1974), unless otherwise noted.
^{b}Fothergill and Fothergill (1970).
^{c}Sen and Whitaker (1973).
^{d}Osuga and Feeney (1968).
^{e}Miller and Feeney (1966).
^{f}Values in parentheses are amides.
^{g}Sulfhydryls -4; disulfide -1.
^{h}Donovan et al. (1970).
^{i}Marshall and Neuberger (1972).
^{j}Miller and Clagett (1973).
^{k}Penasse et al. (1952).

II. Egg Proteins

Most of the proteins in eggs are glycoproteins in which the carbohydrates are either glycosidically attached to seryl or threonyl residues or by an amide linkage to asparaginyl residues (Osuga and Feeney, 1977). Some proteins contain phosphorus and sulfur linked to the seryl or threonyl residues. In addition, lipids are bound to some of the proteins. The shell, the white, and the yolk contain 3, 11, and 17.5% protein, respectively. The MC of the egg white is about 88% and of the yolk 47%; the yolk is rich (33%) in lipids. Egg white is low in lipids (0.02%) and inorganic ions (0.5%). Globular types of proteins, mainly glycoproteins, are soluble. Tables 6.3 and 6.4, and 6.5 present the composition, physical, and biological properties of egg white proteins. Ovalbumin constitutes more than 50% of the egg white proteins; the other main proteins are ovotransferrin (12%) and ovomucoid (11%). Three inhibitors that vary widely in composition, size, and inhibitory action have been isolated and characterized. They are ovomucoid and ovoinhibitor (inhibitors of serine proteases) and ficin inhibitor (inhibitor of thiol proteases). Lysozyme in egg white has a lytic activity. Ovomucin is a large, filamentous, fiberlike polydisperse glycosulfoprotein. It is involved in the deterioration of eggshells. Ovoflavoprotein seems to function as a storage and transport system of riboflavin in the developing embryo. Little is known about the biological role of ovomacroglobulin, a large glycoprotein. Avidin, a minor egg white component, binds biotin and is a toxic constituent.

Yolk proteins (Table 6.6) are composed of several complex macromolecules (glycoproteins, phosphoglycoproteins, lipoproteins, and phosphoglycolipoproteins), which vary widely in properties, including solubility. The proteins–lipoproteins can be fractionated by centrifugations into low density fractions (LDF), heavy granule fractions, and water-soluble fractions. The water-soluble fraction can be subfractionated into livetins and a riboflavin-binding protein. The heavy sedimenting granules contain phosvitin, lipovitellins, and a low density subfraction. The LDF has a micelle-type structure and is composed of a phospholipoprotein having a lipid-rich core with the phospholipid and protein moieties radiating toward the surface (Osuga and Feeney, 1977). The high density fraction (HDF or lipovitellin) is in a complex association with phosvitin in the granules. The HDF has a definite structural integrity and is similar to the globular proteins.

The association forces are hydrophobic. Phosvitin is a phosphoglycoprotein that contains about 70% of the total yolk phosphorus as a monoesterified orthophosphate and exists as a complex with HDF in the yolk granules; it behaves like an acidic polyelectrolyte. The three livetins are acidic proteins with an isoelectric pH range of 4.8 to 5.0. The yolk riboflavin-binding protein is similar to the egg white riboflavin-binding protein; it is a water-soluble globular yolk phosphoglycoprotein.

TABLE 6.4

Physical Properties of Egg White Proteins[a]

Protein	Amount in egg white (%)	pI	Mol wt (gm)	$S_{20,w}$	$D_{20w} \times 10^7$ $cm^2\ sec^{-1}$	$\bar{V}$ (cm^3/gm)	$E_\lambda^{1\%}$
Ovalbumin	54.0	4.5	45,000	3.27	7.67	0.750	$E_{280}^{1\%} = 7.50$
Ovotransferrin	12.0	6.05	76,600	5.05	5.72(Fe)	0.732	$E_{280}^{1\%} = 11.60$
Ovomucoid	11.0	4.1	28,000	2.62	7.70	0.685	$E_{280}^{1\%} = 4.55$
Ovoinhibitor	1.5	5.1	49,000	—	nd	0.693	$E_{278}^{1\%} = 7.40$
Ficin inhibitor[b]	0.05	~5.1	12,700	—	nd	—	$E_{278}^{1\%} = 8.88$
Ovomucin	3.5	4.5–5.0	110,000[c]	6.4(10%)[c]	nd	nd	$E_{277.5}^{1\%} = 9.30$
Lysozyme	3.4	10.7	14,307	1.91	11.20	0.703	$E_{280}^{1\%} = 26.35$
Ovoglycoprotein	1.0	3.9	24,400	2.47	nd	nd	$E_{280}^{1\%} = 3.80$
Ovoflavoprotein	0.8	4.0	32,000	2.76	6.40	0.700	$E_{280}^{1\%} = 15.30$
Ovomacroglobulin	0.5	4.5	900,000 760,000	15.10	1.98	0.745	—
Avidin	0.5	10.0	68,300	4.55	5.98	0.730	$E_{280}^{1\%} = 15.70$

[a]Source: Osuga and Feeney (1974).
[b]Fossum and Whitaker (1968) and Sen and Whitaker (1973).
[c]Reducing conditions used; three peaks during ultracentrifugation: 6.4 S (10%), 2.9 S (5%), and 1.4 S (85%).

TABLE 6.5

Biological Properties of Egg White Proteins[a]

Constituents	Characteristic properties
Ovalbumin	Denatures easily; has four poorly reactive sulfhydryls
Ovotransferrin	Complexes iron ($K_D = 10^{-29}\ M$) and other metals; homologous to serum transferrin; antimicrobial
Ovomucoid	Specific trypsin inhibitor—1:1 complex ($K_D = 1.5 \times 10^{-7}\ M$)
Ovoinhibitor	Inhibitor of serine proteinases: two trypsins and two chymotrypsins simultaneously ($K_D = 4 \times 10^{-8}\ M$, trypsin), subtilisin competes with chymotrypsin; alkaline proteinase competes with elastase. Homologous to blood serum α_2-proteinase inhibitor
Ficin inhibitor	Inhibitor of thiol proteinases: binding site for cathepsin C and a second site for cathepsin B_1, bromelain, papain, and ficin ($K_D = 1.47 \times 10^{-8}\ M$, ficin)
Ovomucin	Viscous; high in sialic acid; inhibitor of virus hemagglutination and clotting of κ-casein by rennin; important in egg deterioration
Lysozyme	Cleaves polysaccharides; antimicrobial; homologous with human lysozyme and milk α-lactalbumin
Ovoflavoprotein	Binds riboflavin $\geqslant$ FMN > FAD in 1:1 complex ($K_A = 7.9 \times 10^8\ M^{-1}$, riboflavin); weakly antimicrobial; homologous with egg yolk and blood serum riboflavin binding protein
Ovomacroglobulin	Strongly antigenic and shows extensive immunological cross-reactivity with ovomacroglobulin of other species
Avidin	Binds 4 biotins/mole avidin; antimicrobial

[a]Adapted from Feeney and Osuga (1976).

TABLE 6.6

Protein Composition of Yolk[a]

Protein	% (of total)				Phosphorus	
	Weight	Solid	Proteins	Lipid	Lipid	Protein
Low density fraction	33.7	65.0	22.0	93.0	19.0	—
Livetin	5.3	10.0	30.0	—	—	—
Phosvitin	2.1	4.0	12.0	—	—	69.0
Lipovitellin	8.3	16.0	36.0	7.0	←12.0→	
Others	2.6	5.0	—	—	—	—
Riboflavin binding protein (YRBP)	—	—	0.4	—	—	some[b]

[a]Modified from Gilbert (1971).
[b]0.2% P or 2 M P/M YRBP.

Useful information has been obtained from fractionation and reconstitution studies on the functional role of egg lipoproteins in sponge cake production. In production of such a cake, the fragile foam produced from egg white must be reinforced and stabilized by egg yolk (Graham and Kamat, 1977; Kamat and Hart, 1973). The HDLP granules are essential in attaining foam stabilization, which is achieved through their contribution to the air cell walls and structural network of the cake. When frozen and subsequently defrosted before use, whole eggs performed in an improved manner in fatless sponge cake production. This is generally attributed to a balanced HDLP to LDLP ratio. Freezing contributes to lipoprotein aggregation and improves stability of foam during baking. Freezing strengthens the binding and complexing capacities of phospholipids to HDLP and enhances their emulsifier properties in effective lipoprotein activity.

III. Muscle Proteins

Protein constitutes 50–95% of the total organic solids in meat, depending on the lipid content. The muscle proteins may be divided into three major classes on the basis of their solubility in aqueous solvents (Table 6.7). The sarcoplasmic proteins are the most soluble and the stroma proteins the least soluble fractions. The sarcoplasmic proteins are globular and form highly viscous solutions with little resistance to shear. Their contribution to meat tenderness is minimal. The

TABLE 6.7

Protein Composition of Vertebrate Muscle[a]

Protein class	Definition
Sarcoplasmic proteins	Those proteins soluble at ionic strengths of 0.1 or less at neutral pH. Constitute 30–35% of total protein in skeletal muscle and slightly more than this in cardiac muscle. Contains at least 100–200 different proteins. Sometimes called myogen.
Myofibrillar proteins	Those proteins that constitute the myofibril. Make up 52–56% of total protein in skeletal muscle but only 45–50% of total protein in cardiac muscle. Although high ionic strength is required to disrupt the myofibril, many of the myofibrillar proteins are soluble in H_2O once they have been extracted from the myofibril.
Stroma proteins	Those proteins insoluble in neutral aqueous solvents. Constitute 10–15% of total protein in skeletal muscle and slightly more than this in cardiac muscle. Includes lipoproteins and mucoproteins from cell membranes and surfaces as well as connective tissue proteins. Although exact percentage composition can vary widely depending on source of the muscle, collagen frequently makes up 40–60% of total stroma protein and elastin may make up 10–20% of total stroma protein.

[a]From Goll et al., 1977.

two connective tissue proteins, collagen and elastin, make up most of the stroma proteins. The stroma proteins are important in food processing because they have several deleterious effects on meat functionality (Goll *et al.*, 1977). They reduce tenderness, depending on the amount and degree of cross-linking among the connective tissue proteins. They decrease the emulsifying capacity of meat, because of their low solubility. They also disturb the water-holding capacity of meat, because of their low content of charged and hydrophilic amino acids. Finally, stroma proteins reduce the nutritive value of meat because of their low proportion of essential amino acids.

Myofibrillar proteins, which constitute the myofibrils, are the largest fraction of proteins of muscle tissue and are intermediate in solubility. They influence the meat's culinary and commercial properties because of their high water-binding capacity (97% of the total) and emulsifying capacity (75–90%). Properties and possible implications of the structure and composition of myofibrillar proteins to meat quality are listed in Tables 6.8 and 6.9. Meat toughness can be resolved into two components: "background" and "actomyosin." Background tough-

TABLE 6.8

Properties of the Myofibrillar Proteins[a]

Protein	% of Myofibril by wt	Mol wt (gm/mole)	Subunit polypeptide composition[b]
Myosin	50–58	475,000	200,000 daltons - two
			20,700 daltons - one[e]
			19,050 daltons - two[e]
			16,500 daltons - one[e]
Actin	15–20	41,785	41,785 daltons - one[e]
Tropomyosin	4–6	70,000	35,000 daltons - one[e]
			32,758 daltons - one[e]
Troponin	4–6	72,000	30,503 daltons - one (TN-T)[e]
			20,864 daltons - one (TN-I)[e]
			17,846 daltons - one (TN-C)[e]
C-protein	2.5–3	140,000	140,000 daltons - one
α-actinin	2–3	206,000	103,000 daltons - two
β-actinin	<1	70,000	?
M-protein[c]	3–5	160,000	160,000 daltons - one
Paramyosin[d]	2–30	220,000	110,000 daltons - two

[a]From Goll *et al.*, 1977.

[b]Subunit polypeptide mass, and composition of myosin, tropomyosin, and troponin differ in "red" (slow-twitch, oxidative) and "white" (fast-twitch, glycolytic) muscles. Figures are given for "white" muscle proteins.

[c]Several reports have indicated that a second M-protein with a 42,000-dalton subunit exists in addition to the M-protein described.

[d]Paramyosin is found only in invertebrate muscles.

[e]Amino acid sequence has been determined.

TABLE 6.9

Molecular Architecture and Biochemical Properties of Myofibrillar Proteins[a]

Attribute of meat	Possible role of myofibrillar architecture and biochemistry
Tenderness	Much of the variation in tenderness is probably determined by integrity of Z-disks and strength and nature of the actin–myosin interaction.
Water-holding capacity	A function of spacings and integrity of the thick and thin filament lattice and possibly also of the extent of the myosin–actin interaction and the high proportion of charged amino acids in the myofibrillar proteins. Over 90% of water-holding capacity due to myofibrillar proteins.
Emulsifying capacity	Related to solubility and structural integrity of the myofibrillar proteins and possibly also to the high proportion of charged amino acids in the myofibrillar proteins. Myofibrillar proteins are responsible for approximately 90% of the emulsifying capacity of meat.
Nutritive value	Myofibrillar proteins contain relatively high proportions of nutritionally essential amino acids.
Cost	Because myofibrillar proteins constitute 35–45% of total organic material in muscle, cost of meat ultimately depends to a large extent on the efficiency with which ingested nutrients are converted to myofibrillar protein.

[a]From Goll *et al.*, 1977.

ness is due to connective tissue and other stromal proteins and actomyosin toughness to myofibrillar proteins. Balance of the two elements combined governs toughness whether perceived by the consumer or measured by instruments (Locker, 1956). Actomyosin toughness is on the average about 10 times as powerful as background toughness in its effect on meat tenderness.

The myofibrillar proteins of vertebrate skeletal muscle are composed of actomyosin (15–20% actin and 50–55% myosin), which is required and suffices for *in vitro* contraction, and of six regulatory proteins. The six are tropomyosin (5–8%), troponin (5–8%), α-actinin (2–3%), β-actinin (0.5–1.0%), component C (2–3%), and M-line proteins (3–5%). The regulatory proteins control actin–myosin interaction and regulate assembly of individual myofibrillar proteins into filaments.

A. Comminuted Meat Products

The manufacture of comminuted processed meat products (such as emulsions, and particulate, sectioned, and formed meats) depends on formation of a functional matrix within the product (Schmidt *et al.*, 1981). Properties of the matrix differ for each class of processed meat and give specific products their characteristic texture and "bite." The properties can be modified in terms of emulsification and/or binding and gelation. The basic structure of an emulsion is a mixture of finely divided meat constituents dispersed as a fat-in-water emulsion, in which the discontinuous phase is fat and the continuous phase is water containing the solubilized protein components. The water-soluble and salt-soluble proteins emulsify fat globules by forming a protein matrix on their surfaces. This stabilizing function is made possible by the reactive groups on the fat–water interface. The proteins are denatured upon contact with the fat globules and form a relatively solid membrane.

Binding capacity of emulsion-type sausages results from denaturation and decreased solubility of the proteins. Once the fat is coated, the emulsion is stabilized. Heating the emulsion coagulates the protein and stabilizes the emulsion so that the protein holds the fat in suspension (Schmidt et al., 1981). Emulsion stability is influenced by the water-holding capacity of the meat; the levels of meat, fat, salt, and additives in the formulation; and the mechanical and heat treatment. Formation of a protein matrix is one of the main factors contributing to creation of a successful emulsion. In products with strong cohesiveness, the matrix is thick and the fat globules are uniform.

The mechanism of binding between chunks of meat is a heat-initiated reaction. Chunks of meat in the raw form show little binding. The binding mechanism is similar to the mechanism acting in the heat-initiated emulsion–stabilization in emulsion sausages. The major difference is that emulsion sausages have no large chunks of meat. In the heat-initiated reaction, a concentrated emulsion forms between meat chunks and acts to bind adjacent meat pieces. The salt-soluble proteins are concentrated between the meat chunks. A protein matrix develops and upon heating acts as a stable binder between meat pieces. Binding between meat chunks involves structural rearrangement of the solubilized meat proteins, which renders them more reactive for protein binding. Orientation of the salt-soluble protein occurs on meat surfaces before and during heat-initiated binding reactions.

The mechanism of binding between meat pieces involves the interaction of thick filaments formed from intact myosin heavy chains in the extracted protein with myofilaments within muscle cells or near the surface. To enable formation of thick filaments at higher temperatures, myosin must be solubilized during mixing before heating. High temperatures unravel the helical portions of proteins into randomly organized chains and produce random cross-links via hydrogen and ionic bonds. In the presence of salt, myosin has the greatest binding power; in the absence of salt, the binding power is enhanced by sarcoplasmic proteins. The sarcoplasmic proteins exert a deleterious effect on the binding power of myosin at high salt concentrations due to the adsorption of denatured sarcoplasmic proteins on the protein structure (Schmidt et al., 1981).

Heat-induced gelation and changes in viscoelastic properties of the protein matrix are related to development of characteristic textures in meat products and to binding between meat pieces and emulsion stabilization. Myosin forms an irreversible gel, which is initiated by heat. The gel has a high water-binding capacity and strong elastic properties. After the gel is formed no syneresis takes place, probably as a result of interaction stabilization. The mechanism of myosin gelation seems to involve formation of bonds that are not ruptured by heat. Heat-induced gelation of myosin is optimum at 60–70°C at pH 6.0 in 0.6 M KCl. An intact myosin molecule has the best functional properties. Properties of heat-induced myosin gels depend on several factors, including the nature of the intact heavy chains.

Heat-denaturation of the myosin rod results in local irreversible conformation changes involving the aromatic amino acids. The ATP-ase active site in myosin plays an important role in its binding properties. Intact and free myosin has the highest functionality both in myosin gels and in meat systems. Functional groups in the rod portion can interact with other myosin molecules and heating induces conformational changes (Schmidt *et al.*, 1981). These interactions are necessary for interaction of functional groups; consequently, a greater number of gel-stabilized bonds will form in a myosin system consisting of nonaggregated molecules. Myosin undergoes a great number of thermal transitions when it is present as single molecules. If the aggregation of myosin is minimized, the degree of intermeshing of the polymer molecules is enhanced and the water-binding capacity and strength of the gel are improved.

B. Modified Muscle Proteins

Brekke and Eisele (1981) reviewed chemical, enzymatic, and combined methods of protein modification to improve functional properties of muscle foods.

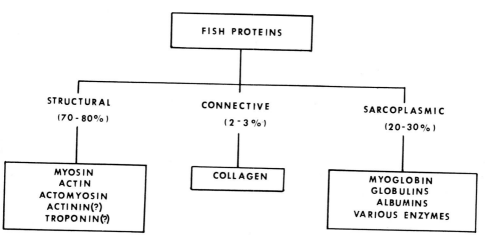

Fig. 6.1. Component proteins of fish muscle. (Courtesy of J. Spinelli, U.S. Department of Agriculture, Seattle, Washington.)

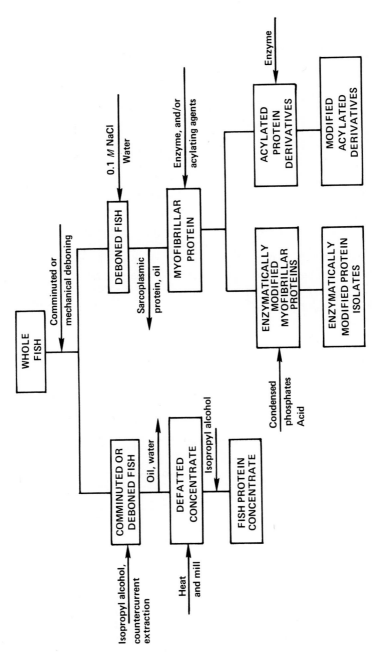

Fig. 6.2. Utilization of whole fish for the preparation of fish protein concentrates, isolates, and derivatives. (Courtesy of J. Spinelli, U.S. Department of Agriculture, Seattle, Washington.)

They described the use of acylation and succinylation processes to improve emulsifying capacity and stability, water-binding, gelation, and cohesion of low functional, low cost muscle proteins. The modifications had little effect on the nutritional value of the products. Selective proteolysis may solubilize the proteins and increase their lipid binding and emulsification.

It has been shown that fish protein can be effectively modified into highly functional proteins (Spinnelli *et al.*, 1977). Component proteins of fish muscle are listed in Fig. 6.1. Unmodified fish myofibrillar proteins are extremely sensitive to conventional processing. They can be modified, however, by enzymatic and/or chemical treatment to yield products with emulsifying potentials surpassing those of the native proteins. Utilization of whole fish for the preparation of fish protein concentrates, isolates, and derivatives is shown in Fig. 6.2; preparation of acylated fish myofibrillar protein is described in Fig. 6.3. The deleterious effects on emulsifying capacity of freeze-drying or isopropanol extraction are compared with the ameliorating modifications induced by enzymatic or chemical

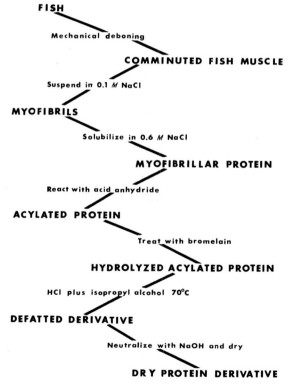

FISH

Mechanical deboning

COMMINUTED FISH MUSCLE

Suspend in 0.1 *M* NaCl

MYOFIBRILS

Solubilize in 0.6 *M* NaCl

MYOFIBRILLAR PROTEIN

React with acid anhydride

ACYLATED PROTEIN

Treat with bromelain

HYDROLYZED ACYLATED PROTEIN

HCl plus isopropyl alcohol 70°C

DEFATTED DERIVATIVE

Neutralize with NaOH and dry

DRY PROTEIN DERIVATIVE

Fig. 6.3. Preparation of acylated fish myofibrillar protein. (Courtesy of J. Spinelli, U.S. Department of Agriculture, Seattle, Washington.)

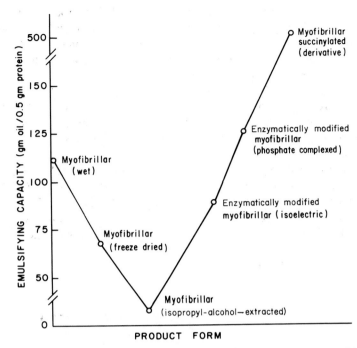

Fig. 6.4. Changes in the emulsifying properties of fish myofibrillar protein induced by various processing techniques. (Courtesy of J. Spinelli, U.S. Department of Agriculture, Seattle, Washington.)

TABLE 6.10

Functionality Differences between Succinylated and Acetylated Fish Protein[a]

Functional property	50–60% Acetylation	50–60% Succinylation
Emulsifying activity: % emulsion with 5 mg/ml of protein	61	97
Emulsifying capacity: gm of oil emulsified with 100 mg of protein	53	132
Gelation concentration of protein for a stable gel: %	4	3
Aeration capacity: volume of foam, ml	550	600
Stability of foam: volume of syneresis, ml	71	32
Water absorption: ml of H_2O absorbed by 1 gm of protein	15	>20

[a]Spinnelli *et al.*, 1977.

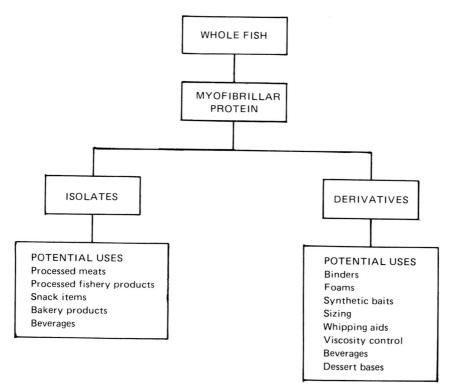

Fig. 6.5. Potential uses of fish protein isolates and derivatives. (From Groniger and Miller, 1974.)

treatments (Fig. 6.4). Table 6.10 lists functionality differences between succiny-lated and acetylated fish protein (Spinnelli *et al.*, 1977). Potential uses of fish protein isolates and derivatives are described in Fig. 6.5.

IV. Plant Proteins

Plant proteins form an intrinsic part of many foods. Although the importance of plant protein is well recognized, the use of concentrated seed protein (CSP) is limited. Some CSP, such as wheat gluten, have been used as food components for several years, but exploration of the potential for the production and use of CSP food components is a comparatively recent development. In 1980 the world protein production totaled 290 million metric tons, of which four-fifths were of plant origin and one-fifth of animal origin (Gassman, 1983). Cereals provided over two-thirds and oilseeds over one-fifth of the plant proteins. Energy costs of protein production (in MJ/kg protein) are 30 for soybeans, 65 for corn (maize),

170 for single-cell protein, about 300 for hen eggs or broilers, 530 for fish, 585 for milk, 590 for pork, and 1300 for beef (Gassman, 1983).

A. Nutrition

As stated in Chapter 5, the proteins of the body are complex molecules constructed from various combinations and proportions of 18 to 22 amino acids. The human body can synthesize 14 amino acids, provided that the necessary precursors are present. However, 8 amino acids cannot be synthesized at rates that allow survival: isoleucine, leucine, lysine, methionine, phenylalanine, threonine, tryptophan, and valine. These 8 amino acids are considered indispensable and must be made available to the body. Nutritional qualities of several proteins are listed in Table 6.11; factors that affect value or acceptability and usefulness of proteins from plant sources are given in Table 6.12 (see Kinsella and Srinivasan, 1981; Satterlee, 1981).

TABLE 6.11

Nutritional Quality of Several Proteins[a]

Species	Limiting amino acids		Biological value (%)	Protein efficiency ratio	Net protein utilization (%)
	Gross deficiency	Marginal deficiency			
Cereals					
Barley	Lys	Ileu, Thr	74	1.6–2.0	59–60
Maize	Lys	Try	53–60	0.9–1.3	49–60
Oats	Lys	Meth, Thr, Ileu	65–75	1.8–2.5	59–66
Rice	Lys	Meth, Thr, Ileu	64–89	1.8–2.3	54–70
Rye	Lys	Phe, Ileu, Try	74	1.6–2.3	—
Sorghum	Lys	Meth, Try, Phe	51–73	0.2–2.0	48–56
Wheat	Lys	Ileu, Try	58–67	0.9–1.7	40–60
Legumes					
Broad bean	Met, Cys	Try	55	—	48
Chick pea	Met, Cys	Try	52–78	1.1–2.2	52–64
Common bean	Met, Cys	Try, Val	45–73	0.0–1.9	31–47
Comon pea	Met, Cys	Try	48–66	0.3–2.2	41–50
Cow pea	Met, Cys	Ileu	49–65	—	35–51
Groundnut	Met, Cys	Ileu, Thr	51–63	1.5–1.8	52–55
Pigeon pea	Met, Cys	Try	46–74	0.7–1.8	52
Soybean	Met, Cys	Val	68–77	0.7–2.4	58–71
Other					
Leaf	Meth	—	59–90	0.5–2.5	55–75
Yeast	Meth	Try	—	0.5–2.0	40–60

[a]From Kinsella and Srinivasan, 1981.

TABLE 6.12

Factors Affecting Nutritive Value or Acceptability and Usefulness
of Proteins from Novel Sources[a]

Factors	Occurrence	Action	Control
Protease inhibitors	Beans, cereals	Inhibits trypsin and chymotrypsin; pancreatic hyperplasia; methionine deficiency	Moist heat
Phytohemagglutinins (lectins)	Beans, (black, kidney) P. vulgaris	Agglutinate erythrocytes; impair absorption	Moist heat
Favism (vicine, convicine)	Broad bean (V. faba)	Hemolytic anemia	Moist heat
Lathyrism	Vetch (Lathyrus sativum)	Neurotoxic, neuromuscular control	Soaking/boiling
Phytic acid	Beans, cereals	Chelation of phosphorus and cations	Milling, moist heat
Gossypol	Cottonseed	Chelation of iron; interaction with protein	Heat moisture, separation
Aflatoxin	Peanuts	Hepatotoxic, carcinogen	Screening
Glucosinolates	Cruciferae	Release toxic thiocyanates, and goitrins	Heat inactivation, extraction
Nucleic acids	Yeast	Uricemia, gout	Hydrolysis
Cell wall	Yeast	Nausea, dizziness	Separation
Oligosaccharides	Beans	Flatulence	Elution, fermentation
Lipoxygenase	Beans	Oxidation of linoleic acid → "off" flavors; carbonyl–protein interactions	Moist heat
Chlorogenic acid	Sunflower	Astringent, bitter; discoloration	Extraction

[a]From Kinsella and Srinivasan, 1981.

The presence of an amino acid in a food does not guarantee that it will be absorbed when the food passes through the gastrointestinal tract. For example, bread crust contains lysine but much of the amino acid is chemically bound to carbohydrate in such a manner that the lysine is not freed by the digestive processes. Therefore, the lysine cannot be absorbed and utilized for protein synthesis. This means that the amino acid spectrum of bread may be misleading in terms of actual nutrional value.

In general, plant proteins are of lower nutritional value than animal proteins.

However, it is possible to form mixtures having values similar to animal products. Furthermore, the indispensable amino acids lysine and methionine, which in general are most limiting in plant proteins, are now available in commercial quantities. Thus it is possible to supplement plant proteins with free amino acids to improve quality significantly.

In the developed countries considerable concern exists about the fact that plant proteins tend to be inferior to animal proteins as sources of amino acids. Although the problem is not serious in developed countries, it may retard research and use of CSP in foods. As far as North America and Western Europe are concerned, two factors should be taken into account. First, the typical diet is mixed and usually contains some animal protein to complement plant protein. Second, diets in Western countries usually contain an excess of protein; thus a slight deterioration in the protein quality of one food is not of major importance.

In many developing countries it is essential to maximize the nutritional quality of the available food protein. Where protein malnutrition occurs, however, the important criterion is the function of quantity multiplied by quality rather than either variable alone. In these cases using CSP as a food component is justifiable, even though it may cause a slight decrease in protein quality of a food. With the availability of amino acids to fortify the CSP, the general trend would be to increase protein quality as well as quantity.

B. Economics

In North America and Western Europe it should be possible to use CSP in the production of foods and thus lower the cost of a high quality, balanced diet. This has particular application in institutional feeding, where cost and product uniformity are of particular importance. Whether the general consumer will purchase CSP foods and make a compensatory reduction in the more expensive animal protein foods is open to debate. In selection of foods for home preparation, cost alone is not always the overriding consideration.

The relatively low cost of CSP foods should make them attractive in many Third World countries, but nutritional education and major changes in sociological attitudes and traditions are needed. Concentrated seed protein foods may be improved by amino acid fortification. For example, the addition of approximately 0.3 kg of L-lysine to 100 kg of wheat flour will increase protein quality by about 45%; thus, in effect, the protein content may be raised from about 14 to 20%. If lysine costs two dollars a kg, then the additional effective protein (6 kg) has cost approximately 10 cents a kg. This is substantially less than any presently available protein in the form of CSP. Such fortification is feasible since it does not change the appearance or taste of cereal-based diets and does not necessitate a change in food habits.

C. Functionality

Concentrated protein can be added to foods to improve appearance, control texture or viscosity, reduce cooking shrinkage, increase shelf life, improve dough handling properties, and meet nutritional requirements.

Soy concentrates are often used as binders and conditioners in meat patties and sausages because of their ability to bind fats and water. Soy isolates are successful as moisture and fat retainers in canned goods subjected to high temperatures and high processing speeds. Isolates are also used in dairy products, sandwich spreads, and snack dips to impart desirable smooth characteristics. Some isolates serve triple functions as emulsifiers, whipping agents, and stabilizers in whipped toppings.

Functional properties also play an important role in the development of new high protein textured products. Protein characteristics enable the products to be spun or extruded in various shapes to provide pleasing texture, eating quality, and appearance. Spun proteins from comparatively inexpensive soybeans can be made to simulate meats, which can be combined with other materials such as wheat gluten, egg albumin, fats, flavors, and colors to make a variety of novel products. Extruded chunks, granules, or bits of soybean fiber have the appearance of ground beef and are used in hamburger patties, chili, meat loaf, or other ground beef products requiring meat extenders (see Chapter 10 and 11).

High protein wheat gluten adds strength to yeast-raised baked goods and to pasta products. Its gas retention and water absorption properties can be used to advantage in achieving the desired textural properties of these foods.

D. Plant Protein Technology

Proteins, as isolates or concentrates, are essential ingredients in many major food processing companies. Their utility is predicated upon the ability of the proteins to perform specific functions beyond nutrition, and the products must have flavors compatible with the end use. Currently, proteins from soybean, wheat (flour), and milk are the only major commercial proteins.

Research has already enabled a reasonable supply and price for methionine and lysine, and other amino acids could be supplied upon demand. Amino acids are classed by the Japanese food industry as "sweet," "bitter," "sour," and "delicious." Sweet amino acids include glycine, alanine, serine, threonine, proline, hydroxyproline, lysine·HCl, and glutamine. Valine, leucine, isoleucine, methionine, phenylalanine, tryptophan, arginine, arginine·HCl, and histidine are bitter. Aspartic and glutamic acids, which are acidic, are regarded as sour; the corresponding sodium aspartate and monosodium glutamate are classed as "delicious."

The sensory characteristics imparted by amino acids are influenced by their configuration. For example, the L-form of monosodium glutamate is classed as "delicious," the D-form as "tasteless." L-Tryptophan is bitter but D-tryptophan is 40 times as sweet as sugar. The racemic mixture of DL-tryptophan is also sweet.

Amino acids also develop flavors upon heating in the presence of sucrose. A list of flavors that arise by heating the amino acid in dextrose at 180°C follows:

Amino acid	Flavor
Glycine	Caramel
Alanine	Caramel
Valine	Chocolate
Leucine	Burnt cheese
Isoleucine	Burnt cheese
Proline	Bread
Hydroxyproline	Cracker
Methionine	Potato
Phenylalanine	Violet
Tyrosine	Caramel
Aspartic acid	Caramel
Glutamic acid	"Butter ball"
Histidine	Corn bread
Lysine	Bread
Arginine	Burnt sugar

The propensity of monomeric amino acids for sweetness is also parlayed into the peptides. For example, the methyl ester of L-aspartylphenylalanine is about 150 times as sweet as sugar.

The principal interest in amino acids is in fortification. L-Lysine is added to gluten breads, to wheat in the milling process, and coated on rice or used in rice pelletizing machines. Isolated proteins should be easily extractable in maximum quantity from low-cost source material, and should be readily and quantitatively recoverable from the extract. Ideally, the recovered protein should be inexpensive, white, and bland, with useful and desired functional qualities. The nutritional qualities of the proteins are generally of minor consideration when applied to the food processing industry, since amino acids could be added as required. Because proteins are expensive ingredients, their use in most food products must be considered in terms of the function they perform. They must also be acceptable with respect to flavor—the indigenous flavor "notes" must be compatible with the final formulation.

Another important consideration is the necessity of aseptic practices: clean, sound, and wholesome raw materials must be used for human consumption and the processing equipment must be sanitary. The production costs of food-grade products are usually much higher than the costs of animal feed.

E. Preparation of Proteinaceous Products

The raw material sources of protein may be converted to different states of matter.

1. They may be used "as is," after proper culling, cleaning, soaking, and so on, prior to cooking. Examples are peas, beans, and parboiled wheat (bulgar).
2. "Flours" may be prepared by comminution depending on the desired physical characteristics of the finished product.
3. Protein concentrates may be prepared by selective removal of soluble materials or by air classification (see Chapter 5).
4. Proteins may be separated from nonprotein substances to yield protein isolates.

Currently, only wheat and soybeans are used on a large-scale basis to prepare isolated proteins. With the exception of wheat gluten and corn zein manufacture, most of the proteins from grains, defatted oilseed meal, or legumes can be extracted almost completely (80%) by alkaline media. However, some of the proteins that are associated with cellular matter require hydrolytic treatment (enzymatic or chemical) to effect solution. It is in the recovery, or "harvesting" of the proteins from the extracted media, that special conditions are required to effect maximum yield of protein. Most plant seed proteins of the storage-type can be precipitated in acidic media.

V. Factors in Production and Use of Concentrated Seed Proteins

The factors that limit the use of plant proteins in human foods are divided into five classes: regulatory, cultural, technological, economic, and biological.

A. Regulation

Many countries have regulatory agencies or groups that determine the types and quantities of materials that may be included in human foods. Countries without such agencies are frequently guided by the United Nations Food and Agriculture Organization and World Health Organization (FAO/WHO) Joint Food Standards Program.

Food legislation is designed to protect the consumer from health hazards and fraud. In general, the successful entry of CSP into the human diet depends upon:

1. Demonstration that the plant protein product is not injurious to health
2. Promulgation of regulations defining the concentrations of nutrients that must be present in foods containing CSP

3. Establishment of acceptable limits for levels of CSP in specific products or product groups
4. Formulation of labeling, packaging, and advertising guidelines to avoid misleading the consumer
5. Development of the technology to allow regulations to be enforced

There is little doubt that these conditions will be met; however, stimulation of research in certain areas is needed to minimize delays. Of particular importance is measurement of the relative amounts of plant and animal proteins in a mixture. This appears to be fundamental to the acceptance of CSP as a meat extender, unless a rigid and expensive system of continual in-plant inspection is used. Also important is the establishment of minimum acceptable levels for the naturally occurring toxins found in some CSP, such as in rapeseed protein preparation, since it is now generally agreed that zero tolerance levels are impractical.

B. Cultural Factors

1. Consumer Attitudes

Historically, humans selected their food from the plants and animals that could be harvested close to home. Because population groups developed in areas varying in available food types, geographic differences in food habits are not surprising. The cultural constraints on the use of CSP in foods vary according to the attitudes of the consumers to which such foods are directed.

2. Substitute Meat Products

Substitute meat products from plant proteins are marketed successfully in the United States. Meat substitutes appeal to several groups besides the poor and the "dollar conscious." Vegetarians are eager to obtain such products and certain religious sects enjoy substitutes for prohibited foods.

3. Developed Countries

In Western countries, there are established precedents for the introduction of substitute and novel foods. When consumers are sophisticated, food processors pay more attention to needs and desires. To utilize large quantities of CSP, government agencies could publicize the safety and merits of such foods, but the onus for persuading the consumer actually to use plant protein foods remains the responsibility of the manufacturer.

4. Developing Countries

The social traditions surrounding the selection, preparation, and consumption of foods are deeply rooted in many cultures and changes in food require major changes in attitudes. Successful ventures are vastly outnumbered by failures. As communications improve and the level of education rises, problems of nonacceptance may decrease. However, it is vital to understand the sociological and cultural changes that may result when basic food traditions are destroyed in a society. This responsibility is too frequently overlooked.

C. Technological Problems

Of course a food product has no nutritional value unless it is eaten, and foods and food components have no commercial value unless they are marketed. Considerable technical effort has been expended in the production and use of CSP, but many problems remain to be solved if potential markets are to be exploited.

1. Processing

The extraction of oil from seeds is well established in many countries. Until recently the seed residue, often very rich in protein, was considered a by-product suitable only for animal feed ingredients or as fertilizers. Consequently, in developing oil extraction processes emphasis was placed on the amount and quality of the oil while the effect of such processes on the seed residues tended to be neglected.

As the formulation of animal feeds became more sophisticated, attention was directed to the improvement of the nutritional values of oilseed residues. Excessive heating can reduce availability of amino acid(s), while insufficient heating allows the survival of certain antimetabolites. These and other considerations have caused many in the oil extraction to change from the expeller to the solvent process. As the demand for plant proteins for food use has increased, oil extraction methods have been subjected to critical examination. The technology to construct factories capable of yielding CSP suitable for human foods is available, but their construction depends upon the existence of a market for products.

Oil extracted seeds are generally unacceptable for human consumption because of their high fiber content. Fiber levels can be controlled to a large extent by dehulling the seeds prior to extraction. An alternative process is the removal of fiber by air classification (see Chapter 5). Dehulling before oil extraction eliminates the need for further processing of the hulls. In addition, the loss of nonhull material is minimized.

Although steam treatment can reduce objectionable flavors, it simultaneously lowers protein solubility and impairs functional properties of seeds. The more highly refined proteins have fewer flavor problems, yet functionality often changes with the degree of isolation. Many seeds contain toxins and antimetabolites, which must be removed or inactivated if seed derivatives are to be used for human food. The control of flavor, toxin, and flatus problems by extraction and concentration of the seeds can be expensive inasmuch as such treatments frequently cause losses of other desirable characteristics. Unless the water-soluble protein is recoverable and utilizable, the cost of the extraction may be prohibitive. Concentration of proteins may cause losses of the first limiting acid(s) and lower biological values. The drying steps associated with extraction and concentration may have a number of adverse effects, such as reduced solubility, denaturation, browning reactions, impaired functional properties, and lowered nutritional value.

2. Plant Proteins

Of the several plant proteins with interesting functional properties, only wheat gluten has been produced and developed because of its functional characteristics. Vital wheat gluten is used to fortify flours that are otherwise unsuitable for bread production, and also serves to strengthen the "hinge" in a sliced hotdog or hamburger bun. Two major problems relating to wheat gluten require solution. Although the term "vital," meaning functional in bread making, is freely used to describe a certain type of gluten, it is not clearly defined. The second question asks whether the production of gluten is feasible on a large scale—is the demand for starch sufficient to justify its production?

Plant proteins are used to produce hydrolysates, which are used to flavor or enhance existing flavors of foods. Although this is a relatively small market in terms of the quantity of protein used, improved flavor technology development is welcomed in the food industry.

The incorporation of CSP into conventional foods can increase protein content and quality. In assessing the use of CSP in raising nutritive value, the effects attainable through fortification with synthetic amino acids must be compared with CSP fortification. In general, amino acid fortification has less effect on the sensory characteristics of foods than does CSP fortification. Considerable effort has been devoted to the development of breads fortified with CSP, but it is questionable whether a wide protein deficiency exists in North America that could be overcome if such bread is made available. Moreover, it is questionable whether North American consumers will pay a premium for higher nutritional value. On the other hand, widely distributed fortified breads could be of enormous value in alleviating malnourishment in developing countries.

Much work with textured vegetable proteins has been directed toward the production of meat analogs or substitutes. In forming such substitutes, spun proteins are bound with materials such as egg albumin. The products, though highly nutritious, are relatively expensive. Substitutes for meat slices and small pieces of cooked meat have been prepared from plant proteins; however, it is not yet possible to make simulated roasts or larger cuts of meat. Meat analogs may be particularly useful in institutions, where low cost and product uniformity are desirable (see also Chapter 11).

It was believed that high protein beverages would serve as readily acceptable nutrient supplements, but most high protein beverages have been commercial failures. Protein beverages represent a relatively new class of foods. The rapidly growing snack food market offers many opportunities for plant proteins. The failure rate of new food products is extremely high, but the acceptance of a novel high protein food may be more easily attainable than acceptance of a substitute food. A completely new food need not face the prejudices and unfavorable comparisons that hound substitute products. Further, in developing new foods, one can exploit the functional characteristics of the proteins, whereas in substitutes such characteristics must frequently be modified to conform to predetermined standards and expectations.

D. Economic Considerations

The type and quantity of CSP that is used in human foods is governed to some extent by economic considerations. Vegetable proteins are cheaper than those of animal origin. The price of protein in a crude product, however, can be misleading; if credit is given for nonprotein material the range in costs would be extended. On the other hand, the potential problems of disposing of the by-product should be given sufficient attention. For example, some loss of sulfur-containing amino acids occurs during the concentration and/or isolation of soy protein. As protein concentration increases, protein quality decreases.

CSP are stable and can be stored at low cost for long periods. They are uniform in composition and waste is minimal. The advantages of CSP must be assessed in the light of the purpose for which they are used. If the objective is to raise nutritional value only, then amino acid fortification may be cheaper than the use of plant proteins.

When CSP are used to extend meat products the important variables include functionality, texture, and flavor. The objective is to prepare an equivalent food, in sensory and biological terms, at a reduced cost. In this case, while the cost of CSP is important, it is not as critical as when nutritional value is the primary concern. Substitution of 20–40% of a meat patty with hydrated CSP (and possibly fat) reduces the cost to the consumer. In institutions the savings could be very significant. The higher uniformity of CSP-extended foods is considered to be another advantage by institutional purchasers. In the production of meat substitutes from CSP, the cost differential between the meat and its analog is proportionally greater than in extended foods, although expensive binders such as egg albumin may reduce the potential margin. As long as texture, flavor, and biological characteristics are acceptable, substitutes appeal to both retail and institutional markets. Recent publicity regarding the adverse health aspects of of animal fat consumption may help stimulate analog consumption, when the preparations are made with vegetable oils and fats. Consumers need to be aware of the fact that whereas 100% of a meat analog may be edible, as little as 30% of a retail raw meat may be consumable. This difference must be recognized if the economic value of the analog is to be calculated and used in the selection of foods. But cost is not the only factor regulating purchasing habits. Food processors are reluctant to risk their share of the market unless they can be certain that product modifications will not materially alter sensory characteristics of foods.

Developing countries tend to be short of food in general, not just protein; consequently, it is not necessarily in their interest to use limited funds to buy plant proteins. The more developed countries often impose tariffs and import duties on processed or analog foods.

E. Biological Factors

Food must be enjoyable to consume. Such enjoyment is the net result of all the sensory values including flavor, texture, odor, and mouth-feel. Unless CSP-containing foods can ensure eating enjoyment at a level equal to that of conventional foods, they will find very limited acceptance by the consumer. In countries where protein malnutrition is not a major problem, the use of CSP to extend animal proteins or to produce substitutes for foods of animal origin is more desirable than the protein enrichment of cereal-based foods. The reason is that the nutrient levels of extended and substitute foods can be made equal, or superior, to those of traditional foods; the use of CSP-containing foods at the expense of traditional foods will not alter nutrient balance. If CSP-containing foods are acceptable and cheaper than traditional foods it is conceivable that a change in dietary habits will result. This in turn may alter nutrient intakes and necessitate other changes in the nutritional specifications for CSP-containing foods.

When CSP are used to prepare substitute foods such as meat analogs, a high level of product uniformity is possible. It is also possible to regulate the levels of cholesterol, saturated fats, energy, and other factors that may influence health. The stresses of medically or self-imposed dietary constraints may be alleviated when the consumer is able to select diet components from a wide array of attractive foods.

If fortification of CSP-containing foods is to be successful, the stability and availability of a supplemental nutrient must be measured. It is not sufficient to add iron to a product if the iron is less available than that in a traditional food or if the iron causes discoloration and/or "off" flavors. Plant proteins are deficient in certain trace minerals, which only now are being recognized as essential for humans. Further, plant proteins are usually associated with phytin, which can sequester mineral elements and reduce their availability as nutrients.

Many plant proteins are associated with toxic factors that must be controlled. For example, vegetable proteins in general are excellent media for the molds that produce aflatoxins, extremely poisonous materials. Organic solvents are widely used to remove the oil from seeds. Unless great care is exercised small amounts of toxic solvent may be left in the pressed cake residue. In preparing CSP for food use, the amount of solvent must be reduced to a safe level. Other external toxins include pesticides and fungicides that may have been applied to the growing crops. Indigestible carbohydrates pass to the large intestine where they are subjected to microbial fermentation; this can lead to excessive gas production. Enzyme inhibitors may interfere with normal digestive processes and cause health problems. Toxins (for example, goitrogens) may be assimilated and interfere with body metabolism.

Other factors that partially constrain the use of CSP include unacceptable flavors and colors. Preparation of CSP also can change biological values. Other unfavorable changes include browning reactions in which lysine is irreversibly bound to carbohydrate, denaturation of protein to the point of reducing digestibility, and amino acid losses. A technology that allows the production of a high quality CSP from one source may not be suitable when applied to other materials. Each protein source presents a specific group of biological problems that may be amenable to a technological solution. Synopses of some of these problems are presented in the following sections.

1. Cereal Grains

It has been argued that cereal grains should be regarded primarily as sources of metabolizable energy rather than as sources of protein. However, cereals provide 80 million metric tons of protein—about one-half of the annual human requirement. This means even a small increase in the amount of protein or improvement in the biological value of that protein could make an important contribution to solving the worldwide problem of protein malnutrition.

According to Miller (1958), the average protein content of corn is only 10.4%; barley and oats average over 13% (Table 6.13). The mean data in Table 6.13 for barley and oats are higher than the more recent data given in Table 6.14 (Whitehouse, 1973).

The protein efficiency ratios (PER) for cereal proteins vary with the crop (Table 6.15). The PER of oat proteins is higher than the PER of other cereal grains; the PER of barley is somewhat higher than the PER of regular corn, wheat, and sorghum (Frey, 1973). The PER of the proteins from different cereals are related to the proportion of the alcohol-soluble prolamin fraction which is very low in lysine, the first limiting amino acid in practically all cereal grains (see discussion of amino acids in next section). Percentages of prolamin fractions in proteins of various cereal grains are shown in Table 6.16.

TABLE 6.13

Protein Percentages in Commercial
Lots of Cereal Grains[a]

| Cereal | Protein (%) (air-dry basis) | | No. of samples |
	Mean	Range	
Barley	13.1	8.5–21.2	1400
Corn	10.4	7.5–16.9	1873
Oats	13.3	7.4–23.2	1850
Rye	13.4	9.0–18.2	112
Sorghum	12.5	8.7–16.8	1160
Wheat	12.0	8.1–18.5	309

[a]From Miller, 1958.

Up to 80% of the protein in the cereal grain is stored in the endosperm. The embryo proteins are rich in albumins and globulins and have a relatively balanced amino acid composition. The endosperm proteins are rich in prolamins and glutelins. The glutelins contain slightly more lysine than do the prolamins. Increasing the protein content of cereals by cultural practices increases primarily the prolamin fraction and thus lowers biological value.

Proportions of amino acids, required by fast-growing chicks, in proteins of six grain crops are given in Table 6.17 (see Whitehouse, 1973). The data in Table 6.17 on net protein utilization by rats indicate that oats and barley are comparable and that both are lower than rice. Table 6.18 summarizes data from several

TABLE 6.14

World Production of Cereal Grain and Cereal Protein (kg × 10⁹)[a]

Item	Wheat	Barley	Oats	Rye	Maize	Sorghum and millet	Paddy rice	All grain crops
Grain[b]	333	131	54	33	251	85	284	1180
Protein (% of fresh weight)[c]	12	10	10.5	12.5	10	11	8	10.2
Protein	40	13	6	4	25	9	23	120

[a]From Whitehouse, 1973.
[b]Food and Agriculture Organization United Nations, 1969.
[c]Adapted from Kent, 1973.

TABLE 6.15

PER of Cereal Proteins in Diets of Rats[a]

Protein source	Protein (%) in ration			
	7.5	9.5	9.0–10.0	8.0–10.0
Barley	1.7	—	1.6	2.0
Corn	1.6	—	1.4	—
Oats	2.1	2.5	1.8	2.2
Rye	2.2	1.8	1.3	—
Sorghum	—	—	0.7	—
Wheat	1.4	1.7	0.9	1.7

[a]From Frey, 1973.

sources on the proportions of albumins, globulins, glutelins, and prolamins in cereal grains and the lysine contents of those fractions.

As stated previously, the role of wheat in cereals is attributed mainly to the unique properties of gluten that can be washed out from milled wheat products. When gluten is processed in such a manner that it retains its viscoelastic properties, the "vital product" can be used to improve the functional properties of baked goods and breakfast cereals. In the Far East, gluten is used as an extender

TABLE 6.16

Origin, Name, and Amount of Different Known Prolamins[a]

Species of Gramineae		Prolamin	
Common name	Scientific name	Name	% of seed protein[b]
Common wheat	*Triticum vulgare*	Gliadin	45
Durum wheat	*Triticum durum*	Gliadin	60
Rye	*Secale cereale*	Secalin	40
Barley	*Hordeum vulgare*	Hordein	40
Oat	*Avena sativa*	Avenin	12
Rice	*Oryza sativa*	—	8
Maize	*Zea mays*	Zein	50
Sorghum	*Sorghum vulgare*	Kafirin	60
Common millet	*Panicum miliaceum*	Panicin	60
Italian millet	*Setaria italica*	—	50
Finger millet	*Eleusine coracana*	Eleusinin	

[a]From Mosse, 1966.
[b]Percent of seed N.

TABLE 6.17

Amino Acid Composition[a] of Grain Proteins[b]

Amino acids essential in some diets	Ideal composition for chicks	Rice[c,f]		Oats[c,d]		Barley[c,d]		Wheat[c,d]		Maize[c,d]		Sorghum[c,e]	
Lysine	5.0	3.5	3.7	4.0	3.7	3.7	3.4	2.6	2.7	2.7	3.0	1.8	2.0
Methionine and cystine	3.5	3.4	3.5	4.8	3.2	4.1	3.5	3.6	3.7	4.6	4.2	3.0	2.3
Threonine	3.5	3.3	4.1	3.6	3.4	3.6	3.7	3.0	2.9	4.0	4.2	3.6	3.0
Isoleucine	4.0	4.5	3.9	4.0	4.6	3.7	3.8	3.4	3.8	3.8	4.0	4.5	3.8
Leucine	7.0	8.0	8.0	7.1	7.0	7.1	6.9	6.8	6.4	10.6	12.0	11.6	13.1
Valine	4.3	5.4	5.7	5.1	5.4	5.3	5.0	4.6	4.3	5.0	5.6	5.4	4.9
Phenylalanine and tyrosine	7.0	10.3	8.5	8.4	8.8	8.6	8.5	7.6	7.8	8.7	8.8	5.2	6.4
Tryptophan	1.0	0.6	1.4	0.9	1.3	1.3	1.4	1.1	1.3	0.7	0.8	0.8	0.7
Histidine	2.0	2.2	2.3	2.2	1.9	2.2	1.9	2.3	2.1	2.6	2.4	2.0	2.1
Arginine	6.0	7.8	7.7	6.1	6.6	5.4	5.0	4.7	4.3	4.3	5.0	3.4	2.7
Net protein utilization by rat (%)		66		59		59		53		51		48	

[a]Grams per 16 gm N.

[b]From Whitehouse, 1973.

[c]Data in first column for each crop from Eggum, 1969.

[d]Data in second column for oats, barley, wheat, and maize from Hughes, 1960.

[e]Data in second column for sorghum (except for tryptophan) from Lykes, 1970.

[f]Data in second column for brown rice (except for tryptophan) from Juliano et al., 1964.

in processed meat and fish goods. Gluten is valued mainly because of its viscoelastic properties; it is also a good emulsifier and strengthener in cheese (Satterlee, 1981). The properties of gluten may be modified to enhance its solubility and water-binding capacity. Thus, for instance, succinylated gluten is relatively water-soluble and phosphorylated gluten binds about 100 times its weight of water.

Corn gluten meal is produced in wet-milling processes. The proteins in the meal are deficient in lysine and tryptophan but are rich in the sulfur-containing amino acids methionine and cystine. A 90% protein commercial corn gluten isolate is only slightly soluble in water but binds three times its weight in water and binds its weight in fat. As in other corn protein preparations, it is low in lysine (1.5 gm/100 gm protein) and tryptophan (0.54 gm/100 gm protein) but rich in methionine plus cystine (4.8 gm/100 gm protein). The isolate can be blended with protein-rich oilseeds (peanuts or soybean) or legumes. Defatted corn germ contains 28–32% protein with a very good balance of essential amino acids and high PER (2.44). It also has excellent water-binding properties.

The amino acid balance of rice protein is acceptable, but the protein content of commercial cultures is low. The limiting amino acid in barley protein is lysine (see Table 6.11). The low lysine level in barley protein is primarily due to the virtual absence of lysine in hordein, the alcohol-soluble protein fraction.

Kernel weight, protein content, and amino acid composition were determined (Pomeranz *et al.*, 1976) for 113 barley cultivars from the United States Department of Agriculture (USDA) World Collection. A close relationship between lysine and glycine was found over different barley types. Decrease in lysine concentration accompanying increase in protein content indicated a curvilinear relation (Fig. 6.6). A similar curvilinear relationship between protein content of barley endosperm and lysine content of the protein was reported for most cereals.

Investigations of Munck *et al.* (1970) have led to the discovery of a high protein, high lysine barley line (Hiproly) with improved nutritional value. The Hiproly barley, C.I. 3947, is of Ethiopian origin; it is an erectoid type with naked, slightly shriveled seeds, and requires a long photoperiod. A sister line to Hiproly, C.I. 4362, has a similar growth habit, but has smoother, heavier seeds. Both lines are high in protein, but C.I. 3947 has substantially more lysine in the protein and a higher nutritional value than C.I. 4362. Kernel weight, protein content, and amino acid composition of C.I. 4362, Hiproly, and average of 113 samples from the USDA Barley World Collection are compared in Table 6.19.

The high lysine, high protein trait markedly reduces the hordein content of barley but the reduction is not as great as the reduction in zein caused by the

TABLE 6.18

Protein Fractions and Their Lysine Content in Some Cereal Grains[a]

Cereal grain	Protein fraction: Soluble in:	Albumin Water[b,c]		Globulin Salt[b,c]		Glutelin Alkali[b,c]		Prolamin Alcohol[b,c]	
Rice		5	4.9	10	2.6	80	3.5	5	0.5
Oats		1	—	78	—	5	—	16	—
Barley (normal)		13	7.9	12	6.3	23	4.8	52	0.8
Barley (Hiproly)		18	8.2	14	6.1	22	4.6	46	0.5
Wheat		5	—	10	—	16	1.9	69	0.6
Maize (normal)		4	3.8	2	6.1	39	3.4	55	0.2
Maize (*opaque*-2)		15	4.1	5	5.2	55	4.7	25	0.1
Sorghum (normal)		8	4.5	8	4.6	32	2.7	52	0.5
Sorghum (160-Cernum)		6	—	10	—	38	—	46	—

[a]From Whitehouse, 1973.
[b]Protein fraction as % of total protein.
[c]Lysine as % of protein fraction.

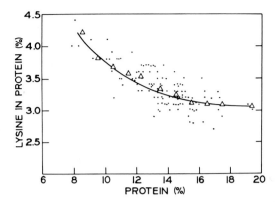

Fig. 6.6. Plot of relationship between percentage of protein and percentage of lysine in protein of 113 barleys from the U.S. Department of Agriculture World Collection. ●, Individual samples; △, averages of percentage of lysine in protein for 1% intervals in protein content. (From Pomeranz, 1975.)

opaque-2 trait in corn. According to Munck (1971), the Hiproly trait increases lysine, aspartic acid, and methionine levels and decreases glutamic acid, cystine, and proline levels. Feeding experiments with rats and mice confirmed the improved nutritional quality of Hiproly lines.

Although naked barley is consumed directly as a food in the Far East, many agrotechnical problems associated with high protein, high lysine barley have yet to be overcome before it becomes a commercially acceptable crop. Similar constraints hold for various corn hybrids with improved nutritional value.

Robbins *et al.* (1971) reported results for a survey of the protein content and amino acid composition of groats for common oat cultivars. The samples contained 12.4–24.4% crude protein (average 17.1%). Chemical analyses indicated that the amino acid composition was superior to that of other cereal grains.

Buckwheat (*Fagopyrum esculentum* Möench) is not a true cereal. It belongs to the Polygonaceae (or buckwheat) family, but, like the cereals, the grain of buckwheat is a dry fruit. The black hulls of the triangular fruit are not suited for human food. Structurally, they have little in common with the bran coats of the cereals. The seed proper (groat) is similar to that of cereals in that it consists of starchy endosperm and oily embryo. Animal feeding experiments have shown that the proteins in buckwheat are a source of high biological value proteins. The proteins of buckwheat have good supplementary value to the cereal grains. Crude protein and 17 amino acids were determined in genetically diverse buckwheats, in buckwheat fractions from a commercial mill, and in the germ and degermed groats (Pomeranz and Robbins, 1972). The buckwheat proteins were particularly rich in lysine (6.1%), and contained less glutamic acid and proline and more arginine and aspartic acid than did cereal proteins. Chemical analyses of the

TABLE 6.19

Kernel Weight, Protein Content, and Amino Acid
Composition[a] of Barleys

Parameter	*C.I. 4362*[b]	*Hiproly*[b]	*World collection*[c]
Kernel weight (mg)	46.7	35.0	43.2
Protein ($N \times 6.25$) (%)	19.3	20.1	13.6
Lysine	3.0	4.3	3.4
Histidine	2.0	2.2	2.1
Ammonia	3.2	2.9	3.3
Arginine	4.0	4.8	4.4
Aspartic acid	5.4	7.4	6.3
Threonine	2.7	3.2	3.1
Serine	3.4	3.7	3.5
Glutamic acid	29.5	25.5	27.4
Proline	15.3	11.9	12.3
Cystine/2	1.1	0.8	1.2
Glycine	3.3	3.8	3.8
Alanine	3.5	4.7	4.1
Valine	4.3	5.1	5.0
Methionine	1.9	2.1	2.6
Isoleucine	3.3	3.6	3.6
Leucine	6.2	6.5	6.5
Tyrosine	2.3	2.3	2.5
Phenylalanine	5.7	5.4	5.2

[a]Grams amino acid per 100 gm amino acids recovered.
[b]From Pomeranz *et al.*, 1974.
[c]From Pomeranz *et al.*, 1976; average of 113 samples from the
USDA Barley World Collection.

buckwheat hydrolysates indicated that the amino acid composition was nutritionally superior to that of cereal grains.

a. Amino Acid Balance The protein reserves in cereal grains are usually made of the following 18 α-amino acids:

1.	Alanine (Ala)	10.	Lysine (Lys)
2.	Arginine (Arg)	11.	Methionine (Met)
3.	Aspartic acid (Asp)	12.	Phenylalanine (Phe)
4.	Cysteine (Cys)	13.	Proline (Pro)
5.	Glutamic acid (Glu)	14.	Serine (Ser)
6.	Glycine (Gly)	15.	Threonine (Thr)
7.	Histidine (His)	16.	Tryptophan (Trp)
8.	Isoleucine (Ile)	17.	Tyrosine (Tyr)
9.	Leucine (Leu)	18.	Valine (Val)

Major systematic tables covering surveys of amino acid contents of foods and feeds were prepared by (a) the U.S. Department of Agriculture (Orr and Watt, 1957); (b) the Commonwealth Bureau of Animal Nutrition (Harvey, 1958); and (c) Revised Tables of the Food and Agriculture Organization of the United Nations (Rao and Odendaal, 1968).

The amino acids Ile, Leu, Lys, Met, Phe, Thr, Trp, and Val are essential in the sense that humans cannot synthesize them *in vivo*. His and Arg are synthesized only partly in the body tissues, usually in amounts insufficient to maintain healthy growth. Cys and Tyr are complementary, respectively, to Met and Phe from which they are formed; insufficient amounts of the former two will deplete the amounts of the latter two.

Several scoring procedures have been recommended as preliminary screening methods for predicting the limiting amino acid or acids and the approximate amino acid balance of a food. The chemical scoring method of Mitchell and Block (1946) does not take into account whether the amino acids are available and whether some of them are in excess to the requirements of the organism. In the Essential Amino Acid Index (EAAI) method of Oser (1951), all essential amino acids are incorporated in the calculations. Hansen and Eggum (1973) developed a model to estimate total amino acid value (TAAV) from amino acid composition. The correlation coefficient between biological value and TAAV for all feedstuffs was 0.74; that coefficient increased to 0.85 when groups of related feedstuffs were analyzed.

Nutritional values of the proteins of wheat, oats, buckwheat, barley, and normal and opaque-2 corn are compared with an egg reference pattern in Table 6.20. Chemical scores for other selected food proteins are given in Table 6.21. The ratio of essential to total amino acids (E : T) in all cereals is lower than that of the egg reference pattern. The E : T ratio in eggs is actually twice as high as would be needed for most efficient use of its essential amino acids. Consequently, the E : T ratio is of limited value in direct evaluation of nutritional adequacy. A more meaningful evaluation can be obtained by calculating the ratio of specific essential amino acids (A) to the total of the essential (TE) amino acids (A : TE).

A comparison of the A : TE ratio with the egg reference pattern yields a chemical score that indicates the limiting amino acids; the lower the value the more limiting the amino acids.

There is presumptive evidence that concentrations of the essential amino acids of egg protein are higher than concentrations required by humans. Isoleucine and methionine concentrations are particularly high, and the use of egg as a reference may overestimate the extent to which those amino acids are limiting and may underestimate the quality of a protein for human use. Consequently, the nutritional value of opaque-2 corn, oat, and buckwheat proteins is probably higher than indicated in the data given in Table 6.20. On the other hand, amino acid

TABLE 6.20

Nutritional Values of Proteins of Cereal Grains Compared with Whole Egg Protein[a]

	Egg reference pattern	Wheat	Oats	Buckwheat	Barley	Corn	
						Normal	Opaque-2
Essential amino acid	3.22	1.99	2.38	E:T values[b] 2.41 A:TE values[c]	2.19	2.65	2.54
Isoleucine	129	122(95)[d]	102(79)	99(77)	105(81)	94(73)	93(72)
Leucine	172	213	194	166	197	328	241
Lysine	125	82(66)	110(88)	158	111(89)	66(53)	116(93)
Tyrosine and phenylalanine	195	243	220	179(92)	208	217	206
Cystine and methionine	107	196	107	106	94(88)	76(71)	81(76)
Threonine	99	93(94)	86(87)	101	97	85(86)	96
Tryptophan	31	41	42	60	40	17(55)	32
Valine	141	150	139	132(94)	148	118(84)	135

[a]Pomeranz, 1975.

[b]Grams essential amino acids per gm total N.

[c]Milligrams specific amino acid per gm of total essential amino acids.

[d]Values in parentheses are A:TE for specific amino acid:A:TE for egg reference pattern × 100. The lowest value under a commodity shows the first limiting amino acid and gives a chemical score.

analyses do not measure one of the most important parameters that determine nutritive value of a food—its digestibility. Chemical scores should be considered primarily as a powerful and convenient screening tool.

While keeping in mind the great potential of cereal grains, one should not forget their limitations. Wheat gluten is a causative agent of some celiac diseases, and it is known to cause allergic reactions through contact, inhalation, and ingestion. Barley causes allergic reactions both in its natural form and as a malt flavoring. Rye also is a known allergen. Oats may be used as a substitute for wheat in diets of allergic individuals. The incidence of allergic reactions to cereal proteins, and especially to those in oats, is low.

2. Legumes

The family Leguminosae includes over 12,000 species in about 500 genera. Of these, only a few are of economic value as foods—peas, beans, lentils, peanuts, and soybeans (Wolff, 1977). Few of the legumes are processed industrially to concentrate the protein fraction. The starch of the mung bean (*Phaseolus aureus*) is used to produce transparent spaghetti-like bean bread; the protein rich residue

TABLE 6.21

Chemical Scores for Selected Food Proteins[a]

Food	Chemical score			Limiting amino acids	Net protein utilization
	Based on FAO pattern 1957	Based on human milk	Based on egg		
Milk (cow's)	80	75	60	Met + Cys	75
Egg	100	90	100	—	100
Casein	80	75	60	Met + Cys	72
Egg albumin	100	80	90	Try	83
Beef muscle	80	80	80	Met + Cys	80
Beef heart	80	80	70	Met + Cys	67
Beef liver	85	85	70	Met + Cys	65
Beef kidney	80	85	70	Met + Cys	77
Pork tenderloin	85	90	80	Met + Cys	84
Fish	70	70	75	Try	83
Oats	80	70	70	Lys	—
Rye	80	90	90	Thr	—
Rice	70	75	75	Lys	57
Corn meal	40	40	45	Try	55
Millet	70	60	60	Lys	56
Kaoliang	70	50	50	Lys	56
White flour	50	50	50	Lys	52
Wheat germ	60	70	65	Met + Cys	67
Wheat gluten	40	40	40	Lys	37
Groundnut flour	60	80	70	Met + Cys	48
Soy flour	70	85	70	Met + Cys	56
Sesame seed	60	50	50	Lys	56
Sunflower seed	70	70	70	Lys	65
Cottonseed meal	70	95	80	Met + Cys	66
Potato	60	85	70	Met + Cys	71
Navy bean	50	50	42	Met + Cys	47
Peas	60	70	60	Met + Cys	44
Sweet potato	80	85	75	Met + Cys	72
Spinach	70	100	90	Met + Cys	—
Cassava	20	50	40	Met + Cys	—

[a]World Health Organization, 1965.

is used to prepare a textured product. The broad bean (*Vicia faba*) can be fractionated into a protein isolate that can be spun into fibrous meat analogs. While field and broad beans are being developed as new protein concentrates and isolates of all legumes, soybean proteins have reached the highest degree of refinement and are added to a variety of foods. The composition of a number of edible legumes are given in Table 6.22. They are rich in protein and have a good balance of essential amino acids (Table 6.23). The various types of soy products that are added to foods are listed in Table 6.24 (see also Chapters 10 and 11).

As stated previously, a major factor limiting the use of soy products is the associated "beany" flavor. Moreover, soybeans, like many other beans, can induce flatulence. The problem arises because indigestible carbohydrates such as raffinose and stachyose pass to the large intestine where they are attacked by microorganisms that produce large quantities of hydrogen and carbon dioxide.

Trypsin inhibitors in raw soybeans are inactivated by moist heat. The hemagglutinins are less well understood but moist heat will inactivate them. The saponins are apparently indigestible and thus pass through the intestinal tract without causing problems. The isoflavones have attracted interest because of their estrogenic activity; whether they represent a significant threat under practical conditions remains to be determined (see also Chapter 10).

Raw soybeans contain a goitrogenic factor, which increases the requirements for iodine. Soybean protein can raise the requirements for vitamin D, vitamin K, vitamin B_{12}, calcium, and phosphorus. Studies with chicks show that isolated soybean protein increases their requirements for manganese, zinc, and copper. These trace mineral difficulties may be associated with the presence of phytic acid in soybean preparations. Since most soy products used for human food are subjected to heat treatment at some stage of their preparation the goitrogenic, vitamin, calcium, and phosphorus effects probably have little significance. Soybean proteins can be allergenic in older children and adults. Ingestion, inhalation of soyflour, and contact can all cause allergic reactions, even if the protein has been exposed to high temperatures.

Heat processed soybean products are considered to contain good quality pro-

TABLE 6.22

Composition of Several Legumes[a,b]

Legume	Protein[c] (%)	Fat (%)	Ash (%)	Fiber (%)	Carbohydrate[d] (%)
Chick peas (Cicer arietinum)	20.6	5.4	2.8	10.3	61
Lentil (Lens esculenta)	29.6	3.1	2.4	3.2	62
Pea (Pisum sativum)	27.9	3.2	2.8	5.9	60
Broad bean (Vicia faba)	31.8	0.9	3.6	8.5	55
Peanut (Arachis hypogaea)	30.0	50.0	3.1	3.0	14
Soybean (Glycine max)	43.9	21.0	4.9	—[e]	30[e]

[a]From Wolff, 1977.
[b]Moisture-free basis.
[c]Kjeldahl N × 6.25.
[d]Measured by difference.
[e]Fiber is included in carbohydrate value.

TABLE 6.23

Essential Amino Acid Contents of Several Legumes[a]

	Chick pea (Cicer arietinum)	Lentil (Lens esculenta)	Broad bean (Vicia faba)	Pea (Pisum sativum)	Soybean (Glycine max)	Peanut (Arachis hypogaea)
Protein content (%)	17.50	24.10	31.8	31.6	61.4	56.9
Amino acids			gm/16 g N			
Arginine	7.98	8.45	10.6	9.2	8.42	—
Histidine	2.57	3.81	2.8	2.5	2.55	—
Isoleucine	4.53	6.30	4.5	4.4	5.10	3.2
Leucine	7.63	10.90	7.7	7.4	7.72	5.9
Lysine	7.72	7.96	7.0	7.7	6.86	3.1
Methionine	1.16	0.70	0.6	1.3	1.56	0.9
Phenylalanine	6.46	6.25	4.3	4.9	5.01	3.8
Threonine	3.86	4.47	3.7	3.8	4.31	2.3
Tryptophan	1.78	1.22	—	1.3	1.28	—
Valine	4.63	5.42	5.2	4.9	5.38	4.1

[a]From Wolff, 1977.

TABLE 6.24

Proximate Analyses and Biochemical Parameters of Soy Flours, Protein
Concentrates, and Protein Isolates[a]

	Full-fat flour[b]	Toasted defatted flour[c]	Protein concentrate[c,d]	Protein isolate[c]
Moisture (%)	3.4	6.5	8.0	4.8
Protein (N × 6.25) (%)	41.0	53.0	65.3	92.0
Crude fat (%)	22.5	1.0	0.3	—
Crude fiber (%)	1.7	3.0	2.9	0.25
Ash (%)	5.1	6.0	4.7	4.0
PER[e]	2.15	2.3	2.3	1.1–1.6
Inactivation of trypsin inhibitors (%)	89	—	—	—
Urease activity, pH change	0.1	—	—	—
Nitrogen solubility index	16	15–25	5	75

[a]From Wolff, 1977.
[b]Experimental sample.
[c]Tech. Serv. Manual, Central Soya Co., Chicago, Illinois.
[d]Prepared by extraction with aqueous alcohol.
[e]PER corrected to casein = 2.5.

tein. The first limiting amino acid is methionine which can be purchased in commercial quantities. The relatively high lysine content has made soybean meal a valuable component of cereal grain diets for monogastric animals such as swine and poultry.

The dry beans most common in the United States are *Phaseolus vulgaris* (navy, pinto, great northern, and kidney beans) and *P. limatus* (lima bean). They are mainly consumed as cooked whole beans (Satterlee, 1981). Dry beans can be fractionated into protein and carbohydrates and the protein fraction can be used as a food protein fortifier and extender. Bean protein concentrates can be produced economically by processing culled beans (wrinkled, split, immature, or off-color). The fractionation can be done by dry methods (sieving and/or air classification; see Chapter 5) or by extraction and isoelectric precipitation. The concentrates are of fair-to-moderate nutritional quality, up to 90% digestible, free of toxic heavy metals and mycotoxins, and generally free of factors that cause flatulence (Satterlee, 1981). Their low level of sulfur-containing amino acids and high lysine content make them well suited for protein enrichment of cereals. They improve the functional properties of both bread and cookies.

About 70% of the total of peanuts produced in the United States are consumed directly as food. Of that amount 55% is peanut butter, 25% roasted peanuts, and 20% candy. In addition, full-fat, partially defatted, and fully defatted peanut grits, flakes, and flours are available commercially. Peanut flours can range from 28% protein in full-fat, to 42% in partially defatted, to 57% in fully defatted flours. Peanut flours are used in cereal-based foods, snacks, meat patties, beverages, ice creams, spreads, frostings, and textured protein products (Satterlee, 1981). Defatted peanut flours, concentrates, or isolates at the 10% level are good functional extenders and retard autooxidative rancidity. Peanut protein concentrates (60–70% protein) are produced by alcohol extraction of carbohydrates, and isolates (at least 90% protein) by an alkaline extraction–acid precipitation procedure. Most oilseeds contain a group of low molecular weight (8,000 to 50,000, 2 S) proteins that are rich in lysine, methionine, and cystine. This group contains 20–25% of the total protein in soybeans but only 5–8% of the peanut proteins. Consequently, the peanut proteins are very deficient in several amino acids and have the lowest nutritional value of common oilseed proteins.

Peanut proteins, and especially isolates, are particularly useful in food formulations in which a bland protein concentrate is needed. They can be used to replace milk proteins (in extended fluid milk products, cheeses, and dry curd milk products), in bread making, and cookie production.

3. Other Protein Sources

Use of sunflower seed proteins has increased in the last decade as a result of breakthroughs in breeding hybrids that contain more than 50% oil and less than 25% hulls. World production of sunflower seed oil ranks second to soybean oil and sunflower seed meal ranks fourth after soy, cotton, and peanut (Gassman, 1983).

Two problems are associated with sunflower protein preparations. The flour causes "off" colors in some foods and the protein isolates are of an unacceptable color as well. Thorough decortication of the seed apparently reduces the undesirable darkening of the protein concentrate, which is due to the oxidation of endogenous phenolic constituents. However, substantial quantities of chlorogenic acid are present in dehulled sunflower seeds. Color remains a problem, but it is encouraging that sunflower protein isolates are free of color problems below pH 7.5. The second concern is that the protein is quite deficient in lysine. This can be overcome by fortification or by using the seeds in combination with proteins containing an excess of this amino acid.

Sunflower seed meals can be extruded alone or in combination with soy meal and used as textured vegetable protein (TVP) to extend beef patties, weiners, and other processed meat products (Satterlee, 1981). Unlike soy proteins, sunflower proteins do not form gels upon heating or treatment with calcium. They form stable foams. The absence of toxic materials in sunflower seed coupled with the very high quality of the oil make this an attractive crop. The protein could be developed for food use while excess material would be readily accepted as an animal feed ingredient.

Rapeseed and mustard proteins are another important oilseed, ranking fifth in world oil production from seeds. They are of particular value because of their worldwide climatic adaptability. A rapeseed protein concentrate contains 57% protein and less than 0.2% glucosinolates, compared with 22% protein and 4% glucosinolates in the intact seed (Satterlee, 1981). Rapeseed glucosinolates are potentially toxic. Once the glucosinolates are removed, the rapeseed proteins have the highest protein nutritional value of plant proteins—a PER of about 3.0. Concentrations of all essential amino acids are above those required by FAO/ WHO recommendations. Textured rapeseed concentrates have better flavor scores and exhibit less shrinkage during frying of extended ground beef than textured soy flour. Adding rapeseed concentrates to weiners improves water and fat binding but impairs texture.

Flaxseed meal is not used in significant amounts in human foods. Small quantities have been used in the preparation of speciality breads and the mucilage extracted from linseed has been used as a mild laxative and to treat peptic ulcers. Linseed meal contains a linamarin, which when degraded by the associated enzyme linase liberates hydrogen cyanide. The enzyme can be inactivated by heating. At present, linseed is not an attractive source of protein for food use, but there is no evidence that its problems could not be overcome.

The soluble potato proteins from starch manufacture can be recovered by direct steam injection to coagulate the proteins. The isolates contain 80–85% protein which has a high biological value (80 compared to 100 in eggs). The protein, however, has the pronounced taste and aroma of cooked potatoes and a gritty texture. To eliminate the undesirable properties in food supplementation, the isolated potato coagulate must be washed to remove the flavor components and ground to a fine powder after drying.

VI. Single Cell Proteins

The idea of single cell protein (SCP) is not new. Torula yeast, obtained from aerobic fermentation of wood waste, was produced and used in Germany as early as World War I. Only recently, however, has protein from single cells been recognized as a source of food supplies independent of agricultural land use. The term "single cell protein" was coined in 1966 by Mateles and Tannenbaum (1968) and is applied to a variety of yeasts, bacteria, fungi, and algae with potential as food or feed ingredients (Tannenbaum, 1977).

Foods of microbial origin are unacceptable to some people who associate microorganisms with disease and decay. On the other hand, yeast has a long and honorable tradition in baking, brewing, and wine making; gourmets pay high prices for truffles, a mushroomlike black fungus; and it is well known that bacterial action changes milk into cheese and yoghurt. Although the industry has been long established for the production of foods (i.e., yeast extracts) and feeds (i.e., torula yeast) of microbial origin, the FAO did not list microbial proteins as a source of unconventional feed, foods, or food additives until after 1955.

Microorganisms have a number of advantages over plants and animals as protein sources (Pomeranz, 1976):

1. They have short generation time. Under optimum conditions, bacteria can double their mass within 0.5–2 hr, yeasts require 1–3 hr, and algae 2–6 hr. The time required to double the mass is 1–5 hr for most microorganisms, 1–4 weeks for most cultivated plants (cereal grains, legumes, tubers), 1 month for poultry, 2 months for swine, and 2–4 months for cattle. Whereas a 500 kg steer produces 0.5 kg of protein per day, 500 kg of bacteria could produce up to 0.5×10^{13} kg of protein per day.

2. Microorganisms can be modified genetically to improve their composition, increase their yield, or develop other desirable properties.

3. The protein content of microorganisms, on a dry-matter basis, is higher than that of most common foods.

4. They can be produced in continuous cultures, are independent of climate, and require small areas and little water.

5. Problems of waste disposal are small.

6. Production of microorganisms can be based on raw materials that are readily available (coal, natural gas) or on waste products that present environmental problems. The latter range from industrial and agricultural wastes to a byproduct of cyclohexane oxidation in nylon manufacture.

Substrates for SCP production are given in Table 6.25. The main requirement for producing protein economically from food wastes is that the waste stream be as concentrated as possible. Digestible solids should be about 1–5% [5,000 to 25,000 ppm biochemical oxygen demand (BOD)]. The volume should be as large as possible; 100,000 to 1 million gallons per day would be adequate. If the waste stream does not meet these requirements, the cost of the product would be prohibitive. In the food industry, it is becoming increasingly important to recover and modify waste. The ultimate aim is to use all the raw material and minimize pollution and loss of so-called waste material. Many streams of so-called waste are actually misplaced resources that are permanently lost. Management must view them as potential sources of salable by-products.

Four groups of microorganisms have been suggested for SCP production: yeasts, fungi, bacteria, and algae. Yeasts generally grow well at 25–40°C and can tolerate high acidity. The species most commonly used in the production of

TABLE 6.25

Substrates for SCP Production

Substrate	Source
Hydrocarbons	Natural gas, crude oil fractions, purified *n*-paraffins
Carbohydrates	Sugars, starch, wood hydrolysates
Alcohols	Methanol and ethanol derived from natural gas and crude oil
Agricultural wastes	Bagasse and pulp, citrus wastes, whey, wood pulp waste
Domestic and industrial wastes	Molasses and other sugar industry wastes, sewage, garbage, food plant wastes
Carbon dioxide	Flue gas, fermentation gas
Recycled animal wastes	Cattle and poultry wastes

yeast food are *Torulopsis utilis* (torula yeast), *Saccharomyces cerevisiae,* and *S. fragilis.* The amino acid composition of torula yeast protein compares favorably with that of the FAO-reference protein, except that it has a lower methionine content.

Fungi grow well over wide ranges of pH, osmotic pressure, temperature, and waste substrates. The methionine and tryptophan contents of fungi are low. The interest in protein obtained from bacteria stems from their high growth rates (e.g., 87 generations per day for *Escherichia coli* and 43 per day for *Pseudomonas fluorescens*) and from their high levels of proteins, which are fairly well balanced in amino acid composition.

Algae are attractive as a potential source of food protein because they use carbon dioxide from the atmosphere and require no carbohydrates. On the other hand, nitrogen is required by algae and cannot be fixed from the air by *Chlorella,* the algae most commonly studied for use as food protein. Large-scale algal cultivation requires an enormous surface area. Still, the effficiency of algal production is attractive. One square meter of an algal culture will yield 7000 gm protein while a good crop of corn yields only 150 gm protein/m². The obstacles to wide acceptance of algae as food are the inefficient digestibility of algal fiber, objectionable flavors, and some gastrointestinal disturbances.

Bacterial SCP is generally comparable to fish meal in that it contains (on a dry-matter basis) 60–70% crude protein; yeast SCP is more like soybean meal, containing 45–55% protein. Mycelial–fungal SCP is somewhat low in protein but can be harvested by simple filtration and does not require costly centrifugation needed for bacteria and yeasts. The chemical composition of microbial cells is affected by the composition of the medium and the cultural conditions. The content of ribonucleic acid increases with growth rate. When nitrogen in the medium is limiting, the cells may accumulate up to 80% fat, on a dry-matter basis. The compositions of some microbial cells are summarized in Table 6.26.

TABLE 6.26

Composition of Microbial Cells[a]

	Filamentous fungi	Algae	Yeast	Bacteria
Nitrogen (%)[b]	5–8	7.5–10.0	7.5–8.5	11.5–12.5
Fat (%)[b]	2–8	7–20	2–6(7–10)	1.5–30(6–20)
Ash (%)[b]	9–14	8–10	5.0–9.5(6–10)	3–7(8–12)
Nucleic acids (%)[b]	variable	3–8	6–12(6–8)	8–16(10–17)

[a]From Kihlberg (1972); values in parentheses from Mauron, 1966.
[b]Percentages on a dry-weight basis.

Only 70–80% of the microbial cell nitrogen is amino acid nitrogen. A considerable part (up to 16%) of the total microbial cell nitrogen is from purine and pyrimidine bases of the nucleic acids and small amounts are from glucosamine, galactosamine, choline, and other cell compounds. Table 6.27 compares the essential amino acid contents of proteins from cereals, eggs, and microbial sources. Compared with egg proteins, microorganisms have a relatively well-balanced amino acid pattern, with the sulfur-containing amino acids in a limiting capacity. Because microorganisms are rich in lysine and threonine, which are low in cereal grains, SCP complements cereal proteins. Microorganisms are rich sources of many vitamins, in particular those from the B group.

TABLE 6.27

Essential Amino Acid Content of Proteins from Eggs, FAO Reference,
Cereals and Microbial Sources[a]

Amino acid	Whole egg	FAO pattern	Barley	Barley spent grain	Oats	A. oryzae on barley	Yeast	Bacterial SCP
Cystine and methionine	5.6	4.2	3.3	3.7	4.1	2.7	1.8–2.7	2.6–3.1
Leucine and isoleucine	15.6	9.0	10.6	13.6	11.3	9.0	12.7–13.8	9.2–10.2
Lysine	6.5	4.2	3.9	3.2	4.2	4.4	6.7–7.6	5.5–6.5
Phenylalanine	5.8	2.8	5.1	5.4	5.3	—	4.5–4.6	2.9–3.5
Threonine	5.1	2.8	3.4	3.0	3.3	3.4	5.0–5.8	4.0–4.1
Valine	7.3	4.2	5.2	5.5	5.3	4.8	4.9–5.9	4.5–4.9
Tryptophan	1.5	1.4	1.4	—	1.6	—	0.8–1.3	0.9–1.0

[a]Expressed in gm/16 gm nitrogen.

A. Safety and Acceptability

Safety and nutritional value are the two major concerns in accepting SCP. Admittedly, microorganisms may produce toxins and sometimes cause nausea and adverse intestinal and skin reactions. However, to say that all SCP is always toxic is as erroneous as to say that it is not. Just as there are toxic and edible mushrooms, there are toxic and nontoxic strains of fungi, yeasts, bacteria, and algae. Actually, toxicity may be a less of a problem than nutritive value. Problems concerning the nutritive value include the indigestiblity of cell wall material and the excessive levels of nucleic acids. The high percentage of nucleic acids in yeast and bacteria is undesirable because in humans it elevates uric acid to levels associated with kidney stones, gallstones, and gout. Consequently, with some organisms extracting nucleic acids may be a necessary step in production of human food. Approval of any novel protein source requires complete information on its toxicity, acceptability, and technological usefulness (incorporation into traditional foods or fabrication of new foods). Evaluation of the product requires appropriate chemical analyses, microbiological examinations, safety evaluations, and studies of protein quality and consumer acceptance.

Much progress has been made in solving SCP problems, though not without creating new ones. There are still several controversial issues on the wholesomeness and the nutritive value of SCP as a food. SCP must undergo stringent evaluation and all reservations on toxic and allergenic effects must be eliminated. Yet even if those reservations are eliminated, two basic problems remain: economics and acceptability. The economics of SCP production will be a key issue in determining its success. The price of SCP products must be able to compete with that of the traditional materials they replace. SCP processing requires large capital investments, and one of the factors affecting processing economics is the size of the operation. To be economically feasible, an SCP manufacturing plant would have to produce at least 100,000 tons of SCP per year. The estimated cost of such a plant would be about 20–70 million dollars.

Other factors that affect the economic feasibility of SCP include the availability and cost of raw materials, water, and land. Economics of SCP production would be markedly improved if microbial cell components other than protein (i.e., nucleic acids) could be used simultaneously or if the microorganisms were grown primarily for the production of some valuable metabolite(s), leaving the cell material as by-product. Thus far, SCP is a rather expensive source of protein. A pound of protein from torula yeast costs about two and one-half times as much as a pound of protein from defatted soy flour and about the same as a pound of protein from a soy protein isolate (see also Chapter 11). To meet the criteria of acceptability, SCP should be unequivocally safe and nutritious and approved by regulatory agencies. It also should have good flavor, texture, and overall consumer acceptance, whether used as a supplement or as an independent food.

REFERENCES

Brekke, G. J., and Eisele, T. A. (1981). The role of modified proteins in the processing of muscle foods. *Food Technol.* **35**(5), 231–234.

Brunner, J. R. (1977). Milk proteins. *In* "Food Proteins" (J. R. Whitaker and S. R. Tannenbaum, eds.), Chapter 7, pp. 175–208. Avi Publ. Co., Westport, Connecticut.

Burrows, V. D., Greene, A. H. M., Korol, M. A., Melnychyn, P., Pearson, G. G., and Sibbald, I. R. (1972). "Food Protein from Grains and Oilseeds—A Development Study Projected to 1980." Report of a Study Group appointed by the Hon. O. E. Lang, Minister Responsible for the Canadian Wheat Board, Ottawa, Canada.

Donovan, J. W., Davis, J. G., and White, L. M. (1970). Chemical and physical characterization of ovomucin, a sulfated glycoprotein complex from chicken eggs. *Biochim. Biophys. Acta* **207**, 190.

Eggum, B. O. (1969). Evaluation of protein quality and the development of screening techniques. *In* "New Approaches to Breeding for Improved Plant Protein." Panel Proceedings, Joint FAO/IAEA Division of Atomic Energy in Food and Agriculture, STI/PUB/212. IAEA, Vienna.

Feeney, R. E., and Osuga, D. T. (1976). Comparative biochemistry of Antarctic proteins. *Comp. Biochem. Physiol. A* **54A**, 281.

Food and Agriculture Organization, United Nations (1969). "Production Yearbook 23." FAO/UN, Rome.

Fossum, K., and Whitaker, J. R. (1968). Ficin and papain inhibitor from chicken egg white. *Arch. Biochem. Biophys.* **125**, 367.

Fothergill, L. A., and Fothergill, J. E. (1970). Thiol and disulphide contents of hen ovalbumin: C-terminal sequence and location of disulphide bond. *Biochem. J.* **116**, 555.

Frey, K. J. (1973). Improvement of quantity and quality of cereal grain protein. *In* "Alternative Sources of Protein for Animal Production," ISBN 0-309-2114-6. Natl. Acad. Sci., Washington, D.C.

Gassman, B. (1983). Preparation and application of vegetable proteins from sunflower seed for human consumption. An approach. *Nahrung* **27**, 351–369.

Gilbert, A. B. (1971). The egg: Its physical and chemical aspects. *In* "Physiology and Biochemistry of the Domestic Fowl" (D. J. Bell and B. M. Freeman, eds.), Vol. 3. Academic Press, New York.

Goll, D. E., Robson, R. M., and Stromer, M. H. (1977). Muscle proteins. *In* "Food Proteins" (J. R. Whitaker and S. R. Tannenbaum, eds.), Chapter 6, pp. 121–174. Avi Publ. Co., Westport, Connecticut.

Graham, G. E., and Kamat, V. B. (1977). The role of egg yolk—lipoprotein in fatless sponge-cake making. *J. Sci. Food Agric.* **28**, 34–40.

Hansen, N. G., and Eggum, B. O. (1973). The biological value of proteins estimated from amino acid analyses. *Acta Agric. Scand.* **23**, 247.

Harvey, D. (1958). "Tables of the Amino Acids in Foods and Feedingstuffs," Tech. Commun. 19. Commonw. Bur. Anim. Nutr., Rowett Inst. Bucksburn, Aberdeenshire, Scotland (Commonw. Agric. Bur., Farnham Royal, Slough, Bucks, England).

Hughes, B. P. (1960). The composition of foods. *Med. Res. Counc. (G.B.), Spec. Rep. Ser.* **297.**

International Dairy Federation (1978). "Characteristics of Products Obtained by Membrane Processes when Applied to Dairy Products," Doc. 106. IDF, London.

Juliano, B. O., Bautista, G. M., Lugay, J. C., and Reyes, A. C. (1964). Rice quality studies on physiochemical properties of rice. J. Agric. Food Chem. 12, 131.

Kamat, V. B., and Hart, C. J. (1973). The contribution of egg yolk lipoproteins to cake structure. F.D. Trade Rev. 43(6), 9–16.

Kent, N. L. (1973). "Technology of Cereals, with Special Reference to Wheat." Pergamon, Oxford.

Kihlberg, R. (1972). The microbe as a source of food. Annu. Rev. Microbiol. 26, 428–466.

Kinsella, J. E., and Srinivasan, D. (1981). Nutritional, chemical, and physical criteria affecting the use and acceptability of proteins in foods. In "Criteria of Food Acceptance," Foster Verlag, Switzerland.

Locker, R. H. (1956). The dissociation of myosin by heat coagulation. Biochim. Biophys. Acta 20, 514.

Lykes, A. H. (1970). "Grain Sorghum in Poultry Nutrition," Tech. Publ. U.S. Feed Grains Council, London.

Marshall, R. D., and Neuberger, A. (1972). Hen's egg albumin. In "Glycoproteins" (A Gottschalk, ed.), Vol. 5B. Amer. Elsevier, New York.

Mateles, R. I., and Tannenbaum, S. R., eds. (1968). "Single-cell Protein." MIT Press, Cambridge, Massachusetts.

Mauron, J. (1966). Plant proteins—a neglected dimension in human nutrition. Int. Z. Vitaminforsch. 36, 362–394.

Miller, D. F. (1958). Composition of cereal grains and forages. Publ. 585. N.A.S.–N.R.C., pp. 1–18.

Miller, H. T., and Feeney, R. E. (1966). The physical and chemical properties of an immunologically cross-reacting protein from avian egg whites. Biochemistry 5, 952.

Miller, M. S., and Clagett, C. O. (1973). Characterization of the glycopeptides of the riboflavin-binding protein. Fed. Proc., Fed. Am. Soc. Exp. Biol. 32, 624.

Mitchell, H. H., and Block, R. J. (1946). Some relationships between the amino acid contents of proteins and their nutritive value for the rat. J. Biol. Chem. 163, 599.

Mosse, J. (1966). Alcohol-soluble proteins of cereal grains. Fed. Proc., Fed. Am. Soc. Exp. Biol. 25, 1663.

Munck, L. (1971). High lysine barley—a summary of the present research development in Sweden. Barley Genet. Newsl. 2, 54.

Munck, L., Karlsson, K. E., Hagberg, A., and Eggum, B. O. (1970). Gene for improved nutritional value in barley seed protein. Science 168, 985.

Orr, M. L., and Watt, B. K. (1957). Amino acid content of food. U.S., Dep. Agric., Home Econ. Res. Rep. 4.

Oser, B. L. (1951). Method for integrating essential amino acid content in the nutritional evaluation of protein. J. Am. Diet. Assoc. 27, 396.

Osuga, D. T., and Feeney, R. E. (1968). Biochemistry of the egg-white proteins of the ratite group. Arch. Biochem. Biophys. 124, 560.

Osuga, D. T., and Feeney, R. E. (1974). Avian egg whites. In "Toxic Constituents of Animal Foodstuffs" (I. E. Liener, ed.). Academic Press, New York.

Osuga, D. T., and Feeney, R. E. (1977). Egg proteins. In "Food Proteins" (J. R. Whitaker and S. R. Tannenbaum, eds.), Chapter 8, pp. 209–266. Avi Publ. Co., Westport, Connecticut.

Penasse, L., Jutisz, M., Fromageot, C., and Fraenkel-Conrat, H. (1952). The determination of carboxy groups of proteins. 2. Carboxy terminal groups of ovomucoids. Biochim. Biophys. Acta 9, 551 (in French).

Pomeranz, Y. (1975). Proteins and amino acids of barley, oats, and buckwheat. *In* "Protein Nutritional Quality of Foods and Feeds" (M. Friedman, ed.), Part 2, pp. 13–78. Dekker, New York.

Pomeranz, Y. (1976). Single-cell protein from by-products of malting and brewing. *Brew. Dig.* **51**(1), 49–55, 60.

Pomeranz, Y., and Robbins, G. S. (1972). Amino acid composition of buckwheat. *Agric. Food Chem.* **20**, 270–274.

Pomeranz, Y., Wesenberg, D. M., Robbins, G. S., and Gilbertson, J. T. 1974. Changes in amino acid composition of maturing Highproly barley. *Cereal Chem.* **51**, 635–640.

Pomeranz, Y., Robbins, G. S., Smith, R. T., Craddock, J. C., Gilbertson, J. T., and Moseman, J. G. (1976). Protein content and amino acid composition of barleys from the world collection. *Cereal Chem.* **53**, 497–504.

Rao, K. K., and Odendaal, P. N. (1968). "New Revised FAO Tables of Amino Acids Content of Foods and Biological Data on Proteins," IDA/68/1. Food Consumption and Planning Branch, Nutr. Div., FAO, Rome.

Robbins, G. S., Pomeranz, Y., and Briggle, L. W. (1971). Amino acid composition of oat groats. *J. Agric. Food Chem.* **19**, 536–539.

Satterlee, L. D. (1981). Protein for use in foods. *Food Technol.* **35**(6), 53, 54, 56, 58, 62, 64, 66–70.

Schmidt, G. R., Mawson, R. F., and Siegel, D. G. (1981). Functionality of a protein matrix in comminuted meat products. *Food Technol.* **35**(5), 235–237, 252.

Sen, L. C., and Whitaker, J. R. (1973). Some properties of a ficin–papain inhibitor from avian egg white. *Arch. Biochem. Biophys.* **158**, 623.

Spinnelli, J., Groninger, H., Jr., Koury, B., and Miller, R. (1975). "Functional Protein Isolates and Derivatives from Fish Muscle," Rep. 5-735. U.S. Dept. of Commerce, Seattle, Washington.

Spinnelli, J., Koury, B., Groninger, H., Jr., and Miller, R. (1977). Expanded uses for fish protein from underutilized species. *Food Technol.* **31**(5), 184–187.

Swaisgood, H. E. (1973). The caseins. *CRC Crit. Rev. Food Technol.* **3**(4), 375–414.

Tannenbaum, S. R. (1977). Single-cell protein. *In* "Food Proteins" (J. R. Whitaker and S. R. Tannenbaum, eds.), Chapter 11, pp. 315–330. Avi Publ. Co., Westport, Connecticut.

Watt, B. K., and Merrill, A. L. (1963). Composition of foods. *U.S., Dep. Agric., Agric. Handb.* **8**.

Whitehouse, R. N. H. (1970). The prospects of breeding barley, wheat and oats to meet special requirements in human and animal nutrition. *Proc. Nutr. Soc.* **29**, 31.

Whitehouse, R. N. H. (1973). The potential of cereal grain crops for protein production. *In* "The Biological Efficiency of Protein Production" (J. G. W. Jones, ed.). Cambridge Univ. Press, London and New York.

Whitney, R. McL., Brunner, J. R., Ebner, K. E., Farrall, H. M., Josephson, R. V., Moor, C. V., and Swaisgood, H. E. 1976. Nomenclature of the proteins of cow's milk. 4th rev. *J. Dairy Sci.* **59**, 785–815.

Wolff, W. J. (1977). Legumes: Seed composition and structure, processing into protein products and protein properties. *In* "Food Proteins" (J. R. Whitaker and S. R. Tannenbaum, eds.), Chapter 10, pp. 291–314. Avi Publ. Co., Westport, Connecticut.

World Health Organization (1965). Protein requirements. *W.H.O. Tech. Rep. Ser.* **301**, 48.

7

Lipids

I. Introduction

Lipids have three important functions in foods: culinary, physiological, and nutritional. Their ability to carry odors and flavors, and their contribution to palatability of meats, tenderness of baked products, and richness and texture of ice cream are examples of culinary functions. Because lipids serve as a convenient means of rapid heat transfer, they have found increasing use in commercial frying operations. Dietary lipids represent the most compact chemical energy available to man. They contain twice the caloric value of an equivalent weight of sugar; they are vital to the structure and biological function of cells. Dietary lipids provide the essential linoleic acid, which has both a structural and functional role in animal tissue; they also provide the nutritionally essential fat-soluble vitamins.

The term "lipid" is used to denote fats and fatlike substances and is synonymous with "lipoids" or "lipins," used in the earlier literature. Lipids are usually defined as food components that are insoluble in water and soluble in organic fat solvents. (Lipids, fats, and oils are used interchangeably in this book.) Lipids are chemical constituents of living organisms, or are derived from such constituents, and most commonly possess fatty acids as part of their moiety. This definition has, however, certain limitations. For example, sterols, squalene, and carotenoids meet the solubility criteria of lipids but contain no fatty acids. On the other hand, gangliosides are soluble in water and alcohol–water mixtures, but insoluble in many organic solvents used to extract lipids from their sources. Despite these limitations, this definition is useful in describing the general characteristic of a class of compounds.

The nomenclature of lipids has been designated by a committee of the Biological Nomenclature Commission, International Union of Pure and Applied Chemistry and the Commission of Editors of Biochemical Journals of the International Union of Biochemistry. The proposed rules concern lipids containing glycerol; sphingolipids; neuraminic acid; fatty acids, long chain alcohols, and amino acid components of lipids; and specific generic terms such as phospholipids and others.

The classification of lipids is difficult because of their heterogeneous nature. The system most commonly used despite its limitation is that proposed by Bloor (1925) as shown in Table 7.1. Molecular structures of major lipid classes are given in Fig. 7.1.

Table 7.2 is shows the classification of lipid groups. Factors that influence the lipid content, extractability, and composition of primary and processed foods are listed in Table 7.3. Fat content and major fatty acid composition of selected foods are summarized in Table 7.4. Table 7.5 shows fatty acid composition of lipids; Tables 7.6 and 7.7 list the physical characteristics of fat raw materials and products.

Foods vary widely in their lipid content and composition. Lard, shortening, and vegetable or animal cooking fats and oil contain almost 100% lipids. The fat content of butter and margarine is over 80%, and of commercial salad dressings 40–70%. Most nuts are rich in lipids (almonds, 55%; beechnuts, 50%; brazilnuts, 67%; cashews, 46%; peanuts, 48%; pecans, 71%; and walnuts up to 64%). The main seeds from which lipids are extracted on a commercial scale include (in addition to peanuts) sesame (50% fat), sunflower (47% fat), hulled safflower (60% fat), and soybeans (18% fat). Among dairy products a wide range of lipids is found. Cottage cheese contains 4%, and cream cheese 38% fat (on an "as is" basis). Fresh fluid cow's milk has 3.7% fat; after drying, 27.5%

TABLE 7.1

Classification of Lipids[a]

Simple lipids—compounds containing two kinds of structural moieties
  Glyceryl esters—these include partial glycerides as well as triglycerides, and are esters of glycerol and fatty acids
  Cholesteryl esters—esters formed from cholesterol and a fatty acid
  Waxes—a poorly defined group which consists of the true waxes (esters of long chain alcohols and fatty acids), vitamin A esters, and vitamin D esters
  Ceramides—amides formed from sphingosine (and its analogs) and a fatty acid linked through the amino group of the base compound. The compounds formed with sphingosine are the most common

Composite lipids—compounds with more than two kinds of structural moieties
  Glyceryl phosphatides—these compounds are classified as derivatives of phosphatidic acid
  Phosphatidic acid—a diglyceride esterified to phosphoric acid
  Phosphatidyl choline—more descriptive term for lecithin which consists of phosphatidic acid linked to choline
  Phosphatidyl ethanolamine—often erroneously called cephalin, a term referring to phospholipids insoluble in alcohol
  Phosphatidyl serine—also erroneously called cephalin
  Phosphatidyl inositol—major member of a complex group of inositol-containing phosphatides including members with two or more phosphates
  Diphosphatidyl glycerol—cardiolipin

Sphingolipids—best described as derivatives of ceramide, a unit structure common to all. However, as in the case of ceramide, the base can be any analog of sphingosine
  Sphingomyelin—a phospholipid form best described as a ceramide phosphoryl choline
  Cerebroside—a ceramide linked to a single sugar at the terminal hydroxyl group of the base and more accurately described as a ceramide monohexoside
  Ceramide dihexosides—same structure as a cerebroside, but with a disaccharide linked to the base
  Ceramide polyhexosides—same structure as a cerebroside, but with a trisaccharide or longer oligosaccharide moiety. May contain one or more amino sugars
  Cerebroside sulfate—a ceramide monohexoside esterified to a sulfate group
  Gangliosides—a complex group of glycolipids that are structurally similar to ceramide polyhexosides, but also contain one to three sialic acid residues. Most members contain an amino sugar in addition to the other sugars. However, not all gangliosides contain amino sugars

Derived lipids—compounds containing a single structural moiety that occur as such or are released from other lipids by hydrolysis
  Fatty acids
  Sterols
  Fatty alcohols
  Hydrocarbons—includes squalene and the carotenoids
  Fat-soluble vitamins, A, D, E, and K

[a]From Bloor, 1925.

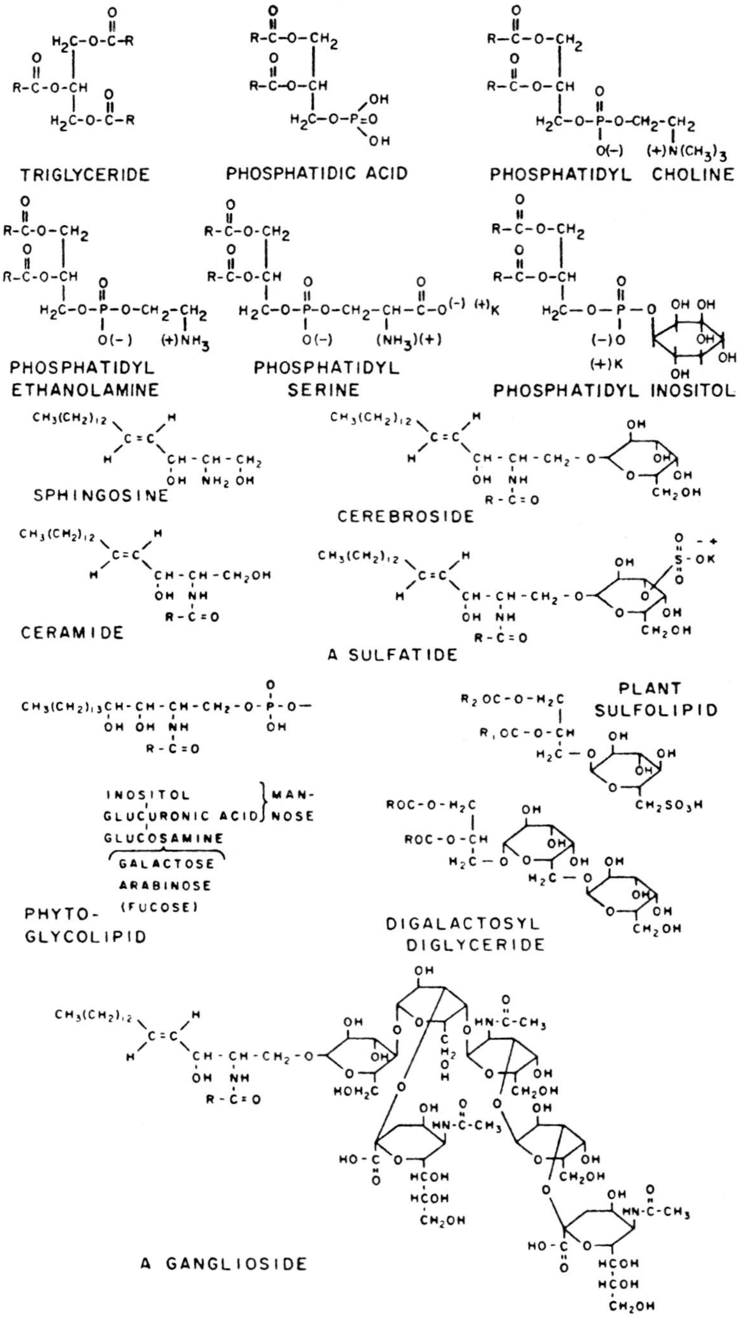

Fig. 7.1. Molecular structures of the major lipid classes. R, a fatty acid residue.

TABLE 7.2

Classification of Lipid Groups[a]

Polarity	Group	Example
Apolar	$H_3C\text{—}(CH_2)_n\text{—}$ $H_3C\text{—}(CH_2)_n$ $\mid$ $R\text{—}CH\text{—}$	Saturated fatty acids and derivatives Cholesterol, Cholesterol esters
	Cycloparaffins	Cholesterol, Cholesterol esters
Polarizable	$H_3C\text{—}(CH_2)_n$ $\mid$ CH $\parallel$ $\text{—}(CH_2)_n\text{—}CH$	Unsaturated fatty acids and derivatives, sphingolipids, carotenoids
Polar (permanent dipole)	$\text{—}C\text{—}O\text{—}$ $\parallel$ O	Triglycerides, cholesterol esters, glycerophosphatides,
	$R\text{—}O\text{—}R^1$	Atherolipids
	$R\text{—}O\text{—}CH$ $\parallel$ $R^1\text{—}CH$	Plasmalogenes
	$R\text{—}OH$	Sphingolipids, mono- and diglycerides, lysophosphatides, hydroxy-fatty acids, inositol phosphatides, sterols
	$\phi\text{—}OH$	Esterogenes
Charged	O $\parallel$ $R\text{—}OSO^-$ $\parallel$ O	Sulfatides
	O $\parallel$ $R\text{—}OSO^-$ $\parallel$ O	Sulfolipids, gallic acids
	O^- $\parallel$ $R\text{—}OPO^-$ $\parallel$ O	Phosphatidic acids
	O^- $\parallel$ $R\text{—}OPO\text{—}R^1$ $\parallel$ O	Glycerophosphatides Sphingomyelins
	$R\text{—}C\overset{\displaystyle O}{\underset{\displaystyle O^-}{\diagup\diagdown}}$	Free fatty acids, gallic acids, phosphatidyl serine
	$\overset{+}{N}$ $\overset{+}{\text{—}NH_3}$	Lecithin, sphingomyelins Phosphatidyl ethanolamine phosphatidyl serine

[a]From Thiele, 1970.

TABLE 7.3

Factors Influencing Lipid Content, Extractability, and Composition
of Primary and Processed Foods[a]

A. Production conditions
  1. Animals, poultry, and fish: breed, species, sex, age, nutrition, season, environment
  2. Plant/vegetable: variety, maturity/ripeness, season, climate
  3. Ingredients used in formulated, imitation, manufactured foods.
B. Processing
  1. Handling prior to processing: crushing, bruising
  2. Types of treatments and methods of processing: aging, blanching, bleaching, canning,
     curing, defatting, degerming, drying, freezing, hydrogenation, milling, nutrification,
     salting, winterizing
C. Storage and packaging conditions: frozen, ambient temperature, relative humidity, packaging
D. Physical state: intact, cut, diced, flaked, ground, milled, minced, mashed, powdered, puffed,
   shredded, sliced, condensed, concentrated, dry, moist, dilute, batter, dough, etc.
E. Cooking method: cooking medium, water/oil, baking, boiling, broiling, braising, deep-fat
   frying, oven frying, breading and cooking, microwave, etc.
F. Designation of portion/cut analyzed
  1. Plant portion: flowers, leaves, stems, roots, pulp with/without seeds, pod with/without
     seeds, seeds, kernels, kernels plus testa, kernels with/without bran layer, etc.
  2. Animal, poultry, and fish: meat cut—typical cuts of beef, pork, sheep with extent of
     trimming; poultry portion—light/dark meats, with/without skin and/or adipose, deboned
     portions, etc.; fish—whole/fillet, shellfish with/without shell, etc.

[a]From Kinsella *et al.*, 1975.

TABLE 7.4

Fat Content and Major Fatty Acid Composition of Selected Foods[a,b]

| | | *Fatty Acids[c]* | | |
| | | | *Unsaturated* | |
Food	*Total fat* (%)	*Saturated[d]* (%)	*Oleic[e]* (%)	*Linoleic[f]* (%)
Salad and cooking oils				
Safflower	100	10	13	74
Sunflower	100	11	14	70
Corn	100	13	26	55
Cottonseed	100	23	17	54
Soybean[g]	100	14	25	50
Sesame	100	14	38	42

TABLE 7.4 (*Continued*)

Food	Total fat (%)	Saturated[d] (%)	Oleic[e] (%)	Linoleic[f] (%)
			Unsaturated	
Soybean, specially processed[h]	100	11	29	31
Peanut	100	18	47	29
Olive	100	11	76	7
Coconut	100	80	5	1
Vegetable fats–shortening	100	23	23	6–23
Table spreads				
Margarine, first ingredient on label[i]				
Safflower (liquid)—tub	80	11	18	48
Corn oil (liquid)—tub	80	14	26	38
Corn oil (liquid)—stick	80	15	33	29
Partly hydrogenated or hydro-				
genated fat	80	17	44	14
Butter	81	46	27	2
Animal fats				
Poultry	100	30	40	20
Beef, lamb, pork	100	45	44	2–6
Fish, raw[j]				
Salmon	9	2	2	4
Tuna	5	2	1	2
Mackerel	13	5	3	4
Herring, Pacific	13	4	2	3
Nuts				
Walnuts, English	64	4	10	40
Walnuts, black	60	4	21	28
Brazil	67	13	32	17
Peanuts or peanut butter	51	9	25	14
Pecan	65	4–6	33–48	9–24
Egg yolk	31	10	13	2
Avocado	16	3	7	2

[a]U.S. Department of Agriculture, 1974.

[b]In decreasing order of linoleic acid content within each group of similar foods.

[c]Total is not expected to equal "total fat."

[d]Includes fatty acids with chains from 8 to 18 carbon atoms.

[e]Monounsaturated.

[f]Polyunsaturated.

[g]Suitable as salad oil; not recommended as cooking oil.

[h]Does not include the isomers of oleic or linoleic acid for which nutritional significance has not been established.

[i]Does not include small amounts of monounsaturated and diunsaturated fatty acids that are not oleic or linoleic.

[j]Linoleic acid includes higher polyunsaturated fatty acids.

TABLE 7.5

Typical Fatty Acid Composition of Fats and Oils[a]

Acid	Common designation	Butter fat (%)	Beef tallow (%)	Lard (%)	Herring oil (%)	Coconut (%)	Palm kernel (%)	Palm (%)	Corn (%)	Cotton (%)	Soya (%)	Cocoa butter (%)	Peanut (%)
Saturated													
Butyric	C4:0	3	—	—	—	—	—	—	—	—	—	—	—
Caproic	C6:0	1	—	—	—	—	—	—	—	—	—	—	—
Caprylic	C8:0	1.5	—	—	—	8	4	—	—	—	—	—	—
Capric	C10:0	3	—	—	—	7	4	—	—	—	—	—	—
Lauric	C12:0	4	—	—	—	48	50	—	—	—	—	—	—
Myristic	C14:0	12	2	1	8	18	16	1	—	1	—	0.5	—
Palmitic	C16:0	25	35	23	12	8.5	8	23	12	24	11	25	11
Stearic	C18:0	9	16	9	—	2.3	2.5	4	2	2	4	35	3
Arachidic	C20:0	—	—	—	—	—	—	—	—	—	—	—	2
Behenic	C22:0	—	—	—	—	—	—	—	—	—	—	—	4
Monounsaturated													
Palmitoleic	C16:1	4	—	—	—	—	—	—	—	—	—	—	—
Oleic	C18:1	—	44	46	—	6	12	37	27	18	25	37.5	46
Diunsaturated													
Linoleic	C18:2	—	2	14	21	2	3	10	57	54	50	2	31
Linoleic	C20:2	—	—	—	28	—	—	—	—	—	—	—	—
Triunsaturated													
Linolenic	C18:3	—	0.4	1	23	—	—	0.3	1	—	8	—	1
Iodine no. (typical)		30	50	73	140	9	17	50	125	110	130	35	98
Slip mp (°C)		27–35	37–43	31–43	—	24	26	40–43	—	—	—	31–33	—

[a] Reproduced with permission from Manley, 1983.

TABLE 7.6

Physical Characteristics of Typical Fat Products[a]

Product	Solid fat index at °C					Melting point, °C (capillary)	Consistency (bloom) @ 23.9°C
	10.0	21.1	26.7	33.3	37.8		
Shortenings							
Cake and icing	28	23	22	18	15	51.1	40
Cake mix	40	31	29	21	15	47.8	75
Coating fat, winter	65	55	45	19	1	39.4	Hard and brittle
Coating fat, summer	67	58	51	31	18	47.8	Hard and brittle
Frying	44	28	22	11	5	42.8	70
Pie crust	33	28	22	10	8	47.8	70
Yeast dough	26	20	12	6	3	44.4	50
Margarines							
Table (premium)	24	12	8	2	0	36.7	15
Table (regular)	28	16	12	3	0	37.8	25
Cake	29	19	17	11	7	44.4	40
Pastry, roll-in	25	21	20	18	15	50.0	80
Puff paste	28	25	24	22	19	51.1	110

[a]From Weiss, 1963.

fat. The fat content of cream ranges from 20% in light coffee cream to 38% in heavy whipping cream. Ice cream contains about 12% fat. A very wide range in fat content is encountered in cereal products: grains contain only 3–5% (but the germ has about 10%), bread 3–6%; most cookies 14–30%, and crackers from 12% in saltines to 24% in chocolate-coated graham crackers. Raw beef carcass trimmed to retail grade contains 16–25% fat; sausages 15–50%; total edible hens

TABLE 7.7

Physical Characteristics of Food Fats—Raw Materials[a]

Fat	Solid fat index (°C)					Melting point, °C (capillary)
	10.0	21.1	26.7	33.3	37.8	
Butter	32	12	9	3	0	36.1
Cocoa butter	62	48	8	0	0	29.4
Coconut oil	55	27	0	0	0	26.1
Lard	25	20	12	4	2	43.3
Palm oil	34	12	9	6	4	39.4
Palm kernel oil	49	33	13	0	0	28.9
Tallow	39	30	28	23	18	47.8

[a]From Weiss, 1963.

and cocks 25%; and herring 11%. The fat ranges from 4% in pink to 16% in chinook salmon. Raw tuna fish contains only 4% fat, but when canned in oil, 21%. The whole edible portion of eggs contains 12% lipids, the yolks alone 29%; after drying the fat content increases to 41% in commercial dried whole eggs, and to 57% in dried yolks. Most fruits and vegetables contain small amounts of lipids (especially when expressed on an ''as is'' basis), but avocado contains 16% lipids, and the lipid content ranges from 10% in giant pickled olives to 36% in salt-dried, oil-coated Greek-style olives. Sweet chocolate contains 35% fat, and bitter or baking chocolate 53%; in dry cocoa powders the fat ranges from 8% in low fat powder to 24% in high fat or breakfast types.

In this chapter, the two general sections on modification of oils and fats and the effects of processing on nutritive values of lipids are followed by four sections on lipids in specific foods: cooking oils, salad oils, and salad dressings; muscle lipids; lipids in cereal products; and lipids in bread making.

II. Modification of Oils and Fats

Oils and fats contain mainly triglycerides (TG), esters of fatty acids, and the trihydric alcohol: glycerol, as shown by the following diagram in which ●, carbon; ○, hydrogen, and ◉, oxygen (courtesy of J. W. E. Coenen). In addition to TG, fats contain a number of minor components such as phosphatides, sterols, steryl esters, fat-soluble vitamins A and D, and tocopherols (which act as antioxidants and have vitamin E activity). Fatty acids in TG differ in chain length from 4 to 24 carbon atoms and in the number of olefinic double bonds from 0 to 6. All natural oils contain significant proportions of at least four fatty acids. This yields 40 different combinations on the three positions of the glycerol molecule, or 40 TG with different chemical and physical properties (Coenen, 1974).

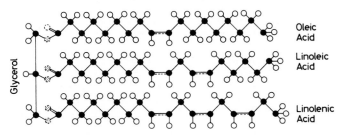

Caloric value of most natural TG shows only small differences. Linoleic acid is of prime importance as an essential fatty acid (EFA) because it cannot be synthesized in the human body and is indispensable for proper functioning. Relatively high levels of linoleic acid in the diet reduce hazards of atherosclerosis; high levels of saturated and probably trans unsaturated fatty acids appear to increase the need for EFA and have the undesirable effect of exacerbating atherosclerosis. Erucic acid, a C_{22} acid with one double bond, occurs in high concentrations in rapeseed oils and has a highly undesirable effect in terms of fatty infiltration and fibrosis of the heart

The flavor of oils is due to small concentrations of volatile breakdown products formed by autocatalytic oxidation of nonvolatile TG. Slight autooxidation may impart unique and delectable flavors (as in olive oil) but generally such flavor is objectionable. This action is termed "reversion" in early stages and "oxidative rancidity" in advanced stages.

Oxidation of fatty acids takes place primarily on methylene groups activated by adjacent double bonds. Consequently, oxidation rate increases with the number of double bonds in the fatty acid chain and with the number of adjacent double bonds, as in polyunsaturated fatty acids. Oxidation is further enhanced by the presence of traces of heavy metals such as copper, iron, and manganese. Oils that are rich in linoleic acid (such as soy, rapeseed, linseed) are less stable than peanut oil, cottonseed, sunflower, palm, and coconut oils. Marine oils with more than three double bonds are most susceptible to oxidation.

Melting and Crystallization

Oils and fats have a melting range rather than a melting point. At room temperature, fats for human consumption are in part liquid and in part crystallized. Component fatty acids of TG govern the extent of crystallization. Cis double bonds and short chain fatty acids lower the melting points. Three factors affect fat structure (Coenen, 1974). First, of the three polymorphic forms (α, β, and β'), β is the densest and the most stable thermodynamically. Second, the more pronounced the mixed crystal formation, the greater the similarity of structure. Third, fast cooling enhances formation of small crystals, which are conducive to a smooth texture. Partial melting during handling or storage may initiate recrystallization and result in increased hardness, a coarse grain, and oil exudation. Fats with a short melting range that completely melt at less than body temperature cause a pleasant cooling effect in the mouth. This is desirable in margarines and essential in confectionery fats used in chocolate and couverture production. The phenomenon is most pronounced in cocoa butter. In all those products, resistance to recrystallization is essential.

Few natural oils and fats meet the stringent requirements of processors and consumers. To meet those requirements various types and degrees of fat modification are available: (a) physical compounding and fractionation without modification of either TG or fatty acids; (b) random or directed interesterification involves reshuffling of fatty acids without modification of the fatty acids; and (c) hydrogenation and isomerization involves modification of fatty acids by elimination, cis–trans isomerization, or shift in double bonds without modification of ester bonds in the triglyceride molecule (Coenen, 1974). All three types of modification influence crystallization and melting properties but only hydrogenation has a significant effect on flavor.

Fractionation can be used to remove an undesirable fraction in order to upgrade the material or to actually isolate a desirable fraction. Examples of removal include treatment of tallow to eliminate fully saturated TG and to produce a lower melting point tallow olein for use in margarine. Another example of elimination is removal of cottonseed stearin or sunflower high melting point waxes through "winterization" (fractionation after storage at low temperatures). Palm oil can be fractionated into a liquid part for use as a frying oil and a stearin part for use in margarines and bakery fats. In addition, fractionation processes can be used to obtain from palm oil a confectionery fat compatible with cocoa butter and different products with narrow and well-defined melting points and textures.

Random esterification is used to modify the solid content index curves of mixtures of TG when a high proportion of high melting glycerides is desirable or to improve resistance to recrystallization. Directed interesterification is used to modify the original TG structure in such a manner that propensity to crystal growth is eliminated.

Hydrogenation is the most drastic modification. Most significantly it alters oil properties; at the same time it makes possible utilization of marine oils. Hydrogenation lowers amounts of EFA and increases levels of saturated and trans unsaturated fatty acids. This may be undesirable from a nutritional standpoint. On the other hand, selective hydrogenation (a) makes it possible to produce from unstable oils stable products with reasonable levels of EFA; (b) eliminates hazards of high degrees of polymerization of deep-fat frying oils; (c) increases stability, (d) eliminates nutritionally objectionable oil components, such as cyclopropenoid sterulic and malvalic acids in cottonseed oil; and (e) makes possible production of a variety of tailor-made products with excellent functional properties.

III. Processing Effects on Nutritive Value
of Lipids

Deteriorative changes in lipids as a result of handling, processing, or storage may be lipolytic, oxidative, polymeric, or degradative. They may involve molecular changes such as saturation of double bonds, changes of position or geometry of double bonds, and changes in fatty acid distribution among the glyceride molecules (Dugan, 1968). Such molecular changes modify the physical properties and utility of lipids and foods containing the lipids. The modifications affect digestibility and energy values and may result in depression of weight and growth; in extreme cases they may cause toxic and carcinogenic effects when ingested. Oxidative rancidity seems responsible for most of the quality and nutritional losses. It involves oxidation at the double bonds on the α-carbon and formation of hydroxy peroxides. The peroxides undergo scission and dismutation reactions to form a wide spectrum of carbonyls, hydroxy compounds, and short chain fatty acids. Hydroxy peroxides have little effect on color, odor, or flavor. Their degradation products, however, lower quality. From a nutritional standpoint, hydroxy peroxides reduce the EFA content; destroy carotenes, vitamin A, and tocopherols; and lower the nutritional value of proteins. In addition to hydroxy peroxides, and their degradation products, oxidized fats contain cyclic and polymeric compounds. They have been shown to depress growth and accelerate deficiency effects of diets marginal in vitamins and protein.

The formation of positional and geometric isomers by hydrogenation of polyunsaturated fatty acids represents a loss of essential fatty acids and in nutritive value. Those isomers, however, may be utilized as a source of energy and inhibit little, if any, growth of test animals when the quantity of EFA in the diet is adequate (Dugan, 1968). Polymer formation is accompanied by increases in viscosity; the increases are greatest with a rise in the unsaturation of the system. Polymers from oxidized cottonseed oil depress growth more than polymers from lard or hydrogenated cottonseed oil. The linolenates form more cyclic compounds than the linoleates.

In general, major changes in nutritional value are reflected in decreases in consumer acceptance. Conversely, processing or use treatments of lipids that result in the most favorable organoleptic properties are also the most favorable from a nutritional standpoint (Dugan, 1968).

IV. Cooking Oils, Salad Oils, and Salad Dressings

Lipids used in the preparation of foods can be divided into two classes based on their consistency at room temperature. The first class includes liquid oils from soybeans, cottonseed, peanuts, olives, or corn. Solids or semisolids, such as shortening, margarine, lard, and coconut oil, make up the second class. The significance of consistency varies with foods. In some it is critical; for example, in salad dressings liquid oil is needed; for margarine, a semisolid consistency is desired. Liquid oils, in general, are suitable for cooking except for those foods that require the production of specific structures, for example, certain cakes or pastries. Liquid oils are unsuitable for the production of plastic products such as cream icings and fillings (Swern, 1982). Oils are used in household cooking and in commercial deep-fat frying of foods that are consumed immediately after frying. Liquid fats are used for frying such foods as potato chips or corn chips that are to be stored for a considerable length of time. Doughnuts require plastic fats for frying, however, because liquid oils give them a greasy appearance.

Cooking oils, with the exception of some marine oils, are of vegetable origin. Marine oils are highly unsaturated and are commonly used after hydrogenation. Generally, bland neutral oil is produced from crude oil of vegetable origin by refining, bleaching, and deodorizing. Salad oils are oils that remain essentially liquid at 4.4°C. To prevent crystallization and turbidity, well-winterized oils and crystallization inhibitors are used. Crystallization inhibitors include lecithin, oxystearin, and mono-, di-, and polysaccharide esters of hydroxy fatty acids.

Salad oils can be prepared from sunflower, safflower, and corn oils after dewaxing. Cottonseed oil is rich in palmitic acid and must be winterized. It is practically impossible to prepare a good salad oil from peanut oil (because of the noncrystallinity of its higher melting point fraction) or from palm oil (because it is rich in palmitic acid). From the standpoints of physical properties and flavor olive oil is uniquely suitable as a cooking and salad oil. Analytical characteristics of typical commercial samples of salad and cooking oils are given in Table 7.8.

As mentioned previously, during storage or cooking bland oils undergo chemical or physical changes because they are sensitive to heat and light. Such changes are catalyzed by trace metals. The primary change that takes place is a reaction of unsaturated sites of the oils with oxygen. At low levels of oxidation flavor reversion takes place. For example, refined soybean oil develops a "beany" or "grassy" flavor. At higher levels, objectionable oxidative flavors are formed. Shelf stability of salad oils depends on many factors, including processing, the nature and amounts of natural and added antioxidants, and storage conditions. The presence of large amounts of linoleic acid generally decreases stability. Cooking oils undergo thermal and degradation changes. Thermally induced oxidative changes result in polymer formation. Such changes are important from nutritional and consumer acceptance points of view.

Most oils contain natural antioxidants, mainly tocopherols. They are often supplemented by synthetic antioxidants, such as butylated hydroxyanisole, butylated hydroxytoluene, tertiary butylhydroquinone, propyl gallate, and ascorbyl palmitate. In addition, citric acid is added to chelate trace metals that accelerate oxidation.

Sucrose polyesters with six or more hydroxy groups esterified with unsaturated fatty acids have the characteristics of conventional cooking and salad oils and affect neither consumer acceptance nor health attributes of products to which they are added. They are not absorbed in the digestive tract and are, therefore, low-calorie substitutes in production of nutrition-oriented salad and cooking oils.

Salad dressings are divided into spoonable (i.e., mayonnaise) and pourable (i.e., French dressings). They have gained wide acceptance as replacements for

TABLE 7.8

Analytical Characteristics of Typical Commercial Samples of Salad and Cooking Oils[a]

	Salad oils				Cooking oils	
	Olive	Corn	Soybean	Winterized cottonseed	Cottonseed	Peanut
Iodine number	85	125	130	112	108	95
Refractive index at 60°C	1.4546	1.4598	1.4600	1.4577	1.4572	1.4550
Free fatty acids, % as oleic	1.5	0.05	0.02	0.02	0.03	0.03
Smoke point (°C)	—	232	232	232	227	227
Keeping quality, AOM hr[b] to peroxide value of 125	20	10	8	10	10	12
Color (Lovibond)[c]	—	40 Y–4.0 R	15 Y–1.8 R	20 Y–2.0 R	20 Y–2.5 R	25 Y–2.0 R
Cold test (hours to cloud at 0°C)	>24	Does not cloud	Does not cloud	20	<1	<1
Cloud point, ASTM (°C)[d]	-5.6	-11.1	-10.0	-4.4	3.3	4.4
Solid point, ASTM (°C)[d]	-10.0	-13.3	-12.2	-6.7	- 2.8	1.1
Titer (°C)	23.0	18.5	21.5	33	36.0	31.3

[a]From Swern, 1982.
[b]Active oxygen method.
[c]Y, yellow; R, red.
[d]Modified by examination of sample at intervals of 1.1°C rather than 2.8°C.

TABLE 7.9

Approximate Composition of Mayonnaise[a]

Ingredient	% by weight
Oil	75.0–80.0
Vinegar (4.5% acetic acid)	10.8–9.4
Egg yolk	9.0–7.0
Sugar	2.5–1.5
Salt	1.5
Mustard	1.0–0.5
White pepper	0.2–0.1

[a]From Swern, 1982.

butter or margarine in sandwich spreads. They have a pleasing flavor, blend well with expensive ingredients to form fillings, and are easily spread, handled, and stored. Mayonnaise is a semisolid food prepared from edible vegetable oil, egg yolk or eggs (fresh, frozen, or dried), vinegar, lemon and/or lime juice, and one or more of the following: salt, sweetener, mustard, paprika, monosodium glutamate, and other seasonings. The finished product contains at least 65% vegetable oil. Mayonnaise is an oil-in-water emulsion, the consistency of which depends to a large extent on the ratio of the aqueous and oil phases and the amount and type of egg solids. Approximate composition of mayonnaise is listed in Table 7.9. Mayonnaise is a semiperishable product. With age it becomes thinner; mechanical shock and temperature extremes accelerate thinning and separation of phases.

Pourable dressings are similar to mayonnaise, except that they contain less oil (minimum 30%). They may contain starch pastes as thickeners, egg yolk solids at a level equivalent to 4% liquid egg yolk, or gums as emulsifiers and dioctyl sodium sulfosuccinate (up to 5% of the weight of the gum).

V. Muscle Lipids

Lipids in avian, aquatic, and mammalian tissues vary widely in quantity and composition. Muscle lipids are associated with, and to a large extent govern, the material's processing and end product properties. These properties include flavor, color, stability, texture, juiciness, protein stability, frozen storage shelf life, emulsion characteristics, and caloric content (Allen and Foegeding, 1981). The potential of lipid oxidation and the development of rancid flavors is one of the constraints in the processing and shelf life properties of muscle foods. On the other hand, lipids and lipid-soluble components are critical to pleasing appearance, texture, flavor, and overall consumer acceptance.

The phospholipid and cholesterol muscle lipids are essential because of their role in the structure of the muscle cell and its organelles and their function. The neutral lipids provide fatty acids for energy metabolism and contribute to the characteristics of the meat. Most of the phospholipid and cholesterol is membrane-associated; neutral lipid is present as microscopic droplets within the muscle cell or in fat cells. As these cells become more numerous, the fat becomes more visible in the muscle cross-section, exhibiting the phenomenon called "marbling."

The content and composition of muscle cells differ within an animal depending on the muscle function (Allen and Foegeding, 1981). Within a species, "light" meat contains less lipid than "dark" meat. This difference is also reflected by the percentages of total lipids that include neutral and phospholipids (Table 7.10). The more aerobic metabolism of red or dark muscle compared with white or light muscle is associated with higher myoglobin levels and higher lipid concentrations; it results in differences in flavor and caloric content.

The fatty acid composition of muscle differs among species. The differences represent significant variables in processing, palatability, and storage characteristics of muscle foods. In all of the nonruminant species dietary fatty acids are important in determining the fatty acid composition of the tissue. Among rumi-

TABLE 7.10

Influence of Muscle and Lipid Content on Quantity of
Intramuscular Neutral Lipids and Phospholipids[a]

Species	Muscle or type	Lipid (%)	Content Neutral lipids (%)	Phospholipids (%)
Chicken	White	1.0	52	48
	Dark	2.5	79	21
Turkey	White	1.0	29	71
	Dark	3.5	74	26
Fish (sucker)	White	1.5	76	24
	Dark	6.2	93	7
Beef	Longissimus dorsi	2.6	78	22
		7.7	92	8
		12.7	95	5
Pork	L. dorsi	4.6	79	21
	Psoas major	3.1	63	37
Lamb	L. dorsi	5.7	83	10
	Semitendinosus	3.8	79	17

[a]From Allen and Foegeding, 1981. Copyright by Institute of Food Technologists.

nant species, it is possible to produce polyunsaturated milk and meat products by feeding the livestock encapsulated oilseed supplements. Data in Table 7.11 summarize differences in unsaturated fatty acid composition of muscle tissue lipids. Fatty acids in fish are highly unsaturated (20–32%). Other species normally have less than 20% polyunsaturated fatty acids in the total intramuscular lipid. Most of the fatty acid is linoleic acid, which is less susceptible to oxidation than fatty acids containing up to six double bonds. Fatty acids in ruminant tissues are the most saturated because of hydrogenation by microorganisms in the rumen. The phospholipids of beef and lamb contain higher percentages of fatty acids with four or more double bonds than pork, but less than chicken. In chicken, beef, pork, and lamb, the neutral or nonpolar lipids contain about 40–50% monoenoic fatty acids and less than 2% of the most highly unsaturated fatty acids. Many of the "off" flavors associated with oxidative rancidity are caused by modifications of phospholipids (Allen and Foegeding, 1981).

According to Hornstein *et al.* (1961), phospholipids do not contribute to desirable meat flavor and in very lean meat may actually cause poor flavor. Some phospholipids (i.e., phosphatidyl serine and phosphatidyl ethanolamine) in beef and pork muscle may produce on oxidation strong "fishy" flavors. Skeletal muscle phospholipids in nonruminant and ruminant species differ only slightly.

One of the problems associated with muscle proteins is their sensitivity to oxidation and associated effects on meat color, functionality, and flavor. The oxidation is governed by modifications of lipids. Lipid oxidation accelerates oxidation of ferrous iron in myoglobin and produces metmyoglobin, a brown-gray pigment that decreases consumer appeal. Hemoprotein and nonheme iron components of meat catalyze lipid peroxidation. Lipid oxidation is a limiting factor in maintaining the quality of mechanically deboned meat. Similar problems are associated with lipid oxidation in mechanically deboned fish. Storage stability of mechanically deboned muscle foods depends on qualitative and quantitative meat composition, processing, equipment, and storage conditions.

Factors that govern the melting point or range of fats are important variables in the manufacture and stability of emulsified meat products. Interaction of lipids with proteins and other muscle components has many implications with regard to characteristics of muscle foods. During the storage of fish, free fatty acids in the muscle tissue increases. This causes protein denaturation and impairs texture and water-holding capacity. The rise in free fatty acids is due to phospholipase A and a lysosomal lipase. Lipolysis is more rapid in dark than in white muscle and triacylglycerol is oxidized more rapidly than phospholipid. Slow freezing or fluctuating temperatures accelerate the release of acid lipase from lysosomes and thereby raise free fatty acid levels in the tissue. During cooking, the fat in meat softens or liquefies and some fat may be lost. At the same time, aromatic compounds are volatilized and the lipids may undergo hydrolytic or oxidative changes. When cooked meat (with the exception of cured meats) is refrigerated

TABLE 7.11

Composition by Class of Intramuscular Unsaturated Fatty Acids as Influenced by Species, Muscle, and Type of Lipid[a]

Species	Muscle source	Lipid type	Composition			
			Monoenoic (%)	Dienoic (%)	Trienoic (%)	>Trienoic (%)
Fish	Marine	Total	20.6–41.8	3.6–6.00	2.1–4.2	8.8–25.0
	Freshwater	Total	23.0–36.2	3.1–11.0	0.7–6.0	8.3–27.7
	Brackish	Total	23.7–37.0	0–6.0	1.3–4.7	14.6–17.3
Fish	White	Total	32.0	3.3	2.8	35.8
	Dark	Total	46.1	4.9	4.9	22.1
Chicken	White	Neutral	41.0	25.5	1.6	0.5
	White	Phospholipid	18.9	19.1	2.7	23.7
	Dark	Neutral	42.2	25.5	1.7	0.5
	Dark	Phospholipid	16.9	21.5	2.1	24.1
Beef	Longissimus dorsi	Triglyceride	49.0	3.0	0.5	0
	L. dorsi	Mono- and diglyceride	45.2	3.2	0.7	0
	L. dorsi	Free fatty acids	33.3	9.3	0.6	5.5
	L. dorsi	Phospholipid	20.0	28.4	2.0	14.0
Pork	L. dorsi	Total	24.4	9.9	1.2	1.0
	L. dorsi	Neutral lipid	51.6	7.7	1.2	0.3
	L. dorsi	Phospholipid	27.2	25.8	0.2	8.2
	Loin	Phosphatidyl choline	19.5	7.1	1.0	0.2
	Loin	Phosphatidyl ethanolamine	20.7	24.7	—	3.0
	Loin	Phosphatidyl serine	18.2	11.3	1.3	7.6
Lamb	L. dorsi	Nonpolar lipid	49.3	3.7	0.8	0
	L. dorsi	Phospholipid	24.5	19.1	0.6	13.6
	L. dorsi	Glycolipid	26.2	29.7	0.8	4.6

[a]From Allen and Foegeding, 1981. Copyright by Institute of Food Technologists.

and then reheated, "warmed-over" flavors result. This condition involves oxidation of intramuscular phospholipids and may be retarded by the presence of natural antioxidants in fresh meat.

VI. Lipids in Cereal Products

A. Wheat Starch Lipids

About 1.1% of wheat starch granules are lipids, which are listed in Table 7.12. About 75% of the starch lipids are lysophosphatides, with lysophosphatidyl choline at 60% as main component. No lecithin is present and lysolecithin should be considered a native, integral starch granule component. Among the starch fatty acids, palmitic acid amounts to 56% of the total, compared with 60% linoleic acid in nonstarch fatty acids. The free fatty acids of starch are 71% saturated. The OH-group in starch lysolecithin is about 75% in the β-position and 25% in the α-position.

Lipids are assumed to be present as inclusion compounds in the starch granule (see Chapter 2) and to be basically unavailable to affect dough processing prior to starch gelatinization. They can, however, influence pasting characteristics (as

TABLE 7.12

Wheat Starch Lipids

Lipid type	Content (% of total lipids)
Lysophosphatides	
Choline	62.3
Ethanolamine	8.5
Serine	4.2
Inositol	1.5
Free fatty acids	11.6
Mono- and diglycerides	0.7
Triglycerides	0.8
Sterols	1.1
Sterol esters	0.7
Galactolipids	1.0
Miscellaneous	7.6
(hydroxy-fatty acids,	
hydrocarbons, unidentified)	

[a]From Acker et al., 1967.

TABLE 7.13

Starch Lipids[a]

Lipids	Percentage of total			
	Wheat	*Rye*	*Barley*	*Oats*
Lysolecithin	62.3	51.9	62.4	51.6
Lysoethanolamine	8.3	n.d.[b]	6.0	5.1
Lysoinositol–phosphatide	1.5	n.d.[b]	3.1	7.0
Free fatty acids	11.6	2.1	4.4	7.7
Total % lipids	1.1	0.5	1.0	1.3

[a]From Acker, 1974; Acker and Becker, 1971.
[b]n.d., not determined.

determined by the amylograph) and overall flour properties in the baked bread. Lipids in wheat, rye, barley, and oat starches are compared in Table 7.13.

B. Wheat Gluten Lipids

Starch-bound lipids are essentially unavailable for interaction with gluten proteins during dough formation (Acker *et al.,* 1967). Lipids are bound to various degrees by gluten fractions and once bound can be extracted only with difficulty; even in lyophilized gluten, polar lipids are practically inextractable and nonpolar lipids only to a limited extent. Gluten lipids are compared with wheat flour (low extraction, white) lipids in Table 7.14; the lipids were extracted with water-saturated butanol at room temperature, separated on a silicic acid column into nonpolar and polar fractions, subfractionated by thin layer chromatography, and assayed densitometrically after spraying. The results were semiquantitative. The following facts are noteworthy: the polar to nonpolar lipid ratios in wheat gluten and flour were about 1.6:1 and 1.8:1, respectively. Flour and gluten are rich in glycolipids and poor in phospholipids; amounts of lysophosphatidyl choline (the main starch lipid) in wheat flour and especially in gluten are low. Whereas the gluten contains more than 95% of the wheat flour's glycolipids, it contains only 53% of the flour's phospholipids. Rye flour lipids did not differ substantially from wheat flour lipids.

According to L. G. Bartolome (Henkel Corp., Minneapolis, Minnesota, personal communication, September 30, 1981), commercial vital wheat gluten (see Chapter 6) contains, on a moisture-free basis, at least 75% protein ($N \times 5.7$), 1.0% ether-extractable fat and 6.5–7% fat by acid hydrolysis. A typical analysis of commercial vital wheat gluten is given in Table 7.15.

TABLE 7.14

Wheat Flour and Gluten Lipids[a,b]

	Content (% of total lipids)	
Lipid type	Flour	Gluten
Total nonpolar	35.6	37.9
Total polar	64.4	62.1
Total phosphatides	19.5	14.0
Phosphatidic acid	2.3	4.8
Phosphatidyl glycerol	4.0	2.7
Choline	2.1	3.6
Ethanolamine	Traces	1.2
Serine	1.2	—
Lysophosphatidylcholine	9.1	1.7
Ethanolamine	0.8	—
Galactolipids, total	19.4	25.1
Monogalactosyl diglyceride	6.0	8.8
Digalactosyl diglyceride	13.4	16.3
Other lipids (free fatty acids, steryl glycosides, glycoside esters, cerebrosides, phytoglycolipids)	25.5	23.0

[a]From Acker et al., 1967.
[b]The total lipid content of the flour was 1.28%, dry-matter basis.

Lipids were determined in flours and gluten from three wheats that varied widely in bread-making potential: Manitoba (good), Jubilar (poor), and Wimax (poor) (Acker, 1974). The results are summarized in Table 7.16. Though some differences are indicated, much more work would be required to determine whether the differences are related to protein contents and/or protein quality (in terms of functional, bread-making properties).

TABLE 7.15

Typical Analysis of Vital Wheat Gluten[a]

Component	Content %
Moisture	5.0–8.0
Protein (dry-basis)	75.0–80.0
Ether-extractable fat (dry-basis)	0.5–1.5
Ash (dry-basis)	0.8–1.2

[a]From "Wheat Gluten—A Natural Protein for the Future—Today," International Wheat Gluten Association, P.O. Box 8193, Shawnee Mission, Kansas 66208.

TABLE 7.16

Lipids in Flour and Gluten from Wheats that Vary Widely
in Bread Making Potential[a]

Component	Manitoba	Jubilar	Wimax
Wheat flour			
Protein (%)	16.8	7.8	9.2
Total lipids (%)	2.05	1.62	1.39
Nonpolar lipids (%)	0.71	0.42	0.38
Polar lipids (%) of which[b]	1.34	1.20	1.01
Monogalactosyl diglyceride			
Digalactosyl diglyceride	13.2	16.3	18.8
Lysolecithin	16.3	18.5	20.5
Lecithin	24.2	24.9	23.5
Wheat gluten	3.9	3.4	3.0
Total lipids (%)	5.70	7.63	7.11
Nonpolar lipids (%)	2.16	1.97	1.98
Polar lipids (%) of which[b]	3.54	5.66	5.13
Monogalactosyl diglyceride	17.7	20.3	21.3
Digalactosyl diglyceride	24.2	25.6	24.3
Lysolecithin	3.8	3.0	3.9
Lecithin	1.2	1.3	1.4

[a]From Acker, 1974.
[b]As percentage of polar lipids.

C. Dry-Milled Maize

The terms "corn" and "maize" are used in this chapter interchangeably. Dry-milled maize can be produced from either degermed or nondegermed grain. In milling nondegermed maize, preferably white dent, the product has a rich oily flavor because of its high fat content. Corn meals with a soft texture are produced on millstones run slowly at low temperatures. Two types of products can be produced: meals—essentially ground whole maize not bolted (not sifted)—and produced in small mills; and meals bolted, with about 5% coarse hull and germ particles removed. Whole or bolted meal has a short shelf life because of the high fat content of the ground material, which has a large surface area and contains active lipases. Proximate compositions of dry-milled maize products are given in Table 7.17.

The typical composition of dry-milled products from degermed maize is listed in Table 7.18. Grits and meal are largely produced from the horny or vitreous endosperm; they contain less than 1.0 and 1.5% fat, respectively. Flour produced by grinding the starchy endosperm contains 2–3% fat from broken germ during

TABLE 7.17

Dry-Milled Maize Products Composition[a,b]

Product	Protein	Fat	CHO (starch)	Fiber	Ash
Whole meal	9.2	3.9	73.5	1.6	1.2
Meal (bolted)	9.0	3.4	74.5	1.0	1.1
Meal (degermed)	7.9	1.2	78.4	0.6	0.5
Grits	8.7	0.8	78.1	0.4	0.4
Flour	7.8	2.6	76.8	0.7	0.8

[a]From Brockington, 1970.
[b]12% moisture basis.

TABLE 7.18

Typical Yields and Analyses of Products from a Degerming-Type Dry-Milled Maize[a]

Products	Yield (%)	Particle size range[b]	Moisture (% wb)[c]	Fat (% db)[d]	Crude fiber (% db)	Ash (% db)	Crude protein (% db)
Maize	100		15.5	4.5	2.5	1.3	9.0
Primary products							
Cereal flaking (hominy grits)	12	−3.5 + 6	14.0	0.7	0.4	0.4	8.4
Coarse grits	15	−10 + 14	13.0	0.7	0.5	0.4	8.4
Regular grits	23	−14 + 28	13.0	0.8	0.5	0.5	8.0
Coarse meal	3	−28 + 50	12.0	1.2	0.5	0.6	7.6
Dusted meal	3	−50 + 75	12.0	1.0	0.5	0.6	7.5
Flour[e]	4	−75 + pan	12.0	2.0	0.7	0.7	6.6
Oil	1						
Hominy feed	35		13.0	6.3	5.4	3.3	12.5
Alternative products							
Brewers' grits	30	−12 + 30	13.0	0.7	0.5	0.5	8.3
100% meal	10	−28 + pan	12.0	1.5	0.6	0.6	7.2
Fine meal	7	−50 + pan	12.0	1.6	0.6	0.7	7.0
Germ fraction[f]	10	−3.5 + 20	15.0	18.0	4.6	4.7	14.9

[a]From Brekke, 1970.
[b]U.S. standard sieve number.
[c]Wet-basis.
[d]Dry-basis.
[e]Break flour.
[f]Yield is distributed between maize oil and hominy feed.

processing. The large surface area and relatively high fat content of maize flour lower its shelf life.

The United States Food and Drug Administration (FDA) has established standards of identity for dry-milled corn products used for food. According to those standards (Code of Federal Regulations, Title 21, part 15), the fat contents of corn meal may not differ more than 0.3% from that of cleaned corn; of bolted corn meal should not be less than 2.25% nor more than 0.3% greater than the fat of cleaned corn; of degerminated corn meal should be less than 2.25%; of corn grits should be not more than 2.25%; and of corn flour may not exceed that of cleaned corn.

Composition of commercial maize products is listed in Table 7.19. The meals and flours are produced from maize ground to typical granulations. The cooked flour hydrates readily in cold water to form a stable paste. The toasted germ is a food-grade product in flake form; the stabilized product contains all the original oil of the germ. The germ cake is an animal feed product from maize germ from which most of the oil has been removed. It is used as a carrier for vitamins and antibiotics in animal feed formulations. Masa harina (yellow regular grind, or yellow coarse grind, or white) is a food-grade product. It is milled from maize that has been steeped, ground, and dried to produce a stable flour for the production of Mexican-style foods.

TABLE 7.19

Composition of Typical Maize Products[a]

Product	Moisture (%)	% Dry-matter basis			
		Protein	Fat	Fiber	Ash
Yellow maize					
Meal[b]	8.2–12.6	8.0–9.0	0.9–2.3	0.3–0.7	0.3–0.7
Flour[b]	8.5–12.1	7.6–8.8	1.4–3.0	0.6–1.0	0.6–1.0
Fine flour	7.1–11.3	6.4–7.4	2.2–3.3	0.04–0.60	0.66–1.12
Cooked flour	6.7–11.5	9.0–9.2	0.6–0.7[c]	—	0.55–0.73
White maize					
Meal[b]	11.0–13.0	8.0–9.0	1.7–2.2	0.5–1.0	0.3–0.7
Flour[b]	8.0–12.0	7.0–8.0	2.0–2.7	0.5–1.0	0.4–0.9
Fine flour	10.0–13.0	8.0–8.8	4.0–5.2	0.5–1.2	0.8–1.0
Toasted corn germ	4.2	17.0	25.4[c]	4.2[d]	7.2
Corn germ cake	5.0 max	14.0 min	3.5 min	8.5 max	—
Masa harina	10.0–12.0	7.0–9.0	3.5–4.5	1.8–2.6	1.2–1.7

[a]Courtesy of Quaker Oats Co., Chicago, Illinois.
[b]From degermed maize.
[c]Ether-extracted.
[d]Dietary fiber = 20.8%.

D. Wet-Milled Maize

The main products of wet-maize milling are starch (unmodified and modified, including syrups, and dextrose) and several coproducts (see Chapter 4). The coproducts, used mainly as feed ingredients, include gluten meal; gluten feed; corn germ meal; and condensed, fermented corn extractives (about 50% solids). Processing maize germ yields refined oil (along with fatty acids from crude oil refining) and corn germ meal (Harness, 1978).

Whereas maize starch contains only 0.04% fat as determined by ether-extraction, total lipids amount to about 0.54%. The lipids are almost entirely associated with amylose, the linear starch fraction, and are predominantly free fatty acids. Commercial maize starch products contain less than 0.1% ether-extractives, typically 0.03% (D. W. Harris, Clinton Corn Processing Co., Clinton, Iowa, personal communication, Sept. 14, 1981).

Coproducts of maize starch wet-milling amount to about one-third of the total output. Except for maize oil and steeping liquor (used in industrial fermentations); the coproducts are mainly sold as animal feed ingredients. Listed in decreasing value, they are corn gluten meal, corn gluten feed, spent germ meal, corn starch molasses or hydrol, steep liquor (condensed corn fermentation extractives), corn bran, and hydrolyzed fatty acids. Composition of the main maize wet-milling feeds is summarized in Table 7.20.

Corn gluten meal is a high protein product, used as a protein balancing ingredient in feed formulations. It is used widely to feed broiler and layer poultry because of its high content of carotenoid pigments. Among the three carotene isomers (β, ζ, and $\beta-\zeta$), only β-carotene has significant vitamin A activity. The dihydroxy xanthophylls are potent pigments for coloring poultry skin and egg yolks. The major isomer, lutein, is slightly superior to zeaxanthin in producing color. The monohydroxy pigments, zeinoxanthin and cryptoxanthin, have less than half the pigmenting value of the dihydroxy pigments. Xanthophyll levels in gluten meal are highest in winter months and drop gradually to half the original value by the end of summer.

The linoleic acid content, on "as is" basis, is 3.2% in corn gluten meal, 2.2% in corn gluten feed, and about 0.5% in corn germ meal (Rapp, 1978). Corn gluten meal is relatively rich in xanthophylls (200–450 mg/kg); 20 mg/kg are present in corn gluten feed and practically none in corn germ meal and concentrated steep water. Corn gluten meal contains 65–145 vitamin A equivalents as retinol (0.15 mg retinol = 5000 IU vitamin A) and 45–65 mg of β-carotene/kg.

E. Maize Oil

Maize contains about 4.5% oil, of which 85% is present in the germ (Reiners, 1978). The germ fraction separated from maize by the wet-milling process con-

TABLE 7.20

Composition of Feeds from Maize Wet-Milling[a]

Component	Corn gluten feed	Corn gluten meal		Corn germ meal	Condensed fermented corn extractives (about 50% solids)
		Guaranteed analysis (%)			
Protein (min)	21.0	60.0	41.0	20.0	23.0
Fat (min)	1.0	1.0	1.0	1.0	0.0
Fiber (max)	10.0	3.0	6.5	12.0	0.0
			Typical analysis (%)		
Fat (average)	2.5	2.5	2.5	1.9	0.0
AOAC range[b]	1.4–3.5	1.0–5.2	1.2–4.4	1.0–2.9	0.0
Total fat[c] (average)	3.8	5.7	4.8	4.6	0.0
Total fat range	2.7–4.7	4.4–7.9	3.6–6.4	4.1–5.3	0.0

[a]Adapted from "Corn Wet-Milled Feed Products." Corn Refiners Association Inc., Washington, D.C., 1975; and S. A. Watson, AACC (American Association or Cereal Chemists, St. Paul, Minnesota) Short Course, April, 1980.

[b]AOAC, Association of Official Analytical Chemists, Washington, D.C.

[c]As determined by extraction with a mixture of chloroform and methanol at a ratio of 4:1. This process is widely used in Europe.

tains about 50% oil and by the dry-milling process about 25% oil. Germ oil can be extracted by a continuous screw press (expeller) to yield a meal with a residual oil content of 7–10%; solvent extraction (directly or following expeller extraction) produces a meal with a residual oil content of 1–3%. About 0.8 kg of oil can be recovered from a bushel of maize (25.4 kg) by solvent extraction of the germ. The products are crude oil and corn germ meal. Crude and refined maize oils are compared in Table 7.21.

TABLE 7.21

Composition of Corn Oil[a]

Component	Crude (%)	Refined (%)
Triglycerides	95.6	98.8
Free fatty acids	1.7	0.03
Phospholipids	1.5	—
Phytosterols	1.2	1.1
Tocopherols	0.06	0.05
Waxes	0.05	—
Carotenoids	0.0008	—

[a]From Reiners, 1978.

Refining maize oil removes free fatty acids, phospholipids, waxes, and carotenoids. The main TG fatty acids are about 60% linoleic acid (all in the cis–cis configuration), 25% oleic acid, and 13.5% palmitic plus stearic acid. The iodine value of refined maize oil produced in the United States shows little variation (125.4–127.6). Oils from African maize may have iodine values as low as 110 and correspondingly reduced ratios of polyunsaturates to saturates (Reiners, 1978).

F. Rice Lipids

Brown rice contains about 1.5–2.5% total fat (Houston, 1972). About 80% of the lipids of brown rice are contained in the bran and polish, and about one-third of lipids in the embryo. In degermed brown rice, about 70% of the total lipids are contained in the outer 8% fraction. The outermost 1.4% milling fraction contains 40% fat, which constitutes about 40% of the total lipids of degermed brown rice.

The lipids of brown rice, bran, and embryo have similar constants: specific gravity 0.91–0.92, ND_{25}^{40} 1.465–1.470, iodine value 95–106, saponification value 177–196 (Juliano, 1972, 1980).

The crude fat content of brown rice and its fractions is on a dry-matter basis:

Fraction	Percent
brown	1.8–4.0
milled	0.2–1.1
bran	14.6–21.7
embryo	15.2–23.8
polish	8.8–15.3

Lipids of all brown rice fractions have similar physical and chemical properties (Juliano, 1980). Lipids are present in the aleurone layer and in the germ in the form of spherosomes (lipid droplets), which are 0.1–10 μm in diameter. They are associated as lipoprotein with protein bodies and starch granules, probably in the membrane fraction. Starch granules contain bound lipids, mainly phospholipids of which lysolecithin is the main fraction.

Most rice lipids are removed with the bran (which contains the germ) and the polish. Commercial bran contains 10.1–23.5% oil and polish 9.1–11.5% oil; the milled rice contains 0.3–0.7% oil (Houston and Kohler, 1970).

Commercial bran oil contains germ oil. The unsaponifiable matter of bran oil consists of about 40% sterols, 25% higher alcohols, 20% ferulic acid esters, 10% hydrocarbons, 2% cholesteryl esters, and small amounts of unidentified compounds. The hydrocarbon fraction is mainly squalene or its isomer.

Sterols in the unsaponifiable fraction include phytosterol, stigmasterol, and dihydrositosterol. Oryzanol, a mixture of ferulate esters of unsaturated triterpenoid alcohols, is a potent antioxidant present at the 1.0–3.0% level in bran oil. Other antioxidants are tocopherols, which have vitamin E activity. Unsaponifiable matter of brown rice contains about 5% total tocopherols, which include α, η, and ζ tocopherols in decreasing amounts. Rice bran oil also contains, depending on the nature and temperature of extractant, various amounts of waxes (up to about 10%).

Neutral fats account for about 85–90% of total bran lipids and 60% of endosperm lipids. The fatty acid composition of brown rice lipids and milled rice lipids is affected significantly by the solvent used to extract the lipids. The main fatty acids, in decreasing amounts, are in brown rice oleic, linoleic, and palmitic; in milled rice linoleic, oleic, and palmitic; and in rice polish, oleic, linoleic, and palmitic.

Nonwaxy rice endosperm contains lysolecithin associated with amylose; in waxy endosperm, palmitic acid is the fatty acid residue.

Fat contents of 241 milled rice samples ranged from 0.19 to 2.73% (average 0.65%) as petroleum ether-extractable material. About 20% of the fatty acids are polyunsaturated, linoleic, and linolenic acids. The oleic to linoleic acid ratio is about 1 : 1 (Kennedy, 1980).

Fat is unevenly distributed within the endosperm; the highest concentration is present in the outer layer and the lowest in the central portion. In a milling study of 12 rice lots, fat in the whole kernel ranged from 0.20 to 0.92%. In the flour passing through a 40-mesh screen, 4.1–11.6% was fat, 17 times as much as in the whole kernel. The flour fraction retained on the 40-mesh sieve contained 4 times as much fat as did the whole kernel, and the residual kernel contained 0.12% fat, about one-fourth that of the original rice.

The composition of fat differs within the endosperm. Neutral fats account for 85–90% of total lipids in the outer layers and for only 60% in the center. Unsaturated fatty acids (oleic and linoleic, generally) show an inverse pattern of distribution. Oleic acid decreases and linoleic acid increases from the outer to the inner layers of the kernel, in both the free fatty acids and the neutral fat fraction (Kennedy, 1980).

G. Sorghum Lipids

Sorghum has a general composition resembling that of maize; wet-milling processes for maize and grain sorghum are basically similar (Watson, 1970).

The object of dry-milling sorghum is to separate the endosperm, germ, and bran while recovering a maximum amount of endosperm (Hahn, 1969). Sorghum germ can be separated from other products and oil extracted from it. High yields of clean grits and minimum flour are desired. By-products of sorghum dry-milling (bran, germ, and ''shorts''—a mixture of bran, germ, and dark flour) are used in the production of hominy feed. Decreasing flour extraction in roller-milling from 90 to 70% decreased oil content of the groats from 2.8 to 2.0%. The oil content decreased from 3.4% in whole sorghum grain to 0.6% in decorticated grain, in which 39.0% of the kernel was abraded; that decrease was accompanied by decreases in fiber (2.2–0.7%), mineral components (1.5–0.4%), and protein (9.6–6.9%), and by an increase in brewers' extract (85.7–97.0%; see Chapter 10).

The impact or attrition degerminators (after grain tempering and dehulling) are effective in germ separation and production (after sieving) of low oil products. Specific gravity separators (after dehulling, impaction, and size classification), can be used to separate the germ and endosperm during dry-milling. Both waxy and nonwaxy grain sorghum can be milled and fractionated. The waxy germ, however, is more difficult to separate, since waxy grits contain more oil than nonwaxy grits. The composition of dry-milled grain sorghum products is summarized in Table 7.22.

TABLE 7.22

Composition of Dry-Milled Grain Sorghum Products[a,b]

| | Percentage of total | | | |
Product	Protein	Oil	Fiber	Ash
Whole grain	9.6	3.4	2.2	1.5
Pearled	9.5	3.0	1.3	1.2
Flour (crude)	9.5	2.5	1.2	1.0
Flour (refined)	9.5	1.0	1.0	0.8
Brewers' grits	9.5	0.7	0.8	0.4
Bran	8.9	5.5	8.6	2.4
Germ	15.1	20.0	2.6	8.2
Hominy feed	11.2	6.5	3.8	2.7

[a]From Hahn, 1969.
[b]"As is" basis.

H. Oat Lipids

Oat hulls contain very little lipid; practically all the fat is in the groats. Oat groats contain higher concentrations of lipids than other cereal grains do. In a comprehensive study of 4000 entries from the World Collection, the free lipid content of oat groats ranged from 3.1 to 11.6%; 90% of the entries contained from 5 to 9% lipid (Brown and Craddock, 1972). Of the fatty acids in oat groats about 20% are palmitic; about 35%, oleic; and more than 35%, linoleic acid (Youngs et al., 1982). TG are the major component of oat lipids (about 40–50%); glycolipids, 8–17%, and phospholipids, 10–20% of the groat lipids. Digalactosyldiglycerides are the major glycolipid component and phos-

TABLE 7.23

Composition of Oats and Oat Fractions[a,b]

	Percentage of total				
Component	Oats	Finished groats	Hulls	Oat shorts	Oat flour, chips, and meal
Protein (N × 6.25)	12.1	15.8	4.2	9.5	15.5
Crude fat	5.1	7.2	1.7	3.2	6.2
Crude fiber	11.0	1.5	32.9	22.0	3.6
Ash	3.4	1.9	6.0	6.7	2.1

[a]From Caldwell and Pomeranz, 1973.
[b]"As is" basis, about 7% moisture.

phatidylcholine is the major phospholipid component of groat lipids. Lysophosphatidylcholine comprises 51.6% of oat starch lipids (Acker and Becker, 1971). Gross composition of oat fractions is summarized in Table 7.23.

Composition of commercial oat products is listed in Table 7.24. The steel oat is an oat that has been hulled and cut to a small size. The steam table product is a dehulled oat that has been steamed and rolled into a thick flake. The regular buckeye rolled product is a dehulled oat that has been steamed and rolled into a large flake. The quick buckeye rolled product is a dehulled oat that has been steamed and rolled into a thin flake for rapid cooking. The flour is a dehulled oat that has been steamed and ground to produce a stable flour in which enzyme activity has been minimized.

TABLE 7.24

Composition of Typical Oat Products[a]

Oat product	Moisture (%)	Dry-matter basis (%)			
		Protein	Fat	Fiber	Ash
Steel oat	9.0–12.0	16.5–18.5	7.0–8.0	1.4–1.8	2.0–2.5
Steam table	9.0–12.0	16.5–18.5	6.0–9.0	1.4–1.8	2.0–2.5
Regular buckeye rolled	9.0–12.0	16.0–18.5	6.0–9.0	1.2–1.8	2.0–2.5
Quick buckeye rolled	9.0–12.0	16.0–18.5	7.0–8.0	1.2–1.8	2.0–2.5
Flour	8.0–12.0	17.0–19.0	7.0–8.0	0.7–1.6	1.8–2.2

[a]Courtesy of Quaker Oats Co., Chicago, Illinois.

I. Barley Lipids

Most barley used in human food is consumed as pot barley or pearl barley. Both are manufactured by gradually removing the hull and outer portions of the barley kernel by abrasive action. The pearling or decortication process used to produce pot barley is merely extended to produce pearl barley; 100 lb of barley normally yields 65 lb of pot barley or 35 lb of pearl barley. Barley flour is a secondary product and polishings are a byproduct of the pearling process. On the basis of decreases in weight and changes in chemical composition, it has been computed that six pearlings remove 74% of the protein, 85% of the fat, 97% of the fiber, and 88% of the mineral ingredients contained in the original barley.

Table 7.25 compares the gross composition of barley products obtained in pearling barley. The composition of commercial barley products is listed in Table 7.26. Chester and portage are creamy white, pearled barley products; both are used in human and pet food applications as thickeners and fillers. About 90–95% and 80–90% is retained on U.S. sieve no. 8 for chester and portage barley products, respectively. The quick cooking product is a creamy white, pearled barley that has been steamed and rolled into a flake for quick cooking (100% of the flakes is retained on U.S. sieve no. 8). The quick cooking barley is used as a major ingredient in dry soups and as a thickener when a quick cooking product is required. The flakes are a creamy white, pearled barley that has been steamed and rolled into thin flakes. Barley flakes provide a less chewy texture than oat flakes. Barley flakes can be used as an ingredient in granola products, or in specialty breads to provide texture. Barley flour is milled from barley grain that has been pearled, steamed, and ground to produce a stable product in which enzyme activity has been minimized. Only 0–2% is retained on U.S. sieve no. 2,

TABLE 7.25

Composition of Milled Barley and Products of Barley Milling[a]

Product	Percentage of total					
	Moisture	Protein	Fat	N-free extract	Crude fiber	Ash
Dehulled barley	12.5	10.6	1.7	72.1[b]	1.6	1.5
Pearls	12.5	7.8	1.0	76.2	1.4	1.1
Pearling dust	12.5	9.5	1.4	74.3	0.8	1.5
Feed meal	12.0	12.5	3.0	64.0	5.0	3.5
Bran	10.5	14.0	3.5	57.1	10.0	4.9
Husks	10.4	3.6	1.0	49.2	28.6	7.2

[a] Adapted from Rohrlich and Bruckner, 1966; "as is" basis.
[b] In hulled barley.

TABLE 7.26

Composition of Typical Barley Products[a]

Barley product	Moisture (%)	% Dry-matter basis			
		Protein	Fat	Fiber	Ash
Chester	9.0–10.0	11.0–12.0	2.0–2.5	1.0–1.5	0.7–1.3
Portage	9.0–10.0	11.0–12.0	1.0–1.5	1.0–1.5	0.7–1.3
Quick cooking	9.0–10.0	10.0–12.0	1.0–1.5	0.5–1.0	0.7–1.3
Flakes	10.0–12.9	10.0–12.0	1.0–1.5	0.5–1.0	0.7–1.3
Flour	—	13.0–15.0	1.9–2.9	1.2 max	1.0–2.0

[a]Courtesy of Quaker Oats Co., Chicago, Illinois.

and 70–80% pass U.S. sieve no. 100. It can be used as a thickener, stabilizer, binder, or protein source for baby foods, malt beverages, prepared meats, and pet foods.

The main products of malting are malt and sprouts. The main byproducts of brewing are brewers' spent grains, spent hops, and yeast (Pomeranz, 1973); their composition is summarized in Table 7.27.

Lipids impair the foam, flavor, and taste of beer as it ages. Beer foam is basically proteinaceous in nature and relatively stable. Major differences in foam stability are caused by substances that interact with the proteins. Thus, for instance, isohumulones in combination with metal ions interact with the proteins and improve "head" retention. Interaction of lipids with the proteins is usually damaging. Apparently, when a lipid is added to a beer its initial form differs

TABLE 7.27

Average Composition of Raw Materials and By-Products of the Brewing Industry[a]

	Percentage of total					
	Moisture	Protein	Fat	Crude fiber	Ash	N-free extract
Malt	7.7	12.4	2.1	6.0	2.9	68.9
Malt sprouts	7.6	27.2	1.6	13.1	5.9	44.6
Brewers' dried grains	7.2–7.7	21.1–27.5	6.4–6.9	15.3–17.6	3.9–4.2	39.4–42.9
Hops[b]	12.5	17.5	18.7[c]	13.2	7.5	27.5
Spent hops	6.2	23.0	3.6	24.5	5.3	37.4
Yeast	4.3	50.0	0.5	0.5	10.0	34.7

[a]From Leavell, 1942; "as is" basis.
[b]From Luers, 1950.
[c]Total ether-extract; includes nonlipid components; additional components, 3% tannins.

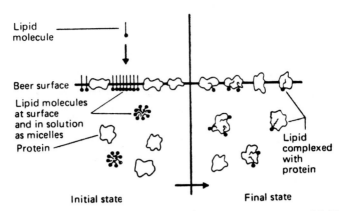

Fig. 7.2. Diagrammatic representation of the proposed initial and final states of lipid material when added to beer. (From Roberts *et al.*, 1978.)

from its final form (Fig. 7.2). The magnitude of the effect of lipids on beer foam depends on the concentration and on the physical state of the lipid in the beer. Roberts *et al.* (1978) examined effects of several fatty acids and glycerides on beer foam. In all cases, lipids added to beer a short time before foaming had a much more negative effect than did the same concentration of lipid added some time previously. The reason for foam collapse in the first instance was due to interaction of the lipid with the proteinaceous bubble wall surface, modification of that surface, and the resultant collapse.

VII. Fats and Oils in Bakery Goods

A. Amounts of Fats Required in Formula Balances

The requirement of fats in formula balance depends on type of bakery goods (Table 7.28). In general, chemically leavened doughs or batter systems require larger quantities of fats than do yeast-leavened doughs. Pie crust dough requires the largest quantity of fats, whereas sponge or angel food cakes and crepes or wafers require no additional fats. A complete formula balance for each type of baked goods has been described by Chung and Pomeranz (1984).

Frying fats contribute a large portion of total fat content of some products such as doughnuts and fried fruit pies. Doughs for cake or yeast-raised doughnuts require 6 or 7–16% fat (flour weight), respectively (Table 7.28), whereas finished doughnuts contain much more lipid due to additional fat absorbed during frying. Absorbed fats from frying contribute 50–63 and 35–44% of total fats for cake or yeast-raised doughnuts, respectively.

1. Chemically Leavened Doughs

Shortening is used in cookie production as a tenderizing ingredient. Though most commercially available shortenings can be suitably applied to cookie production, each type exerts a variable effect on the spreading factor of the cookie dough and on the cookies' surface characteristics. The major problems encountered with fats do not relate so much to their specific properties as to their improper use. Bakers often increase the level of shortening without due regard to

TABLE 7.28

Fat Requirements in Formula Balances[a]

Baked goods	Added fats (gm/100 gm flour)
Chemically leavened doughs	
Pie crust	50–75
Cookies	
Deposit	50–60
Wire-cut	25–60
Rotary	18–35
Baking powder biscuits	10–25
Batter systems	
Cakes	
Pound	30–70
Layer	20–55
Chiffon	25–50
Sponge/angel food	0
Waffles/pancakes	25–30
Crepes/wafers	0
Yeast-leavened doughs	
Breads	
Whole wheat	4–6
White premium	8
White pan	2–3
Pumpernickel	1
Rolls and buns	8–10
Sweet goods	12
Miscellaneous	
Doughnuts	
cake	6
raised	7–16
Crust mix (fried pies)	30
Crackers	8–15
Pretzels	2–3

[a]Summarized from data by Chung and Pomeranz, 1984.

the proper balance with other ingredients, in attempting to improve the richness of a cookie formula. The result may be a dough that tends to smear and lacks adequate cohesion for machine processing. When creaming is an essential part of cookie dough mixing, hydrogenated vegetable shortening or interesterified lard is generally superior to regular lard or fluid shortening.

The relationship of sugar and fat enrichment in biscuit and cookie recipes is shown in Fig. 7.3. The ratios of fat, sugar, and water in biscuit and cookie doughs are given in Fig. 7.4. In Fig. 7.3, sugar and fat are calculated per 100 parts of flour. In Fig. 7.4, the sum totals of fat, sugar, and water are adjusted to 100 and the ratios are plotted independent of flour. The two lines (20°C and 40°C) indicate saturated sugar levels at the two temperatures; dough recipes below these two lines contain some crystalline sugar prior to the temperature rise

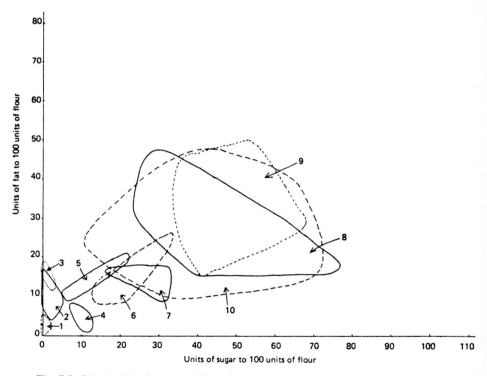

Fig. 7.3. Relationship of sugar and fat enrichment in cookie and biscuit recipes. By areas: 1, bread, pizza, and crispbread; 2, water biscuits and soda crackers; 3, cream crackers; 4, cabin biscuits; 5, savory crackers; 6, semisweet/hard–sweet; 7, "Continental" semisweet; 8, short doughs (molded); 9, wire-cut types; 10, short dough (sheeted). (Reproduced with permission from *Technology of Biscuits, Crackers, and Cookies* by D. J. R. Manley (1983) and published by Ellis Horwood, Ltd, Chichester, England.)

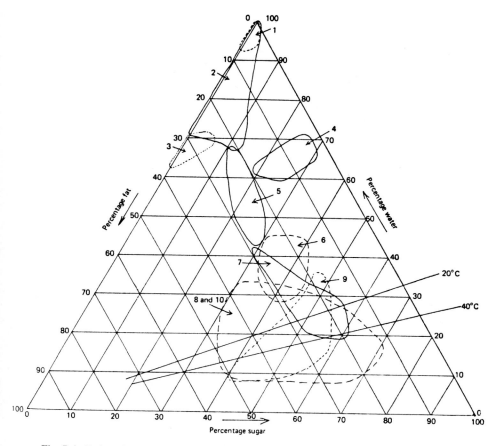

Fig. 7.4. Ratios of fat, sugar, and water in biscuit and cookie doughs. By areas: 1, bread, pizza, and crispbread; 2, water biscuits and soda crackers; 3, cream crackers; 4, cabin biscuits; 5, savory crackers; 6, semisweet/hard–sweet; 7, "Continental" semisweet; 8, short doughs (molded); 9, wire-cut types; 10, short dough (sheeted). (Reproduced with permission from *Technology of Biscuits, Crackers, and Cookies* by D. J. R. Manley (1983) and published by Ellis Horwood, Ltd, Chichester, England.)

in the oven. Table 7.29 shows how different parameters or properties change as the recipe becomes enriched with fat and sugar.

Pie crust requires 50–75 parts of shortening (Table 7.28): obviously shortening is the most expensive ingredient in pie dough. Shortening plays a significant role in developing the final structure, texture, and appearance of the crust. The type of shortening employed in pie production is largely determined by the personal preference of the baker. Regular lard has been used for pie baking because of its excellent shortening characteristics, its desirable plasticity at low

TABLE 7.29

Property Changes in Recipe as Fat and Sugar Are Added[a]

| | Crackers | Semisweet | Short | | Soft |
			High fat	High sugar	
Moisture in dough	30%	22%	9%	15%	11%
Moisture in biscuit	1–2%	1–2%	2–3%	2–3%	3+%
Temperature of dough	30–38°C	40–42°C	20°C	21°C	21°C
Critical ingredients	Flour	Flour	Fat	Fat and sugar	Fat and sugar
Baking time	3 min	5.5 min	15–25 min	7 min	12+ min
Oven band type	Wire	Wire	Steel	Steel	Steel

[a]Reproduced with permission from Manley, 1983.

temperatures, and its ability to disperse in a manner that promotes flakiness. It also contributes to the flavor of the finished crust. Since the late 1960s blends of hydrogenated vegetable shortening and lard or margarine have been used extensively in pies.

2. Batter Systems

The formula balance in cake batter systems differs from those in yeast-raised doughs and varies with the type of product: sugar levels are high and exceed those of flour, while shortening and milk levels fluctuate. Pound cake batter requires 30–70 parts shortening whereas true sponge and angel cake batters require no shortenings (Table 7.28). A modern pound cake formula is much leaner than that of the original old-fashioned pound cake, whose name originated from a formula that required 1 lb each of flour, sugar, butter, and eggs.

The three basic functions of fats in cake baking are (a) entrapment of air during the creaming process, resulting in the proper aeration or leavening of the batter and finished cake; (b) lubrication of the protein and starch particles to break the continuity of the gluten and starch structure that forms the crumb and to tenderize it; and (c) emulsification and holding of considerable amounts of liquid to increase and extend softness of the cakes. Hydrogenated plastic shortenings have been commonly used in cake batters because of their capability to retain air. Since the 1930s the fluid shortenings have gained popularity over the plastic shortenings. The fluid shortenings contain aerating emulsifiers (surfactants) and are widely used in baked products prepared from aerated batters, including all types of layer cakes, pancakes, waffles, muffins, doughnuts, pound cakes, quick breads, coffee cakes, and sweet rolls (see also Chapter 9).

Margarines can be used for cheese or chiffon cake. Because margarines do not have the creaming capacity of modern emulsifying shortenings, they are not highly functional in high sugar ratio layer cake production unless the formulas are supplemented with dry powdered emulsifiers. Margarines can replace shortenings in cake formulas rich in whole egg and/or egg yolk (such as pound cake), however, because egg yolk is an excellent emulsifier (due to its high lecithin content).

3. Yeast-Leavened Doughs

Among yeast-raised baked goods, breads are rather lean products, dinner rolls and buns are intermediate, and sweet goods are rich in fat. White pan bread in the United States typically is made using 2–3% fat. Fat content varies somewhat in production of specialty breads. In the production of white premium bread or 100% whole wheat bread, generally higher fat levels are employed than for white pan bread. The pumpernickel formula or white hearth (sour French) formula requires less fat.

Shortening, or fat, is one of the essential ingredients in bread making. In commercial bakeries, shortening is added (a) to facilitate dough handling and processing; (b) to improve slicing properties; (c) to prolong shelf life and enhance keeping qualities; (d) to increase loaf volume; and (e) to improve crumb grain uniformity and tenderness. See Table 7.30 for a comparison of white pan bread formulas.

For years the fat most commonly used was lard, although other types of shortenings were also used in bread making. Application of fluid shortening in breads began in the early 1960s. At present, a third generation of fluid bread shortening is available. In addition to shortening effects, the added surfactant(s) impart other functional properties. Fluid shortenings, which are liquid and pumpable, have many advantages over the plastic shortenings.

Yeast-raised sweet doughs contain relatively high levels of shortening (Table 7.28). Some sweet dough formulas contain as much as 20–25 parts each of sugar, shortening, and eggs. Margarines have gained steadily since late 1960s as primary fats in the production of sweet, Danish, and puff pastry because they (a) contribute to rich and natural flavor and aroma; (b) have excellent shortening power; (c) are availabile in plastic ranges that adapt to a broad range of processing conditions; (d) contribute to the rich color to baked foods; and (e) are economical. Margarine is available in several tailor-made forms ranging in melting point and plasticity from waxy, puff pastry types to completely fluid products.

Danish pastry doughs differ from regular sweet doughs in that they have specially rolled-in fat, that is, incorporated by a series of folding and rolling operations. The fat in the rolled-in sweet dough is augmented by 15–25 parts of a

TABLE 7.30

Comparison of White Pan Bread Formulas[a]

Ingredient	Formula[b]	
	Old (1964)	Revised (1980)
Flour	100	100
Water	Variable	65.5
Sweeteners		
Sucrose (granulated)	9.2	0
High fructose corn syrup	0	6.2
Corn syrup	0	1.2
Fats		
Lard	2.6	0.6
Shortening (vegetable)	0.7	0
Soybean oil	0	1.7
Surfactants		
Emulsifier/dough strengthener	0.35	0.75
Miscellaneous dough conditioner	0.35	0.50
Enzymes (protease)	0	0.25
Dairy-type products		
Nonfat dry milk	2.50	0
Soy–whey blend	0	2.20
Yeast	2.70	2.75
Salt	2.20	2.10
Yeast food	0.60	0.50
Mold inhibitor	0.20	0.20
Malt	trace	0
Bread produced from formula	158.40	160.79

[a]From Schnake, 1981.
[b]Parts by weight based on 100 parts flour.

rolled-in fat which may be butter, margarine, or a vegetable or compound shortening.

4. Other Baked Goods

The two kinds of doughnuts are chemically raised cake types and yeast-raised types. Most bakers have adapted commercial mixes to the production of both cake and yeast-raised doughnuts.

Frying fats have an important role in doughnut production. Two phenomena are involved in deep-fat frying: (a) absorption, which involves that part of fat that seeps into the product to become a part of the crust and which is mostly involved in variation in fat usage; and (b) adsorption, which refers to fat that merely adheres to the surface of the crust rather than becoming an integral part of it.

Among these two phenomena, absorption plays an important role in enhancing flavor, improving mouth-feel and eating quality, and extending shelf life. Doughnuts with a below minimal level of fat absorption have a pastry texture and those with too much absorbed fat impart a greasy flavor and mouth-feel.

Various types of frying fats affect absorption. Fat absorption is highest with hydrogenated lard and lowest with a blend of lard and tallow. Many other factors influence fat absorption besides type of fat and frying temperature, for example, temperature of doughnut batter or dough, degree of mixing, and amount of moisture added to a doughnut mix.

Shortening level is much higher for the crust mix of fried fruit pies than that for the doughnut mixes, but only about one-half of the amount in the regular pie crust formula. High grade hydrogenated vegetable shortening is required for a crust formula to produce dough that will sheet well and will not shrink during processing, filling, or frying. Because of these special physical requirements for fried pie dough, it has not been possible to make a flaky fried pie crust to resemble that of a baked dessert pie.

Most soda cracker-type products require less shortening than do rich specialty crackers, such as butter-thin, milk, or round. For butter-thin crackers total fat addition is 15% (7% shortening and 8% butter) of flour weight. Many types of chemically leavened crackers are sprayed with hot oil as they emerge from the oven. In the past, coconut oil was used almost exclusively but now several lightly hydrogenated vegetable oils are used. Oil levels range from 5 to 20%.

Pretzel (_Brezel,_ a traditional German snack food) doughs require low levels of fats, usually vegetable shortenings.

B. Trends in Usage of Fats and Oils in the Baking Industry

Utilization of shortening in bread and cake baking in the United States decreased from 1972 to 1977 by 22% (from 433 million to 336 million kg) and of lard by 54% (from 121 million to 55.3 million kg). Utilization of shortening in all baked goods including cookies and crackers was 644 million kg (34% of total) in 1972 and 525 million kg (27% of total shortening consumption) in 1977.

The trends are to reduce total fat contents and to replace the plastic fats in baked goods with fluid shortenings (Smith, 1979). These changes bring about significant nutritional advantages because of substantial reductions in fat in bakery products and modification of fatty acid composition of the base oil. Since fluid shortenings can be prepared with unhydrogenated or only partially hydrogenated vegetable oils as base stocks, they can be manufactured to retain the EFA in an unaltered form or altered only slightly. A further advantage of fluid shortenings involves direct utilization of liquid vegetable oil in combination with surfactants in bread and cake production.

Several factors are responsible for the trends just described: (a) availability and cost of edible oil base stocks; (b) bakers' need to reduce raw material costs; (c) bulk handling requirements; and (d) consumer preference of "all vegetable" and "polyunsaturated" labeling. Making this trend workable is predicated on the utilization of fat-based food surfactants. A combination of certain surfactants with oil can replace effectively the functionality of plastic shortening or fats in baking (see Chapter 9).

The use of surfactants in enhancing shortening response or in making the oil functional is based on the basic multifaceted properties of surfactants in wheat products. Surfactants are used in bread making because they soften the crumb and can be used as antistaling agents due to amylose- and/or amylopectin-complexing. With surfactants, the level of soluble starch from bread crumb decreases. Surfactants complex with wheat proteins and/or nonwheat proteins and can be used as dough conditioners or as loaf volume improvers. Surfactants also spare or replace shortening and effectively replace flour lipids. They can be used in specialty breads such as high fiber, high germ, or whole wheat bread and facilitate production of acceptable bread from nonwheat flours or starches.

1. Changes in Baking Formulations

Major changes in the past decade were in levels of sweeteners, fats, surfactants, and dairy-type products. Total fat levels dropped from 3.3 parts to 2.3 parts and at the same time total amounts of surfactants increased from 0.7 to 1.25 parts (almost 80% increase). Lard consumption in bread production decreased almost 80% (from 2.6 to 0.6 parts) and soybean oil replaced vegetable shortening and most of lard in the white pan bread formula (Table 7.30).

White dinner rolls, and sandwich, hamburger, or frankfurter buns call for 8–10 parts of shortening, which is about three to four times as much as needed for white pan bread formula. The trend in production of rolls and buns is also to reduce shortening levels to 5 parts by use of new types of fluid bread shortenings; a blend of partially hydrogenated soybean oil (88–84%) and a mixture of surfactants (12–16%).

Similar trends have been shown for usage of fats and oils in production of other bakery goods. The trends include reduction of total fat contents and replacement of plastic shortening or lard by vegetable oils in combination with suitable surfactants.

Emulsifying surfactants have always played important roles in cake production. In selecting the right cake emulsifier combination, 41% (based on flour weight) of plastic shortening could be replaced by 17% soybean oil for devil's-food cake and 21% plastic shortening by 14% soybean oil for white cake. A proper cake emulsifier can make nonfunctional liquid oil as functional as a plastic fat, which can retain air. This new trend undoubtedly will change the types and amounts of fat in cake formulations.

Use of emulsifiers in cookie making improves spread ratio, cookie surface, texture, and machinability, and also allows reduction of shortening by 20–50% in formulas for rotary and wire-cut cookies. Emulsifiers are used in cookies and crackers only to a limited degree, however, probably due to bakers' reluctance to change the successful traditional formula.

VIII. Native Wheat Flour Lipids in Baking

Wheat and wheat flour lipids are a minor constituent of major functional importance in bread making. The literature on wheat lipids up to 1970 was reviewd by Mecham (1971) and Pomeranz (1971), on cereal lipids from 1969 to 1976 by Morrison (1978) and on the functionality of wheat flour lipids in bread making by Chung *et al.* (1977, 1978). Chung and Pomeranz (1981) reviewed findings on wheat and flour lipids from 1975 to 1981.

Lipids affect bread making in several ways (Pomeranz, 1971). During progressive stages in the baking process, the lipids may (a) modify gluten structure at the mixing stage; (b) catalyze oxidation of sulfhydryl groups; (c) catalyze polymerization of proteins through a process that involves lipid peroxidation; (d) act as lubricants; (e) improve gas retention by sealing gas cells; (f) prevent interaction between starch granules during gelatinization; (g) lend some structural support to the gluten; (h) retard water transport from proteins to starch; (i) retard starch gelatinization; and (j) act as antistaling agent. It is possible that some of these effects (e.g., retardation of starch gelatinization and the antistaling effect) are due to the same mechanism.

Total wheat flour lipids contain about equal amounts of nonpolar and polar components. Wheat flour lipid composition is shown schematically in Fig. 7.5. TG are a major component of nonpolar lipids; digalactosyl diglycerides of glycolipids; and lysophosphatidyl cholines and phosphatidyl cholines of phospholipids. Differences in solubility provide a convenient and useful means of separating wheat flour lipids into major categories: free and bound. Free lipids can be extracted with nonpolar solvents such as petroleum ether. For extraction of bound (mainly to protein) lipids, polar solvents such as water-saturated butanol or a mixture of chloroform–methanol–water are required. Lipids extracted by petroleum ether are arbitrarily defined as free and those lipids extracted by water-saturated butanol, following petroleum ether-extraction, are defined as bound.

The free lipids can be fractionated according to their elution from a silicic acid column. The 70% free lipids can be eluted with chloroform and form what is arbitrarily called the "nonpolar" fraction containing TG as a major component. The residual 30% free lipids can be eluted from the column with a more polar solvent, such as methanol; they constitute a mixture of free polar lipids. Among

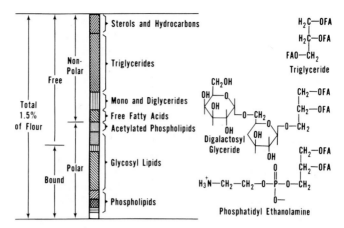

Fig. 7.5. Types and amounts of lipids found in wheat flour. (Courtesy of J. S. Wall.)

the free polar lipids, about two-thirds are glycolipids containing digalactosyl diglycerides as a major component, and one-third are phospholipids as a major component.

About 0.6–1.0% of bound lipids can be extracted from flour with water-saturated butanol after petroleum ether-extraction. Bound lipids contain about 30% nonpolar and 70% polar lipids. Bound polar lipids are rich in phospholipids with lysophosphatidyl choline as a major phospholipid component. Glycolipids are richer in free than in bound polar lipids, but the actual amounts of both glycolipids and phospholipids are higher in bound than in free polar lipids.

A. Glycolipids

Our knowledge of the chemistry, biosynthesis, and metabolic pathways of complex lipids in animal, plant, and microbial cells is expanding rapidly. Intensive investigations of glycolipids in many laboratories throughout the world indicate ubiquitous occurrence of such complex lipids and their high concentration in specific tissues. The results suggest that the glycolipids have a fundamental role in the living cell.

Glycolipids are of great interest to the cereal chemist for several reasons. Historically, certain types of glycolipids were first isolated from wheat flour. From the practical standpoint, glycolipids constitute a major portion of polar wheat flour lipids. Glycolipids of wheat flour are essential in producing bread of acceptable quality. In addition, natural and synthetic glycolipids can improve the loaf volume, crumb grain, and softness retention of high protein bread.

Glycolipids, as the name implies, are complexes of carbohydrates and lipids. Glycolipids generally combine the polar features of polyols with the lipophilic behavior of long aliphatic chains. Thus, some glycolipids, while showing considerable solubility in lipid solvents, also form aqueous solutions. The combination of polar and nonpolar properties makes it attractive to speculate that glycolipids may be structural features of lipid–aqueous interfaces. Noncharged surface-active glycolipids interact with structural proteins by hydrophobic bonding. Together with other complex lipids they may be responsible for ionic interactions between lipid micelles and proteins in the chloroplast.

1. Types of Glycolipids

The four main types of glycolipids are glycosyl ceramides, phytoglycolipids, complex glycolipids of microorganisms, and glycosyl glycerides. Formulas of the main types of glycolipids are given in Fig. 7.1.

Glycosyl ceramides are a family of sphingolipid derivatives in which the amino group of a sphingosine is amide-linked to a fatty acid (forming a sphingolipid), and the primary hydroxyl group is glycoside-linked to a molecule of sugar (usually galactose) or to an oligosaccharide. The compounds are widely distributed in animal tissues, each organ containing a dominant type of glycolipid. The simplest sphingolipid is ceramide, which has been isolated from brain and yeast. More complicated derivatives of ceramides are cerebrosides, which occur abundantly in nervous tissue. They have a carbohydrate (generally galactose) attached to the primary hydroxyl group of ceramide, and may contain either a 2-hydroxy fatty acid or a nonhydroxy fatty acid. Gangliosides, glycosphingolipids containing sialic acid (N-acylneuraminic acid), have been isolated from many tissues. Gangliosides are associated with the microsomes, which are considered to be composed of endoplasmic reticulum and ribonucleoprotein particles. Gangliosides associated with the endoplasmic reticulum may be important in membrane structure and function.

Phytoglycolipids are unique among complex glycolipids because they have the structural features of a glycolipid and a phosphatide. The oligosaccharide is attached to the long chain base through phosphatidyl inositol. Some of the isolated glycolipids have unique immunochemical properties, some show specific forms of toxicity, and some are present in a few bacterial strains and can be used for taxonomical purposes.

Glycosyl glycerides consist of mono- and digalactosyl diglycerides and of plant sulfolipid. Figure 7.1 shows the structure of a galactolipid in which two galactose molecules are linked to a diglyceride unit. Plant sulfolipid is distinct from animal sulfatides in that it is related to the glycosyl glycerides rather than sphingolipids; and it contains a sulfonic acid group (RSO_3H) and not a sulfuric acid group ($ROSO_3H$), which characterizes the cerebroside sulfates or sulfatides. The sulfolipid appears to be mainly concentrated in the lamellar membranes of the chloroplasts. Although little or no sulfolipid occurs in seeds, its molar concentration in photosynthetic tissues of most plants is $1–6 \times 10^{-3}$. Clover and alfalfa are good sources of the sulfolipid. Highest concentrations were found in marine red algae. Sulfolipid and its derivatives comprise the most concentrated anionic sugar compounds in plants. As salts of a very strong sulfonic acid, they are anionic under all conditions. The plant sulfolipid combines the lipophilic properties of two fatty acid esters and an extremely hydrophilic moiety. The amphiphatic molecule, therefore, is expected to exhibit excellent surfactant properties.

Galactosylglycerides have been found in a wide variety of plant, animal, and microbial sources. The concentration of galactolipids in chloroplasts is remarkably high. The galactolipids probably are universal constituents of photosynthetic tissue. While TG are generally the main components of seed lipids, the seeds of *Briza spicata,* a member of the grass family, have a unique composition. The seeds contain 20% of lipid that is semisolid and quite unusual. The lipid contains 49% digalactosyldiglycerides, 29% monogalactosyldiglycerides, and little, if any, conventional TG. The main fatty acids are palmitic, oleic, and linoleic.

B. Wheat Flour Lipids in Bread Making

Small amounts of polar wheat flour lipids substantially improve loaf volume, crumb grain, and freshness retention of bread, as stated earlier. The effects are much more pronounced in defatted flour than in untreated flours. The improvement is mainly due to galactolipids. Storage of damp wheat at elevated temperatures is accompanied by rapid dissappearance of glycolipids and phospholipids. Similarly, severe damage to bread-making potentialities of stored wheat flour is accompanied by almost complete breakdown of lipids rather than changes in the gluten proteins, starch, or water-solubles. Loaf volume potential and crumb grain of damaged flour could be restored by adding polar lipids. The involvement of polar lipids in bread making is also indicated by changes in extractability during dough making. The free polar lipids (including galactolipids) in flour become bound during dough mixing and cannot be extracted with nonpolar solvents such as petroleum ether.

1. Synthetic Glycolipids in Bread Making

Several dough improvers are used by the baking industry. They include, in addition to mono- and diglycerides, various derivatives of glycerides, glycerol, or fatty acids. The compounds are designed to improve one or more of the following: dough strength; ingredient and processing tolerance; and volume, texture, and overall quality of the baked bread. It is desirable that the compounds also retard the crumb firming rate without causing excessive initial bread softness. The improvement involves interaction between the gluten and the starch fractions of wheat flour (see also Chapter 9).

Sucrose esters or sucroglycerides are synthesized by esterification of fatty acids or natural glycerides with sucrose. Sucrose esters were originally designed to provide non-ionic surfactants that can be easily broken down and to eliminate problems in extremely slow biodegradation of detergents in sewage disposal. Sucrose esters can be useful in food products because they are not toxic. They usually do not interfere with the action of preservatives and bioactive products, are low foaming, have a moderate surfactant activity, and are completely and readily biodegradable. The reaction of sucrose esters of fatty acids and glycerides with starches and starch fractions is similar to that of various substances with antistaling activity in bread. It has been suggested that sucrose esters are absorbed on the surface of wheat starch granules and form an insoluble complex which has an antistaling capacity.

Commercially available sucroglycerides counteract the deleterious effects of soy flour on loaf volume. The effects of sucroglycerides increase with an increase in hydrophilic–lipophilic balance (i.e., with decrease in number and chain lengths of fatty acids attached to the sucrose molecule). Sucroglycerides and glycolipids from wheat flour and *Briza spicata* also improve bread baked from wheat flour enriched with defatted cottonseed flour, fish protein concentrate, sesame seed flour, and food-grade yeast.

2. Lipid Shortening Effects

In practice, the most significant contribution of native flour lipids to bread making is that they are essential to bring out optimum shortening effects. This is illustrated in Figs. 7.6 and 7.7. Chung *et al.* (1981) extracted flour free lipids with petroleum ether (PE) and total lipids with 2-propanol (2-PrOH). 2-PrOH extracted all of free lipids and most of lipids bound to proteins. Extracted and reconstituted flours were baked with various levels (% of flour weight) of commercial shortening. Loaf volume (LV) increased, in general, as shortening level increased for all PE-defatted flours as well as for the unextracted control flour (Fig. 7.6). The LV of breads baked with flour defatted with PE and then reconstituted with polar lipids were the most uniform, probably because polar lipids, especially in the absence of nonpolar lipids, were potent enough to raise LV to nearly maximum potential, even with no shortening at all. However, a small amount of shortening (0.375%) improved crumb grain of polar lipid-reconstituted bread.

For the flour defatted with 2-PrOH and then reconstituted with polar lipids, the LV response to shortening was similar (Fig. 7.7) to that of PE-defatted and polar lipid-reconstituted flour (Fig. 7.6). However, shortening had a detrimental effect on LV of 2-PrOH-defatted flour and its nonpolar lipid-reconstituted flour (unlike

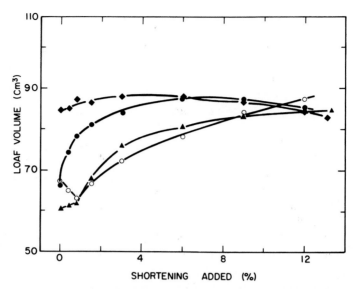

Fig. 7.6. The effect of shortening level (% of flour weight) on loaf volume of bread baked from 10 gm (14% moisture) of unextracted control flour (●), flour defatted with petroleum ether (○), flour defatted with petroleum ether and then reconstituted with polar lipids (♦), or flour defatted with petroleum ether and then reconstituted with nonpolar lipids (▲). The overall standard deviation of four replicates was 1.27 cm³. (From Chung *et al.*, 1981.)

the improving effect of high shortening levels on the PE-defatted flour and its nonpolar lipid-reconstituted flour). LV of 2-PrOH-defatted flour decreased significantly with shortening up to 6% and then leveled off. At all shortening levels, LV was lowest for the flour containing only nonpolar lipids.

Insofar as LV and crumb grain were concerned, 18 mg of free polar lipids could be replaced by 900–1200 mg (9–12%) of shortening, whereas 56 mg of total polar lipids could not be replaced by any level of shortening. High shortening levels required to restore optimum LV and crumb grain of PE-defatted or nonpolar lipid-reconstituted flours were usually accompanied by significant decreases in water absorption and/or substantial increases in mixing time—both undesirable in commercial baking. Thus, the ability of shortening to replace flour free lipid components is great and yet limited.

The shortening effect was reviewed by Bell *et al.* (1977), who posited two mechanisms of shortening effects: chemical and/or physical. The true chemical effect would involve lipid oxidation. The chemical mechanism was considered to be nonexistent (or at least insignificant) in bread making. The physical effects of lubrication, sealing, foam formation, involvement of hydrogen and hydrophobic bonds, and delayed carbon dioxide release were reviewed. Bell *et al.* (1977) showed that the rate of carbon dioxide release was faster in doughs baked without

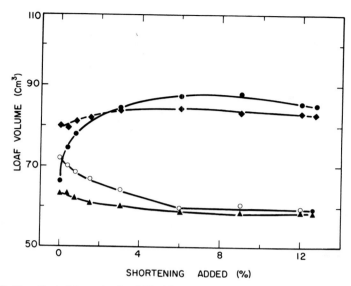

Fig. 7.7. The effect of shortening level (% of flour weight) on loaf volume of bread baked from 10 gm (14% moisture) of unextracted control flour (●), flour defatted with 2-propanol (○), flour defatted with 2-propanol and then reconstituted with polar lipids (◆), or flour defatted with 2-propanol and then reconstituted with nonpolar lipids (▲). The overall standard deviation of four replicates was 1.47 cm³. (From Chung *et al.*, 1981.)

shortening than in doughs baked with shortening. They suggested that the difference in rate of release might explain the response of flour to added shortening. They postulated that LV increases when sufficient solid shortening components are present in a free form in the dough. The free components are especially important in oven spring (the initial rise in loaf volume at the early stages of baking).

MacRitchie and Gras (1973) emphasized that if baking formulations include shortening or other lipid additives, the effect of the natural flour lipid may be obscured. On the other hand, it is difficult to determine the effects of adding lipids to an untreated flour because of the presence of natural flour lipids. In discussing the role of flour lipids in bread making, we must keep in mind that LV and crumb grain are affected by many factors. Both wheat flour lipids and shortening contribute to the final size and quality of the bread, but probably by different mechanisms; interactions between lipids and shortening do take place and depend on the quality of lipids present in the flour.

Interactions between flour lipids and shortening also depend on the quantity of lipids added. Shortening responses varied with the level of lipid supplementation and with the type of flour supplemented, that is, defatted or untreated. In general, shortening responses are slightly increased when lipid fractions are added to the untreated flour than to the defatted flour. The shortening response rises with increasing levels of the nonpolar monogalactosyldiglyceride and phospholipid fractions added to the defatted flour, but decreases with higher levels of fractions that are rich in digalactosyl diglyceride. Yet, in untreated flour the effects of supplementation levels of flour lipids on shortening response are either masked or even reversed, as in the case of phospholipids.

The third factor involved in shortening response is wheat flour quality. Recent studies have shown that the shortening response is much higher in good-quality than in poor-quality bread-making flours (Table 7.31). Addition of 3% shortening to the poor flour was detrimental to LV instead of beneficial. On the other hand, shortening was beneficial when added to Skelly B-defatted poor flour and detrimental when added to the defatted good bread-making flours. The results indicated that the shortening effect is influenced by or dependent on various flour components, including lipids, and also on the quality and quantity of lipids added to the defatted or untreated flours, that is, on total amounts and types of lipids present in the dough.

3. Surfactant–Lipid Shortening

Lipid-related surfactants have several functional properties in bread making: as antistaling agents, dough modifiers, shortening sparing agents, and improvers for production of high-protein breads.

TABLE 7.31

Effects of Flour Quality and Lipid Removal on Loaf Volume Response
to Shortening in Baked Bread from 10 gm Untreated
or Skelly B-Defatted Flours

	Loaf volume (cm³)		
	Shortening (%)		Response to shortening (ΔLV)[a]
Flour	0	3	
Untreated original flour			
Good A	73.3	91.7	+18.4
Good B	64.5	81.0	+16.5
Poor	51.6	47.3	−4.3
Skelly B-defatted flour			
Good A	71.8	71.7	−0.1
Good B	71.0	67.5	−3.5
Poor	46.3	53.2	+6.9

[a]$\Delta LV = LV_{(s)} - LV_{(o)}$ where $LV_{(s)}$ and $LV_{(o)}$ are, respectively, loaf volume (LV) of bread baked with or without 3% shortening.

DeStefanis *et al.* (1977) reported that very little binding of the surfactants (crumb softeners) sodium stearoyl lactylates, succinylated monoglycerides, and monoglycerides by the major flour components occurred at the sponge stage. The additives were firmly bound to the gluten proteins during dough mixing and strongly bound to the starch by complexing with both the amylose and amylopectin fractions in bread. Based on a study that employed model systems, it was concluded that two concurrent phenomena occurred during baking. First, the bonds between the gluten proteins and the additive weakened progressively (protein denaturation) as the dough temperature increased. As starch gelatinized above 50°C, the additives weakly bonded to proteins readily formed a strong complex with starch; thus, the bonds were translocated from the proteins to the starch.

Chung *et al.* (1976) studied the effects on baking properties of replacing wheat flour lipids and 3% shortening with several commercial mixtures of mono-, di-, and triesters of sucrose, which varied in hydrophilic–lipophilic balance (HLB) (see also Chapter 9). The sucrose esters were separated into PE-soluble and PE-insoluble fractions. The fractions were characterized by thin layer chromatography, and the characterized components were compared with flour lipid components and standard lipids. The beneficial effect of sucrose esters on LV and texture in bread with PE-defatted flour increased with increased HLB values. A sucrose ester with an HLB of 1.0 only partially replaced flour lipids, whereas an ester with an HLB of 9.0 was a better replacement than the one with an HLB of 1.0; and two sucrose esters with an HLB of 14.0 functionally replaced both flour lipids and 3% shortening. Removing the PE-soluble fractions enhanced the beneficial effects of the two sucrose esters with an HLB of 14.0. Bake tests and thin layer chromatography showed that the most effective component(s) of the sucrose esters has R_f values close to those of digalactosyldiglyceride from flour.

4. Contribution of Surfactants

Studies on surfactants (sucrose monopalmitate, ethoxylated monoglycerides, and sodium stearoyl lactylates) as a replacement for wheat flour lipids and the three-way contribution of surfactants in the presence of wheat flour lipids and/or shortening were reported by Chung *et al.* (1978). PE extracted 1% total free lipids (0.7% nonpolar and 0.3% polar) and 2-PrOH extracted 1.36% total lipids including free and bound lipids (0.73% nonpolar and 0.63% polar). PE-defatted or 2-PrOH-defatted flours were baked with (a) total, nonpolar, or polar wheat flour lipids, and (b) equivalent amounts of nonionic sucrose monopalmitate, ethoxylated monoglycerides (both having an HLB value of 14.0), or anionic sodium stearoyl lactylate (with an HLB value of 9.0) alone or in combination with wheat flour lipid fractions.

For a positive shortening response, flour lipids were required; adding 3%

shortening to untreated flour increased LV; adding shortening decreased LV for PE-defatted flour and for 2-PrOH-defatted flour. Restoring the free lipids to PE-defatted flour partially restored the shortening effect on LV; for a positive shortening response with 2-PrOH-defatted flour, the extracted total lipids had to be restored. When surfactants were added to the defatted flour, the LV–shortening response was, in general, positive for the PE-defatted flour and negative for the 2-PrOH-defatted flour. The presence of shortening was essential for increasing the LV of breads from PE-defatted flour supplemented with 0.7% nonpolar lipids plus 0.3% surfactants to replace free polar lipids. For 2-PrOH-defatted flour, LV rose with an increase in surfactants, indicating interactions between surfactants and shortening. Shortening produced a satisfactory crumb grain in bread from untreated flour and improved grain in bread baked from PE-defatted or 2-PrOH-defatted flours that were reconstituted with the extracted flour lipids (total, nonpolar or polar); shortening had no effect, however, in breads from defatted flours that were supplemented with surfactants alone.

Surfactants, especially nonionic sucrose monopalmitate or ethoxylated mono-glycerides (HLB of 14.0), effectively performed some functions of the free lipids extracted from wheat flour with PE; they could have a shortening sparing effect; and all surfactants added with the extracted free polar lipids increased LV over that of the untreated flour baked with shortening. However, surfactants alone neither completely replaced total 2-PrOH-extracted lipids plus shortening nor replaced the extracted polar lipids in bread making; a positive synergetic effect on LV of all three surfactants with polar lipids was most significant for a combination of sodium stearoyl lactylate and 2-PrOH polar lipids. The surfactants exerted negative synergetic effects when added with the extracted nonpolar lipids in breads without shortening. Although a combination of surfactant and nonpolar flour lipids exerted a positive synergetic effect in bread with shortening, the surfactants were less effective than polar flour lipids in counteracting the deleterious effects of nonpolar lipids. The lower effectiveness of the surfactants were most notable in flour from which polar (especially bound) lipids were absent. The three surfactants were not equally effective as replacements for polar flour lipids. Whereas the two nonionic surfactants (sucrose monopalmitate and ethoxylated monoglycerides) were better replacements for free polar lipids, rich in nonionic digalactosyldiglycerides and monogalactosyldiglycerides, the anionic sodium stearoyl lactylate was a slightly better replacement for the 2-PrOH-extracted polar lipids (free plus bound), which contained more phospholipids and highly polar glycolipids of sucrose and raffinose than the free polar lipids alone.

Effects of surfactant on overall bread-making quality or in replacing lipids (especially polar) extracted from the wheat flour are evidently quite complex. The optimum balance between the HLB of the surfactant and the number of charged groups as well as degree of polarity depends on the quantity and quality of the lipids that are to be replaced or supplemented, the presence of shortening, and, in the production of high protein breads, the quantity and nature of the protein-rich additives.

REFERENCES

Acker, L. (1974). Die Lipide des Getreides. Ihre Zusammensetzung und ihre Bedeutung. *Getreide, Mehl Brot* **7**, 181–187.

Acker, L., and Becker, G. (1971). Neuere Untersuchungen über die Lipide der Getreidestärken. Part II. Die Lipide verschiedener Stärkearten und ihre Bindung an die Amylose. *Stärke* **23**, 419–424.

Acker, L., Schmitz, H. J. and Hamza, J. (1967). Getreidechemikertagung. *Arbeitsgem. Getreideforsch,* pp. 30–34.

Allen, C. G., and Foegeding, E. A. (1981). Some lipid characteristics and interactions in muscle foods. A review. *Food Technol.* **35**(5), 253–257.

Bell, B. M., Daniels, D. G. H., and Fisher, N. (1977). Physical aspects of the improvement of dough by fat. *Food Chem.* **2**, 57–70.

Bloor, W. R. (1925). Biochemistry of fats. *Chem. Rev.* **2**, 243–300.

Brekke, O. L. (1970). Corn dry milling industry. *In* "Corn: Culture, Processing, Products" (G. E. Inglett, ed.), Chapter 14, pp. 262–291. Avi Publ. Co., Westport, Connecticut.

Brockington, S. F. (1970). Corn dry milled products. *In* "Corn: Culture, Processing, Products" (G. E. Inglett, ed.), Chapter 15, pp. 292–306. Avi Publ. Co., Westport, Connecticut.

Brown, C. M., and Craddock, J. C. (1972). Oil content and groat weight of entries in the world oat collection. *Crop Sci.* **12**, 514–515.

Caldwell, E. F., and Pomeranz, Y. (1973). Industrial uses of cereals: Oats. *In* "Industrial Uses of Cereals" (Y. Pomeranz, ed.), Chapter 12, pp. 393–411. Am. Assoc. Cereal Chem., St. Paul, Minnesota.

Chung, O. K., and Pomeranz, Y. (1981). Recent research on wheat lipids. *Baker's Dig.* **55**(5), 38–50, 55, 96, 97.

Chung, O. K., and Pomeranz, Y. (1984). Fats and oils as functional ingredients in the baking industry–Nutritive value. *In* "Advances in Modern Human Nutrition" (J. J. Kabara, ed.), Vol. 3. Chem-Orbital Publ. Co., Park Forrest South, Illinois (in press).

Chung, O. K., Pomeranz, Y., Goforth, D. R., Shogren, M. D., and Finney, K. F. (1976). Improved sucrose esters in breadmaking. *Cereal Chem.* **53**, 615–626.

Chung, O. K., Pomeranz, Y., Finney, K. F., and Shogren, M. D. (1977). Defatted and reconstituted wheat flours. II. Effects of solvent type and extracting conditions on flours varying in breadmaking quality. *Cereal Chem.* **54**, 484–495.

Chung, O. K., Pomeranz, Y., and Finney, K. F. (1978). Wheat flour lipids in bread-making. *Cereal Chem.* **55**, 598–618.

Chung, O. K., Pomeranz, Y., Jacobs, R. M., and Howard, B. G. (1980). Lipid extraction conditions to differentiate among hard red winter wheats that vary in breadmaking. *J. Food Sci.* **45**, 1168–1174.

Chung, O. K., Shogren, M. D., Pomeranz, Y., and Finney, K. F. (1981). Defatted and reconstituted wheat flours. VII. The effects of 0–12% shortening (flour basis) in bread making. *Cereal Chem.* **58**, 69–73.

Coenen, J. W. E. (1974). Modification of oils and fats. *In* "Contribution of Chemistry Food Supplies" (I. Morton and D. N. Rhodes, eds.), pp. 15–54. Butterworth, London.

DeStefanis, V. A., Ponte, J. G., Jr., Chung, F. H., and Ruzza, N. A. (1977). Binding of crumb softeners and dough strengtheners during breadmaking. *Cereal Chem.* **54,** 13–24.

Dugan, L. R., Jr. (1968). Processing and other stress effects on the nutritive value of lipids. *World Rev. Nutr. Diet.* **91,** 181–205.

Hahn, R. R. (1969). Dry milling of grain sorghum. *Cereal Sci. Today* **14,** 234–237.

Harness, J. (1978). II. Corn wet milling industry in 1978. *In* "Products Corn Refining Industry in Foods," pp. 7–10. Corn Refiners Assoc., Inc., Washington, D.C.

Hornstein, I., Crowe, P. F., and Heimberg, M. J. (1961). Fatty acid composition of meat tissue lipids. *J. Food Sci.* **26,** 581–584.

Houston, D. F. (1972). Rice bran and polish. *In* "Rice Chemistry and Technology" (D. F. Houston, ed.), Chapter 11, pp. 272–300. Am. Assoc. Cereal Chem., St. Paul, Minnesota.

Houston, D. F., and Kohler, G. O. (1970). "Nutritional Properties of Rice." Natl. Acad. Sci., Washington, D.C.

Juliano, B. O. (1972). The rice caryopsis and its composition. *In* "Rice Chemistry and Technology" (D. F. Houston, ed.), Chapter 2, pp. 16–74. Am. Assoc. Cereal Chem., St. Paul, Minnesota.

Juliano, B. O. (1980). Properties of rice caryopsis. *In* "Rice Production and Utilization" (B. S. Luh, ed.), Chapter 10, pp. 403–438. Avi Publ. Co., Westport, Connecticut.

Kennedy, B. M. (1980). Nutritional quality of rice endosperm. *In* "Rice Production and Utilization" (B. S. Luh, ed.), Chapter 11, pp. 439–469. Avi Publ., Co., Westport, Connecticut.

Kinsella, J. E., Posati, L., Weinrauch, J., and Anderson, B. (1975). Lipids in foods: Problems and procedures in collating data. *CRC Crit. Rev. Food Technol.* **5,** 299–324.

Leavell, G. (1942). Brewers and distillers by-products and yeast in livestock feeding. *U.S. Dept. Agric., Bur. Anim. Ind., Bull.* **58.**

Luers, H. (1950). "Die Wissenschaftlichen Grundlagen von Malzerei und Brauerei." Huber, Nuremberg, West Germany.

MacRitchie, F., and Gras, P. (1973). The role of lipids in baking. *Cereal Chem.* **50,** 292–302.

Manley, D. J. R. (1983). "Technology of Biscuits, Crackers, and Cookies." Ellis Horwood, Ltd., Chichester, England.

Mecham, D. K. (1971). Lipids. *In* "Wheat: Chemistry and Technology" (Y. Pomeranz, ed.), 2nd ed., pp. 393–451. Am. Assoc. Cereal Chem., St. Paul, Minnesota.

Morrison, W. R. (1978). Cereal lipids. *Adv. Cereal Sci. Technol.* **2,** 221–348.

Pomeranz, Y. (1971). Composition and functionality of wheat flour components. *In* "Wheat Chemistry and Technology" (Y. Pomeranz, ed.). *Monogr. Ser.—Am. Assoc. Cereal Chem.* **3** (rev.), 585–674.

Pomeranz, Y. (1973). Industrial uses of barley. *In* "Industrial Uses of Cereals" (Y. Pomeranz, ed.), pp. 371–392. Am. Assoc. Cereal Chem., St. Paul, Minnesota.

Rapp, W. (1978). *In* "III. Co-products. Products Corn Refining Industry in Foods," pp. 11–17. Corn Refiners Assoc., Inc. Washington, D.C.

Reiners, R. A. (1978). *In* "IV. Corn oil. Products Corn Refining Industry in Foods," pp. 18–21. Corn Refiners Assoc., Inc., Washington, D.C.

Roberts, R. T., Keeney, P. J., and Wainright, T. (1978). The effects of lipids and related materials on beer foam. *J. Inst. Brew.* **84,** 9–12.

Rohrlich, M., and Bruckner, G. (1966). "Das Getreide und seine Verarbeitung," 2nd ed. Parey, Berlin.

Schnake, L. D. (1981). Revised white pan bread marketing spreads. *In* "Wheat Outlook and Situation," pp. 11–14. Econ. Res. Serv., U.S. Dept. Agric., Washington, D.C.

Smith, W. M. (1979). Evolution of fluid bread shortenings. *Baker's Dig.* **53**(4), 8–10.

Swern, D. (1982). Cooking oils, salad oils, and salad dressings. *In* "Bailey's Industrial Oil and Fat Products" (D. Swern, ed.), Vol. 2, Chapter 5, pp. 315–341. Wiley, New York.

Thiele O. W. (1970). Classification of lipid groups. *Dtsch. Med. J.* **21,** 266–273.

U.S. Department of Agriculture (1974). *Agric. Inf. Bull. (U.S. Dep. Agric.)* **361.**

Watson, S. A. (1970). Wet milling process and products. *In* "Sorghum Production and Utilization" (J. S. Wall and W. M. Ross, eds.), Chapter 17, pp. 602–626. Avi Publ. Co., Westport, Connecticut.

Weiss, T. J. (1963). Fats and oils. *In* "Food Processing Operations" (M. A. Joslyn and J. L. Heid. eds.), Vol. 2. Avi Publ. Co., Westport, Connecticut.

Youngs, V. L., Peterson, D. M., and Brown, C. M. (1982). Oats. *Adv. Cereal Sci. Technol.* **5,** 49–105.

8

Enzymes

Enzymes are proteins that catalyze chemical reactions in living cells. They are generally composed of about 200 to 1000 amino acid residues covalently linked in a sequence regulated by the cell's genetic code. The amino acid sequence governs nonbonded covalent interactions, such as hydrogen-bonds, and thereby the three-dimensional conformation. The uniqueness of the conformation gives the enzymes both their catalytic activity and specificity. The active site of an enzyme is directly involved in binding and catalytic action. The substrate is the substance acted on by the enzyme during the biological modification. Some enzymes depend on their conformation only for their activity. Others require additional cofactors. The latter may be metallic ions or complex organic co-enzymes (i.e., nucleotides, vitamins of the B group, pentoses, or others). Co-enzymes are generally carriers of electrons or specific functional groups or atoms that are transferred during enzymatic reactions. The metal ions may serve as a bridge between the enzyme and substrate or as a catalytic group (Skinner, 1975).

Enzymes are named by adding the suffix "ase" to the name of the enzyme substrate or reaction: for example, urease acts on urea and glucose oxidase oxidizes glucose to gluconic acid. Additionally, the suffix "in" is added to the name of the enzyme source: papain is obtained from the papaya plant. In 1955, an international commission on enzymes was established to develop a systematic nomenclature and classification scheme for enzymes. The system devised by the commission divides enzymes into six major classes based on the type of reactions they catalzye: hydrolases, transferases, oxidoreductases, isomerases, lyases (which catalyze group removal or addition to double bonds), and ligases (which join molecules at the expense of high energy bonds).

TABLE 8.1

Classification of Enzymes Significant in Food and in the Food Industry[a]

Trivial name	Systematic name	Enzyme Commission (EC) number	Reaction (as significant in food material)
Oxidoreductases			
Glucose oxidase	β-D-Glucose:O$_2$ oxidoreductase	1.1.3.4	β-D-Glucose + O$_2$ → D-glucono-δ-lactone + H$_2$O$_2$
Phenolase (polyphenol oxidase)	o-Diphenol:O$_2$ oxidoreductase	1.10.3.1	2 o-Diphenol + O$_2$ → 2 o-quinone + 2 H$_2$O
Ascorbic acid oxidase	L-Ascorbate:O$_2$ oxidoreductase	1.10.3.3	2 L-ascorbate + O$_2$ → 2 dehydroascorbate + 2 H$_2$O
Catalase	H$_2$O$_2$:H$_2$O$_2$ oxidoreductase	1.11.1.6	H$_2$O$_2$ + H$_2$O$_2$ → O$_2$ + 2 H$_2$O
Peroxidase	Donor:H$_2$O$_2$ oxidoreductase	1.11.1.7	Donor + H$_2$O$_2$ → oxidized donor + 2H$_2$O
Lipoxidase (lipoxygenase)	—	1.99.2.1	Unsaturated fat + O$_2$ → a peroxide of the unsaturated fat
Hydrolases			
Lipase	Glycerol ester hydrolase	3.1.1.3	Triglyceride + H$_2$O → glycerol + fatty acids
Pectin methylesterase	Pectin pectyl-hydrolase	3.1.1.11	Pectin + n H$_2$O → pectic acid + n MeOH
Chlorophyllase	Chlorophyll chlorophyllido-hydrolase	3.1.1.14	Chlorophyll + H$_2$O → phytol + chlorophyllide
Phosphatase (acid or alkaline)	Orthophosphoric monoester phosphohydrolase	3.1.3. (1,2)	An orthophosphoric monoester + H$_2$O → an alcohol + H$_3$PO$_4$

298

α-Amylase	α-1,4-Glucan 4-glucanohydrolase	3.2.1.1	Hydrolysis of α-1,4-glucan links ⎰ Internal random hydrolysis
β-Amylase	α-1,4-Glucan maltohydrolase	3.2.1.2	⎱ Successive maltose units removed
Glucoamylase	α-1,4-Glucan glycohydrolase	3.2.1.3	Successive glucose units removed
Cellulase	β-1,4-Glucan 4-glucanohydrolase	3.2.1.4	Hydrolyses β-1,4-glucan links in cellulose
Amylopectin-1,6-glucosidase (R-enzyme)	Amylopectin 6-glucanohydrolase	3.2.1.9	Hydrolyses α-1,6-glucan links in amylopectin
Polygalacturonase	Polygalacturonide glycanohydrolase	3.2.1.15	Pectic acid + $(x - 1)$ $H_2O \rightarrow x$ α-D-galacturonic acid
Maltase (α-glucosidase)	α-D-Glucoside glucohydrolase	3.2.1.20	Maltose + $H_2O \rightarrow 2$ α-D-glucose
Lactase	β-D-Galactoside galactohydrolase	3.2.1.23	Lactose + $H_2O \rightarrow$ α-D-glucose + β-D-galactose
Invertase (sucrase)	β-D-Fructofuranoside fructohydrolase	3.2.1.26	Sucrose + $H_2O \rightarrow$ α-D-glucose + β-D-fructose
Pepsin	—	3.4.4.1	
Rennin	—	3.4.4.	
Trypsin	—	3.4.4.4	
Chymotrypsin	—	3.4.4.5	
Elastase	—	3.4.4.7	
Papain	—	3.4.4.10	Hydrolysis of peptide linkages
Chymopapain	—	3.4.4.11	
Ficin	—	3.4.4.12	
Bromelain	—	3.4.4.c	
Bacterial protease	—	3.4.4.16	
Fungal protease	—	3.4.4.17	
Collagenase	—	3.4.4.19	

[a]From Eskin et al., 1971.

299

Altering the pH or temperature optima can disrupt the conformation of the enzyme and may result in loss or reduction of enzyme activity or stability. Stability denotes the length of time over which an enzyme remains catalytically active. The catalytic activity of enzymes is often expressed in terms of specific activity. Specific activity is the number of enzymatic units per mg of enzymatic protein. One unit is the amount of enzyme that catalyzes the transformation of 1 μmole of substrate per min under standardized conditions. In 1972 the Commission on Biochemical Nomenclature and Enzyme Nomenclature recommended a new unit of enzymatic activity, the katal. One katal equals 6×10^7 traditional enzyme units and is defined as the amount of activity that converts one mole of substrate per sec (see Appendix).

Industrial enzymes are derived from plant, animal, or microbial sources. The cells of the enzyme source are disrupted or broken down by physical and/or chemical methods—and an extract of the enzyme is prepared. For isolation of some enzymes, the enzyme is separated from the bulk of the liquid. From these solutions, concentrated or even purified preparations can be made. The production of high purity enzyme preparations requires fractional precipitation, differential adsorption or elution, chromatographic separation, electrophoretic separation, dialysis, crystallization, freeze-drying, or a combination of these methods (Skinner, 1975).

Enzymes in foods can be viewed from three perspectives: (a) as indigenous food enzymes; (b) as added enzyme concentrates or isolates; and (c) as enzymes produced by microorganisms and present either as contaminants or added as cultures (Acker, 1962, Fox, 1974, Whitaker, 1972; 1974). Many beneficial effects result from the activity of endogenous enzymes in foods; often they are overshadowed by deleterious effects. Consequently, much of the early work on food enzymology was concerned with the inactivation of indigenous enzymes by heat processing, with reduction of their activity by cold storage, or with inactivation of microbial contaminants by drying to safe moisture levels.

Added enzymes are important in food technology because of the roles they play in the composition, processing, and shelf life of foods (Underkofler, 1980; Whitaker, 1980). Some naturally occurring enzymes are highly desirable (e.g., amylases in sweet potatoes produce pleasing textures and flavors); others are undesirable (e.g., lipases produce rancidity and polyphenol oxidase activity results in browning). Sometimes enzyme activities are determined to measure adequacy of processing—for example, phosphatase in milk pasteurization and catalase and peroxidase in vegetable blanching.

Table 8.1 classifies enzymes significant in foods and the food industry. A brief survey of commercial applications of enzymes in food production is given in Table 8.2. Tables 8.3 and 8.4 detail enzymes and enzyme products for food processing.

In milk, lipase and phosphatase are destroyed by pasteurization; peroxidase and xanthine oxidase are destroyed by sterilization at 115°C for 15 min—a small

TABLE 8.2

Commercial Applications of Enzymes in Food Production[a]

Processing difficulty or requirement	Enzyme function	Enzyme used	Enzyme source
Milling and baking			
High dough viscosity	Catalyzes hydrolysis of starch to smaller carbohydrates by liquefaction, thus reducing dough viscosity	Amylase	Fungal
Slow rate of fermentation	Accelerates process	Amylase	Fungal
Low level of sugars resulting in poor taste, poor crusts, and poor toasting characteristics	Converts starch to simple sugars such as dextrose, glucose and maltose by saccharification, thus increasing sugar levels	Amylase	Bacterial
Staling of bread	Enables bread to retain freshness and softness longer	Amylase	Bacterial
Mixing time too long for optimum gas retention of doughs	Reduces mixing time and makes doughs more pliable by hydrolyzing gluten	Protease	Fungal
Curling of sheeted dough for soda crackers as dough enters continuous cracker ovens	Prevents curling of sheeted dough	Protease	Fungal
Poor bread flavor	Aids in flavor development	Lipoxidase Protease	Soy flour Fungal
Off color flour	Bleaches natural flour pigments and lightens white bread crumbs	Lipoxidase	Soy flour
Low loaf volume and coarse texture	Hydrolyzes pentosans	Pentosanase	Fungal
Meats			
High fat content	Aids in removal or reduction of fat content	Lipase	Fungal
Upgrade meat	‚Hydrolyzes muscle protein and collagen to give more tender meat	Protease	Ficin (figs) Papain (papaya)
Serum separation of fat in meat and poultry products	Produces liquid meat products and prevents serum separation of fat in meat products and animal foods	Protease	Fungal
High viscosity of condensed fish solubles	Reduces viscosity while permitting solids levels over 50% without gel formation of the condensed fish solubles	Protease	Fungal

(continued)

TABLE 8.2 (*Continued*)

Processing difficulty or requirement	Enzyme function	Enzyme used	Enzyme source
Protein shortage	Prepare fish protein concentrate	Protease	Fungal/bacterial
Distilled beverages			
Thick mash	Thins mash and accelerates saccharification	Amylase	Bacterial Malt
Chill haze	Chillproofs beer	Protease	Fungal Bacterial Papain
Low runoff of wort	Assists in physical disintegration of resin and improves runoff of wort	Amylase	Fungal Malt
Fruit products and wines			
Apple juice haze	Clarifies apple juice	Pectinase	Fungal
High viscosity due to pectin	Reduces viscosity by hydrolyzing the pectin	Pectinase	Fungal
Slow filtration rates of wines and juices	Accelerates rate of filtration	Pectinase	Fungal
Low juice yield	Facilitates separation of juice from the fruit, thus increasing yield	Pectinase	Fungal
Gelled purée or fruit concentrate	Prevents pectin gel formation and breaks gels	Pectinase	Fungal
Poor color of grape juice	Improves color extraction from grape skins	Pectinase	Fungal
Fruit wastes	Produces fermentable sugars from apple and grape pomace	Cellulase	Fungal
Sediment in finished product	Helps prevent precipitation and improves clarity	Pectinase	Fungal
Syrups and candies			
Controlled level of dextrose, maltose, and higher saccharides	Controls ratios of dextrose, maltose, and higher saccharides	Amylase	Bacterial Fungal Malt
High viscosity syrups	Reduces viscosity	Amylase	Bacterial Fungal
Sugar loss in scrap candy	Facilitates sugar recovery from scrap candy by liquefaction of starch content	Amylase	Bacterial
Filterability of vanilla extracts	Improves filterability of vanilla extracts	Cellulase	Fungal
Miscellaneous			
Poor flavors in cheese and milk	Improves characteristic flavors in milk and cheese	Lipase	Fungal

TABLE 8.2 (*Continued*)

Processing difficulty or requirement	Enzyme function	Enzyme used	Enzyme source
Tough cooked vegetables and fruits	Tenderizes fruits and vegetables prior to cooking	Cellulase	Fungal
Inefficient degermination of corn	Produces efficient degermination of corn	Cellulase	Fungal
High set times in gelatins	Reduces set times of gelatin without significantly altering gel strength	Protease	Fungal
Starchy taste of sweet potato flakes	Increase conversion of sweet potato starch	Amylase	Fungal Bacterial
High viscosity of precooked cereals	Reduces viscosity and allows processing of precooked cereals at higher solid levels	Amylase	Fungal Bacterial

[a]From Pulley, 1969.

percentage of α-amylase and *p*-diamine oxidase survives, however. The presence of phosphatase is used to determine adequacy of pasteurization. The technique can detect 0.1% raw milk in pasteurized milk or a drop of 2°C in temperature or a slightly inadequate length of pasteurization time. Lipases are undesirable because they may lead to rancidity when fresh milk is cooled too fast, when raw milk is agitated or homogenized; or when milk foams or is exposed to excessively high temperatures.

After an animal carcass is dressed, it is chilled. Within about 24 hours, the fat solidifies and rigor mortis sets in. This involves hardening and shortening of the muscles and toughening of the meat. Thereafter, proteolytic enzymes (cathepsins) soften the fibers at a rate that depends on the temperature of storage. Best results are obtained at temperatures slightly above freezing. Such changes are accompanied by development of desirable flavors in properly age-tenderized meat (ses also Chapter 6).

Cereal grains harvested at moisture levels below 14% exhibit slow respiratory changes. These changes are greatly enhanced at elevated moisture levels. Moisture-rich fruits and vegetables are much more perishable than cereal grains. To minimize loss, and damage, and other changes in quality, they must be stored under proper conditions of temperature, ventilation, humidity, and so on. Yet, total inactivation of respiratory enzymes is undesirable as spoilage occurs much more rapidly in dead than in live tissue. The unwanted changes can be reduced by freezing, drying, and canning. In all these preservation methods, the vegetables are blanched (heated in boiling water or steam) to destroy native enzyme systems prior to preservation. Adequacy of blanching is determined by measuring residual peroxidase or catalase. Blanching is also effective in inactivating

TABLE 8.3

Some Commercial Enzyme Preparations Used in Food Processing[a]

Trivial name	Takamine brand	Classification[b]	Source	Systematic name[c]	IUB no.[c]
α-Amylases	Clarase series	Carbohydrase	*Aspergillus oryzae*	α-1,4-glucan 4-glucanohydrolase	3.2.1.1
	Fungal amylases	Carbohydrase	*Aspergillus oryzae*	α-1,4-glucan 4-glucanohydrolase	3.2.1.1
	HT amylase series	Carbohydrase	*Bacillus subtilis*	α-1,4-glucan 4-glucanohydrolase	3.2.1.1
	Tenase	Carbohydrase	*Bacillus subtilis*	α-1,4-glucan 4-glucanohydrolase	3.2.1.1
Cellulase	Cellulase	Carbohydrase	*Aspergillus niger*	β-1,4-glucan glucanohydrolase	3.2.1.4
Glucoamylase	Diazyme	Carbohydrase	*Aspergillus niger*	α-1,4-glucan glucohydrolase	3.2.1.3
Pectinase[d]	Spark-L	Carbohydrase	*Aspergillus niger*	Poly-α-1,4 galacturonide glycanohydrolase	3.2.1.15
				Pectin pectylhydrolase	3.1.1.11

Glucose oxidase	Dee O	Oxidoreductase	*Aspergillus niger*	β-D-glucose: oxygen oxidoreductase	1.1.3.4
Catalase	Catalase L	Oxidoreductase	Bovine liver	Hydrogen peroxide: hydrogen peroxide oxidoreductase	1.11.1.6
Lipases	Pancreatic lipase	Lipase	Animal pancreatic tissues	Glycerol-ester hydrolase	3.1.1.3
	Lipase powders	Lipase	Edible forestomach tissue of calves, kids, and lambs	Carboxylic-ester hydrolase	3.1.1
Proteases	HT proteolytic	Protease	*Bacillus subtilis*	Peptide hydrolase	3.4
	Fungal protease	Protease	*Aspergillus oryzae*	Peptide hydrolase	3.4
	Bromelain	Protease	Pineapples: *Ananas comosus, Ananas bracteatus* (L)	Peptide peptidohydrolase	3.4.4.24
	Papain	Protease	Papaya: *Carica papaya* (L)	Peptide peptidohydrolase	3.4.4.10

[a]Courtesy of Marschall Division, Miles Laboratories, Elkhart, Indiana.
[b]Classification according to the First Supplement to the Food Chemicals Codex, 2d edition, 1974.
[c]Enzyme nomenclature, 1964 Recommendations of International Union of Biochemistry, Elsevier, Amsterdam, 1965.
[d]A mixture of polygalacturonase and pectin methylesterase.

TABLE 8.4

Uses of Some Commercial Enzyme Preparations[a]

Product	Description	Major Food Uses	Suitable operating conditions	Principal enzymes
Carbohydrases				
Clarase series	A series of fungal enzymes (primarily α-amylases) characterized by both dextrinizing and saccharifying action on starch. Hydrolyze starch more extensively than most α-amylases, resulting in the formation of substantial quantities of maltose and low levels of glucose.	Fruit juices: Facilitates filtration by converting starch to soluble dextrins of lower viscosity, and eliminates starch hazing problems Chocolate syrups: Converts cocoa starch to dextrins and sugars, reducing viscosity and insuring a stable syrup Syrups: Hydrolyzes acid and enzyme liquefied syrups to intermediate or high maltose syrups. In combination with Diazyme enzyme produces high conversion and high fermentable noncrystallizing syrups.	Effective pH range: 4.0–6.6 Effective temperature range: 20–60°C	α-amylase Phosphatase Protease
Fungal amylase	A fungal α-amylase for flour supplementation at the mill or bakery. Rapidly dextrinizes and progressively saccharifies gelatinized starch, yielding substantial quantities of maltose and glucose.	Baking: Results in increased gas production, improved crust color, improved moistness and keeping quality of the crumb and additional flavor contributed by caramelized sugar in the crust	Effective pH range: 4.0–6.6 Effective temperature range: 20–60°C	α-amylase Protease Phosphatase

	Description	Applications	Effective conditions	Enzyme(s)
Tenase	A liquid bacterial α-amylase for economical starch liquefaction at high temperatures. Displays thermal stability and is compatible with batch and continuous starch liquefaction processes.	Syrups: Liquefies starches in dual-enzyme processes to produce dextrose syrups, crystalline dextrose, high fermentable dextrose, high fermentable syrups and high maltose syrups. Distilling: Economically replaces malt for grain mash liquefaction	Effective pH range: 5.0–7.5. Effective temperature range: 20–90°C	α-amylase
HT-amylase series	A series of powdered bacterial α-amylases characterized by an ability to liquefy and dextrinize starch polymers at temperatures above that of starch gelatinization. Hydrolysis products are primarily dextrins with small quantities of glucose and maltose. There is a corresponding rapid decrease in viscosity of starch slurries with a limited increase in fermentable and reducing sugars.	Brewing: Rapidly and completely liquefies adjuncts and removes starch haze. Candies: Aids in scrap candy recovery. Cereals: In the production of precooked cereals, reduces the viscosity of the cereal prior to roll drying. Chocolate syrups: Converts cocoa starch to dextrins and sugars, reducing viscosity and insuring a stable syrup. Distilling: Used for effective mash liquefaction. Starches: Reduces viscosity in gelatinized starch slurries for easier handling or to prepare specific viscosity products	Effective pH range: 5.0–7.5. Effective temperature range: Up to 90°C	α-amylase, Protease
Cellulase	A fungal cellulase system primarily active on soluble forms of cellulose (Cx—component) with minor ac-	Fruit juices: Aids in extraction and clarification of juice from citrus fruits. Catalyzes hydrolysis of grape and ap-	Effective pH range: 3.0–5.0. Effective temperature range: 20–60°C	Cellulase, Hemicellulase, Pectinase

(continued)

TABLE 8.4 (*Continued*)

Product	Description	Major Food Uses	Suitable operating conditions	Principal enzymes
	tivity on highly ordered forms (C₁—component). Hemicellulase component of the system rapidly reduces the viscosity of several plant gums.	ple pomace to fermentable sugars. Essential oils and spices: Increases yields of essential oils and other plant extracts.		
Diazyme	A glucoamylase which removes successive glucose units from the non-reducing ends of starch polymers. Unlike α-amylase and β-amylase, can hydrolyze both the linear and branched glucosidic linkages of starch, amylose, and amylopectin. Essentially transglucoside-free. Hydrolyzes starch polymers to glucose in quantitative yields	Distilling: Economically and compatibly replaces malt in existing processes. Dextrose: Hydrolyzes corn, wheat, potato, and other starches to dextrose, producing a final syrup with excellent color, taste and low ash. Brewing: Provides a continuous supply of fermentable sugar by converting nonfermentable dextrins to fermentables, thereby increasing alcohol yields and reducing the carbohydrate content of beer Fermentation: Converts starch to dextrose, which serves as a growth medium or nutrient in industrial fermentation processes.	Effective pH range: 3.5–5.0 Effective temperature range: 30–60°C	Glucoamylase

Spark-L	A fungal pectic enzyme system designed for efficient de-polymerization of naturally occurring pectins. Extremely flexible, and effective for a wide variety of plant processing conditions, including a broad range of pH and temperature.	Starches: Applicable in the production of vinegar, yeast, and other starch-based products Fruit juices: Improves total juice yield and produces clear juice of exceptionally high quality from apples, cranberries, grapes and other fruits Wines: Increases free-run and total juice yield. Resulting wines are clear with excellent color and body. Jams and jellies: Used to remove pectin prior to gel standardization Fruits and vegetables: Hydrolyzes undesirable gel formations in fruit and vegetable purees and extracts	Effective pH range: 2.5–5.5 Effective temperature range: 5–65°C.	Cellulase Hemicellulase Protease Pectin methylesterase Polygalacturonase
Proteases HT-Proteolytic	A bacterial protease system extremely effective for protein hydrolysis over the neutral and alkaline pH range. Primarily a neutral protease, also demonstrates significant liquefying α-amylase activity.	Proteins: Hydrolyzes plant and animal proteins where a significant level of hydrolysis and lower molecular weight peptides are desired Baking: Produces excellently flavored, quality crackers and cookies in a fraction of	Effective pH range: 6.0–8.5 Effective temperature range: 20–55°C	α-amylase Neutral protease Alkaline protease

(continued)

TABLE 8.4 (*Continued*)

Product	Description	Major Food Uses	Suitable operating conditions	Principal enzymes
Fungal protease	A fungal system of endo- and exo- peptidases capable of catalyzing the hydrolysis of a wide range of peptide bonds. Primarily an acid protease, fungal protease is active over a broad pH range, and demonstrates significant saccharifying α-amylase activity.	the normal processing time. Improves handling of pizza doughs Baking: Improves grain, texture and compressibility of bread crumb, reduces mixing requirements of sponge doughs and increases loaf volume and color Proteins: Hydrolyzes and modifies plant and animal protein under acid conditions Meats: Major component in meat tenderizing formulations	Effective pH range: 4.0–7.5 Effective temperature range: 20–50°C	α-amylase Protease
Bromelain	A mixture of proteases, isolated from the pineapple plant, capable of hydrolyzing both plant and animal proteins to peptides and amino acid. Depending upon hydrolysis conditions, bromelain can hydrolyze protein substrates to large or small peptides.	Meats: A useful ingredient in tenderizer formulations Baking: Produces excellently textured sugar wafers, waffles and pancakes. Improves handling of pizza doughs. Fish: Used to process inedible and scrap fish for fish meal, oil and solubles Proteins: Reduces the viscosity of protein solutions used in the manufacture of protein hydrolysates. Eliminates protein hazes.	Effective pH range: 4.0–9.0 Effective temperature range: 20–65°C	Protease

Papain	A potent protease preparation derived from papaya latex by a purification procedure designed to activate all latent proteolytic activity. Papain has a broad substrate specificity which allows it more extensively to hydrolyze proteins. It is further characterized by excellent stability at elevated temperatures.	Brewing: Stabilizes and chillproofs beer. Meats: Tenderizes meat. Proteins: Ideal for protein modifications, either for modifying the physical properties of dispersed protein, or for increasing dispersibility	Effective pH range: 6.0–8.0 Effective temperature range: 20–75°C	Protease
Dee O Series	A glucose oxidase–catalase system available in differing purities, catalyzes the oxidation of glucose in the presence of molecular oxygen. In contrast with chemical antioxidants, glucose oxidase offers continuing protection because it is not oxidized during catalysis.	Eggs: Stabilizes egg solids by desugaring whites, yolks or whole eggs Beverages: Removes oxygen from carbonated and still beverages, thereby increasing shelf life Salad dressings: Removes oxygen to prevent undesirable flavor changes in oxygen sensitive foods	Effective pH range: 4.5–7.0 Effective temperature range: 30–60°C	Glucose oxidase Catalase
Catalase L	A beef liver catalase stabilized in a liquid formulation. Decomposes hydrogen peroxide into molecular oxygen and water over a broad pH range.	Milk: Rapidly and safely destroys excess hydrogen peroxide in milk after that substance has been used to destroy harmful microorganisms in the milk Eggs: In desugaring eggs, decomposes hydrogen peroxide to liberate oxygen required by glucose oxidase to cata-	Effective pH range: 6.5–7.5 Effective temperature range: 5–45°C	Catalase

(continued)

TABLE 8.4 (Continued)

Product	Description	Major Food Uses	Suitable operating conditions	Principal enzymes
		lyze the oxidation of glucose to gluconic acid		
		Sterilization: Removes residual peroxide after its use in destroying salmonella and staphylococci in food products such as eggs and starchy preparations		
Pancreatic lipase	A porcine pancreas-derived lipase capable of hydrolyzing insoluble fats and fatty acid esters to yield diglycerides and monoglycerides as intermediate products. Prolonged hydrolysis results in glycol formation with liberation of free fatty acids.	Eggs: Improves whipping properties of egg albumin Fats: Solubilizes or modifies fats or fatty acid esters. Breaks down fats acting as emulsifiers.	Effective pH Range: 5.5–9.5 Effective temperature range: 20–50°C	Lipase α-amylase Protease
Lipase powders	A series of lipase powders derived from glandular edible tissue. Lipase powders specifically hydrolyze milk fat, liberating a reproducible ratio of short chain fatty acids.	Dairy: Used in the manufacture of Italian types of cheese.	Effective pH range: 5.5–9.5 Effective temperature range: 20–50°C	Lipase
Specialty enzymes Brew Ñ zyme	A blend containing primarily α-amylase and protease. Designed specifically to replace	Brewing: Economically replaces a portion of the malt in the mashing operation to	Effective pH range: α-amylase: pH 5.0–7.5	α-amylase Protease β-glucanase

Name	Description	Application	Operating conditions	Enzyme
	a portion of the brewers malt in the production of wort.	produce a wort of equal character and quality	Protease: pH 5.5–7.0 Effective temperature range: α-amylase: 70–75°C. Protease: 45–55°C	α-amylase Glucoamylase Protease
Dextrinase A	A fungal amylase system for the production of 62–65 DE noncrystallizing syrups with excellent glucose–maltose ratios.	Syrups: Converts 38–42 DE acid syrups to 62–65 DE noncrystallizing syrups with 35–40% both glucose and maltose	Effective pH range: 5.0–5.3 Effective temperature range: 50–55°C	
Hemicellulase CE-100	An enzyme preparation that demonstrates a high degree of specificity for a class of polysaccharides designated as galactomannans such as locust bean gum, guar gum, soybean hull gum, and coffee gum.	Coffee: Hydrolyzes coffee gums which cause liquid coffee concentrates to gel	Effective pH range: 3.5–6.0 Effective temperature range: 30–65°C	Cellulase Hemicellulase
Marzyme	A milk coagulant produced by the pure culture fermentation of the organism *Mucor miehei*.	Cheese: Economically replaces rennet in the production of a wide variety of cheeses.	Operating conditions depend on specific type of cheese being produced	Protease
Tendrin	A blend of proteases, testing. Hydrolyzes protein to a proper point, avoiding over-tenderization.	Meat tenderizers: Compatible with most other ingredients in powdered or liquid tenderizers.	Effective pH range: 4.5–7.5 Effective temperature range: 40–60°C	Protease

[a]From Marschall Division, Miles Laboratories, Elkhart, Indiana.

heat-sensitive polyphenol oxidases, which are responsible for enzymatic browning of cut surfaces when they are exposed to oxygen.

Industrial enzymes are produced from animal and plant tissues and from microbial sources: molds, yeasts, and bacteria (Beckhorn *et al.*, 1965). Advantages of enzymatic preparations in food processing include (a) specificity; (b) controlled action at biological temperatures and pH; (c) high efficacy; and (d) ease of inactivation at the end of reaction (Schultz, 1960; deBecze, 1970; Villadsen, 1972; Wieland, 1972; Wingard, 1972; Reed, 1975; Wiseman, 1975; Pintauro, 1979; Birch *et al.*, 1980).

I. Major Classes of Enzymes and Their Applications

A. Carbohydrases

Carbohydrases hydrolyze polysaccharides or oligosaccharides. Starch-splitting enzymes include amylases of three types (see Fig. 2.41):

$$\text{starch} \xrightarrow{\text{α-amylase}} \text{dextrins} + \text{maltose}$$

$$\text{starch} \xrightarrow{\text{β-amylase}} \text{maltose} + \text{dextrins}$$

$$\text{starch} \xrightarrow{\text{glucoamylases}} \text{glucose}$$

The α-amylases are basically dextrinogenic enzymes that hydrolyze $\alpha\text{-}(1 \rightarrow 4)$ linkages at random. They rapidly liquefy large starch molecules and after prolonged action produce maltose. They cannot hydrolyze $\alpha\text{-}(1 \rightarrow 6)$ linkages. They produce large amounts of low molecular weight dextrins, maltose, maltoligosaccharides, and small amounts of glucose and of the trisaccharide panose (which has the original $\alpha\text{-}(1 \rightarrow 6)$ linkage). The α-amylases are from fungal, bacterial, and plant sources. They vary widely in their thermostability: bacterial amylases are most thermostable, fungal amylases are most thermolabile, and plant amylases are intermediate. The β-amylases are produced by higher plants (cereals and sweet potatoes). This class is a saccharifying enzyme. The β-amylases produce only maltose by splitting it progressively off the nonreducing end of linear chains. If they act on amylopectin, their activity ceases at the $\alpha\text{-}(1 \rightarrow 6)$ linkage and produce a β-limit dextrin. Glucoamylase (also called amylogucosidase) is formed mainly by fungi, and is a saccharifying enzyme. It produces only glucose by progressively hydrolyzing glucose units from the nonreducing end.

Amylases have many uses; for example, in the production of sweet syrups (see

Chapter 2), in baking, in saccharification of fermentation mashes in distilling and brewing industries, and in starch modification or removal (Pomeranz, 1966). The four most important disaccharide-splitting enzymes (all of microbial, mainly yeast, origin) are

$$\text{maltose} \xrightarrow{\text{maltase}} \text{glucose } + \text{ glucose}$$

$$\text{sucrose} \xrightarrow{\text{invertase}} \text{glucose } + \text{ fructose}$$

$$\text{lactose} \xrightarrow{\text{lactase}} \text{glucose } + \text{ galactose}$$

$$\text{melibiose} \xrightarrow{\text{melibiase}} \text{glucose } + \text{ galactose}$$

Maltase is used very little on an industrial scale. Invertase is used in the production of artificial honey and invert sugar for noncrystalline jams, to prevent crystallization of concentrated molasses, and in removal of sucrose from foods. Invertase is important in the manufacture of confections, liqueurs, frozen desserts, and chocolate-coated liquid center candies. Hydrolysis of lactose by α-D-galactosidase increases sweetness, solubility, and osmotic pressure.

Lactase can be used to prevent lactose crystallization in ice cream, as lactose imparts a grainy or sandy texture. It is also used in whey concentrates for animal feeds, in frozen milk concentrates, in bread making, and in cases of lactose intolerance. Cleavage of lactose to glucose and galactose increases the sweetness, and produces a fermentable sugar (in panary fermentation) and two reducing sugars that contribute to browning reactions. Fungal melibiase, an α-galactosidase, is finding increased applications in hydrolyzing raffinose to galactose and sucrose.

Pectic substances (see also Chapter 3) are structural polysaccharides in the middle lamella and the primary cell wall of higher plants (Rombouts, 1981). They are basically α-D-(1 → 4)-galacturonans partially esterified with methyl groups (see Chapter 3). L-Rhamnose residues are linked to the main galacturonan chain and hemicellulose side chains may be bound to either galacturonate or to rhamnose residues. In the middle lamella, primary cell wall cellulose fibers are bound through a hemicellular–pectin complex into a relatively rigid network. Fresh fruits contain 0.2–4.5% pectin. The richest sources are citrus fruits; excellent sources of pectin are lemon rind and pulp (about 25–30%) and sugar beet pulp.

According to Rombouts (1981) there are two main groups of pectic enzymes: deesterifying (pectin esterases) and chain-splitting (depolymerases). The pectinesterases deesterify pectins to produce methanol and pectates. Plant pectinesterases have a single chain, zipperlike action; fungal pectinesterases have a multiple chain, random action. The depolymerases split the glycosidic bonds either by hydrolysis (glycosidases) or by β-elimination (lyases). Polygalacturonases are produced by several fruits, many fungi, and some yeasts and bacteria. Pectate lyases are produced by bacteria, protozoans, and some phytopathogenic fungi. The basic structures of galacturonan and the points of attack of pectin depolymerases on their substrates are given in Fig. 8.1 (Rombouts and Pilnik, 1978).

According to Underkofler (1980), pectic enzymes can be divided into those that act on pectin and those that act on pectic acid (see Chapter 3). Those that act on pectin are methylesterases, polymethylgalacturonases (endo and exo types), and pectin lyases (endo and exo types). Those that act on pectic acid are polygalacturonases (endo and exo types) and pectic acid lyases (endo and exo types). The enzymes are found in plant tissues, mainly of fruits, and in microorganisms, mainly fungi. Pectinolytic enzymes have two functions in food processing. Modification of pectin characteristics is the first function. Natural pectins vary in degrees of esterification between 20 and 80%. High methoxyl pectins (75% esterification) require high sugar levels (above 50%) for gelation. Low methoxyl pectins in the presence of Ca^{2+} can accomplish such gelation at 30% sugar levels. Naturally occurring low methoxyl pectins, for example, from sugar beets, are highly acylated and ineffective in gelation. Pectins can be de-esterified by acid or alkali treatment, or as stated earlier, by pectinolytic enzymes. Second, the enzymes can increase juice yields and dehaze or clarify juices, such as apple juice (Fox, 1974). According to Underkofler (1980), pectic enzymes are used to increase juice yields, clarify juices, reduce viscosity of fruit purees and concen-

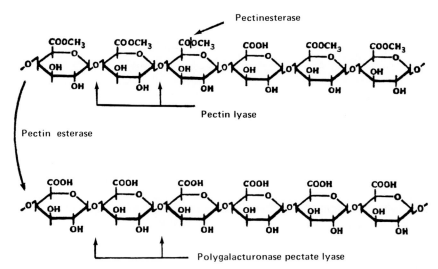

Fig. 8.1. Basic structures of galacturonan and the points of attack of pectinesterase and pectin depolymerases on their substrates. (From Rombouts and Pilnik, 1978.)

TABLE 8.5

Application of Commercial Pectolytic Enzymes in Fruit and Vegetable Juice Technology[a]

Clarification of fruit juices, e.g., apple juice; depectinized juices can also be concentrated without gelling and without developing turbidity
Enzyme treatment of pulp of soft fruit, red grapes, citrus, and apples, for better release of juice (and colored material); enzyme treatment of pulp of olives, palm fruit, and coconut flesh to increase oil yield
Maceration of fruits and vegetables (disintegration by cell separation) to obtain nectar bases and baby foods
Liquefaction of fruits and vegetables, to obtain products with increased soluble solids content (pectolytic and cellulolytic enzymes combined)
Special applications: preparation of clouding agents from citrus peel, cleaning of peels for use in "candying" and marmalade production, recovery of oil from citrus peel, depectinizing citrus pulp wash

[a]From Rombouts and Pilnik, 1978

trates, produce low sugar jellies, improve extraction of juice and color of wines, digest fruit peel, and recover citrus oil. (See Table 8.5.)

Cellulases and hemicellulases are used in the production of glucose from cellulosic wastes (for dextrose, SCP, and alcohol, and other fermentation products), in agar recovery from seaweeds, in upgrading commercial feeds, in processing dehydrated vegetables and fruits, in tenderization of fibrous vegetables, and in removal of soybean seed coat (Underkofler, 1980).

A considerable number of proteases of plant, animal, and microbial origin are used by the food industry. Generally, commercial preparations are mixtures of several proteolytic enzymes. Some of the uses include dairy, baking, brewing, tenderizing meat, production of fish solubles and oriental condiments. Rennin is used to coagulate milk in cheese production. Cheese is manufactured essentially as follows (see also Chapter 11):

$$\text{casein} \xrightarrow[\text{microbial culture}]{\text{rennin}} \text{paracasein} \xrightarrow{\text{Ca}^{2+}, 30°C} \text{clot (whey)} \xrightarrow[\text{adjustment}]{\text{pH and temperature}}$$

$$\text{"green cheese"} \xrightarrow[\text{bacterial enzymes}]{\text{rennin}} \text{ripe cheese}$$

Most proteolytic enzymes can clot milk but may dissolve the curd by hydrolyzing the insoluble protein. The proteolytic action of rennin in the curd is limited. The mechanism of curd formation involves limited proteolysis of κ-casein to form a soluble glycopeptide and para-κ-casein. The clotted paracasein then enhances the precipitation of other casein fractions which have interacted with calcium ions released during hydrolysis of κ-casein.

The main source of commercial rennet has been the stomach mucosa of calves.

In recent years, vegetable (ficin) and microbial (mainly fungal) rennets have become increasingly popular (often for economic or religious reasons).

Proteolytic enzymes are used to make the gluten in tough ("bucky") doughs more pliable, reduce mixing time, and improve overall bread processing, quality, and consumer acceptance. The use of microbial proteases is widespread in cracker, biscuit, and cookie doughs to reduce processing times, improve quality of end products, and reduce amounts of expensive ingredients (such as shortening). Proteases in the brewing industry are used to eliminate (or at least reduce) "chill haze" in beer cooled to temperatures below 10°C. Papain is most commonly used for this purpose, although any proteolytic enzyme active at pH 4.5 can be used. Besides papain, pepsin, ficin, bromelain, and fungal and bacterial proteases may be employed. The amount of papain required to accomplish effective chillproofing (about 8 ppm) does not adversely affect the functional properties of beer, that is, body, foam formation and retention, or taste.

Meat is tender immediately postmortem but toughness is considerably increased with the formation of actomysin at the onset of rigor mortis. Meat again becomes tender after aging for about 1 week at 3°C, during which the main change is disintegration of the Z-line. During this process, there is little change in the solubility of the connective tissue or of actomysin (Fox, 1974). Proteases are used to tenderize beef. Generally, the plant proteases (such as papain, bromelain, or ficin) used to loosen connective tissues, mainly collagen and elastin, are combined with bacterial and/or fungal proteases to affect muscle fibers. In recent years, some attention has been given to tenderization of poultry and ham (see Chapter 6).

Enzymatic hydrolysates of proteins in soybeans or wheat gluten have bland, pleasant flavors. Enzymatic hydrolysis of milk proteins may produce bitter tastes, which must be removed by carboxypeptidases or aminopeptidases.

Lipases are of limited application in foods. Applications include development of flavors in dairy products (ice creams, cheeses, margarine, and creams) and chocolate confections, and improvement of egg white whipping properties. The main sources of lipases are animal (pancreatic or pregastric tissue) or microbial. The role and action of lipases depend on the surface area of emulsions upon which they act. In addition, most lipases show specificity with regard to chain length, degree of saturation, and position of fatty acids. Some lipases may impart soapy or rancid flavors; some are useful in manufacture of special products (Italian-type cheeses, butter oils, creams); and some require very tight control to avoid aberrant and atypical flavors. Lipolyzed creams can be used in certain candies, confections, snacks, and sauces. Lipolyzed butter products enhance the flavors of cheese dips, sauces, dressings, soups, gravies, baked products, and sweets. Free fatty acids in modified butterfat-containing products can vary widely in their effects on chocolate confections: small amounts enhance the natural flavor, intermediate amounts impart buttery flavors, and high levels impart cheesy flavors (Underkofler, 1980).

B. Oxidoreductases

Oxidoreductases vary widely in their effects on foods. Some cause deteriorative changes such as excessive browning by polyphenoloxidases, destruction of ascorbic acid by ascorbic acid oxidase, or development of objectionable oxidative flavors by peroxidase. Others may improve food quality; for example, the lipoxygenases, glucose oxidases, and catalases. Lipoxygenase from minimally heat-treated soybeans can be used to bleach carotenoid pigments in wheat flour. This enzyme also affects the functional properties of wheat flour doughs (mixing, bread crumb, grain, and flavor). Glucose oxidase of fungal origin can oxidize glucose to gluconic acid. In this reaction, both glucose and oxygen are removed from the system. Powdered eggs deteriorate during storage because of nonenzymatic browning reactions between glucose and proteins. As a result of this reaction, whipping properties and foam stability of the eggs are reduced and at advanced stages of browning, the egg products are discolored, have "off" flavors, and their protein solubility is markedly lowered. The undesirable changes can be reduced by removing glucose from egg albumin or whole eggs prior to drying, with glucose oxidase.

Glucose oxidase can be used to remove oxygen from canned and bottled beer and thereby improve the stability of flavor. It can remove oxygen from apple wine, thereby preventing growth of certain microorganisms and "off" flavor development. Similar improvement can be observed in preventing color and flavor deterioration of canned or bottled soft drinks, especially citrus drinks. Removal of oxygen also may be important in reducing hazards of corrosion in canned acid foods.

Removal of oxygen from mayonnaise slows color changes and rancidity development. Glucose oxidase incorporated into a water-impermeable, oxygen-permeable package can be placed into a hermetically sealed container along with a dry food product, such as whole milk powder, roasted coffee, white cake mix, or active dry yeast. In this case, the glucose oxidase acts as oxygen scavenger and protects the dry or dehydrated food from oxidative deterioration. Catalases of animal, bacterial, and fungal sources are employed in conjunction with glucose oxidase to remove traces of hydrogen peroxide. Catalase converts hydrogen peroxide to water and oxygen. Catalase can be used in cold sterilization or preservation of milk by hydrogen peroxide in cheese production.

C. Production of Food Flavors

Fungal phosphodiesterases can be used to produce 5'-nucleotide flavor enhancers, such as 5'-inosine monophosphate and 5'-guanosine monophosphate. These compounds accentuate "meaty" flavors and are used in canned vegetables, soups, sauces, gravies, and so on. Some flavorless precursors can be modified by enzymatic reactions to restore, enhance, or develop flavor components in foods. Finally, enzymes are used to remove some flavor components or to stabilize them and prevent undesirable modifications.

II. Immobilized Enzymes

Usefulness of enzymes in food systems has increased dramatically as a result of progress in stabilization, isolation, cofactor regeneration, and especially through advances in immobilization. Enzymes can be immobilized on water-insoluble materials while retaining their catalytic activity (Weetall, 1974; Wingard *et al.*, 1976; Chang, 1977; Konecny, 1977). Water-insoluble derivatives of enzymes are of great practical and theoretical value. They can be readily added to or removed from a reaction mixture, permitting close control of the reaction, automation, and reuse of the immobilized enzyme. These materials can be used for continuous processing. Immobilized enzymes are of theoretical value because like many enzymes in living cells they are attached to a matrix and often have characteristics that differ from the soluble form of the enzyme (Dunlap, 1975; Hultin, 1974; Zaborsky, 1973).

Several methods are available for attaching an enzyme to a water-insoluble matrix. They are depicted diagrammatically in Fig. 8.2 (Weetal, 1974). Enzymes can be adsorbed on charcoal, organic polymers, glass, mineral salts, metal oxides, and various siliceous materials. The most common materials for entrapment are polymers of synthetic (mainly polyacrylamide) or natural origin. Enzymes have been encapsulated in membranes of various polymers; the selective membrane retains the enzymes and permits entry of low molecular weight substrates. Enzymes can be attached to ion-exchange resins by electrostatic interactions. Polymerization with bifunctional agents is the basis of immobilization by cross-linking. Enzymes prepared by both adsorption and cross-linking are highly stable. The enzymes can be adsorbed to colloidal silica and cross-linked with glutaraldehyde. Enzymes can be copolymerized with maleic anhydride and ethylene that was previously reacted with ethylenediamine. And, finally, enzymes can be attached to organic polymers; the covalent attachment offers the most permanent method of immobilization. Immobilization of enzymes changes their characteristics. The most important factor that determines the value of immobilized enzymes for commerical uses is their thermal and operational stability. Some derivatives show decreased stability and others show increased stability. Immobilization also changes pH profiles and kinetics. Weetall (1974) listed several important applications for immobilized enzymes: in hydrolysis of proteins to peptides and amino acids, in cheese manufacture, conversion of corn starch to dextrose and of dextrose to fructose, clarification of wine and fruit

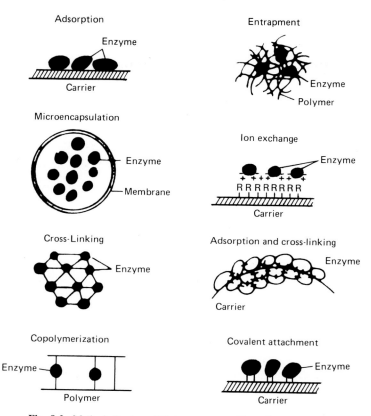

Fig. 8.2. Methods for immobilizing enzymes. (From Weetall, 1974.)

juices, removal of glucose from egg white, chillproofing beer, inversion of sucrose with invertase, stabilization of milk, and utilization of new and inexpensive raw materials or by-products.

According to Olson and Richardson (1974), the activity of enzymes almost invariably decreases as a result of immobilization. They found a number of successful applications for immobilized enzymes: resolution of racemic mixtures of amino acids employing aminoacylase conjugates, glucose isomerase, invertase, cleavage of lactose to glucose and galactose, several amylolytic enzymes, use of proteases in chillproofing beer, continuous coagulation of milk, inhibition of oxidative rancidity of milk, flavor modification to destroy naringin (the bitter component of grapefruit juice), conversion of adenosine to inosine, and continuous synthesis of L-amino acids (see also Kilara and Shahani, 1979). The development of immobilized enzymes for production of high fructose corn syrup was described by Carasik and Carroll (1983), and current and potential uses of immobilized enzymes were reviewed by Hultin (1983).

Appendix: Assay Unit Definitions

**Enzyme
Classification**   **Unit Definitions**

Carbohydrase

α-Amylase
Maltose assay   One maltose unit (MU) is that amount of enzyme which will
                                       liberate 1 mg of reducing sugar as maltose in 30 min
                                       under the conditions of the assay.

Determination of liquefying   One modified Wohlgemuth unit (MWU) is that amount of
  amylase (modified   enzyme which will dextrinize 1 mg of soluble starch to a
  Wohlgemuth method)   definite size dextrin in 30 min under the conditions of the
                                       assay.

α-Amylase assay (method of   One SKB unit is that amount of enzyme which will dex-
  Sandstedt, Kneen and Blish,   trinize 1 gm of β-limit dextrin to a definite size dextrin in
  see Chapter 2)   1 hr under the conditions of the assay.

Assay of bacterial amylase used   One bacterial amylase unit (BAU) is that amount of enzyme
  in desizing   which will dextrinize 1 mg of soluble starch to a definite
                                       size dextrin per min under the conditions of the assay.

Determination of bacterial   Thermostability is expressed as percent relative activity
  α-amylase thermostability   retained with time under the conditions of the assay.
Cellulase
Viscometric cellulase assay   One cellulase unit (CU) is that amount of enzyme which will
                                       produce a change in the relative fluidity of 1 in a defined
                                       sodium carboxymethyl cellulose substrate in 5 min under
                                       the conditions of the assay.

Glucoamylase
  Diazyme assay   One diazyme unit (DU) is that amount of enzyme which will
                                       liberate 1 gm of reducing sugar as glucose/hr under the
                                       conditions of the assay.

72-hour maltose digestion assay   A glucoamylase is considered transglucosidase-free when the
  (transglucosidase assay)   resulting specific rotation of its maltose digestion is less
                                       than the resulting rotation of a maltose digestion con-
                                       ducted with a glucoamylase standard known to be trans-
                                       glucosidase-free.

Pectinase
  Apple juice depectinizing   Apple juice depectinization units (AJDU) are determined by
  assay   correlating depectinization time of a defined apple juice
                                       substrate by the unknown pectinase with depectinization
                                       time by a pectinase standard of known activity.

Pectinmethylesterase assay   One pectinmethylesterase unit (PMEU) is that amount of
                                       enzyme which will liberate 1 μmole of titratable carboxyl
                                       groups/min under the conditions of the assay.

Viscometric polygalactouronase   One polygalacturonase unit (PGU) is that amount of enzyme
  assay   which will cause a change in relative fluidity of 0.1/sec in
                                       a defined sodium polypectate substrate under the condi-
                                       tions of the assay.

Enzyme Classification

Oxidoreductase
Glucose Oxidase

Manual assay method for glucose oxidase

Catalase
Keil assay

Exhaustion method for catalase

Lipase

Lipases
Lipase assay

Esterase assay

Protease

Proteases
Colorimetric Northrop assay

Spectrophotometric detergent protease assay

Half-hour colorimetric hemoglobin assay

Viscometric protease assay

Spectrophotometric neutral protease assay

Acid protease assay

Unit Definitions

One glucose oxidase unit (GOU) is equivalent to that amount of enzyme which will cause the uptake of 10 mm^3 oxygen/min under the conditions of the assay.

One Keil unit (KU) is that amount of enzyme which will decompose 1 gm of 100% hydrogen peroxide in 10 min under the conditions of the assay.

One exhaustion unit (EU) is that amount of enzyme which will decompose 300 mg of hydrogen peroxide under the conditions of the assay.

One lipase unit (LU) is that amount of enzyme which will liberate 1 meq of fatty acid in 2 hr under the conditions of the assay.

One FCC esterase unit (FCC EU) is that amount of enzyme which will liberate 1.25 micromoles of butyric acid per 1 min under the conditions of the assay.

One Northrop unit (NU) is that amount of enzyme which gives 40% hydrolysis of 1 liter of the casein substrate in 60 min under the conditions of the assay

One detergent unit (MDU) is that amount of enzyme which will liberate 10 nmole of tyrosine/min under the conditions of the assay.

One hemoglobin unit (HU) is that amount of enzyme which will liberate 0.0447 mg of nonprotein nitrogen in 30 min under the conditions of the assay.

One viscometric protease unit (VU) is that amount of enzyme which will produce a relative fluidity change of 0.01/sec in a defined gelatin substrate under the conditions of the assay.

One neutral protease unit (NPU) is that amount of enzyme which will liberate 1 μmole of tyrosine/min under the conditions of the assay.

One acid protease unit (APU) is that amount of enzyme which will produce a change in absorbance (ΔA) of 1 at 660 nm under the conditions of the assay.

(*continued*)

Enzyme Classification	**Unit Definitions**
Spectrophotometric assay for bromelain	One bromelain tyrosine unit (BTU) is that amount of enzyme which will liberate 1 μmole of tyrosine/min under the conditions of the assay.
Papain assay	One NF papain unit (NF PU) is that amount of enzyme which will liberate the equivalent of 1 microgram of tyrosine/hr under the conditions of the assay.

Formulations and Specialty Enzymes

Dextrinase evaluation procedure	The dextrinase ratio is the percent reducing sugar divided by the percentage glucose of a starch solution hydrolyzed under the conditions of the assay.
Viscometric hemicellulase assay	One hemicellulase unit (HCU) is that amount of enzyme which will produce a relative fluidity change of 1 in a defined locust bean gum substrate/5 min under the conditions of the assay.
Colorimetric lactase assay	One lactase unit (LU) is that amount of enzyme which will liberate 1 μmole of *o*-nitrophenol/min under the conditions of the assay.
Colorimetric pentosanase assay	One xylanase unit (XU) is that amount of enzyme which will liberate 1 μmole of reducing sugar as xylose/min under the conditions of the assay.

Source: Marshall Division Miles Laboratories, Elkhart, Indiana.

REFERENCES

Acker, L. (1962). Enzymic reactions in foods of low moisture content. *Adv. Food Res.* **11**, 263–330.

Beckhorn, E. J., Labbee, M. D., and Underkofler, L. A. (1965). Production and use of microbial enzymes for food processing. *Agric. Food Chem.* **13**, 30–34.

Birch, G. G., Blakebrough, N., and Parker, K. J., eds. (1980). "Enzymes and Food Processing." Appl. Sci. Publ., London and New York.

Carasik, W., and Carroll, J. O. (1983). Development of immobilized enzymes for production of high-fructose corn syrup. *Food Technol.* **37**(10), 85–91.

Chang, T. M. S., ed. (1977). "Biomedical Applications of Immobilized Enzymes and Proteins," Vol. 1. Plenum, New York.

Commission on Biochemical Nomenclature and Enzyme Nomenclature (1972). "Recommendations of the International Union of Pure and Applied Chemistry and the International Union of Biochemistry." Elsevier, Amsterdam.

deBecze, G. I. (1970). Food enzymes. *CRC Crit. Rev. Food Technol.* **1**, 479–518.

Dunlap, R. B., ed. (1975). "Immobilized Biochemicals and Affinity Chromatography." Plenum, New York.

Eskin, N. A. M., Henderson, H. M., and Townsend, R. J. (1971). "Biochemistry of Foods." Academic Press, New York.

Fox, P. F. (1974). Enzymes in food processing. *In* "Industrial Aspects of Biochemistry" (B. Spencer, ed.), pp. 213–239. Fed. Eur. Biochem. Soc., London.

Hultin, H. O. (1974). Characteristics of immobilized multienzymic systems. *J. Food Sci.* **39,** 647–652.

Hultin, H. O. (1983). Current and potential uses of immobilized enzymes. *Food Technol.* **37**(10), 66, 68, 72, 74, 76–78, 80, 82, 176.

Kilara, A., and Shahani, K. M. (1979). The use of immobilized enzymes in the food industry: A review. *CRC Crit. Rev. Food Sci. Nutr.* **10,** 161–169.

Konecny, J. (1977). Theoretical and practical aspects of immobilized enzymes. *Surv. Prog. Chem.* **8,** 195–251.

Olson, N. F., and Richardson, T. (1974). Immobilized enzymes in food processing and analysis. *J. Food. Sci.* **39,** 653–659.

Pintauro, N. D. (1979). "Food Processing Enzymes: Recent Developments." Noyes Data Corp., Park Ridge, New Jersey.

Pomeranz, Y., (1966). The role of enzymes additives in breadmaking. *Brot Gebaeck* **20,** 40–45.

Pulley, J. E. (1969). Enzymes simplify processing. *Food Eng.* **41**(2), 68–71.

Reed, G., ed. (1975). "Enzymes in Food Processing," 2nd ed. Academic Press, New York.

Rombouts, F. M. (1981). Pectic enzymes, their biosynthesis and roles in fermentation and spoilage. *Proc. Int. Yeast Symp., 5th, 1980* Sect. 90, pp. 585–592.

Rombouts, F. M., and Pilnik, W. (1978). Enzymes in fruit and vegetable juice technology. *Process Biochem.* **13**(8), 9–13.

Schultz, H. W., ed. (1960). "Food Enzymes." Avi Publ. Co., Westport, Connecticut.

Skinner, K. J. (1975). Enzymes technology. *Chem. Eng. News* **53**(33), 22–29, 32–41.

Underkofler, L. A. (1980). Enzymes. *In* "Handbook of Food Additives" (T. E. Luria, ed.), 2nd ed., Vol. 2, pp. 57–124. CRC Press, Boca Raton, Florida.

Villadsen, K. J. S. (1972). Production and application of enzymes within the food industry. DE-CHEMA-*Monogr.* **70,** 135–174.

Weetall, H. H. (1974). Immobilized enzymes; analytical applications. *Anal. Chem.* **46**(7), 602A–604A, 607A, 608A, 610A, 612A, 615A.

Whitaker, J. R. (1972). "Principles of Enzymology for the Food Industries." Dekker, New York.

Whitaker, J. R., ed. (1974). "Food Related Enzymes," Adv. Chem. Ser. 136. Am. Chem. Soc., Washington, D.C.

Whitaker, J. R. (1980). Some present and future uses of enzymes in the food industry. *In* "Enzymes. The Interface Between Technology and Economics" (J. P. Danehy and B. Wolnak, ed.). Dekker, New York.

Wieland, H. (1972). "Enzymes in Food Processing and Products." Noyes Data Corp., Park Ridge, New Jersey.

Wingard, L. B., ed. (1972). "Enzyme Engineering." Wiley, New York.

Wingard, L. B., Katchalski-Katzir, E., and Goldstein, L., eds. (1976). "Applied Biochemistry and Bioengineering," Vol. 1. Academic Press, New York.

Wiseman, A., ed. (1975). "Handbook of Enzyme Biotechnology." Wiley, New York.

Zaborsky, O. (1973). "Immobilized Enzymes." CRC Press, Cleveland, Ohio.

II

Engineering Foods

9

Additives

I. Standards for Additives

The term "food additive" indicates a substance added to a natural food and is considered to have an innately perjorative connotation (Wodicka, 1980). Most people see food additives as unnecessary and probably harmful. Yet food additives are used for many good and valid reasons. Reduction of variations in quantity and quality of seasonably supplied agricultural produce, ease of processing, improvement of nutritional value, protection from chemical and microbiological damage during transportation and storage, and better consumer acceptance and eating quality are a few justifications for the use of food additives. Over 100 countries have joined the Food and Agriculture Organization (FAO) and World Health Organization (WHO) of the United Nations to establish an organization to set food standards, known as the Codex Alimentarius Commission (1975). The goal of the commission is to develop a body of food standards to govern international trade. According to the definitions used by the commission,

> "Food additives" means any substance not normally consumed as a food by itself and not normally used as a typical ingredient of the food, whether or not it has nutritive value, the intentional addition of which to food for a technological (including organoleptic) purpose in the manufacture, processing, preparation, treatment, packing, packaging, transport or holding of such food results, or may be reasonably expected to result,

(directly or indirectly) in it or its byproducts becoming a component of or otherwise affecting the characteristics of such foods. The term does not include "contaminants" or substances added to food for maintaining or improving nutritional qualities.

The Commission's Committee on Food Additives is responsible for endorsing or establishing maximum levels for individual food additives. The general principles for the use of food additives, as adopted by the Ninth Session of the Codex Alimentarius Commission, are as follows:

1. All food additives, whether actually in use or being proposed for use, should have been or should be subjected to appropriate toxicological testing and evaluation. This evaluation should take into account, among other things, any cumulative, synergistic, or potentiating effects of their use.

2. Only those food additives should be endorsed, which so far as can be judged on the evidence presently available, present no hazard to the health of the consumer at the levels of use proposed.

3. All food additives should be kept under continuous observation and should be reevaluated, whenever necessary, in the light of changing conditions of use and new scientific information.

4. Food additives should at all times conform with an approved specification (e.g., the Specifications of Identity and Purity recommended by the Codex Alimentarius Commission).

5. The use of food additives is justified only where they serve one or more of the purposes set out below and only where these purposes cannot be achieved by other means which are economically and technologically practicable and do not present a hazard to the health of the consumer:

    a. To preserve the nutritional quality of the food; an intentional reduction in the nutritional quality of a food would be justified in the circumstances dealt with in (b) and also in other circumstances where the food does not constitute a significant item in a normal diet

    b. To provide necessary ingredients or constituents for foods manufactured for groups of consumers having special dietary needs

    c. To enhance the keeping quality or stability of a food or to improve its organoleptic properties, provided that this does not so change the nature, substance, or quality of the food as to deceive the consumer

    d. To provide aids in the manufacture, processing, preparation, treatment, packing, transport, or storage of food, provided that the additive is not used to disguise the effects of the use of faulty raw materials or of undesirable (including unhygienic) practices or techniques during the course of any of these activities

6. Approval or temporary approval for the inclusion of a food additive in an advisory list or in a food standard should:

    a. As far as possible be limited to specific foods for specific purposes and under specific conditions

   b. Be at the lowest level of use necessary to achieve the desired effect

   c. As far as possible take into account any acceptable daily intake, or equivalent assessment, established for the food additive and the probable daily intake of it from all sources. Where the food additive is to be used in foods eaten by special groups of consumers, account should be taken of the probable daily intake of the food additive by consumers in those groups.

Standards relating to the carry-over of additives into foods, as adopted by the Eleventh Session of the Codex Alimentarius Commission (paragraph 121, AL-INORM 76/44) are as follows:

1. For the purposes of the Codex Alimentarius, the "carry-over principle" applies to the presence of additives in food as a result of the use of raw materials or other ingredients in which these additives are used. The principle does not apply to the labeling of such food or the presence of contaminants.

2. The principle applies to all Codex standards, unless otherwise specified in such standards.

3. The presence of an additive in food, through the application of the "carry-over principle," is generally permissible if:

   a. The additive is permitted in the raw materials or other ingredients (including additives) by an applicable Codex standard or under any other acceptable provision which takes into account the health requirements of food additives

   b. The amount of the additive in the raw material or other ingredient (including additives) does not exceed the maximum amount so permitted

   c. The food into which the additive is carried-over does not contain the additive in greater quantity than would be introduced by the use of the ingredients under proper technological conditions or manufacturing practice

   d. The additive carried-over is present at a level that is nonfunctional, that is, at a level significantly less than that normally required to achieve an efficient technological function in its own right in the food

4. An additive carried-over into a particular food in a significant quantity or in an amount sufficient to perform a technological function in that food as a result of the use of raw materials or other ingredients in which this additive was used shall be treated and regarded as an additive to that food, unless the Codex Commodity Committee responsible, in conjunction with the Codex Committee on Food Additives, decides otherwise.

5. The appropriate Codex Commodity Committee, in conjunction with the Codex Committee on Food Additives, shall decide the specific cases to which the "carry-over principle" shall not apply, particular attention being paid to cases where no relevant Codex standard applicable to the ingredient exists.

6. The appropriate Codex Commodity Committee, in conjunction with the Codex Committee on Food Additives, shall establish an overall limit for an additive which is added to the food intentionally and which is also carried-over in an ingredient.

In the United States, interstate commerce of foods (other than meat) is governed by the Federal Food, Drug, and Cosmetic Act of 1938 as amended in 1958 with regard to food additives. The definition of food additives of the Food and Drug Administration (FDA) reads as follows:

> "Food additives" include all substances not exempted by section 201 (s) of the act, the intended use of which results or may reasonably be expected to result, directly or indirectly, either in their becoming a component of food or otherwise affecting the characteristics of food. A material used in the production of containers and packages is subject to the definition if it may reasonably be expected to become a component, or to affect the characteristics, directly or indirectly, or food packed in the container. "Affecting the characteristics of food" does not include such physical effects, as protecting contents of packages, preserving shape, and preventing moisture loss. If there is no migration of a packaging component from the package to the food, it does not become a component of the food and thus is not a food additive. A substance that does not become a component of food, but that is used, for example, in preparing an ingredient of the food to give a different flavor, texture, or other characteristics in the food may be a food additive.

Part and parcel of the definitions is the concept of "generally recognized as safe" (GRAS). Following the Food Additives Amendment of 1958, the FDA published a list of GRAS materials (21 CFR 182) and several other amendments, including one from 1976 (Code of Federal Regulations), which reads

> The food ingredients listed as GRAS in Part 182 of this chapter or affirmed as GRAS in Part 184 or §186.1 of this chapter do not include all substances that are generally recognized as safe for their intended use in food. Because of the large number of substances the intended use of which results or may reasonably be expected to result, directly or indirectly, in their becoming a component or otherwise affecting the characteristics of food, it is impracticable to list all such substances that are GRAS. A food ingredient of natural biological origin that has been widely consumed for its nutrient properties in the United States prior to January 1, 1958, without known detrimental effects, which is subject only to conventional processing as practiced prior to January 1, 1958, and for which no known safety hazard exists, will ordinarily be regarded as GRAS without specific inclusion in Part 182, Part 184, or §186.1 of this chapter.

Thus, substances and processes used in the United States prior to 1958 are assumed to be GRAS, unless there is some question about their safety. There is, however, no final or definitive list of GRAS additives and the FDA reserves the right to reevaluate the status of

1. GRAS substances modified by processes developed since 1958 which may reasonably be expected to alter their composition
2. Substances, otherwise GRAS, which have had significant alteration of composition by breeding or selection after January 1, 1958, when that

change may reasonably be expected to alter the nutritive value or the concentration of toxic constituents

3. Distillates, isolates extracts and concentrates of extracts of GRAS substances
4. Reaction products of GRAS substances
5. Substances, not of natural biological origin, including those purporting to be chemically identical to GRAS substances
6. Substances of natural biological origin consumed for other than nutrient properties

In 1979, the FDA published a list of the major functions of the following 32 categories of food additives:

1. Anticaking and free-flowing agents
2. Antimicrobial agents
3. Antioxidants
4. Colors and coloring adjuncts
5. Curing and pickling agents
6. Dough strengtheners
7. Drying agents
8. Emulsifiers and emulsifier salts
9. Enzymes
10. Firming agents
11. Flavor enhancers
12. Flavoring agents and adjuvants
13. Flour-treating agents
14. Formulation aids
15. Fumigants
16. Humectants
17. Leavening agents
18. Lubricants and release agents
19. Nonnutritive sweeteners
20. Nutrient supplements
21. Nutritive sweeteners
22. Oxidizing and reducing agents
23. pH control agents
24. Processing aids
25. Propellants, aerating agents, and gases
26. Sequestrants
27. Solvents and vehicles
28. Stabilizers and thickeners
29. Surface-active agents
30. Surface-finishing agents
31. Synergists
32. Texturizers

In the United States, criteria for food additives are stated in 21 CFR 1720.22 ("Safety to be considered") and in 21 CFR 170.20 ("General principles for evaluating the safety of food additives"). Those regulations are designed to assure that "a food additive for use by man will not be granted a tolerance that will exceed 1/100th of the maximum amount demonstrated to be without harm to experimental animals." Levels are translated, generally, from experimental animals to man in terms of mg additive per kg body weight. The criteria apply to food or color additives that are legally GRAS on the basis of scientific evidence. In the United States, food additives fall into three categories: substances tolerated by special authorization granted prior to 1958, substances considered GRAS by experts, and additives as such.

The 1958 Food Additives Amendment contains a highly controversial clause known as the "Cancer Clause" or the "Delaney Clause." It was introduced by Representative J. Delaney (New York) and states that no additive shall be deemed safe if it is found to induce cancer when ingested by man or animal. This applies only to food additives and has been interpreted to mean that no substance (including coloring agents) may be added at any low level even if it induces cancer only when added at an extremely high level. This clause has generated much discussion and controversy.

The European Economic Community (EEC) has no uniform definition of food additives; different nations use different definitions, terminology, and standards. The EEC is attempting to formulate a uniform definition of food additives (EEC, 1982).

Kermode (1972) of the FAO/WHO Food Standards Program estimated that as many as 2500 food additives were in existence by 1972. According to Pilnik (1973), the distinction between additives and food ingredients is basically a matter of semantics. Food additives involve the following functions: nutritive value, sensory qualities, keeping properties, physical characteristics related to end-use, performance of a food or ingredients, and processing. Table 9.1 lists reasons for using additives, additives used, and foods in which such additives are used.

TABLE 9.1

Food Additives[a]

Purpose	Additives	Foods
To impart and maintain desired consistency— emulsifiers distribute tiny particles of one liquid in another to improve texture homogeneity and quality. Stabilizers and thickeners give smooth uniform texture and flavor and desired consistency.	Lecithin, mono- and diglycerides, gum arabic, agar, methyl cellulose	Baked goods, cake mixes, salad dressings, frozen desserts, ice cream, chocolate milk, beer

TABLE 9.1 (*Continued*)

Purpose	Additives	Foods
To improve nutritive value—medical and public health authorities endorse this use to eliminate and prevent certain diseases involving malnutrition. Iodized salt has eliminated simple goiter. Vitamin D in dairy products and infant foods has virtually eliminated rickets. Niacin in bread, corn meal, and cereals has eliminated pellagra in the southern states.	Vitamin A, thiamine, niacin, riboflavin, ascorbic acid, vitamin D, iron, potassium iodide	Wheat flour, bread and biscuits, breakfast cereals, corn meal, macaroni and noodle products, margarine milk, iodized salt
To enhance flavor—many spices and natural and synthetic flavors give a desired variety of flavorful foods such as spice cake, gingerbread, and sausage.	Cloves, ginger, citrus oils, amyl acetate, carvone, benzaldehyde	Spice cake, gingerbread, ice cream, candy, soft drinks, fruit-flavored gelatins and toppings, sausage
To control acidity or alkalinity—leavening agents are used in the baking industry in cakes, biscuits, waffles, muffins, and other foods. Similar additives make fruits and potatoes easier to peel for canning; others neutralize sour cream in making butter.	Potassium acid tartrate, tartaric acid, sodium bicarbonate, citric acid, lactic acid	Cakes, cookies, quick breads, crackers, butter, process cheese, cheese spreads, chocolates, soft drinks
To maintain appearance, palatability, and wholesomeness—by delaying deterioration of food due to microbial growth or oxidation. Food spoilage caused by mold, bacteria, and yeast is prevented or slowed by certain additives. Antioxidants keep fats from turning rancid and certain fresh fruits from darkening during processing when cut and exposed to air.	Propionic acid, sodium and calcium salts of propionic acid, ascorbic acid, butylated hydroxyanisole, butylated hydroxytoluene	Bread, cheese, syrup, pie fillings, crackers, fruit juices, frozen and dried fruit, margarine, lard, shortening, potato chips, cake mixes
To give desired and characteristic color—to increase acceptability and attractiveness by correcting objectionable natural variations.	FDA-approved colors, such as annatto, carotene, cochineal, chlorophyll	Confections, bakery goods, soft drinks, cheese, ice cream, jams, and jellies
To mature and bleach—and to modify gluten to improve baking, to improve appearance of certain cheeses and to meet the desire for white flour by changing the natural yellow pigments.	Chlorine dioxide, chlorine, potassium bromate and iodate	Wheat flour, certain cheeses
Other functions—such as humectants to retain moisture in some foods and to keep others, including salts and powders, free-flowing.	Glycerol, magnesium carbonate	Coconut, table salt

[a]Courtesy of the Manufacturing Chemists Association, Inc., Washington, D.C. Additives are listed roughly in order of quantity used.

Detailed discussion of all groups of additives is outside the scope of this book. In terms of functional properties, the three most important groups are texturizers (such as polysaccharides, discussed in Chapter 3), enzymes (see Chapter 8), and emulsifiers. Emulsifiers are discussed in this chapter, with special emphasis on lecithin.

II. Food Emulsifiers

A. Dispersion Systems

Emulsions represent one of nine two-phase systems which result from intimate mixing of two substances in the solid–liquid–gas states of aggregation. Emulsions are disperse two-phase systems: an intimate mixture of two immiscible liquids. Food emulsions are complex rather than simple systems. They can be defined as finely to coarsely dispersed systems of two or more phases in which both the disperse and the dispersing phase are in a liquid to semisolid form. Under normal conditions both phases are immiscible or only slightly miscible. Composition of the emulsified food components determines the typical structure of the emulsified food.

The dispersed liquid is the inner, disperse, or open phase; the dispersing liquid is the outer, coherent, or closed phase. For a short period both liquids may permeate each other so uniformly that no inner or outer phase can be distinguished. Eventually, the system tends toward a state of equilibrium in which one liquid forms the closed phase.

When oil and water are made into an emulsion, the oil can be dispersed in water or the water can be dispersed in oil. If oil is finely dispersed in water, the oil is present in fine droplets and forms the disperse phase and water is coherent and forms the closed phase, and an oil-in-water emulsion (o/w) is formed. The basic character of this emulsion is set by the water. Since the outer phase is water, o/w emulsions can be diluted with water. Typical examples of o/w emulsions are homogenized milk and ice cream. If, however, water is dispersed in oil, the water becomes the inner phase and the oil the outer phase and a water-in-oil emulsion (w/o) is obtained. Because the oil is the outer phase, a w/o emulsion behaves like an oily system and can be diluted with oil. Margarine is a typical example of a w/o emulsion. In some polyphase emulsions, such as o/w/o, the inner phase is an o/w emulsion and the outer phase is oil.

The particle size of the disperse phase in emulsions governs its appearance. Macroemulsions with a particle size range of 100–0.5 μm in diameter appear milky due to the effect of scattered light caused by the particles of the dispersed phase. Microemulsions with a particle size range of 0.5–0.01 μm in diameter vary from nearly transparent with a bluish tinge to optically clear systems when the particles are smaller than the light wavelength. Particle size and distribution also influence the consistency of an emulsion. Emulsions with small particles are more viscous than emulsions that contain the same amount of dispersed phase in the form of large particles. Emulsions with different particle sizes, for example, polydisperse emulsions, have less viscosity than emulsions with a uniform particle size, such as monodisperse emulsions.

In the preparation of emulsions, one liquid is dispersed in another. By forming many small droplets, the surface of the dispersed liquid is enlarged considerably. If, for example, a drop of oil of 1 ml volume (spherical surface of 4.83 cm² and diameter of 1.24 cm) is dispersed in 2.39×10^{11} droplets with a diameter of 2×10^{-4} cm (2 μm), the surface increases to 30,000 cm² or 6210 times (Schuster, 1981). These small droplets have considerably more energy than a coherent oil phase. This energy, called "surface tension," is directed parallel to the surface and prevents the dispersion of the oil phase. Work must be done to overcome the surface tension. The required work (W) is proportional to the increase in surface (dA) and the surface tension (δ): $W = dA \times \delta$. Conversely, work can be reduced considerably by lowering the surface tension.

In borderline cases, energy can be augmented mechanically or physicochemically by adding an emulsifier. In practice, a combination of both methods is used. If the phases are present as solids, thermal energy must be used because the phases that are to be emulsified must be present in liquid form before emulsification can take place. Whereas microemulsions are stable, macroemulsions are thermodynamically unstable. Sedimentation, aggregation, and coalescence decrease the stability of an emulsion. Whereas sedimentation and aggregation are reversible, coalescence is irreversible and results in the breakup of the emulsion.

A good emulsion should be adequately stable to withstand mechanical, thermal, and temporal influences. The stability of an emulsion depends on such factors as (a) degree of dispersion of the inner phase, (b) quality of the interfacial layer, (c) viscosity of the outer phase, (d) phase to volume ratio, and (e) specific gravity of the phases.

B. Emulsifier Properties

Emulsifiers facilitate formation of finely dispersed systems. They lower the interfacial tension between two immiscible phases and thereby facilitate interfacial work, or emulsification. Emulsifiers develop steric or electrical barriers at the interface and hinder the dispersed particles from flowing together. This property is also characteristic of stabilizers. The main function of emulsifiers is the stabilization of emulsions. Stabilization is more important than primary dispersion, as the latter also can be attained by mechanical means.

Emulsifiers should accumulate at the interface between two phases. They must be surface active and interfacial active. In addition, they must either charge the particles so that they repel each other or form a stable, often highly viscous or even protective, layer around the particles. Emulsifiers are characterized by their ambiphilic structure. The emulsifier molecule consists of a lipophilic–hydrophobic part and a hydrophilic–lipophobic part (Fig. 9.1). Modern techniques of organic synthesis make it possible to manufacture a great number of tailored emulsifiers. Possible constructions are listed in Fig. 9.2. The lipophilic (nonpolar) part is formed by straight, branched chain, or cyclic hydrocarbons; the formation of the hydrophilic (polar) part involves a great number of possibilities. Of the wide spectrum of potential emulsifiers, only those that are physiologically and toxicologically safe are permitted, provided they have technological advantages in the production of foods. Table 9.2 is a compilation of emulsifiers certified for use in the United States, and those proposed for food use by the EEC, and which passed toxicological evaluation by the FDA and the FAO/WHO Joint Expert Committee on Food Additives.

Emulsifiers can be classified according to (a) their electrochemical charge in aqueous systems; (b) their dissolving properties; (c) the relation of hydrophilic to lipophilic groups; and (d) the functional groups of which they are composed. Emulsifiers behave in water electrochemically depending on their structure. Emulsifiers in which the surface activity is bound to the salt form dissociate in water to ions. They are classified according to the ion that is primarily responsi-

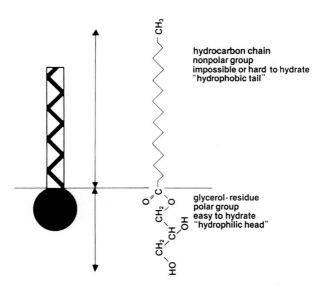

Fig. 9.1. Emulsifier ambiphilic structure. Top portion, the lipophilic area; lower portion, the hydrophilic area of an emulsifier molecule. (From Schuster and Adams, 1984.)

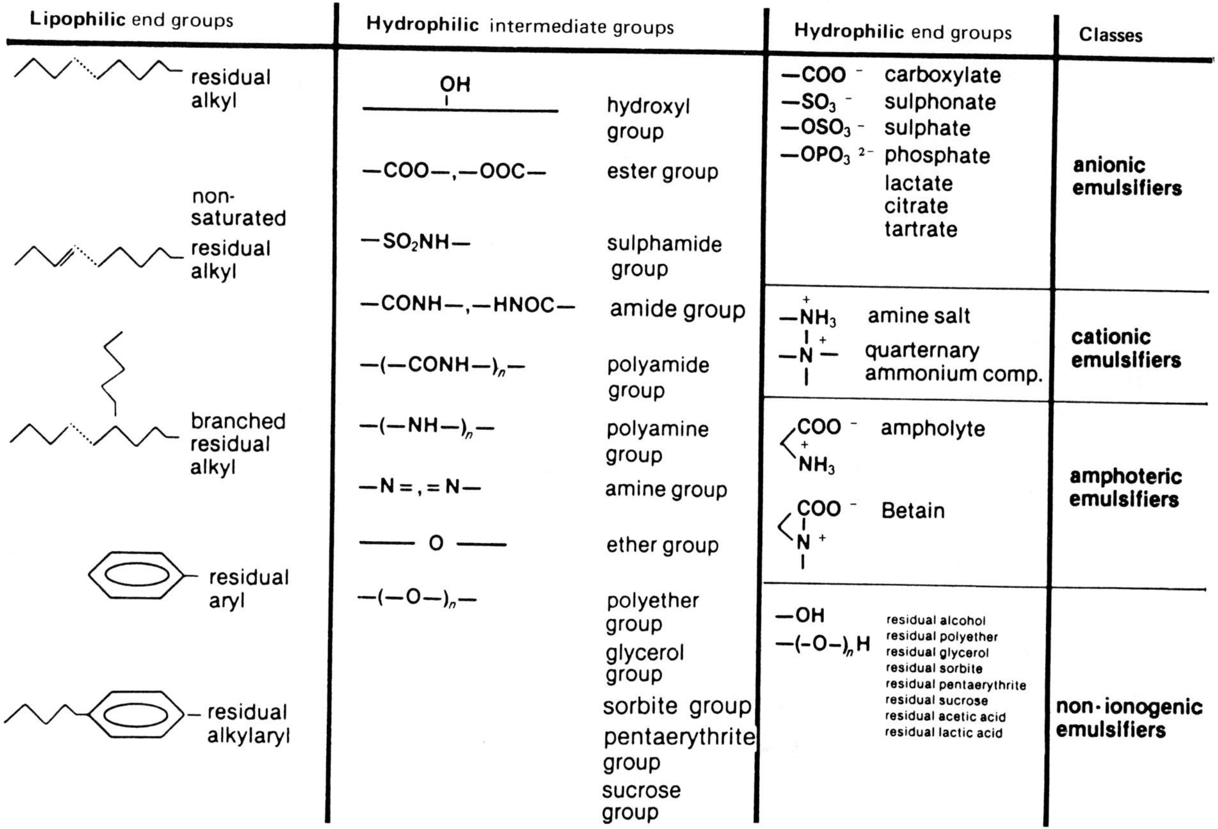

Lipophilic end groups	Hydrophilic intermediate groups		Hydrophilic end groups	Classes
residual alkyl	hydroxyl group	$\overset{\mid}{\underset{\mid}{OH}}$	$-COO^-$ carboxylate $-SO_3^-$ sulphonate $-OSO_3^-$ sulphate $-OPO_3^{2-}$ phosphate lactate citrate tartrate	anionic emulsifiers
non-saturated residual alkyl	ester group	$-COO-,-OOC-$		
	sulphamide group	$-SO_2NH-$		
	amide group	$-CONH-,-HNOC-$	$-\overset{+}{N}H_3$ amine salt $-\overset{+}{\underset{\mid}{N}}-$ quarternary ammonium comp.	cationic emulsifiers
branched residual alkyl	polyamide group	$-(-CONH-)_n-$		
	polyamine group	$-(-NH-)_n-$	$\underset{\overset{+}{N}H_3}{\overset{COO^-}{<}}$ ampholyte	amphoteric emulsifiers
	amine group	$-N=,=N-$		
residual aryl	ether group	$-\!\!-\!O\!-\!\!-$	$\underset{-\overset{\mid}{\underset{\mid}{N}}-}{\overset{COO^-}{<}}$ Betain	
residual alkylaryl	polyether group glycerol group	$-(-O-)_n-$	$-OH$ $(-O-)_nH$	non-ionogenic emulsifiers
	sorbite group pentaerythrite group sucrose group		residual alcohol residual polyether residual glycerol residual sorbite residual pentaerythrite residual sucrose residual acetic acid residual lactic acid	

Fig. 9.2. Possible tailor-made emulsifiers. (From Schuster, 1981.)

TABLE 9.2

Emulsifiers Acceptable in the United States or Europe[a]

| Emulsifiers permitted in bread and fine baked products by the U.S. Food and Drug Administration (FDA) | | Emulsifiers proposed for use in food by the European Economic Community (EEC) Council Directive 80/597/EEC | | Toxicological evaluation | | | Special remarks for application in the United States | Abbreviation |
| | | | | | JECFA[b] | | | |
Designation	Code of federal regulations no.	Designation	EEC no.	FDA	Acceptable daily intake (mg/kg)	Report no.		
Lecithin	182.1400	Lecithins	E 322	GRAS[c]	Not specified	17, 18	In bread, milk bread, whole wheat bread, buns, rolls (GMP)[a]	LC
Hydroxylated lecithin	172.814	—	—	—	—	24	As above	HLC
Salts of fatty acids (aluminum, calcium, magnesium, potassium, sodium)	172.863	Sodium, potassium, and calcium salts of fatty acids	E 470	—	Not specified	17, 18	As a binder, emulsifier, and anti-caking agent in accordance with GMP (not in bread)	SFA
Mono- and diglycerides of edible fats or oils, or edible fat-forming acids	182.4505	Mono- and diglycerides of fatty acids	E 471	GRAS	Not specified	17, 18	In bread, milk bread, whole wheat bread, buns, rolls (GMP)	MDG
Acetylated mono-glycerides	172.828	Acetic acids esters of mono- and di-glycerides of fatty acids	E 472a	—	Not specified	17, 18	May be safely used in or on food in accordance with GMP (not in bread)	AMG
Glyceryl lactoesters of fatty acids	172.852	Lactic acid esters of mono- and di-glycerides of fatty acids	E 472b	—	Not specified	17, 18	May be safely used in food as emulsifiers and plasticizers in accordance with GMP (not in bread)	LMG

Name	Synonym	21 CFR	E number	Status	Limitation	FAS	Use	Abbreviation
Diacetyl tartaric acid esters of mono- and diglycerides of edible fat-forming fatty acids	Mono- and diacetyl tartaric acid esters of mono- and diglycerides of fatty acids	182.4101	E 472e	GRAS	0–50 (tartaric acid maximum 30)	17, FAS[e] no. 5	In bread, milk bread, whole wheat bread, buns, rolls (GMP)	DATEM
Succinylated monoglycerides	—	172.830	—	—	New specifications are being prepared	24	0.5% by weight of flour in bread, rolls, buns (GMP)	SMG
Monosodium phosphate derivatives of mono- and diglycerides of edible fats or oils, or edible fat-forming fatty acids	—	182.4521	—	GRAS	—	—	Not in bread	MP-MG
Ethoxylated mono- and diglycerides (polyoxyethylene mono- and diglycerides of fatty acids)	—	172.834	—	—	—	—	0.5% by weight of flour in yeast-leavened bakery products, 0.5% by weight of the dry ingredients in cakes and cake mixes	EMG
Stearyl monoglyceridyl citrate	—	172.755	—	—	—	—	As an emulsion stabilizer in or with shortenings containing emulsifiers	SMGC
Lactylated fatty acid esters of glycerol and propylene glycol	—	172.850	—	—	—	—	May be safely used in food in accordance with GMP (not in bread)	LGPE
Propylene glycol mono- and diesters of fats and fatty acids	Propane 1,2-diol esters of fatty acids	172.856	E 477	—	0–25 (as propylene glycol)	17, FAS no. 5	In bread, milk bread, whole wheat bread, buns, rolls (GMP)	PGME
Succistearin (stearoyl propylene glycol hydrogen succinate)	—	172.765	—	—	—	—	In or with shortenings and edible oils intended for use in cakes, cake mixes, and pastries (GMP)	SCS

(continued)

TABLE 9.2 (*Continued*)

Emulsifiers permitted in bread and fine baked products by the U.S. Food and Drug Administration (FDA)		Emulsifiers proposed for use in food by the European Economic Community (EEC) Council Directive 80/597/EEC		Toxicological evaluation			Special remarks for application in the United States	Abbreviation
					JECFA[b]			
Designation	Code of federal regulations no.	Designation	EEC no.	FDA	Acceptable daily intake (mg/kg)	Report no.		
Polyglycerol esters of fatty acids	172.854	Polyglycerol esters of fatty acids	E 475	—	0–25	17, FAS no. 5	May be safely used in food as emulsifiers (GMP) (not in bread)	PGE
—	—	Sucrose esters of fatty acids	E 473	—	0–10 (Dimethylformamide should not be detectable)	24, 20	—	SUE
—	—	Sucroglycerides	E 474	—	0–10 (Dimethylformamide maximum 10 ppm)	24, 20	—	SUG
Sorbitan monostearate	172.842	Sorbitan monostearate	E 491	—	0–25	17, FAS no. 5	0.61% on dry weight basis in cake and cake mixes (not in bread)	SMS
Polysorbate 60 (polyoxyethylene sorbitan monostearate)	172.836	—	—	—	0–25	17, FAS no. 5	0.5% by weight of flour in yeast-leavened bakery products, 0.46% on dry weight basis in cake and cake mixes	PS 60
Polysorbate 65 (polyoxyethylene sorbitan tristearate)	172.838	—	—	—	0–25	17, FAS no. 5	0.32% of the cake or cake mix, on a dry weight basis (not in bread)	PS 65

Name	No.	Chemical name	E number	Range	Reference	Usage	Abbrev.
Polysorbate 80 (polyoxy-ethylene sorbitan monooleate)	172.840	—	—	0–25	17, FAS no. 5	1% of the weight of the finished shortening (not in bread)	PS 80
Lactylic esters of fatty acids	172.848	—	—	—	—	In bakery mixes, bakery products, pancake mixes, when standards of identity do not preclude their use (GMP)	SLA
Calcium stearoyl-2-lactylate	172.844	Calcium stearoyl-2-lactylate	E 482	0–20	17, FAS no. 5	0.5% by weight of flour in yeast-leavened bakery products and prepared mixes for yeast-leavened bakery products	CSL
Sodium stearoyl lactylate	172.846	Sodium stearoyl-2-lactylate	E 481	0–20	17, FAS no. 5	0.5% by weight of flour in baked products, pancakes, waffles (including the mixes for these products)	SSL
Sodium stearoyl fumarate	172.826	—	—	—	—	0.5% by weight of flour in yeast-leavened bakery products. 1% by weight of flour in non-yeast-leavened bakery products	SSF
—	—	Stearoyl tartrate	E 483	0–500	FAO Nutrition Series A, B, C	—	STA
Copolymer condensates of ethylene oxide and propylene oxide	172.808	—	—	—	—	0.5% by weight of flour in yeast-leavened bakery products for which standards of identity do not preclude such use	CEP

[a]From Schuster and Adams (1984).
[b]Joint Expert Committee on Food Additives (Food and Agriculture Organization/World Health Organization).
[c]GRAS, generally recognized as safe.
[d]GMP, good manufacturing practices.
[e]FAS, food additives series.

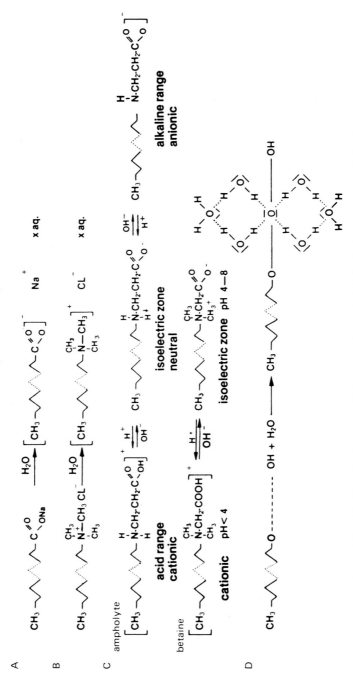

Fig. 9.3. Classification of emulsifiers according to their behavior in water. A, Anionic; B, cationic; C, amphoteric; D, non-ionic emulsifiers. (From Schuster, 1981.)

ble for their surface activity (see Fig. 9.3). In anionic emulsifiers the anion is responsible and in cationic emulsifiers the cation is responsible for their surface activity. Amphoteric emulsifiers have a different dissociation behavior. As inner salts in the isoelectric zone, they show no dissociation. They behave non-ionically, because the charges are balanced intramolecularly. If the pH value is changed by adding alkali or acid, the intramolecular salt form is eliminated and the emulsifier assumes either a positive or a negative charge. The position of the isoelectric zone and its extension depend on the structure of the amphoteric compound (Schuster, 1981). The isoelectric zone of betains, for example, comprises a wide pH range from about 4.0 to 8.0. Nonionic compounds do not dissociate in aqueous systems and form no ions.

From a practical point of view, we can classify emulsifiers into four groups: anionic, cationic, amphoteric, and nonionic. Figure 9.4 shows a schematic diagram of the interfacial activity of emulsifier classes as related to the pH of a system. Anionic emulsifiers form negatively charged organic ions in water and are effective in the neutral to alkaline pH range, depending on the neutralization range; they lose their efficiency in the acid range. Cationic emulsifiers form positively charged ions in water. They are effective in the acidic range, that is, below the neutralization range of the organic bases. Amphoteric emulsifiers form positively and negatively charged ions in water and are effective at all pH values (although in different forms) but are sometimes ineffective in the isoelectric zone. The nonionic emulsifiers, which form no ions in water, are usually independent of the pH value. They are compatible with all emulsifiers.

Another property of anionic and cationic emulsifiers is electroneutral salt formation, which yields large molecular electroneutral salts. They have no sur-

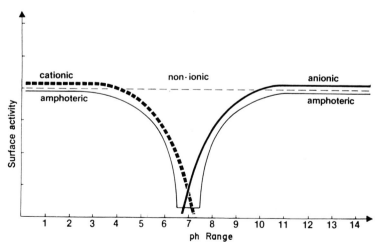

Fig. 9.4. Interfacial activity of emulsifier classes as a function of pH. (From Schuster, 1981.)

face activity. Consequently, these emulsifiers are incompatible and cancel each other. The electroneutral salt formation is depicted in Fig. 9.5. Emulsifiers can be roughly classified according to their solubility in hydrophilic (polar) and lipophilic (nonpolar) solvents. Strongly hydrophilic emulsifiers are normally water-soluble and strongly lipophilic emulsifiers are oil-soluble.

Equilibrium of the hydrophilic–lipophilic balance (HLB) is a decisive factor in efficacy of emulsifiers. In a homologous series in which the hydrophilic equilibrium changes, a point is reached at which the hydrophilic properties are so balanced that an optimum efficiency is attained for a particular application. This equilibrium was quantified by Griffin (1956), who rated the emulsifiers on a scale of 1 to 20. Lipophilic emulsifiers have a low HLB, and hydrophilic emulsifiers have a high HLB. The change from lipophilic to hydrophilic, on this HLB scale, occurs at number 10. HLB values originally were stipulated on the basis of empirical tests. Today, HLB values of emulsifiers of known structure and molecular weight can be determined by calculation. Calculations are based on the premise that the HLB value is a function of the ratio of weight of the hydrophilic part to the total molecule of the emulsifier (interfacially active substance). Figure 9.6 shows calculated HLB values of emulsifiers. It is relatively easy to predict with confidence the HLB values of non-ionic emulsifiers. Prediction of HLB values of emulsifiers whose hydrophilic nature is not based on nonionic hydrophilic groups, but on other functional ionic hydrophilic groups, is more difficult. The strength of interaction, rather than percentage of functional groups, is decisive for the hydrophilic nature of emulsifiers. Davies (1957) demonstrated a relationship between HLB and definable physicochemical properties. From this relationship he developed a method for calculation of HLB values. Calculated and empirical HLB values agreed very well. According to Manley (1983), lipophilic emulsifiers (HLB range of 3 to 6) include monoglycerides, glycerol lactopalmitate, propylene glycol monostearate, sorbitan esters, and triglycerol distearate. The important amphophilic emulsifiers (HLB range of 8 to 14) are diacetyltartaric acid esters, polyoxyethylene sorbitan esters, and sucrose esters. Soaps, lecithin, and monolaurates are hydrophilic emulsifiers with an HLB range of 14 to 18.

In summary, the HLB value provides an efficient indication of potential ap-

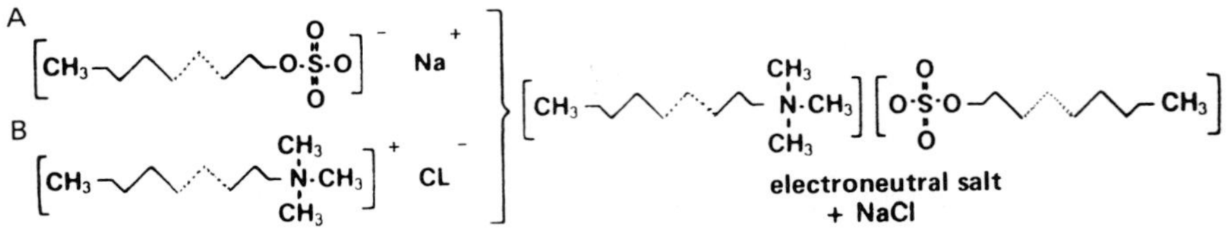

Fig. 9.5. Electroneutral salt formation of inversed charged emulsifiers. A, Anionic; B, cationic. (From Schuster, 1981.)

$$HLB = \frac{H_M}{T_M} \times 20$$

H_M = hydrophilic molecular part
T_M = total molecular weight

$$HLB = \frac{H}{5}$$

H = hydrophilic part [%]

$$HLB = 20\left(1 - \frac{SV}{AV}\right)$$

SV = Saponification value
AV = Acid value

Percentage hydrophilic groups	Percentage lipophilic groups	HLB value	behavior in water
0	100	0	not dispersed
10	90	2	not dispersed
20	80	4	poor dispersion
30	70	6	poor dispersion
40	60	8	milky dispersion
50	50	10	stable milky dispersion
60	40	12	transparent clear dispersion
70	30	14	transparent clear dispersion
80	20	16	clear colloidal solution
90	10	18	clear colloidal solution
100	0	20	clear colloidal solution

field of application

- 1–3: antifoaming agents
- 3–6: **water/oil emulsifiers**
- 7–9: wetting agents
- 8–18: **oil/water emulsifier**
- 13–15: washing agents
- 15–18: solubilizers

Fig. 9.6. Calculated hydrophilic–lipophilic balance (HLB) values of emulsifiers. (From Schuster, 1981.)

plication of emulsifiers in pure binary systems. In complex systems, however, such as foods containing carbohydrates, proteins, and lipids that interact with each other, the value of HLB is more limited.

1. Interfacial Activity

Due to their lipophilic/hydrophilic structure, emulsifiers orientate at the interfaces of binary systems. This can be demonstrated for a simple water foam by the system water/air and an emulsifier with a balanced hydrophilic/hydrophobic equilibrium (HLB of about 10). At first, the molecules orientate at the surface, since the hydrophilic groups are attracted and the lipophilic groups are repelled by water. At the same time, surface tension decreases considerably (Fig. 9.7). After adding more emulsifier, the surface of the water is completely covered and a monomolecular layer of emulsifier is formed. Up to this point, the reduction of surface tension runs parallel to the increase in emulsifier concentration. If the concentration of emulsifier exceeds a certain point, the surface tension decreases very little. This point, called the critical concentration of micelle formation, is the point at which the surface of the water is completely covered with a monomolecular layer. If more emulsifier is added, the emulsifier molecules find no free space at the surface and are pushed under the surface of the water. This affects the hydrophobic forces of the emulsifier, and a thermodynamically unsta-

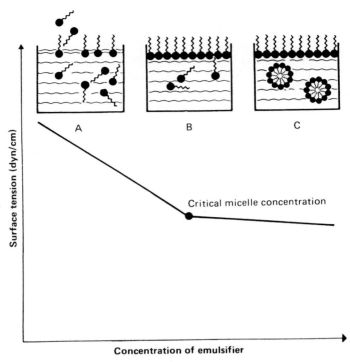

Fig. 9.7. Mechanism of surface activity of emulsifiers. A, Phase 1—molecules concentrate at the surface; B, phase 2—molecules crowd into the interior; C, phase 3—micelles form. (From Schuster, 1981.)

ble system results. As soon as sufficient emulsifier molecules penetrate the water, they orientate to form a more thermodynamically stable condition. The lipophilic parts associate and the hydrophilic parts orientate toward the water and micelles are formed (Fig. 9.8). Because of the interlocking of the lipophilic emulsifier groups, the cores of the micelles contain pure lipophilic and pure hydrophobic zones. At the same time, a circle of hydrophilic groups forms the exterior of the micelles. Depending on the composition of the emulsifier and other parameters, micelles can be lamellar, spherical, or rod-shaped, and may be of varying sizes.

The intramolecular polarity generated by hydrophobic and lipophilic groups enables emulsifiers to form a film between the water and oil so that the emulsifiers orientate toward the water with their hydrophilic part while the lipophilic part penetrates the oil. This results in an interaction between the water molecules and the hydrophilic part and the oil molecules and the lipophilic part of the emulsifier. This interaction changes the interfacial tension. By reducing the interfacial tension, two originally immiscible liquids can be made into homogeneous emulsions.

The strength of interaction between the hydrophilic area of the emulsifier and water is responsible for the type of emulsion (Schuster, 1981). If the interaction is strong, the surface tension of the water is reduced and approaches zero. The water ceases to form droplets and becomes the outer phase of the emulsion; the oil is dispersed in the form of fine droplets and becomes the inner, open phase. If interaction between water and the hydrophilic area of the emulsifier is weak, the surface tension of the water is not greatly reduced and a w/o emulsion is formed.

The solubility of an emulsifier in the phases governs the type of an emulsion. The phase in which the emulsifier is most soluble, becomes the outer one. Hydrophilic emulsifiers, which dissolve little in the oil phase, diffuse quickly into the phase interface. The marked interaction with water or high water-solubility enables the hydrophilic group to penetrate the water portion and form a stable hydrophilic surface film. The emulsifier film is overelongated on the hydrophilic side and the oil phase is enclosed. Lipophilic emulsifiers, which dissolve easily in the oil phase, diffuse slowly into the interface. They adhere closely to the oil phase with their lipophilic parts and do not enter the aqueous phase. Interaction of the lipophilic parts with the oil phase is stronger so that the film of the emulsifier is overelongated at the lipophilic side and the water is enclosed.

According to Davies (1957), when the coalescence speed of the aqueous droplets is higher than that of the oil droplets, the water is in the outer phase and an o/w emulsion is formed. When the coalescence speed of the oil droplets is

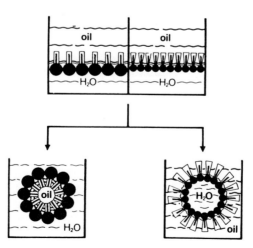

Fig. 9.8. Mechanism of the function of emulsifiers. Left, oil/water emulsion; right, water/oil emulsion. (From Schuster, 1981.)

higher than that of the aqueous droplets, a w/o emulsion results. After formation of an emulsion, it should be stabilized by arrangement of the emulsifier at the interface and the formation of steric or electrostatic barriers (see Fig. 9.9). Excess emulsifier also produces a stabilizing effect by forming liquid-crystalline mixed phases in the interfacial film covering the emulsion droplets. These interfacial films are comparable to barrier membranes of biological systems.

As stated previously, emulsifiers can be classified according to the charge of hydrophilic groups in water: ionic [anionic, cationic, and amphoteric (ampholytes and betains)], non-ionic, solubility in hydrophilic and lipophilic solvents, lipophilic groups, and HLB. Additional considerations in classification are crystal form and size, and arrangement of the emulsifier molecule during interaction with water. Regular fat is polymorphic, that is, it can exist in more than one crystalline form (see Chapter 7). These are known as α, β′, and β forms. When fats are melted and rapidly cooled, they generally crystallize in the α form. This form has the lowest melting point, is the most unstable, and converts to the β′ or β form. Crystal reversion can be retarded by incorporating a suitable emulsifier with crystal-modifying capacity, such as sorbitan esters of fatty acids. Sorbitan esters are used to prevent "bloom" formation on the surface of chocolates containing cocoa butter substitutes. Bloom is the result of reversion of crystalline

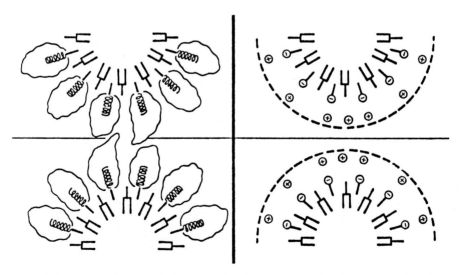

Fig. 9.9. Diagram of the stabilization of emulsions by means of a hydrate layer and electric double layer. Left, hydrate layer, formation of a steric barrier by the addition of water to the hydrophilic chain; right, electric double layer, formation of electric barrier by orientation of the anionic emulsifiers to the interface. ⊐⊏—, Triglyceride; ⟨⟨⟨⟨⟨⟨—, non-ionic polyglycerol ester; ⊖—, anionic tenside. (From Schuster, 1981.)

fat. Similarly, the α-tending capacity of propylene glycol esters, acetylated monoglycerides, and lactylated monoglycerides stabilizes monoglycerides in the α-form. Those emulsifiers are used to stabilize sponge cakes and other aerated foods. The α-tending emulsifiers also form a protective α-crystalline membrane of lipid droplets in high fat cakes and thereby enhance their aeration.

Many food products are manufactured in the form of emulsions that are subsequently aerated fo a foam, for example, dairy cream, imitation cream, spray-dried toppings, and various whipped products. In such products, emulsion stability is attained through the action of proteins. The emulsifiers are used to improve whipping rate, stiffness, volume, and stability of foam. These emulsifiers are of the α-tending type and enhance agglomeration of fat globules to promote formation of stable and acceptable foams. For cakes and sponge products containing little or no fat, emulsifiers are used to facilitate aeration and allow a faster whipping rate of the cake batter and a more even cell structure and higher volume of the finished cake. The most effective emulsifiers in these products are distilled monoglycerides, which are commonly added as an aqueous dispersion in α-crystalline gel form. Free-flowing powders, prepared by spray-drying distilled monoglycerides with other emulsifiers, can be added directly to sponge and other cake doughs.

The interaction of emulsifiers with water is determined by their free energy (G_0), change in enthalpy (δH_0), and change in entropy (δS_0). Solubility of emulsifiers in water may be expressed thermodynamically as

$$\delta G_0 = \delta G_0^{sol} + \delta G_0^h$$

where δG_0^{sol} and δG_0^h are energy contributions of the hydrophilic and hydrophobic moieties, respectively.

A summary of properties and values that allow the classification of emulsifiers is given in Table 9.3. Functional properties of emulsifiers and usage of emulsifiers in specific food applications are listed in Table 9.4 and 9.5. Fig. 9.10 gives formulas of important emulsifiers and Tables 9.6 and 9.7 are guides to some of the functions of those emulsifiers.

In practice, it is often advantageous to use mixtures or combination of emulsifiers because in general, they guarantee stable emulsions using low emulsifier levels. By combining hydrophilic and lipophilic emulsifiers, a mixed product can be tailored to the fat phase that is to be emulsified (Schuster, 1981). Emulsifier complexes are formed and the lipophilic emulsifiers act as "lipophilic stabilizers" in a basically hydrophilic complex. This pseudocomplex results from secondary valency bonds of the lipophilic emulsifier. The complexes are in an ordered state and produce stable interfacial films. The following emulsifiers can be used as lipohilic stabilizers: lecithin, monodiglycerides, acetic and lactic acid esters, and sorbitan fatty acid esters.

TABLE 9.3

Summary of Properties and Values of Emulsifiers Listed in Table 9.2[a]

Designation	Lipophilic part[b]	HLB[c] value	Solubility[d] Oil	Solubility[d] Water	Polymorphism phase behavior[e,f]	Surface film[g,h]	Mesomorphic phase[i]	Micelle formation	Abbreviation
Lecithin	RFA	3–4	s	d	—	I	Lamellar	Reverse micelle	LC
Hydroxylated	RFA	7–10	s	d	—	I	—	Micelle	HLC
Salts of fatty acids									
Sodium	RFA	16–18	i	s	—	I	Hexagonal I	Micelle	
Potassium	RFA	16–18	i	s	—	I	Hexagonal I	Micelle	
Magnesium	RFA	3–5	d	i	—	S	Lamellar	—	SFA
Calcium	RFA	3–5	d	i	—	S	Lamellar	—	SFA
Aluminum	RFA	3–5	d	i	—	S	Lamellar	—	SFA
Mono- and diglycerides	RFA	2.8–3.8	s	d	$\text{Melt} \underset{T_K}{\overset{}{\rightleftharpoons}} \alpha\text{ Form} \underset{T_M}{\searrow} \beta\text{ Form}$	S / I	Lamellar / Cubic / Hexagonal II / Gel / Disperse	Micelle / Reverse micelle	MG, DG
Acetylated MG	RFA	2.5–3.5	s	i	$\text{Melt} \underset{T_K}{\overset{}{\rightleftharpoons}} \alpha\text{ Form} \underset{T_M}{\rightleftharpoons}$	S / I	—	Reverse micelle	AMG
Glyceryl lactoesters of fatty acids	RFA	3–4	s	i	$\text{Melt} \underset{T_K}{\overset{}{\rightleftharpoons}} \alpha\text{ Form} \underset{T_M}{\rightleftharpoons}$	S	—	Reverse micelle	LMG

Substance		HLB			Thermal behavior		Mesophase	Association structure	Abbreviation
Diacetyl tartaric acid esters of MG and DG	RFA	8–10	s	d	$\text{Melt} \xrightleftharpoons[T_M]{T_K} \alpha\text{ Form}\longrightarrow$	I	Lamellar Disperse	Micelle	DATEM
Succinylated MG	RFA	5–7	s	d	$\text{Melt} \xrightleftharpoons[T_M]{T_K} \alpha\text{ Form}\longrightarrow \beta\text{ Form}$	I	Lamellar Disperse	Micelle; Reverse micelle	SMG
Monosodium phosphate derivatives of MG and DG	RFA	—	—	—	—	—	—	—	MPMG
Ethoxylated MG and DG	RFA	12–13	s	s	—	I	Hexagonal I	Micelle	EMG
Stearyl monoglyceridyl citrate	RFA, RFAL	—	—	—	—	—	—	Reverse micelle	SMGC
Lactylated fatty acid esters of glycerol and propylene glycol	RFA	2.5–3.5	s	i	$\text{Melt} \xrightleftharpoons[T_M]{T_K} \alpha\text{ Form}$	—	—	Reverse micelle	LGPE
Propylene glycol mono- and diesters	RFA	1.5–3.0	s	i	$\text{Melt} \xrightleftharpoons[T_M]{T_K} \alpha\text{ Form}$	S	—	Reverse micelle	PGME
Succistearin	RFA	4–6	s	i	—	—	—	—	SCS
Polyglycerol esters	RFA	4–14	s, s	d, s	$\text{Melt} \xrightleftharpoons[T_M]{T_K} \alpha\text{ Form}$	I	Lamellar Hexagonal II Disperse Gel	Micelle; Reverse micelle	PGE
Sucrose esters	RFA	3–16	d, i	d, s	—	—	—	—	SUE
Sucroglycerides	RFA	3–8	d	d	—	—	—	—	SUG
Sorbitan monostearate	RFA	4.7–5.7	s	d	$\text{Melt} \xrightleftharpoons[T_M]{T_K} \alpha\text{ Form}$	S	Lamellar Disperse	Reverse micelle	SMS

(continued)

TABLE 9.3 (Continued)

Designation	Lipophilic part[b]	HLB[c] value	Solubility[d]		Polymorphism phase behavior[e,f]	Surface film[g,h]	Mesomorphic phase[i]	Micelle formation	Abbreviation
			Oil	Water					
Polysorbate 60	RFA	14–15	s	s	—	I	Hexagonal I	Micelle	PS 60
Polysorbate 65	RFA	10–11	s	d	—	I	Hexagonal I	Micelle	PS 65
Polysorbate 80	RFA	14–15	s	s	—	I	Hexagonal I	Micelle	PS 80
Lactylic esters of fatty acids	RFA	5–7	s	d	—	I	—	Reverse micelle	SLA
Calcium stearoyl lactate	RFA	7–9	s	d	—	S	—	Reverse micelle	CSL
Sodium stearoyl lactate	RFA	18–21	s	d		I	Lamellar Hexagonal II Disperse	Micelle Reverse micelle	SSL
Sodium stearoyl fumarate	RFAL	—	d	i	—	—	—	—	SSF
Stearoyl tartrate	RFAL	—	s	d	—	—	—	—	STA
Copolymer condensates of ethylene and propylene oxide	RPO	—	d i	s	—	—	—	—	CEP

[a]From Schuster and Adams (1984).

[b]RFA, Residual fatty acid; RFAL, residual fatty alcohol; RPO, residual polypropylene glycol.

[c]Hydrophilic–lipophilic balance.

[d]s, Soluble; i, insoluble; d, dispersible.

[e]Values from Flack and Krog (1970).

[f]T_K, Temperature of crystallization; T_M, temperature of melting.

[g]Values from Schuster (1983).

[h]I, Instable; S, stable.

[i]Values from Krog and Lauridsen (1976) and Rosevear (1968).

TABLE 9.4

Characteristics of Emulsifiers[a]

Functions	Examples
Emulsification—Combination of oil and water in a compatible dispersion	All food products containing substantial amounts of both oil and water
Antistaling—complexing action on starch, to reduce firming of the crumb	Most baked goods
Texture modification—complexing action on starch to reduce clumping, and improve consistency and uniformity	Macaroni, dehydrated potatoes, breads, and cakes
Aeration or foaming—to initiate or control gas-in-liquid dispersions	Toppings, icings, cakes, and convenience desserts
Emulsion stabilization—to improve the stability or quality of an emulsion	Nonstandardized dressings, frozen desserts, other vegetable dairy products
Solid fluidization—uniform suspension of solid particles in a liquid	Confectionery coatings and chocolates, fluidized shortenings
Crystal modification—modification of polymorphic form, and size and rate of growth in fat crystals	Shortenings, coatings, and peanut butters
Palatability improvement—emulsification of a lipid system to enhance eating quality	Icings, coatings
Agglomeration—controlled coagulation of fat particles in a liquid	Frozen desserts
Defoaming—de-emulsification of gas-in-liquid emulsions usually on top of liquid systems	Syrups, yeasts
Wetting—reduction of interfacial tension between liquid and solid surfaces to cause the liquid to spread more quickly and evenly over surfaces	Convenience foods such as spray-dried dessert mixes, coffee whiteners, and instant breakfasts
Solubilizing—improving ability of liquid-in-liquid dispersions to form clear solutions	Colors, flavors, perfumes
Stickiness and tackiness reduction	Candies, chewing gum

[a]Courtesy of Durkee Industrial Foods Group, Cleveland, Ohio.

C. Emulsifying Capacity of Food Components

Egg yolk is a traditional emulsifier. Its emulsifying properties stem from its phosphatides and membrane-forming lipoproteins (see Chapter 6). Another food with good emulsifying properties is skim milk. Liquid skim milk binds considerable amounts of fat or oil while it forms stable o/w emulsions. Although skim milk contains no emulsifiers, it is rich in proteins. These proteins interact with the fats and/or oils and form a natural emulsifier complex of lipoproteins. The emulsifying properties of proteins result from the interaction of nonpolar, hydrophobic side chains with hydrocarbon chains of TG through hydrophobic interac-

TABLE 9.5

Emulsifiers in Specific Food Applications[a]

Emulsifier type[b]	Consumer shortening[c] (%)	Cake mix[c] (%)	Bread, rolls and buns[d] (%)	Cakes (baker's)[e] (%)	Sweet goods[d] (%)	Icings and fillings[c] (%)	Frozen desserts (%)	Vegetable-type whipped topping (%)	Frozen whipped topping (%)	Coffee whitener, liquid (%)	Coffee whitener, powdered (%)	Peanut butter (%)	Margarine (%)	Confectionery coatings (%)
Monodiglycerides	1.5–4	8–12	0.4–0.5	2–8	10–20	4–6	0.1–0.3	0.1–0.2	—	0.25–1.0	2–5	1–4	0.5	0.25–0.4
Monoglycerides (distilled)	—	—	0.1–0.3	—	7–15	3–5	0.1–0.25	—	—	0.25–1.0	2–5	0.25–2	—	1–3
Monoglycerides (hydrated)	—	—	0.25–0.75	—	—	—	—	—	—	—	—	—	—	—
Polysorbate 60[f]	0.5–2	1–3	0.15–0.35	0.5–2	0.5–1	1–2	—	—	—	0.1–0.3	1–2	—	—	—
Polysorbate 65	—	—	—	—	—	—	0.05–0.1	—	—	—	—	—	—	—
Polysorbate 80	—	—	—	—	—	—	0.05–0.1	0.1–0.3	0.4–0.6	0.25–0.5	1–3	—	—	0.1–0.4
SMS	—	2–4	—	2–4	—	—	—	—	—	—	—	—	—	—
PGME	—	4–12	—	2–12	—	2–6	—	—	—	—	—	—	—	—
GLE	—	4–8	—	2–8	—	—	—	0.1–0.2	0.15–0.30	—	—	—	—	0.2–3.0
ATMG	—	—	0.4–0.5	—	—	—	—	—	—	—	—	—	—	—
EMG	—	—	0.25–0.5	—	0.25–0.5	—	—	—	—	—	—	—	—	—
Ca-st-2-L	—	—	0.25–0.5	—	0.25–0.5	—	—	—	—	—	1–3	—	—	0.5–1.0

Emulsifier													
Na-st-2-L	—	0.25-0.5	—	—	—	—	—	—	—	0.1-0.3	1-3	—	—
Succ. mono.	—	0.25-0.5	—	—	—	—	0.1-0.2	—	0.1-0.2	—	—	—	—
3-1-S	—	—	2-4	—	—	—	0.1-0.2	0.2-0.3	0.2-0.5	2-5	—	—	—
6-2-S	—	—	—	2-4	—	—	—	—	—	—	—	—	—
3-1-O	0.1-0.3	—	—	—	—	—	—	—	—	—	—	0.5-1.0	—
Lecithin and derivatives	—	—	—	—	—	—	—	—	—	—	—	0.5	0.25-0.4
Hydrated blends (wet-basis)	—	0.25-1	0.5-1	—	0.1-0.25	—	—	—	—	—	—	—	—
GMS-Polysorbate 80 80/20	—	—	—	—	0.1-0.25	—	—	—	—	—	—	—	—
GMS, Polysorbate 65 60/40	—	2-4	—	—	—	—	—	—	—	—	—	—	—
SMS, Polysorbate 60 60/40	—	2-4	—	—	—	—	—	—	—	—	—	—	—
SMS, Polysorbate 60 70/30	—	0.25-0.5	0.25-0.5	—	—	0.1-0.2	—	0.2-0.5	—	—	—	—	0.2-0.5
SSF	—	—	—	—	—	—	—	—	—	—	—	—	—

[a] From Nash and Brickman (1972).

[b] Abbreviations: SMS, sorbitan monostearate; PGME, propylene glycol monoesters; Ca-st-2-L, calcium stearyl-2 lactylate; Succ. mono., succinylated monoglycerides; 6-2-S, hexaglycerol distearate; ATMG, acetylated tartrated esters of mono- and diglycerides; GLE, glycerol lactoesters; EMG, ethoxylated monoglycerides; Na-st-2-L, sodium stearyl 2-lactylate; 2-1-S, triglycerol monostearate; 3-1-O, triglycerol monooleate; GMS, polysorbate 80 80/20 (blend of 80% glycerol monostearate, 20% polysorbate 80); and SSF, sodium stearyl fumarate.

[c] On a shortening basis.

[d] Flour basis.

[e] Refer to Federal Bread Standards, Code of Regulation 21, Foods and Drugs, Parts 100 to 199.

[f] Polysorbate 60 is to be used in conjunction with other emulsifiers.

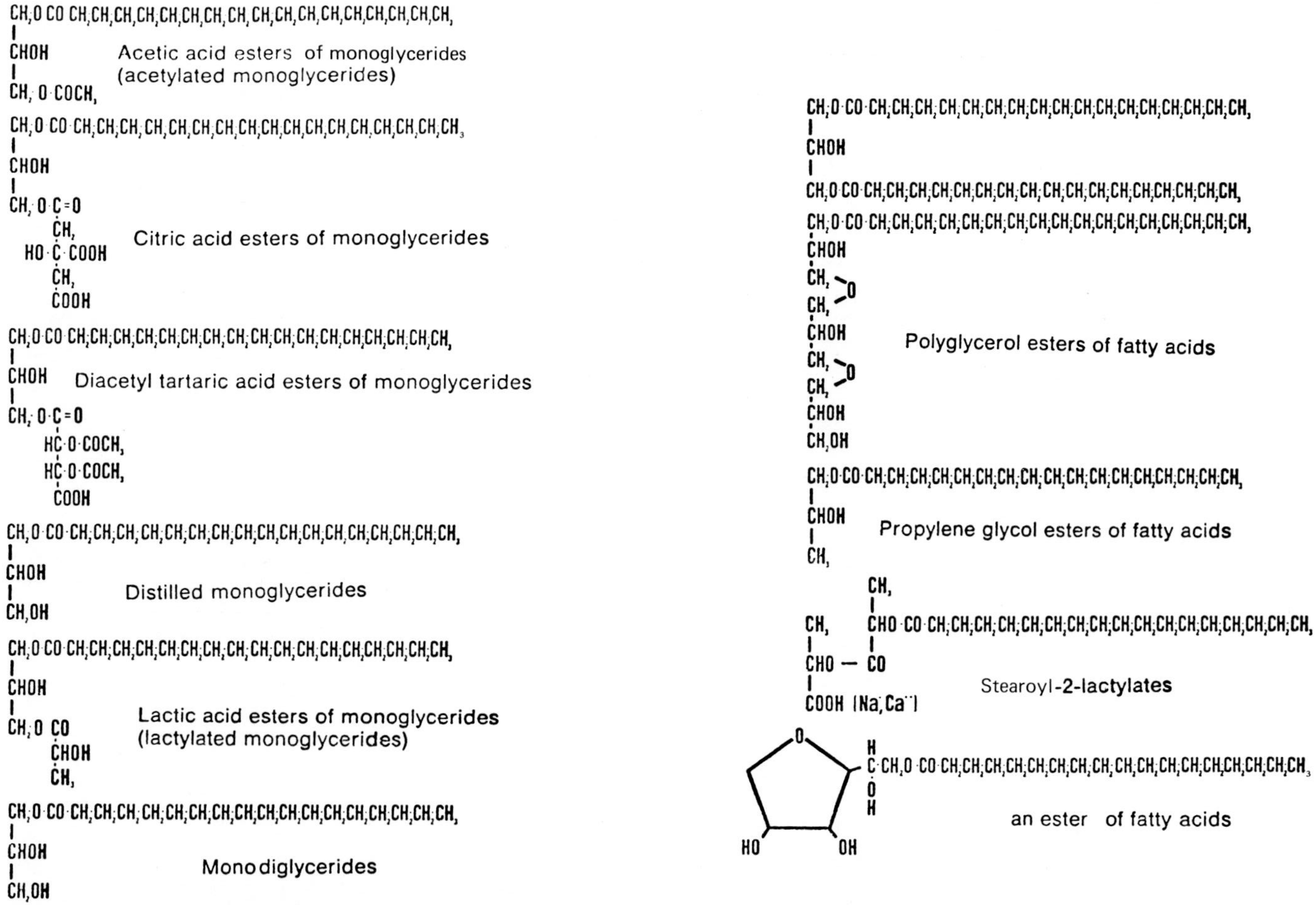

CH₂O CO CH₂CH₂CH₂CH₂CH₂CH₂CH₂CH₂CH₂CH₂
|
CHOH   Acetic acid esters of monoglycerides
| (acetylated monoglycerides)
CH₂ O COCH₃

CH₂O CO CH₂CH₂CH₂CH₂CH₂CH₂CH₂CH₂CH₂CH₂CH₂CH₂CH₃
|
CHOH   Citric acid esters of monoglycerides
|
CH₂ O C = O
      |
      CH₂
  HO·C·COOH
      |
      CH₂
      COOH

CH₂O CO CH₂CH₂CH₂CH₂CH₂CH₂CH₂CH₂CH₂CH₂CH₂CH₂CH₃
|
CHOH   Diacetyl tartaric acid esters of monoglycerides
|
CH₂ O·C = O
      |
   HC·O·COCH₃
      |
   HC·O·COCH₃
      |
      COOH

CH₂O CO CH₂CH₂CH₂CH₂CH₂CH₂CH₂CH₂CH₂CH₂CH₂CH₂CH₂CH₂CH₃
|
CHOH   Distilled monoglycerides
|
CH₂OH

CH₂O CO CH₂CH₂CH₂CH₂CH₂CH₂CH₂CH₂CH₂CH₂CH₂CH₂CH₃
|
CHOH   Lactic acid esters of monoglycerides
| (lactylated monoglycerides)
CH₂ O CO
      |
      CHOH
      |
      CH₃

CH₂ O CO CH₂CH₂CH₂CH₂CH₂CH₂CH₂CH₂CH₂CH₂CH₂CH₂CH₃
|
CHOH   Monodiglycerides
|
CH₂OH

CH₂O CO CH₂CH₂CH₂CH₂CH₂CH₂CH₂CH₂CH₂CH₂CH₂CH₂CH₂CH₃
|
CHOH

CH₂O CO CH₂CH₂CH₂CH₂CH₂CH₂CH₂CH₂CH₂CH₂CH₂CH₂CH₂CH₂CH₃
|
CH₂O CO CH₂CH₂CH₂CH₂CH₂CH₂CH₂CH₂CH₂CH₂CH₂CH₂CH₂CH₂CH₃
|
CHOH   Polyglycerol esters of fatty acids
|
CH₂
   ╲O
CH₂
|
CHOH
|
CH₂
   ╲O
CH₂
|
CHOH
|
CH·OH

CH₂O CO CH₂CH₂CH₂CH₂CH₂CH₂CH₂CH₂CH₂CH₂CH₂CH₂CH₂CH₂CH₃
|
CHOH   Propylene glycol esters of fatty acids
|
CH₃

CH₃   CHO CO CH₂CH₂CH₂CH₂CH₂CH₂CH₂CH₂CH₂CH₂CH₂CH₂CH₂CH₂CH₃
|
CHO — CO   Stearoyl-2-lactylates
|
COOH [Na,Ca⁻¹]

H
|
C·CH₂O CO CH₂CH₂CH₂CH₂CH₂CH₂CH₂CH₂CH₂CH₂CH₂CH₂CH₂CH₃   an ester of fatty acids
‖
O
|
H

Fig. 9.10. Formulas of important emulsifiers. (Courtesy of Grinsted Products Inc.. Kansas City. Kansas.)

TABLE 9.6

Functional Effect of Emulsifiers in Food Products[a]

Emulsifier	Stability of o/w and w/o emulsions	Starch complexing in bread, pasta, and potato products	Protein interaction: dough conditioning effect in yeast-leavened goods	Aerating and stabilizing effect in topping fats, whipped desserts, and bakery products
Acetic acid esters	None	None	None	Very good (α-tending)
Citric acid esters	Very good (o/w and w/o)	Slight	Slight	Slight
Diacetyl tartaric acid esters	Very good (o/w)	Good	Very good	None
Distilled monoglycerides				
Saturated	Good (w/o)	Very good	Slight	Very good (α-crystal form)
Unsaturated	Very good (w/o)	Slight	Slight	Slight
Lactic acid esters	Good (w/o)	Slight	None	Very good (α-tending)
Mono- di-glycerides	Good (w/o)	Slight	Slight	Good
Polyglycerol esters	Very good (o/w)	Slight	Slight	Good
Propylene glycol esters	None	None	None	Very good (α-tending)
Stearoyl-2-lactylates (sodium salts)	Very good (o/w)	Good	Good	Good
Sorbitan esters	Slight (o/w)	None	None	Slight

[a]Courtesy of Grinsted Products Inc., Kansas City, Kansas.

tions (Fig. 9.11). A large interfacially active molecular complex is formed and the interfacial tension is lowered in the system of oil and water. The emulsifying effects of proteins depend on their structure and are influenced by the pH value, salt level, concentration, and temperature of the interacting system. Proteins can interact also with other components of a food, such as carbohydrates and emulsifiers.

D. Stabilization of Emulsions

Stabilizers are added to unstable systems to avoid undesirable physicochemical changes of the food state or—in the context of this discussion—to increase

TABLE 9.7

Emulsifiers in Foods[a]

Emulsifiers	Functions
Bakery and other starch products	
Bread, rolls, bread improvers	
Cold dispersible emulsifiers based on distilled monoglycerides	Crumb softeners, volume enhancers, dough conditioners
Sodium stearoyl-2-lactylates	
Distilled monoglycerides	
Diacetyl tartaric acid	
Esters of monoglycerides	
Cake and sponge improvers	
Distilled monoglycerides	Ensure good fat distribution in batter, aid moisture retention in finished cakes, facilitate mechanical handling, improve aeration and volume, better texture, and shelf life
Blends of distilled monoglycerides and α-tending emulsifiers	
Lactic acid esters	
Propylene glycol esters of fatty acids	
Polyglycerol esters of fatty acids	
Pasta products, snacks	
Sodium stearoyl-2-lactylates	Improve texture, reduce stickiness, aid extrusion, impart crispness
Distilled monoglycerides	
Potatoes (dehydrated flakes, granules)	
Sodium stearoyl-2-lactylates	Improve texture and whip, reduce stickiness
Distilled monoglycerides	
Margarine and related products	
Bakery compound	
Distilled monoglycerides	Stabilize liquid emulsions prior to cooling, provide a fine and stable water dispersion in the bakery compound
Monodiglycerides	
Margarine (table)	
Distilled monoglycerides	Stabilize liquid emulsion prior to cooling, provide a fine and stable water dispersion in the margarine
Monodiglycerides	
Margarine (frying)	
Citric acid esters of monoglycerides	Impart excellent frying properties and minimize spattering
Blends of monoglycerides and special soy lecithin	
Margarine (puff pastry)	
Specially prepared emulsifiers	Improve plasticity of the margarine and produce a flakier puff pastry
Peanut butter	
Distilled monoglycerides	Inhibit oil separation
Monodiglycerides	
Shortening	
Distilled monoglycerides	Excellent creaming properties in the dough resulting in increased cake volume and uniform structure
Monodiglycerides	

TABLE 9.7 (*Continued*)

Emulsifiers	*Functions*
Dairy products	
Cultured milk	
Stabilizer blends	Stabilize protein content
Ice cream	
Combined and integrated emulsifier and stabilizer products	Improve creaminess, "stand-up," dryness, "melt-down," texture, consistency, and overrun control
Milk drinks	
Integrated emulsifiers and stabilizers	Improve fat dispersion, increase heat stability, enhance palatability
Mousse	
Emulsifier and stabilizer blends	Improve aeration and stability
Sorbet and sherbet	
Stabilizer blends	Provide stable overrun, impart refreshing mouth-feel and improve texture
Multifoods	
Caramels, toffees	
Distilled monoglycerides	Better fat distribution, improved chewing properties, reduction of stickiness
Monodiglycerides	
Chocolate, confectionery, coatings	
Sorbitan esters of fatty acids	Bloom retardation and improvement of palatability
Coatings for nuts, raisins	
Acetic acid esters of monoglycerides	Protection against loss of moisture and fat oxidation
Coffee whiteners	
Sodium stearoyl-2-lactylates	Improve dispersion and whitening
Distilled monoglycerides	
Frozen whipped dairy cream and toppings	
Sodium stearoyl-2-lactylates	Used together these products enhance aeration and prevent syneresis and shrinkage of the foam
Lactic acid esters of monoglycerides	
Mayonnaise and salad dressings	
Stabilizer blends	Improve emulsification, regulate viscosity, enhance flavor
Recombined and filled cream	
Distilled monoglyceride specialties	Improve aeration, enhance whippability, stabilize cream during storage
Whipped toppings	
Acetic acid esters of monoglycerides	Improves aeration and foam stabilization, enhance whippability
Lactic acid esters of monoglycerides	
Propylene glycol esters of fatty acids	

[a]Courtesy of Grinsted Products Inc., Kansas City, Kansas.

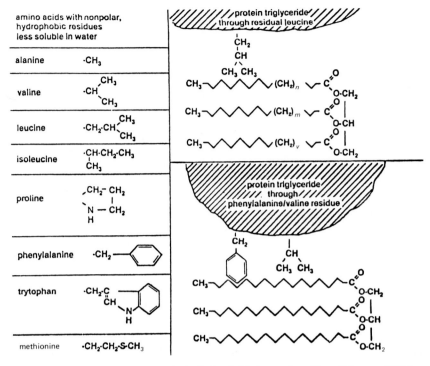

Fig. 9.11. Formation of lipoprotein by hydrophobic interaction. (From Schuster, 1981.)

stability of emulsions. Stabilization of an emulsion involves increasing the viscosity of the outer phase. Many stabilizers interact with the emulsified particles by attaching in a complex manner to the particle that is to be stabilized. They cover the droplets without penetrating directly into the inner phase, and intensify the electrostatic charge or the solvate cover and inhibit aggregation and coalescence. Hydrocolloids (see Chapter 3) exert these effects to a high degree in o/w emulsions (Schuster, 1981). Their emulsion-stabilizing power is due to the formation of a three-dimensional network that forms in the continuous phase as viscosity increases. Some hydrocolloids have hydrophobic zones and show weak interfacial-active properties. Hydrocolloids can associate with emulsifiers and form stable interfacial films.

Hydrocolloids are macromolecular hydrophilic substances. They dissolve, or disperse or swell in water and form viscous solutions, pseudogels, or gels. These substances are mainly polysaccharides. Most hydrocolloids are thread-shaped macromolecules with intramolecular interaction forces that allow different bonds between the hydrocolloids and other emulsion components. Used as stabilizers for food emulsions, hydrocolloids are part of the food and are governed by the same legal criteria as emulsifiers. The EEC allows several compounds as emulsion stabilizers in foods; they are listed in Table 9.8. The materials shown in Table 9.8 are stabilizers for o/w emulsions. Only microcrystalline cellulose and related cellulose powders are suitable for use as stabilizers of w/o emulsions.

TABLE 9.8

Composition and Properties of Stabilizers for Emulsified Food as Stipulated in EEC Directive 80/957/EEC[a]

EEC no.	Name	Composition	Form in water[b]	Ionogenicity	Interfacial activity
E 406	Agar	Seaweed extracts D-galactose β-(1-4), 3,6-anhydro-L galactose-(1-3), + sulfate acid ester groups	hv	Anionic	−
E 401	Sodium	D-mannuronic acid β-(1-4)	lv–hv	Anionic	−
E 402	Potassium alginate	L-guluronic β-(1-4)			
E 403	Ammonium				
E 405	Propane-1,2-diol alginate	Propane-1,2-diol ester of alginate	lv–mv	Non-ionic	+
E 407	Carrageenan	D-Galactose, 3,6-anhydro-D-Galactose + sulfate acid ester groups	mv–hv	Anionic	−
E 414	Acacia or gum arabic	Tree exudates L-arabinose, D-galactose, L-rhamnose, D-glucuronic acid	lv	Anionic	+
E 413	Tragacanth	D-galactose, D-xylose, D-glucuronic acid	hv	Anionic	−
E 412	Guar gum	Seed gums D-mannose β-(1-4), D-galactose-(1-6) branches	hv	Non-ionic	−
E 410	Locust bean gum	D-mannose β-(1-4), D-galactose-(1-6) branches	hv	Non-ionic	−
E 415	Xanthan gum	Fermentation product D-glucuronic acid β-(1-2), D-glucose β-(1-4), D-mannose, D-glucose β-(1-2) or β-(1-3) branches	hv	Anionic	−
E 460	Microcrystalline or powdered cellulose	Cellulose derivatives D-glucose β-(1-4)	insoluble	Non-ionic	−
E 466	Carboxymethylcellulose		lv–hv	Anionic	−
E 465	Ethylcellulose	Ethylether of cellulose	lv–hv	Non-ionic	+
E 461	Methylcellulose	Methylether of cellulose	lv–hv	Non-ionic	+

[b]From Schuster (1981).

[b]lv, Low viscous; mv, medium viscous; hv, high viscous.

III. Lecithins

"Lecithin" is the commercial or popular name for a naturally occurring mixture of similar compounds more accurately identified as phosphatides or phospholipids. The principal components of the natural mixture are phosphatidylcholine, phosphatidylethanolamine, inositol phosphatides, and related phosphorus-containing lipids. The chemical structures of the three major phosphatides are given in Fig. 9.12.

Lecithin is nature's principal surface-active agent. It is found in all living cells of animal and vegetable origin. The highest concentrations generally occur in animal products. For example, fresh egg yolk contains 8–10% phosphatides

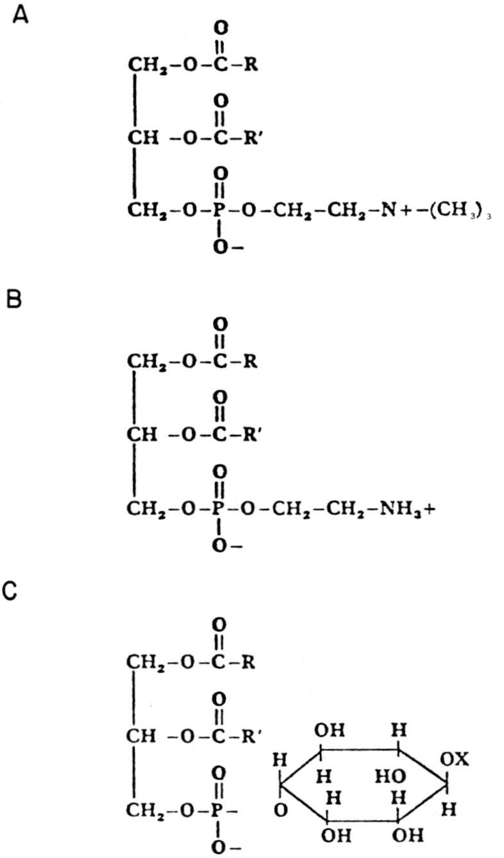

A

$$CH_2-O-\overset{\overset{\text{O}}{\|}}{C}-R$$
$$CH-O-\overset{\overset{\text{O}}{\|}}{C}-R'$$
$$CH_2-O-\underset{\underset{\text{O}-}{|}}{\overset{\overset{\text{O}}{\|}}{P}}-O-CH_2-CH_2-N+-(CH_3)_3$$

B

$$CH_2-O-\overset{\overset{\text{O}}{\|}}{C}-R$$
$$CH-O-\overset{\overset{\text{O}}{\|}}{C}-R'$$
$$CH_2-O-\underset{\underset{\text{O}-}{|}}{\overset{\overset{\text{O}}{\|}}{P}}-O-CH_2-CH_2-NH_3+$$

C

$$CH_2-O-\overset{\overset{\text{O}}{\|}}{C}-R$$
$$CH-O-\overset{\overset{\text{O}}{\|}}{C}-R'$$
$$CH_2-O-\underset{\underset{\text{O}-}{|}}{\overset{\overset{\text{O}}{\|}}{P}}-$$

Fig. 9.12. Chemical structures of three major phosphatides: A, Phosphatidyl choline; B, phosphatidyl ethanolamine; C, phosphatidyl inositol. X, one or more phosphate groups linked to one or more sugar molecules; R, R', fatty acid constituents.

TABLE 9.9

Composition of Important Pure Lecithins (Oil-Free)[a]

	Lecithin type (%)			
	Soybean	*Egg*	*Rapeseed*	*Safflower*
Phosphatidyl choline (lecithin)	18.5–21	73	30–32	32–39
Lysophosphatidyl choline (lysolecithin)	1–2	5–6	3	1–2
Phosphatidyl ethanolamine (ethanolaminecephalin)	12–16	15	30–32	14–17
Lysophosphatidyl ethanolamine (lysoethanolaminecephalin)	1–2	2–3	3	2
Phosphatidyl serine (serinecephalin)	3–4	—	—	—
Phosphatidyl inositol (inositol phosphatide)	11–14	1	14–18	21–27
Lysophosphatidyl inositol (lysoinositol phosphatide)	1	—	1	1
Sphingomyelin	—	2–3	—	—
Plasmalogen	—	1	—	—
Phytoglycolipids	13	—	10	—
Phosphatide acids	2	—	1	—
Other lipids, carbohydrates	25–30	—	2–8	15–28

[a]Courtesy of Lucas Meyer Co., Hamburg, West Germany.

which play a role in the remarkable emulsifying properties of the product. The major commercial source of lecithin is the soybean, which contains 0.3–0.6% phosphatides. In humans and in animals the phosphatides are concentrated in the vital organs, such as the brain, liver, and kidney; in vegetables they are concentrated in the seeds, nuts, and grains. Lecithins from various sources vary widely in composition (Tables 9.9 and 9.10).

A. Types, Properties, and Combinations

Crude commercial soybean lecithin has the following representative composition: 16% phosphatidylcholine; 14% phosphatidylethanolamine; 10% phosphatidylinositol; 35% soybean oil; and 25% miscellaneous compounds (water, sugars, sterols, other phosphatides). The crude dried product obtained from degumming of soybean oil contains 28–33% oil. This product can be used to prepare a variety of grades of lecithin by adding bleaching agents (such as H_2O_2) to lighten the tan-to-brown color, by adding fatty acids or oil to increase fluidity, by removing the oil to increase the phosphatide content, or by separating the oil-

TABLE 9.10

Fatty Acid Composition of Lecithins[a]

| | Type of Lecithin (%) | | | |
	Soybean	Cottonseed	Egg	Groundnut
Palmitic acid	17–21	32–34	35–37	7
Stearic acid	4–6	2–3	9–15	2
Oleic acid	12–15	25–27	33–37	38–41
Linoleic acid	53–57	39–42	12–17	17–19
Linolenic acid	6–7	3–4	0.5	—
Arachidic acid	—	—	—	20–22
Arachidonic acid	—	—	3.7	—

[a]Courtesy of Lucas Meyer Co., Hamburg, West Germany.

free lecithin into alcohol-soluble and alcohol-insoluble fractions (Cowan and Wolf, 1974; Wolf and Sessa, 1978). Different types of commercial lecithins are described in Table 9.11. Separation of an acetone-soluble fraction increases the amount of phosphatides in the acetone-insoluble fraction by decreasing the amount of triglycerides. Further fractionation can separate the alcohol-soluble phosphatidyl choline and phosphatidyl ethanolamine from the alcohol-insoluble inositol phosphatides. The alcohol-soluble fraction gives stable o/w emulsions; the alcohol-insoluble fraction gives stable w/o pan-grease (release of baked or fried foods from the pan) agents. Composition and properties of commercial

TABLE 9.11

Different Grades of Commercial Bleached Lecithin[a]

| | Properties of different grades[b] | | | |
Characteristics	A	B	C	D
Consistency	Plastic	Fluid	Fluid	Fluid
Acetone-Insoluble (%)[c]	67–72	62–64	54–60	62–64
Moisture (%)	1.1	0.75	0.75	0.75
Benzene-Insoluble				
(max %)	0.1	0.1	0.1	0.05
Acid value	25	32	40	32
Centipoises at 26.6°C	—	15,000	7,000	15,000

[a]From Wolf and Sessa (1978).

[b]Grades with A, B, C, and D can result from different amounts of bleaching with hydrogen peroxide.

[c]Actual percentage of phosphatides—remainder primarily oil or additives like fatty acids.

TABLE 9.12

Composition and Properties of Commercial Lecithin Fractions[a]

Property	Oil-free lecithin	Phosphatidyl choline fraction	Phosphatidyl inositol fraction
Solubility in			
Alcohol	Dispersible	Soluble	Insoluble
Water	Dispersible	Dispersible	Dispersible
Oil	Soluble	Soluble	Soluble
Choline fraction (%)	30	60	4
Cephalin fraction (%)	30	30	29
Inositol fraction (%)	32	2	55
Oil (%)	3	4	4
Type of emulsifier	o/w or w/o	o/w	w/o

[a]From Wolf and Sessa (1978).

lecithin fractions are described in Table 9.12. Hydrogen/peroxide, besides bleaching, can also hydroxylate to impart hydrophilic properties, improve moisture retention, and contribute to the formation of stable o/w emulsions.

The National Soybean Processors Association (1974) classifies lecithin into six grades, depending on viscosity and color (Table 9.13). Phosphatide content is

TABLE 9.13

Specifications for Commercial Lecithins[a]

Analysis	Natural		Single-Bleached		Double-Bleached	
	Fluid	Plastic	Fluid	Plastic	Fluid	Plastic
Acetone insoluble (%, min)	62.0	65.0	62.0	65.0	62.0	65.0
Moisture (%, max)	1.0	1.0	1.0	1.0	1.0	1.0
Benzene insoluble (%, max)	0.3	0.3	0.3	0.3	0.3	0.3
Acid value (max)	32.0	30.0	32.0	30.0	32.0	30.0
Color, Gardner (max)[c]	10.0	10.0	7.0	7.0	4.0	4.0
Viscosity poises (25°C, max)[d]	150.0	—	150.0	—	150.0	—
Penetration m (max)[e]	—	22.0	—	22.0	—	22.0

[a]From National Soybean Processors Association (1974).

[b]By toluene distillation for 2 hr or less (American Oil and Chemical Society method Ja 2-56).

[c]On a 5% solution in mineral oil.

[d]By any appropriate conventional viscometer, or by AOCS Bubble Time Method Kz 6-59, assuming density to be unity. Fluid lecithin with a viscosity less than 75 poises may be considered a premium grade.

[e]Using Precision cone 73525, Penetrometer 73510; sample conditioned 24 hr at 25°C.

specified in terms of acetone-insolubility; product clarity and purity, and benzene-insolubility. Plastic lecithins are converted to fluid forms by adding 2–5% fatty acids or carriers such as soybean oil (Wolf and Sessa, 1978). Sullivan and Szuhaj (1975) distinguished between natural lecithins and those refined or chemically modified (Table 9.14). The refined ones are obtained either by custom blending or by alcohol fractionation; the modified ones are hydroxylated or acetylated lecithins. Hydroxylated lecithins are prepared by treating the phosphatides with peracids or hydrogen peroxide in the presence of a water-soluble aliphatic carboxylic acid. Hydroxylated lecithins are readily dispersed in water and have excellent fat-emulsifying properties; they are approved for food application in the United States. The amino group of phosphatidyl ethanolamine acetylates when treated with acetic anhydride. Introducing a substituent on the positively charged portion of the zwitterionic phosphatidyl choline converts it to a negatively charged lecithin with improved solubility and o/w emulsifying properties (Wolf and Sessa, 1978).

Lecithins that contain more than 67% acetone-insoluble compounds or moieties disperse with difficulty in aqueous media. Solubility is imparted by

TABLE 9.14

Classification of Soybean Lecithins[a]

I. Natural
    A. Plastic
        1. Unbleached
        2. Single-bleached
        3. Double-bleached
    B. Fluid
        1. Unbleached
        2. Single-bleached
        3. Double-bleached
II. Refined
    A. Custom blended natural
    B. Oil-free phosphatides
        1. As is
        2. Custom blended
    C. Fractionated oil-free phosphatides
        1. Alcohol-soluble
            a. As is
            b. Custom blended
        2. Alcohol-insoluble
            a. As is
            b. Custom blended
III. Chemically modified

[a]From Sullivan and Szuhaj; cited by Wolf and Sessa, 1978.

reaction with polyhydric alcohols and epoxy compounds, hydroxylation of unsaturated fatty acid constituents, fractionation, or compounding with dispersing agents. Oil-free lecithin is more susceptible to oxidative deterioration than is natural lecithin. Possibly the small amounts of tocopherols in natural lecithins act as antioxidants (Wolf and Sessa, 1978). Antioxidant activity of commercial lecithin is probably due to phosphatidyl ethanolamine as an effective scavenger of metal traces. The granular form has appreciably more surface area than do the thick fluid or plastic forms. Exposure of the increased surface to oxygen accelerates deterioration and causes rancid or bitter flavors. These flavors are also generated by heating above 40°C. Heat discoloration of phospholipids added to oils and fats to prevent spattering can be prevented by adding a compound that forms carbon dioxide in heating, such as $NaHCO_3$.

Lysolecithins, which have a more hydrophilic character than lecithin, show stronger o/w emulsifying properties. Hydrogenated phosphatides have a lighter color, greater resistance to oxidative rancidity, and less odor and flavor. They are, however, more difficult to disperse in aqueous systems and have poorer emulsifying properties than do natural phosphatides (Wolf and Sessa, 1978).

The main function of phosphatides is to emulsify fats. The long chain fatty acid moieties contribute hydrophobic properties; those properties are counterbalanced by the polar or hydrophilic character of the phosphate moiety. In an o/w system the phosphatide components concentrate at the oil/water interface. The polar, hydrophilic parts of the molecules are directed toward the aqueous phase and the nonpolar, hydrophobic (or lipophilic) parts are directed toward the oil phase. Concentration of phosphatides at the oil–water interface lowers the surface tension and makes it possible for emulsions to form. Once the emulsion is formed, the phosphatide molecules at the surface of the oil or the water droplets act as barriers that prevent the droplets from coalescing and thus they stabilize the emulsion (Wolf and Sessa, 1978).

Instant dispersibility in water generally requires chemical modification. It is possible to attain the desirable properties in a blend of soybean lecithin and ethoxylated monoglycerides. In baking formulations that call for soy flour, lecithinated (usually at a level of 15%) low fat or high fat flours can be used. Lecithin is often incorporated into solid or fluid shortenings used in baking.

Many researchers have demonstrated the synergistic effects in bread making of lecithin in combination with mono- and diglycerides and other surface-active agents. The use of lecithin in combination with monoglycerides (a) improves quality characteristics of the raw materials; (b) optimizes technical processing; (c) reduces shortening requirements; and (d) improves overall quality of final products, including freshness retention and nutritive value. Hydroxylated lecithins are particularly valuable in bakery products because of an apparent synergy with mono- and diglycerides, in addition to their high dispersibility in water systems (in contrast to oil solubility of most lecithins).

The following product modifications and combinations were described in several publications from Lucas Meyer Co., Hamburg, Germany: (a) improved water dispersibility through formation of N-acylphospholipids; (b) improved water dispersibility through use of high levels of lysophosphatides; (c) interaction with amylose as coemulsifier for monoglycerides; (d) defatted soy lecithin, 95% powder as emulsifier and dispersant; (e) spray-dried combination of standard soy lecithin and milk solids (mainly lactose) as emulsifier, dispersant, wetting agent; and (f) spray-dried combination of modified partial glycerides and milk solids for flour treatment and assurance of uniformity and high quality of bread-making wheat flours.

B. Use in Foods

1. Chocolate and Confections

Lecithin is used extensively in the cocoa processing industry. In cocoa powder and instant drink powder mixes, the wetting and suspension properties of lecithins are of great importance. For manufacture of chocolate and coatings, the interfacial activity of lecithin serves to reduce the proportion of cocoa butter, the most expensive ingredient, without impairing the flow and melting properties. Lecithin provides a physicochemical link between two immiscible phases, the cocoa fat and the sugar solution.

During a process called conching, powdered and crumbly milled material is converted into a flowable suspension of sugar, cocoa, and milk powder particles in cocoa butter. Mechanical input is expended to separate the agglomerates formed during rolling and to coat the individual particles so formed with a sheath of fat, in order to disperse them in the cocoa butter. The longer the duration of the conching, the better the flow characteristics. The decreasing content of water as hydrophilic component and the progressive coating of solid particles are responsible for the net improvement of flow properties during conching. The addition of small amounts (0.1–0.5%) of lecithin reduces the viscosity of chocolates and enables a 3–8% saving in cocoa butter. The amount of lecithin required to attain optimum flow properties decreases as the fat content increases. On the other hand, the amount of lecithin required to produce optimum lowering of viscosity increases as the particle size of the solid components decreases. Unlike cocoa butter and other fats, lecithin performs a number of functions. It facilitates fusion in molten chocolate compositions and improves distribution of the fatty phase and flowability of the multicomponent system due to lecithin's wetting and emulsifying properties. Lecithins are heat-sensitive, and they retard evaporation of water and volatilization of water-vapor-volatile components. Lecithin should not be added, however, until after the melted material is cooled to below 60°C. For the best distribution of small amounts of lecithin in the chocolate suspension, the lecithin can be added as a lecithin–cocoa butter mixture. The best distribution of the surface-active lecithin is obtained by continuously and intensively mixing the melting material for 30 min after lecithin addition. The liquefying action of lecithins increases considerably as the mixture proceeds from cocoa (single component system) thrugh cocoa–sugar (two component system) to cocoa–sugar–milk (three component system). Lecithins reduce the frictional forces between the interfaces of the cocoa mix, the sugars, and the proteinaceous components, and thereby decrease the yield point and viscosity.

Without added lecithins cocoa powders have poor wetting properties. Cocoa powder containing added 1.0–1.5% lecithins wets rapidly on sprinkling over cold water or milk. A certain amount of water is required to achieve good wetting properties in instant powders to align the hydrophilic and hydrophobic groups on the surface of the particles. A high correlation exists between the specific area of lecithins distributed in the solid phase and its wetting properties. Instant drink powders are multicomponent mixes, which consist of sugars, cocoa powder, nutrient salts, flavors, and vitamins. The components, except for cocoa powder, are blended in a mixer to a homogenate while lecithin is incorporated through spraying. Subsequently, the cocoa powder is incorporated into the mixture and the remainder of lecithin is sprayed directly into it as a finely divided aerosol. To further improve the wetting characteristics and to accelerate dissolution of the sugar, instant drink powders are often agglomerated by a thermal finishing treatment in a moist atmosphere. Moistening the surface during the fluidized-bed treatment facilitates formation of a fine porous agglommerate of adhering particles. The consistency of soft milk caramels, toffees, fudges, and similar products is governed by their water content (5–8%) and their fat content (at least 6%). These two components govern the chewing and biting properties and require a special lecithin for homogeneous bonding. Satisfactory emulsion and proper consistency of soft caramels depend on the presence and contributions of lecithins to reduce fat separation and loss of water. Recrystallization of sugar, in the absence of lecithin, leads to hard products with objectionable textures.

Automation and mechanization of the food industry make it mandatory to eliminate contamination of food components by mineral lubricants. Lecithins can be employed as lubricants and "slip" agents. Even the formation of a hard sugar deposit on mechanical pouring heads can be combated effectively through the use of aqueous lecithin emulsions.

2. Flour, Bread, and Pastry Doughs

Lecithins play an important role in the cereal processing industry. Besides a shortening-sparing effect, they improve rheological properties (in terms of stability), and enhance overall bread quality, including shelf life and consumer acceptance. Processing lecithins into powdered and free-flowing products by bonding them to suitable carriers makes possible direct addition of lecithins to

TABLE 9.15

Guide to Some Lecithin Uses[a]

Uses	Typical action	Suggested concentration (%)	Results and remarks
Animal feeds (other than pet foods)	Emulsifier, wetting agent	~1–3	Promotes and stabilizes emulsions. Provides instant wetting of powdered formulations
Baking	Modify the gluten characteristics of flour, emulsifier, antioxidant, wetting agent	0.1–0.3 based on flour in bread 0.5–1 based on shortening in pies	Improve the "shortening effect"; reduce emulsifier cost; benefit flavor, shelf life, texture, and moisture retention
Cake mixes	Modify the gluten characteristics of flour, emulsifier, antioxidant, wetting agent	1–3 based on shortening in cakes plus other emulsifiers	Same as in baking. Allows rapid wetting with milk or water. Generally used in combination with monoglycerides
Candy	Emulsifier	~1 based on shortening	Used in fat-containing candy, e.g., caramels, brittle, nougats, and taffies. Aids mixing of sugar, fat, and water to prevent greasiness, graining, and streaking

Application	Function	Amount (%)	Comments
Chocolate	Reduces viscosity by wetting and dispersing action	0.3–0.5	More effective and lower priced than cocoa butter. The higher level may be used to improve moisture tolerance as in ice cream coatings, etc.
Ice cream		0.001–0.1 of final weight	Improves smoothness, prevents "sandiness" on storage; aids dispersion of chocolate.
Instant foods	Wetting, dispersing, emulsifying	0.5–3	Reduces use of egg yolk and stabilizes fat. Aids reconstitution by improving wettability and dispersion of solids and fats (i.e., dry beverage mixes, etc.)
Margarine	w/o emulsifier, anti-spattering agent, browning agent	0.15–0.5	Concentration depends on degree of emulsion stability desired and grade of lecithin used
Shortening	Emulsifier, dispersant	0.5–1	Produces a more complete and uniform blending of the shortening. Increases the shortening action

[a]Courtesy of Central Soya, Chemurgy Div., Chicago, Illinois.

flours. Adding lecithin is important in the manufacture of ready-mixed and specialty flours. Many specialty flours for yeast-leavened doughs (for rolls, rusks, and so on) contain lecithin as gluten-stabilizing components, as antiaging agents, and as antioxidants to retard fat spoilage. Liquid lecithin can be incorporated into specialty flours along with the fat component.

In fat-containing bakery goods of the confection, rusk, and pastry types, the function of lecithins is to facilitate fat dispersion. Lecithins improve yeast-leavened pastry products and promote suppleness and stability in sponge cake mixes and creamed dough mixes. In short-type doughs, blighting is prevented. Lecithins contribute to the layering of puff pastries and the flakiness of Danish pastries. The egg content of brioches can be reduced when lecithin is used. Lecithin balances out differences in qualities of flours used to produce biscuits, crackers, wafers, rusks, and the like. Biscuit and cracker doughs are drier and more supple and the baked products are more uniform in size, shape, and quality. The most important role of lecithins in production of cookies is probably the fact that they facilitate rapid blending and incorporation of all components. This reduces mixing time and prevents the development of doughs that are too tough for cookie making. Lecithins represent an important quality-stabilizing factor in the production of alimentary pastes.

3. Fats and Oils

The main problem in margarine production is emulsification. Prior to the use of lecithin, egg yolk was used as emulsifier. Microscopic examination of margarine reveals a network of fat particles in which droplets of water are embedded. Each kilogram of margarine has an internal surface area of about 600 m^2. Such an extended interfacial system can be stabilized only by an emulsifier that reduces the surface tension between water and fat. Lecithin in margarine acts as an emulsifier and as a membrane-forming agent. In addition, it acts to limit spattering properties of margarine. Spattering, the rapid explosive evaporation of water, takes place when finely divided water particles coalesce to form large droplets. Lecithins envelop the water particles with a protective membrane and in this way reduce hazards of water coalescence. Fine dispersion of water in oil is of fundamental significance in ensuring acceptable consistency and spreadability of margarines. From a nutritional perspective, it is becoming more and more important to increase the content of polyunsaturated fatty acids in the American diet, yet these fatty acids are susceptible to oxidation. Lecithin has a limited antioxidative effect.

4. Milk and Milk Products

In unprocessed milk, fat is present as protein-enveloped globules. The fat globule membrane is rich in phosphatides and contains practically all the phosphorus-containing lipids in cow's milk; it constitutes 0.8–1.0% of the total milk fat. Lipoproteins of the milk fat globule membrane, formed from phosphatides and proteins, prevent rapid creaming of milk fat; they also stabilize the foam structure of whipped cream and protect butter from oxidation. In addition, the lipoproteins affect the antispattering and browning properties of butter. Besides these functions, lecithins fulfill the functions of emulsifiers during the preparation of reconstituted milk products. Lecithin rich in phosphatidyl choline are used to produce fat-containing imitation milk products that are instantly dispersible. These lecithins also produce butterfat emulsions that contain 60–70% water.

Whole milk contains 0.2–5.0% free fat, which is distributed over the surface of individual particles of milk powders as a fine layer. The monomolecular or oligomolecular layer makes the surface strongly water-repellent, especially in cold water. There is a strong surface tension between the fat layer and cold water, so that wetting is inadequate. In addition, the hydrophobic action of the surface prevents the intrinsically hydrophilic milk protein components from swelling and dissolving. Hydrophilic lecithin reduces the interfacial tension drastically and improves the solubility of milk powder. A brief summary of the action and results of lecithin used in foods is presented in Table 9.15.

REFERENCES

Code of Federal Regulations (1976). Title 21 CFR 182. Office of the Federal Register, U.S. Govt. Printing Office, Washington, D.C.

Codex Alimentarius Commission (1975). "Procedural Manual," 4th ed. Food and Agriculture Organization of the United Nations, Rome.

Codex Alimentarius Commission (1979). "Guide to the Safe Use of Food Additives," 2nd Ser., CAC/FAL, 5/1979/. FAO/WHO, Rome.

Cowan, J. C., and Wolf, W. J. (1975). Soybeans. In "Encyclopedia of Food Technology" (A. H. Johnson and M. L. Peterson, eds.), pp. 818–828. Avi Publ. Co., Westport, Connecticut.

Davies, J. T. (1957). A quantitative kinetic theory of emulsion type. Proc. Int. Congr. Surf. Act., 2nd, 1957, Vol. 1, pp. 426–428.

European Economic Community (EEC) (1982). "Draft Proposal for a Council Directive on the Approximation of the Laws of the Member States on the Labeling of Food Additives Sold as Such by Retail for Manufacturing Purposes," Working Doc. III/991/82-DE. European Economic Community, Commission of the European Communities, Directorate for Internal Market and Industrial Affairs, Brussels.

Flack, A., and Krog, N. (1970). The functions and applications of some emulsifying agents commonly used in Europe. Food Trade Rev. 40, 27–33.

Food and Agriculture Organization/World Health Organization (FAO/WHO Joint Expert Committee on Food Additives) (1974). "Evaluation of Certain Food Additives," Tech. Rep. Ser. 557. FAO, Rome.

Food and Agriculture Organization/World Health Organization (FAO/WHO Joint Expert Committee on Food Additives) (1978a). "Specifications for Identity and Purity: Food Colours, Enzyme Preparations and Other Food Additives," No. 7. FAO, Rome.

Food and Agriculture Organization/World Health Organization (FAO/WHO Joint Expert Committee on Food Additives) (1978b). "Specifications for Identity and Purity: Thickening Agents, Anticaking Agents, Antimicrobials, Antioxidants, Emulsifiers," No. 4. FAO, Rome.

Food and Drug Administration (1976). "Federal Food, Drug, and Cosmetics Act, as Amended." U.S. Govt. Printing Office, Washington, D.C.

Griffin, W. C. (1956). Clues to the surfactant selection offered by the HLB system. *Off. Dig., Fed. Paint Varn. Prod. Clubs* **28,** 446–455.

Kermode, G. O. (1972). Food additives. *Sci. Am.* **226**(3), 15–21.

Krog, N., and Lauridsen, B. J. (1976). Food emulsifiers and their associations with water. *In* "Food Emulsions" (S. Friberg, ed.), pp. 67–139. Dekker, New York.

Manley, D. J. R. (1983). "Technology of Biscuits, Crackers, and Cookies." Ellis Horwood, Ltd., Chichester, England.

Nash, N. H., and Brickman, L. M. (1972). Food emulsifiers—Science and art. *J. Am. Oil Chem. Soc.* **49**(8), 457–461.

National Soybean Processors Association. (1974). "Yearbook and Trading Rules." Washington, D.C.

Pilnik, W. (1973). Food additives—Recent developments in the food industry. *Gordian* **73**(5), 208–214; (6), 252–256; (7/8), 299–302, 305–307.

Rosevear, F. B. (1968). Liquid crystals: The mesomorphic phases of surfactant compositions. *J. Soc. Cosmet. Chem.* **19,** 581–594.

Schuster, G. (1981). Manufacture and stabilization of food emulsions. Lecture presented at the INSKO Training Course, Helsinki, Finland.

Schuster, G. (1983). Untersuchungen zum Grenzflaecheverhalten von Emulgatoren. *Seifen-Oele-Fette-Wachse* **109,** 3–9.

Schuster, G., and Adams, W. F. (1984). Emulsifiers as additives in bread and fine baked goods. *Adv. Cereal Sci. Technol.* **6,** 139–287.

Sullivan, D. R., and Szuhaj, B. F. (1975). Commercial lecithin: Types, properties and uses. Paper presented at a meeting of the American Oil and Chemical Society, Dallas, Texas.

Wodicka, V. O. (1980). Legal considerations of food additives. *In* "Handbook of Food Additives" (T. E. Furia, ed.), 2nd ed., Vol. 2, pp. 1–12. CRC Press, Inc., Boca Raton, Florida.

Wolf, W. J., and Sessa, D. J. (1978). Lecithin. *In* "Encyclopedia of Food Science" (A. H. Johnson and M. L. Peterson, eds.), pp. 461–467. Avi Publ. Co., Westport, Connecticut.

Some Traditional Foods

This chapter reviews the role of components in four well-established foods: dairy, wheat flour, malt, and soybean products. The first section covers dairy ingredients. Next, investigations designed to determine the role of wheat flour components in bread making are reviewed. Modifications and transformations of malt components in beer production are discussed in the third section, which describes changes from barley to beer. The last section concerns the composition and potential of soybean products.

I. Dairy Ingredients*

Many dairy-based ingredients provide emulsifying, stabilizing, whipping, water-absorbing, browning, crystallizing, and flavor-enhancing properties to foods. Their use often improves both flavor and nutritional value. A less obvious benefit of utilizing dairy based ingredients is the possible reduction in the number of necessary ingredients. By properly selecting dairy-based ingredients, one may obtain the required functional proteins, fats, carbohydrates, vitamins, and minerals in one ingredient without the use of single purpose additives.

The common names established for dairy ingredients are "milk" for milk, concentrated milk, reconstituted milk, and dry whole milk; "skim milk" or "nonfat milk" for skim milk, concentrated skim milk, reconstituted skim milk, and nonfat dry milk; "buttermilk" for sweet cream buttermilk, concentrated sweet cream buttermilk, reconstituted sweet cream buttermilk, and dried sweet cream buttermilk; "whey" for whey, concentrated whey, reconstituted whey, and dried whey, "cream" for cream, reconstituted cream, dried cream, and plastic cream; and "butterfat" for butteroil and anhydrous butterfat. Dairy products for which a standard of identity has been established are labeled by that established name.

A. Nutrition

Milk, from which all dairy products are produced, is a rich source of nutrients. Depending on the composition of the specific dairy product, considerable nutritional enrichment may be obtained by the use of dairy-based ingredients.

1. Milk Proteins

The essential amino acid profiles of the major milk proteins and selected other proteins and FAO standards are presented in Table 10.1.

The protein efficiency ratio (PER) represents the weight gained per gram of protein consumed by weaning rats and is a commonly used biological assay. In the United States, PER is used to determine the percentage of the recommended daily allowance (US-RDA) of protein supplied by a specific food. If the PER of the protein in a food is equivalent or greater than casein, 45 gm of that protein provides the US-RDA. If the PER is less than casein, 65 gm of that protein are

---

*This section is a condensation and update of three publications on dairy-based ingredients: *Dairy-Based Ingredients for Food Products* by A. G. Hugunin and N. L. Ewing, *Milk-Derived Ingredients for Confectionery Products* by A. G. Hugunin and R. K. Nishikawa, and *A Fresh Look at Dairy-Based Ingredients for Processed Foods* by A. G. Hugunin and S. M. Lee. The reviews were published by Dairy Research, Inc., a division of United Dairy Industry Association, Rosemont, Illinois.

TABLE 10.1

Essential Amino Acid Composition of Food Proteins[a,b]

Food protein	Threonine	Valine	Leucine	Isoleucine	Lysine	Methionine	Phenylalanine	Tryptophan	Histidine[c]	Arginine[c]	Cysteine
Whole milk	4.6	7.1	12.1	6.7	7.4	2.8	5.5	1.4	2.2	3.7	0.8
Whey	5.8	6.1	12.3	5.8	10.3	2.3	3.9	2.6	2.2	—	2.3
Casein	4.5	7.4	10.0	6.4	8.1	3.3	5.4	0.96	3.0	3.9	0.4
Soybean meal	3.9	5.3	8.0	6.0	6.8	1.7	5.3	1.4	2.9	7.3	1.5
Egg albumin	4.9	7.5	8.8	6.4	7.2	4.2	6.0	1.2	2.4	6.0	2.4
Meat, beef	4.4	5.1	7.8	5.2	8.6	2.7	3.9	1.0	3.3	6.5	1.3
Whole wheat	3.3	4.3	7.0	4.0	2.7	2.5	5.1	1.2	2.1	4.3	2.7
Whole rice	3.8	6.2	8.2	5.2	3.2	3.4	5.0	1.3	1.7	7.2	1.6
Whole corn	3.7	5.3	15.0	6.4	2.3	3.1	5.0	0.6	2.5	4.8	1.5
Potatoes, white	6.9	5.3	9.6	3.7	8.3	2.5	5.9	2.1	2.2	5.0	0.6
FAO standard	2.8	4.2	4.8	4.2	4.2	2.2	2.8	1.4	—	—	2.0

[a]In grams of amino acid per 100 grams of protein.
[b]Courtesy of Dairy Research, Inc., Rosemont, Illinois.
[c]Histidine and arginine may be important in the nutrition of infants and children.

TABLE 10.2

Protein Efficiency Ratio of Milk
and Other Food Proteins[a]

Food protein	Adjusted PER (casein = 2.5)
Corn, germ flour	2.2
Wheat, gluten	0.8
Soy	2.2
Peanut	1.8
Torula yeast	1.7
Egg, hen	2.6
Nonfat dry milk solids	2.7
Caseinate, sodium	2.6
Whey protein	3.2
Casein	2.5

[a]Courtesy of Dairy Research, Inc., Rosemont, Illinois.

required to provide the US-RDA. The PER of major milk proteins and other selected proteins are presented in Table 10.2.

Data in Tables 10.1 and 10.2 indicate that milk proteins have excellent nutritional quality. The excess quantities of lysine are important in fortification because this is the limiting amino acid in wheat and rice proteins. Milk proteins, particularly whey proteins, also contain adequate concentrations of tryptophan, which is the limiting amino acid in corn. The use of milk proteins to improve protein quality is very feasible. Soluble forms of milk protein concentrates are available for beverage products. These products are bland and do not upset delicate flavor balances in most products. Because they have unique functional properties, they may impart improved body or texture to the products they fortify.

2. Vitamins

Milk contains at least 12 water-soluble vitamins and four fat-soluble vitamins, as shown in Table 10.3. Although the quantities present in 1 quart of fluid milk (132 gm of dry whole milk) are less than that of the US-RDA, all essential vitamins are present in fresh fluid milk. During processing some of the vitamins are destroyed, and when milk is fractionated disproportionate quantities of vitamins are detected in the various fractions.

The majority of milk marketed for fluid consumption is fortified with vitamin D; low fat milk is fortified with vitamin A. Manufacturing-grade milk used for ingredients usually is not vitamin fortified. Of the naturally occurring fat-soluble vitamins, vitamin A is present in the most abundant quantity. It has excellent

TABLE 10.3

Vitamin Content in Milk Products[a]

Milk product	Vitamin A (IU/100 gm)	Vitamin E (mg/kg)	Thiamine (mg/kg)	Riboflavin (mg/kg)	Niacin (mg/kg)	Pantothenic acid (mg/kg)	Vitamin B_6 (mg/kg)	Biotin (mg/kg)	Folic acid (mg/kg)	Vitamin B_{12} (mg/kg)	Vitamin C (mg/kg)
Whole milk											
Fluid	150	1.0	0.43	1.74	0.93	3.39	0.60	0.03	0.059	0.0042	10.5
Dried	1110	7.5	3.3	15.5	7.3	27.3	3.9	0.30	0.50	0.026	81
Skim milk											
Fluid	9	—	0.4	1.7	0.86	3.6	0.41	0.016	—	0.0037	19
Dried	68	0.4	3.6	18.9	10.6	38.8	4.4	0.27	0.60	0.034	98
Butter	2917	2.4	0.03	2.3	0.5	2.3	0.04	—	—	—	—
Cheddar cheese	1169	—	0.30	5.0	0.92	2.7	0.74	0.022	0.095	0.013	—
Cottage cheese	291	—	0.26	3.3	0.92	2.2	0.35	0.020	0.30	0.0085	—
Whey (dried)	50	—	3.7	23.4	9.6	47.3	4.0	0.37	0.89	0.021	—

[a]Courtesy of Dairy Research, Inc., Rosemont, Illinois.

stability during pasteurization, evaporation, and drying. As is true with all of the fat-soluble vitamins, 90% or more of this vitamin remains in the fat fraction when milk fat is separated. Vitamins D, E, and K are present at relatively low concentrations in unfortified milk. Vitamin K is heat-sensitive and may be partially destroyed during milk processing, but vitamins D and E have excellent heat stability.

The water-soluble vitamins are primarily distributed in the serum phase of milk and are concentrated in low fat dairy products. Vitamin B_1 (thiamine) is present at significant concentrations in freshly secreted milk. During heat processing, some destruction of B_1 may occur. Up to 50% loss has been reported in instant nonfat dry milk. Thiamine is relatively stable when stored in dry products, but some loss may occur during storage of fluid products. Riboflavin is resistant to heat provided exposure to light is minimal. Sunlight, and to a lesser extent artificial light, destroys riboflavin, and the rate of destruction is directly related to temperature. No loss of riboflavin occurs during storage of dry products.

Although a quart of milk contains less than 10% of the US-RDA of niacin, milk is one of the most effective foods in preventing pellagra (niacin deficiency). The antipellagra properties of milk are attributed to the availability of the niacin and the high concentration of tryptophan, which humans can metabolize to niacin. Niacin is stable during all reasonable dairy processing and is not destroyed by light.

Vitamin B_6 (pyridoxine) is present in significant concentrations in fresh milk, but 20–30% can be destroyed during production and storage of condensed and sterilized milk products. Under proper manufacturing conditions and during storage of dry products, the majority of vitamin B_6 is retained.

Pantothenic acid, biotin, and folic acid are essential to humans. Estimated daily requirements are 5–10 mg pantothenic acid, 150–300 μg biotin, and 400 μg folic acid. A quart of milk supplies 35–65%, 10–20%, and 14% of those vitamin requirements, respectively. Pantothenic acid and biotin are relatively stable when exposed to heat and light and no detectable loss occurs in dried milk stored for a year. Folic acid is sensitive to heat in the presence of oxygen. Losses of 40–50% of the folic acid have occurred during sterilization of milk. Less significant losses occur during pasteurization, evaporation, and storage in sunlight.

A quart of fresh milk can supply up to 80% of the required vitamin B_{12}. During processing, variable quantities of this vitamin are destroyed. Reconstituted nonfat dry milk contained 20% less vitamin B_{12} than the fluid whole milk. The effect of heat on vitamin B_{12} is increased in the presence of oxygen and may be linked to the destruction of vitamin C.

Vitamin C (ascorbic acid) is present at a high concentration in freshly secreted milk, but processed milk products contain very little vitamin C.

3. Minerals

The mineral content of milk, more specifically the calcium content, is perhaps the most important nutritional property of dairy products. The National Research Council recommends the following daily intake of calcium: 0.8 gm for adults, 1.2 gm for pregnant women and 1.2 gm for children 11–18 years of age. Whole milk contains approximately 0.9 gm calcium/100 gm of milk solids.

Phosphorus is the second most important mineral in milk. The recommendations for phosphorus are based on the calcium requirements and vary with age. The calcium to phosphorus ratio in the diet should be 1:1 for adults and growing children, and 1.5:1 for infants up to 1 year of age. The calcium to phosphorus ratio in cow's milk is 1.2:1.

A quart of milk (132 gm of dry whole milk solids) contains 33% of the US-RDA of magnesium, 3% of the US-RDA of iron and 4% of the US-RDA of iodine. Sodium, potassium, chlorides, citrate, and sulfur are also present in milk.

4. Carbohydrates

Lactose is the principal carbohydrate and the major component in milk. Certain individuals are unable to metabolize lactose. Utilization of lactose requires the presence of the enzyme lactase (β-D-galactoside galactohydrolase, EC 3.3.1.23) in the small intestine. Lactase cleaves the disaccharide into glucose and galactose, which enter the blood stream and are metabolized. In individuals lacking the enzymes, lactose moves on to the large intestine where it is fermented by bacteria, producing acids and carbon dioxide. Milk solids in which lactose is nearly completely hydrolyzed are now commercially available.

5. Milkfat

The cholesterol content per gram of product in butter is less than half that of whole egg and three to four times that of chicken and fish. In most applications in which milkfat would be used as an ingredient, the final concentration is 1–10%. The cholesterol content contributed by milkfat in these products is 2.8–28.0 mg/100 gm. Milkfat imparts unique functional, flavor, and aesthetic characteristics not easily duplicated by other fats. These properties may improve consumer acceptance of a product.

B. Functional Properties

The functional properties of individual milk components used as an ingredient are related to the composition of the final product and processing conditions which may be used.

1. Milk Proteins

Approximately 80% of the proteins present in milk are classified as caseins. Caseins are very diverse in the functional properties, which include water-binding, gelling, whipping, fat-emulsifying, whitening, and stabilizing properties. Caseins and casein-containing milk products are used in breakfast cereals, baked goods, comminuted meats, synthetic meat products, powdered toppings, desserts, coffee whiteners, simulated whipping creams, imitation milk, instant breakfast beverages, puffed snacks, and nondairy or imitation cheeses.

Whey proteins are proteins that remain in the serum after casein is precipitated. No commercial attempts are made to fractionate these proteins; the functional properties of whey protein concentrates relate to the total effect of these proteins. Whey proteins have low water absorption properties, permitting high concentrations without excessive viscosity. Undenatured whey proteins are soluble in both acid and alkaline solutions. Ionic strength has little effect on solubility except at lower pH values. Whey proteins exhibit good emulsifying properties over a broad pH range. They have been used to produce foams, but the percentage of foam overrun and the foam stability are influenced by variations in heating temperature, redox potential, pH, and fat content.

Whey proteins have the ability to coagulate and form complexes with other proteins. This apparently results from sulfhydryl and disulfide interactions. The resulting complexes often have functional properties different than those of whey proteins. When heated with other proteins such as casein, whey proteins contribute to water retention. A gel structure produced by protein–protein complex may be responsible.

2. Lactose

The functional properties of lactose are considerably different than those of other sugars, such as sucrose. Lactose has a low relative sweetness. Depending on the concentration, two to four times as much lactose is required to produce the sweetness of sucrose. Therefore, large quantities of lactose may be used to provide functional properties without causing excessive sweetness.

Most uses for lactose depend upon its ability to facilitate browning, stabilize proteins, induce crystallization, and enhance flavor. Because of its relatively low sweetness, lactose is used to increase the bulk of solids and improve mouth-feel, texture, and/or viscosity. Lactose may improve the texture of some products such as canned fruits and vegetables, bakery products, cream filling, and toppings. In the crystalline state, lactose is nonhydroscopic and free-flowing, making it a desirable dispersing agent.

3. Milk Lipids

The major advantage of milkfat is its delicate, pleasant flavor. This flavor is retained through baking and other heat treatment. Milkfat is used most often in premium formulations for bakery products, confections, and vegetable products.

Triglycerides make up 95–96% of the total milk lipids. The major fatty acids are 16:0, 18:0, and 18:1, which includes approximately 70% of the total fatty acids. The less prominent short chain fatty acids are primarily responsible for the flavor of milk. Milk fats do not exhibit sharp melting points; rather the range is from 30 to 41°C.

Phospholipids make up 0.8–1.0% of the total lipids in whole milk. In skim milk and buttermilk the phospholipid to fat ratio is much higher, whereas in high fat dairy products the ratio is somewhat lower. Like all phospholipids, those in milk have excellent emulsifying properties.

C. Dairy-Based Products

The fractionation of milk into various dairy-based ingredients is shown in Fig. 10.1. The average compositions of various dairy ingredients are presented in Table 10.4.

1. Whole Milk

Whole milk is expensive. Because of its high concentration of water, it is often difficult to use in formulations, expensive to transport, and its storage stability is limited.

2. Condensed and Skim Milk

Condensed milk products offer the advantage of easily specifying fat and solids concentration and special heat treatments. Because of their reduced water content, they are more economical to transport and store. The major disadvantage of condensed products is the limited storage life and the necessity of refrigerated storage. Sweetened condensed milk has declined in popularity because of improved technology in production of dry products. Although sweetened condensed milk is resistant to bacterial growth, some molds and sugar-fermenting yeasts may grow in bulk-handled products. The high sugar concentration may be a disadvantage in some applications; however, large quantities are still being used by the baking, ice cream, and confectionery industries.

3. Dry Whole Milk

United States Department of Agriculture (USDA) standards for grades of dry whole milk identify three grades, U.S. Premium, U.S. Extra, and U.S. Standard. Dry whole milk is produced either by spray- or roller-drying processes. Spray-drying is the most common: the product has higher solubility and fewer scorched particles.

Dried whole milk products are easier to use in many formulations, less expensive to transport, and resistant to microbial growth. An oxidized or "tallowy"

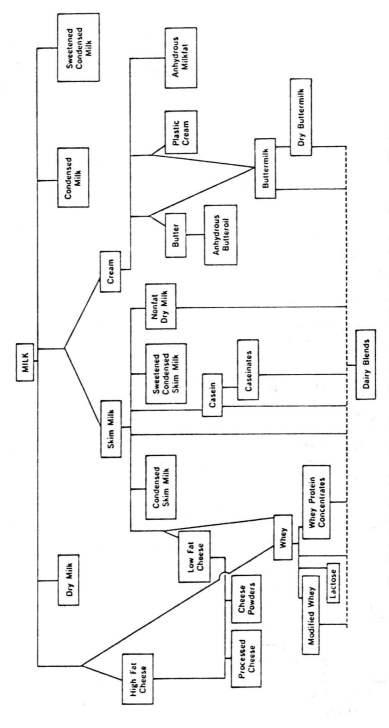

Fig. 10.1. Dairy-based ingredients from milk. (Courtesy of Dairy Research, Inc., Rosemont, Illinois.)

TABLE 10.4

Typical Composition of Dairy-Based Ingredients[a]

Ingredient	Moisture (%)	Protein (%)	Fat (%)	Lactose (%)	Ash (%)	Calcium (%)	Phosphorus (%)
Fluid whole milk	87.4	3.5	3.5	4.9	0.7	0.10	0.09
Fluid skim milk	90.5	3.6	0.1	5.1	0.7	0.12	0.09
Sweetened condensed milk[b]	27.1	8.1	8.7	11.4	1.8	0.26	0.21
Sweetened condensed skim milk[c]	28.4	10.0	0.3	16.3	2.3	0.25	0.20
Dried whole milk	2.0	26.4	27.5	38.2	5.9	0.91	0.71
Nonfat dry milk	3.0	35.9	0.8	52.3	8.0	1.31	1.02
Dried buttermilk	2.8	34.3	5.3	50.0	7.6	1.25	0.97
Cream, light	73.0	2.9	19.3	4.2	0.6	0.10	0.08
Cream, light whipping	62.9	2.5	30.5	3.6	0.5	0.08	0.06
Cream, heavy whipping	57.3	2.2	36.8	3.2	0.5	0.07	0.05
Plastic cream	18.2	0.7	80.0	1.0	0.1	0.03	0.02
Butter[d]	16.5	0.6	80.5	0.4	2.5	0.02	0.02
Spray-dried butter	2.0	—	80.0	—	—	—	—
Anhydrous butteroil	0.2	0.3	99.5	—	—	—	—
Cheddar cheese[e]	37.0	22.0	32.0	0–2.1	3.7	0.70	0.50
Spray-dried cheddar cheese	3.0	—	—	—	—	—	—
Cottage cheese curd[f]	79.0	16.9	0.4	2.7	0.8	0.09	0.05
Casein	7	88	1.0	—	4.0	—	—
Caseinate	3–5	90–94	0.7–1.0	—	6–7	—	—
Dried whey	4.5	12.9	1.1	73.5	8.0	0.65	0.59
Whey protein concentrates	2	20–60	2–9	18–60	3–18	—	—
Lactose	0.5	0.1	0.1	99.0	0.2	—	—

[a] Courtesy of Dairy Research, Inc., Rosemont, Illinois.
[b] Added sugar, 44.3%.
[c] Added sugar, 42.0%.
[d] Added salt, 0–2.3%.
[e] Added salt, 1.6%.
[f] Added salt, 1.0%.

flavor is a common defect in stored dry whole milk. The onset of these flavors is delayed by packaging the milk under nitrogen or carbon dioxide and the addition of antioxidants has proved to be partially acceptable. Although flavor problems restrict use of dry whole milk, large quantities are used in confections, chocolate products, and soup mixes. The other flavors of these products usually overshadow any flavor that may develop in dry whole milk.

4. Nonfat Dry Milk

Nonfat dry milk is the most popular dairy-based ingredient because of its excellent flavor, functional properties, nutritional value, and convenient, storage-stable form.

USDA standards and ADMI standards for grades of nonfat dry milk identify "Extra" and "Standard" nonfat dry milk produced by spray- or roller-drying. Nonfat dry milk is also assigned a Heat Treatment Classification. This classification depends on the undenatured serum proteins present in the product:

1. U.S. High-Heat. The finished product shall not exceed 1.5 mg undenatured whey protein nitrogen per gram of nonfat dry milk.
2. U.S. Low-Heat. The finished product shall not show less than 6 mg undenatured whey protein nitrogen per gram of nonfat dry milk.
3. U.S. Medium-Heat. The finished product shall show undenatured whey protein nitrogen between the levels of "high-heat" and "low-heat."

To improve solubility, nonfat dry milk may be instantized. This process involves slightly wetting the particles while they are suspended in an air stream and forcing the sticky-surfaced particles together to form porous aggregates. These aggregates are then redried and ground. The resulting products are easily dispersed and solubilized.

High-heat nonfat dry milk is essential to prevent LV depression in bakery products. The high water absorption properties of high-heat nonfat dry milk are also advantageous in confections and comminuted meats. Low-heat nonfat dry milk is preferred for beverage applications because of its better flavor and solubility. Medium-heat nonfat dry milk is used in products (such as ice cream) in which water absorption and flavor are important.

5. Dried Buttermilk

Dried buttermilk is similar in composition to nonfat dry milk, but contains a higher concentration of fat. Much of this fat is phospholipids, which have good emulsifying and whipping properties. The fat content may reduce storage stability; it is recommended that dry buttermilk be used as rapidly as possible.

6. Cream

Cream is prepared by the centrifugal separation of the lower density fat fraction of milk. The FDA requires that cream labeled "heavy whipping" have a minimum of 36% fat; "light whipping" 30–36% fat; and "light," "coffee," or "table" cream 18–30% fat. Specialized treatments such as heating and homogenization may be specified as they affect the viscosity of the cream.

Cream is subject to microbial deterioration and must be kept under refrigeration. A process has been developed for producing sterilized cream with extended shelf life.

7. Plastic Cream

Plastic cream is prepared with a special type of separator or repeated separation to achieve a product containing 80% milkfat. It is similar to butter in consistency and must be handled in the same manner as butter. In contrast to butter, the product is an o/w emulsion. Because of the reduced water content, plastic cream is cheaper to transport and store. It may be frozen to increase storage life but some destabilization of the emulsion may occur.

8. Butter

Butter is manufactured from cream by a churning process. In the course of churning, the mechanical agitation destabilizes the fat-in-water emulsion and results in production of granules of butterfat approximately 0.25 in. diameter. After the aqueous buttermilk is drained from the fat, the fat is further worked to subdivide and disperse the remaining water droplets and to produce the smooth consistency of butter.

USDA standards for grades of butter identify four grades: U.S. Grade AA, U.S. Grade A, U.S. Grade B, and U.S. Grade C. The grade of butter first is determined on the basis of classifying the flavor characteristics and then the characteristics of body, color, and salt content.

Butter can be transported at a relatively low cost and stored for several months at −23 to 29°C. Butter prepared from good quality fresh cream and properly stored will maintain excellent flavor. It does not have the excellent whipping properties of cream, but apparently the use of emulsifiers can improve whipping properties of butter.

9. Cheese

Natural ripened cheeses differ according to source of milk, processing procedure, and location of production. Most cheeses are produced by use of microbial cultures to yield the desired flavor and acidity. A rennin-type enzyme is used to coagulate the casein.

Spray-dried cheddar cheese powders are produced by comminuting and slurrying cheese in water to obtain 35–45% total solids, adding sodium citrate or disodium phosphate to stabilize the emulsion, and then homogenizing and spray-drying the slurry. Another method of drying cheese is to slurry comminuted cheese in skim milk or whey to facilitate drying. Flavoring, salt, and coloring may be added prior to drying. Spray- and foam-dried cheese powders are soluble and stable at room temperature for 9 months when packed under nitrogen. Although spray-dried cheese is quite expensive, it finds applications in bakery products, dry mixes, and snack foods.

Coagulation of the protein in fresh, unripened cheeses is largely dependent upon the acidity progress. When sufficient acidity has developed to coagulate the casein, the curd is cut and cooked. Whey is drained off and the soft curd is packaged for sale. Unripened cheeses are high in moisture and have bland flavors. They must be refrigerated and have a short shelf life. Their relatively concentrated casein is an advantage in some food applications.

10. Caseins

Edible caseins are produced in a manner similar to that of unripened cheese. Skim milk is heated to approximately 27°C and the pH value is lowered to 4.4–4.7 through the production of lactic acid by lactic-acid-producing bacteria or by the direct addition of lactic, sulfuric, or hydrochloric acid. The skim milk is then rapidly heated to 43–46°C to form a casein curd. The curd is broken to allow the whey to escape, and the whey is drained. The curd is repetitively washed to remove traces of remaining acids, salts, and whey proteins; pressed to remove excessive moisture; milled to reduce particle size and increase surface area; and finally dried. It is necessary to temper dried casein to allow hardening and promote even distribution of moisture throughout the batch. The casein may then be ground to the desired particle size. Very little edible dry casein is used in food products except in protein fortification of cereals and breads.

11. Caseinates

Caseinates are salts of casein. They are manufactured by preparing an aqueous colloidal suspension of acid-coagulated casein by the addition of alkali to achieve pH 6.7. The suspension is pasteurized and spray-dried. The type of caseinate salt produced is determined by the alkali used. Sodium and calcium caseinates are most common, being produced from casein and either sodium hydroxide or calcium hydroxide. Caseinates have good storage stability and impart many unique functional properties. Because caseinates are considered a food chemical derived from milk, they may be legally used in nondairy products such as coffee whiteners, and whipped toppings.

12. Whey

Whey is that fraction of skim milk remaining after the coagulation and separation of casein. Most processed whey in the United States is identified as sweet whey, a by-product of ripened cheese production. Lesser amounts of acid whey resulting from the production of cottage and cream cheese are processed. Whey is available on the market at a low cost. Food processors use whey as a source of crude lactose, milk solids, milk protein, or total solids.

13. Lactose

Lactose is generally produced from cheese whey or the permeate fraction produced by ultrafiltration of whey. The whey or permeate are concentrated to 50–60% solids and the lactose is crystallized by slowly cooling the supersaturated solution. Lactose crystals are centrifugally separated and sprayed with water to remove adhering liquor. This crude lactose contains up to 2% protein, ash, and lipids combined. Applications include animal feed, and as a substrate for fermentation products.

Edible and USP grades of lactose are produced by redissolving and refining the crude lactose. Refining techniques include heat precipitation of proteins, treatment with activated carbon to decolorize, and ion exchange or electrodialysis to reduce the ash content. After the refining process, lactose may be recrystallized and dried or simply spray-dried.

Lactose is available in different crystalline forms. The most common is α-lactose monohydrate or simply α-hydrate. Although these crystals are hard and not readily soluble, they are stable. The β-lactose crystals are sweeter and more readily soluble than α-hydrate. Since all forms of lactose in solution gradually achieve a β:α ratio of 62.25:37.75, there is usually no advantage in purchasing β-lactose. Anhydrous lactose "glass" (amorphous noncrystalline glass) may be produced by rapidly drying lactose solutions. This form of lactose is hydroscopic and may take up moisture from the air to form a crystal lattice.

14. Dried Dairy-Based Blends

Various blends of dairy and nondairy ingredients have been developed and marketed. Nonfat dry milk, whey, whey protein concentrates, caseinates, and buttermilk solids may be blended with nondairy proteins, carbohydrates, and fats. The blends are usually formulated for specific food applications. Some are produced by dry blending ingredients; others are produced by specially processing solutions of the ingredients and spray-drying. Most of these blends are designed to replace a more expensive ingredient in a specific food, but some blends have unique functional properties.

D. Applications

1. Processed Meats

Binding the fat and water in comminuted meats is important to the emulsion stability. Nonfat dry milk is the ingredient of choice, effectively providing the desired functionality without impairing the meaty flavor. Numerous studies demonstrate attributes of dairy-based ingredients as binders in meats. In cured meats, combinations of lactose with glucose reduce sweetness and prevent softening of meat tissue.

2. Synthetic Meat-Type Products

Casein can be spun into fibers and bonded with a starch matrix. The resulting product is similar in appearance to meat. The starch matrix melts when subjected to cooking temperatures but the synthetic meat-type product may be used on salads or as a condiment.

Whey proteins have the unique ability to complex with other proteins in heated solutions and may be used as a binder in production of textured vegetable protein. Up to 50% of the egg albumin commonly utilized to bind spun soy fibers now is replaced by whey protein concentrate. Whey protein concentrates are the major constituent in a textured synthetic product containing fat and chicken flavor. This product hydrates when boiled and imitates the flavor and texture of chicken meat in soups.

3. Sauces, Gravies, and Soups

Butter and cream impart a unique desirable flavor to sauces and soups. Skim milk and nonfat dry milk add a bland but pleasant flavor and produce a milky white color. Many food processors use whey products in soups, sauces, and gravies. Whey yields a slight milky flavor; it also enhances and balances desired flavor. When added to foods, the effect of lactose is similar to that of monosodium glutamate. Whey proteins are less likely to adhere to the sides of cooking utensils, minimizing the requirements for agitation. Whey stabilizes sauces and gravies subjected to freeze–thaw stress and prolonged high temperatures on steam tables.

Whey is a major ingredient in dry mixes; it is an inexpensive nonhygroscopic functional filler easily dispersed in hot or cold solutions. The lactose present in whey has the ability to absorb and retain flavors and aromas, thereby improving flavor stability during processing and storage. Alternatively, pure lactose can be added to the mix prior to drying to capture and retain flavors, or it may blended into the mix after drying to carry added flavors, enhance flavors, control sweetness, or serve as a disperser to prevent clumping and promote free-flowing properties. Other ingredients available for dry mixes include instant nonfat dry milk, spray-dried cream, and cheese powders.

4. Vegetables and Fruits

Usage of lactose in jams, jellies, and preserves has been limited because of the inherent problems of lactose crystallization.

5. Beverages

Protein fortification with milk proteins is easily accomplished because their high PER enables the manufacturer to use the US-RDA base of 45 gm. Many dairy-based ingredients are important sources of calcium. The flavor of beverages fortified with milk-derived ingredients is unique, but not objectionable.

Citrus-flavored dairy-based beverages have been developed to create nutritional carbonated soft drinks. In numerous hot cocoa mixes, whey solids are used to replace nonfat milk solids. Lactose enhances natural chocolate flavor and makes possible a substantial reduction in expensive cocoa concentration. The addition of lactose to chocolate drinks imitates the rich mouth-feel normally produced by butterfat. This could be an advantage in dry hot cocoa mixes in which the use of cream is limited by its flavor instability.

Dairy ingredients have been utilized in nutritional beverages for populations of developing countries. Whey proteins are particularly advantageous because of the reasonable cost, acceptable flavor, and high nutritional value of the proteins.

6. Cereals and Pasta Products

Milk-derived proteins increase the protein content and improve the protein quality of cereal grain products. For example, in macaroni, an increase in protein content from 13 to 20% raised PER from 0.8 to 2.5. The detectable differences in flavor and texture were not objectionable and no process modifications were required. Resiliency and improved sauce-clinging tendency in canned spaghetti were gained by the addition of 3% whey protein concentrate to semolina-flour spaghetti noodles.

Dried whey is an nutritious additive for breakfast cereals, and the lactose contributes to browning reaction. Casein, caseinates, and whey protein concentrates are economical sources of protein for breakfast cereals. The addition of modified whey in which lactose has been enzymatically hydrolyzed to glucose and galactose permits reduction in sugar content without sacrificing sweetness. Lactose can be used as a cereal coating to form an opaque, anhydrous, non-hygroscopic vapor barrier to atmospheric moisture. Such lactose could provide sweetness, enhance nutrition, and absorb added food colorings. Sweetened condensed milk has found application as a nutritious, flavorful binder in granola-type products.

Bacterial and fungal lactases were added to bread doughs containing several lactose-containing substances in studies by Pomeranz (1964). Under conditions of bread production, hydrolysis of lactose did not supply adequate amounts of fermentable sugars. Other studies demonstrate that it is possible to use commercial products of dry milk solids in which lactose has been enzymatically almost completely hydrolyzed to glucose and galactose in the production of acceptable bread (Shogren *et al.*, 1979). If 4 parts of milk solids per 100 parts wheat flour are used in the production of such bread, the contribution of glucose to panary fermentation must be supplemented by sucrose and/or malt to provide adequate levels of fermentable sugars. For use of lactase-hydrolyzed milk solids, the milk must be heat-treated under conditions comparable to those for regular milk powder used in bread production.

7. Snack Foods

Numerous snack foods are formulated with dairy ingredients. They include extruded nutritional snacks produced from a corn–soy–sodium caseinate base containing 17% protein, and puffed snacks made from skim milk curd and whey protein concentrate containing 25% protein.

Lactose is beneficial in color development of fried snack foods. For example, potato slices rinsed in 0.5% lactose solution before frying develop a golden-brown color.

Cream cheese, sour cream, and processed cheese are the major base for dips and spreads. Increasing the concentration of nonfat milk solids has the effects of improving appearance and spreading properties. Butterfat contributes to the overall flavor and smooth creamy texture of dips and spreads.

8. Imitation Food Products

Most of the successful imitation food products, coffee whiteners, and vegetable fat toppings depend upon dairy ingredients to provide the desired properties and flavor. For example, caseinates have excellent whipping properties for toppings. They also form a protective protein film around the fat particles, which prevents clumping of the fat and loss of stability. Coffee whiteners contain caseinates to stabilize the emulsions and rely upon the whitening effect of caseinates to produce the desired appearance. Imitation beverages and hot cocoa bases are prepared with whey, vegetable fats, corn syrups, and caseinates. Caseinates and whey, particularly demineralized whey, produce a milky flavor in these otherwise bland products, and contribute to the emulsion stability. Some food processors take advantage of the synergistic effect of caseinates with synthetic emulsifiers and the stability of the encapsulating layers to heat or freezing. This attribute facilitates the production of a wide range of spray-dried or frozen products that retain the original emulsion characteristics.

Many frozen desserts are prepared with dairy ingredients. Typical products contain milk solids (nonfat) from whey and nonfat dry milk or skim milk. Low-calorie frozen desserts contain skim milk solids, whey solids, and caseinates. The caseins and whey proteins present in frozen desserts are critical to the whipping properties of the original mix and the smooth texture and flavor of the final frozen product.

The unique properties of milk proteins are exemplified in nondairy cheese-type products. The protein sources utilized to give the desired texture are sodium and calcium salts of casein. These proteins are essential for the desired melting properties and texture associated with true cheese.

Simulated human milks for infants have become popular. Simulated human milk can be prepared by blending skim milk with electrodialyzed whey. Technology is available to produce infant formulas with ratios of casein to whey protein, lactose, and minerals similar to the ratios in human milk. Some manufacturers change the fat so that the fatty acid composition is the same as in human milk.

9. Confections

Dairy products are traditional ingredients in confections. Their original use may have resulted from availability, but the flavors, textures, and colors that are the expected characteristics of many confections result directly from dairy ingredients.

a. Milk Proteins. Caseins and whey proteins, the two major classes of proteins in milk, have different functions in candy. Both proteins have emulsifying properties that aid in blending formula ingredients, and both absorb on the surface of fat particles to improve emulsion stability, but caseins bind more than twice their weights of water and produce a drying effect in confections. Undenatured whey protein binds little water, but water absorption can be increased by heat-denaturing whey proteins. Whey proteins have better foaming properties than caseins.

Many of the beneficial aspects of milk proteins occur during cooking of confections. Milk proteins undergo denaturation during heating, gradually unfolding into long fibrous structures that interact and bind together. The network of interconnected thread-like molecules imparts the final texture to the confection. Caseins produce a firm chewy body that is neither sticky nor tough. Whey proteins form a softer coagulum with less resilience than that produced by casein.

Milk proteins contribute to the final color and flavor of confections. Caseinates are largely responsible for the white color of milk and this whitening effect is transferred to most food products that include significant concentrations of casein.

b. Lactose. Lactose or milk sugar can have an important effect on the flavor, color, and texture of confections. The flavors depend on the temperature and duration of cooking, the acidity of the mix, and the presence of free amino groups of proteins. Caramelization of lactose occurs at temperatures above the melting point, producing yellow and brown pigments and an array of flavorful and fragrant aldehydes, ketones, alcohols, acids, and reductones. More important is the interaction of lactose with free amino groups and the resulting Maillard browning reaction. This reaction is greatly accelerated by high temperatures and the presence of moisture.

Sweetness is a major sensory attribute in confections. Lactose is the least sweet of the principal sugars used in confections and can be added to establish flavor balance. Another significant property of lactose is its ability to hold flavors, odors, and pigments. The odors, pigments, and flavors are absorbed on the surface as lactose crystals form. The flavors, colors, and odors are therefore retained until the lactose is dissolved during consumption.

Lactose can greatly affect the texture of confections—lactose is much less soluble than sucrose or dextrose. Depending on the temperature, one of two types of lactose crystals form. Above 93°C β-anhydride crystals form, but as the candy mass cools and during storage α-anhydride crystals redevelop. The α-anhydride crystals are harder and less soluble than the β-form, and if they reach appreciable size they result in a gritty or sandy texture. Complications often result because a supersaturated solution of lactose may remain stable for several weeks before crystallization commences. Supersaturated solutions of lactose may form a solid mass termed ''lactose glass.'' This monocrystalline form is stable at room temperature and imparts softness, elasticity and freshness of body to candies. Lactose in the amorphous form is hygroscopic. If sufficient moisture is absorbed to dilute the lactose concentration, recrystallization to less desirable forms may result.

Lactose contributes to the enhancement of flavor of many confections. It decreases vapor pressure of the water in confections, which retards moisture loss and prolongs freshness. Addition of lactose to a mixed solution of sucrose and corn syrup can increase the viscosity. The resulting effect on the body and texture of the confection is similar to that observed when the ratio of corn syrup to sugar is increased. Finally, controlled crystallization of lactose yields desirable body and texture and prevents deleterious ''graining'' during storage. Three methods of controlling crystallization are maintaining the lactose concentration below about 3%; adding sufficient inhibitory ingredients such as corn syrup, protein, and fat to prevent crystallization; and forming fine crystals by forced crystallization at the time of pouring or at the time fondant is added.

c. Milk Lipids. The major attribute of milkfat in confections is the flavor it imparts. The fat in fresh fluid milk is bland; however, during processing of confectionery products desirable flavors can be developed by several mechanisms. Triglycerides are the predominant lipids in milkfat; TG can be broken down into flavorful free fatty acids and glycerol by enzymes already present in milk or by the heat of processing. Storage at fluctuating temperatures, or blending of hot and cold milk and physical agitation to disrupt the protective fat globule membrane accelerate the enzymatic hydrolysis. The result is an increase in free fatty acids, including butyric, caproic, and capric acids, which impart the rancid flavor notes often considered desirable in strong-flavored candies, particularly milk chocolates.

Milk lipids also influence the mixing of ingredients, the viscosity of the mix, and the texture and appearance of the final confection. Fats act as a lubricant and the phospholipids present in milk lipids also act as emulsifiers. They facilitate the mixing of ingredients for confectionery products and help stabilize the fat globules in an even, well-dispersed phase throughout the confection. Milkfat may coat other emulsified particles, such as proteins, and reduce stickiness. In the finished product, the semifluid fat interfaces provide shear planes, which impart plasticity to the confection. Candy coatings with unacceptable "waxy" mouth-feel can be modified to an acceptable texture by addition of milkfat.

d. Salts. High concentrations of milk salts or imbalance of milk salts can adversely affect flavor in confections. At lower levels, milk salts may aid in producing a well-rounded milk flavor and may enhance other flavors.

II. Wheat Flour Components in Bread Making

This section reviews investigations designed to determine the role of wheat flour components in bread making (Pomeranz, 1973a, 1980a,b,c). The investigations involved fractionation of the components without impairing their functional properties, appropriate characterization of the separated fractions (and, whenever possible, synthesis of purified components), and performance tests. Those studies (combined with examination of dough and bread by light microscopy and scanning and transmission electron microscopy) have established the roles of gluten proteins, water-solubles, and lipids in bread making. The studies offer information about the interaction among various flour components and the significance of the complexes formed in bread making. In addition, they stimulate research on making new proteins functional and compatible with well-established wheat flour proteins, thereby enhancing consumer acceptance of nutritionally improved, protein-enriched bread.

A. From Wheat to Flour

Wheat is cultivated in most countries on all continents, and while about 30,000 wheat varieties belonging to 14 species are grown throughout the world, only about 1000 varieties are of commercial significance. Most of the varieties grown for bread production belong to the species *Triticum aestivum* (common wheat). Flour from common wheat is best suited for breads acceptable to consumers and bakers in most Western countries.

Before wheat can be used in the production of yeast-leavened bread, it must go through several mechanical and chemical changes. To understand how wheat is made into flour—the main raw material in bread making—it is necessary to learn about wheat structure. The grain, or kernel, of wheat is a one-seeded fruit with a coat, the bran, adhering to the seed. Fig. 10.2 shows longitudinal and cross-sections of a kernel of wheat and Table 10.5 outlines the nutritional charac-

a Kernel of Wheat

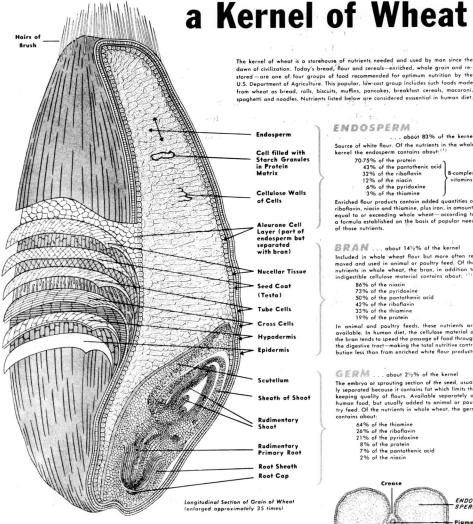

The kernel of wheat is a storehouse of nutrients needed and used by man since the dawn of civilization. Today's bread, flour and cereals—enriched, whole grain and restored—are one of four groups of food recommended for optimum nutrition by the U.S. Department of Agriculture. This popular, low-cost group includes such foods made from wheat as bread, rolls, biscuits, muffins, pancakes, breakfast cereals, macaroni, spaghetti and noodles. Nutrients listed below are considered essential in human diet.

Hairs of Brush

Endosperm

Cell filled with Starch Granules in Protein Matrix

Cellulose Walls of Cells

Aleurone Cell Layer (part of endosperm but separated with bran)

Nucellar Tissue

Seed Coat (Testa)

Tube Cells

Cross Cells

Hypodermis

Epidermis

Scutellum

Sheath of Shoot

Rudimentary Shoot

Rudimentary Primary Root

Root Sheath

Root Cap

Longitudinal Section of Grain of Wheat (enlarged approximately 35 times)

ENDOSPERM

. . . about 83% of the kernel

Source of white flour. Of the nutrients in the whole kernel the endosperm contains about: [1]

70-75% of the protein
43% of the pantothenic acid
32% of the riboflavin   } B-complex
12% of the niacin   } vitamins
6% of the pyridoxine
3% of the thiamine

Enriched flour products contain added quantities of riboflavin, niacin and thiamine, plus iron, in amounts equal to or exceeding whole wheat—according to a formula established on the basis of popular need of those nutrients.

BRAN . . . *about 14½% of the kernel*

Included in whole wheat flour but more often removed and used in animal or poultry feed. Of the nutrients in whole wheat, the bran, in addition to indigestible cellulose material contains about: [1]

86% of the niacin
73% of the pyridoxine
50% of the pantothenic acid
42% of the riboflavin
33% of the thiamine
19% of the protein

In animal and poultry feeds, these nutrients are available. In human diet, the cellulose material of the bran tends to speed the passage of food through the digestive tract—making the total nutritive contribution less than from enriched white flour products.

GERM . . . *about 2½% of the kernel*

The embryo or sprouting section of the seed, usually separated because it contains fat which limits the keeping quality of flours. Available separately as human food, but usually added to animal or poultry feed. Of the nutrients in whole wheat, the germ contains about:

64% of the thiamine
26% of the riboflavin
21% of the pyridoxine
8% of the protein
7% of the pantothenic acid
2% of the niacin

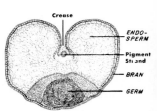

Crease

ENDO-SPERM

Pigment Strand

BRAN

GERM

Cross Section View

As a product group in the U.S. Department of Agriculture's recommended Daily Food Guide for good nutrition—bread, flour and cereals make a generous contribution to human requirements for the B-vitamins—thiamine, niacin and riboflavin—and the mineral, iron. They also help fill daily needs for protein and calcium. The other three food groups are: milk and milk products; meats, poultry, fish,

eggs and dry lentils; fruits and vegetables. Nutritionists advise eating a variety of foods from each of the four groups every day to obtain all the nutrients necessary for adequate diet. With fresh, frozen, canned and prepared foods readily available, it is easy for everyone to satisfy nutritional requirements by following the Daily Food Guide.

Fig. 10.2. Longitudinal and cross-sectional views of a kernel of wheat (×20). (Courtesy of the Wheat Flour Institute, Chicago, Illinois.)

TABLE 10.5

Nutrients Available in Endosperm, Bran,
and Germ of Wheat[a]

Nutrients	Endosperm (%)	Bran (%)	Germ (%)
Protein	70–75	19	8
Pantothenic acid	43	50	7
Riboflavin	32	42	26
Niacin	12	86	2
Pyridoxine	6	73	21
Thiamine	3	33	64

[a]Percentage of amounts in whole grain.

teristics of some of its tissues. Scanning electron microphotograph (SEM) views in Fig. 10.3 give a three-dimensional image of some of the tissues shown in Fig. 10.2.

The endosperm, which forms about 83% of the kernel, is the source of white flour and contains 70–75% of the kernel's protein. The bran, forming about 14% of the kernel, is included in whole wheat flour, but more often is removed and used in animal or poultry feed. Because the cellulosic material of the bran cannot be digested and tends to speed the passage of food through the human digestive tract, the total nutritive contribution of whole wheat flour is less than that of enriched white flour products. The germ, forming about 3% of the kernel, is the embryo or sprouting tissue of the seed. It is usually separated out because it contains oil, which limits the keeping quality of flours. Although the germ is available as human food, it is usually added to animal or poultry feed.

The steps involved in wheat-to-flour production are wheat selection, blending, conditioning, milling, and maturing. The milling process consists of a gradual pulverization, first between corrugated rolls and then between smooth, or reduction, rolls (Fig. 10.4). Tables 10.6 and 10.7 compare the composition of wheat tissues and the composition of flours varying in milling extraction, a term used to denote the percentage of the wheat contained in the flour after milling. The flour itself, as noted, comes mainly from the starchy endosperm: the rest of the kernel (mainly bran, part of the aleurone, and germ) is used for animal feed.

In the production of white flour in a modern mill, the object is to separate the endosperm—the central part of the grain—from the bran and germ. A partial mechanical separation of these closely adhering structures is possible because of differences in their physical properties. Because of its high fiber content bran is tough, whereas the starchy endosperm is friable. The germ, with its high oil content, readily flakes when passed between smooth rolls. Because the structures

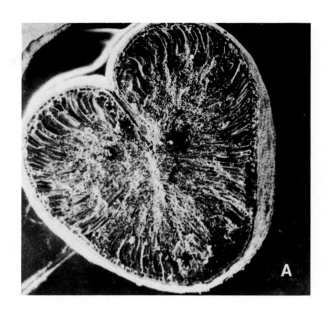

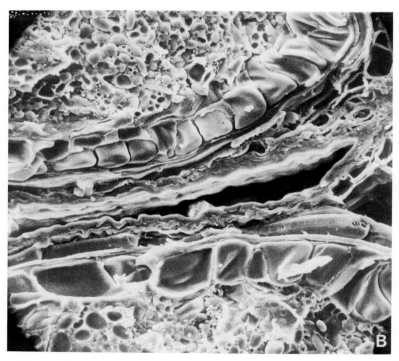

Fig. 10.3. Scanning electron microphotographs of A, cross section of whole wheat kernel (×12); B, section through the crease (×175); C, central starchy endosperm (×730); D, aleurone layer (×850); E, high magnification of aleurone grains (×8500).

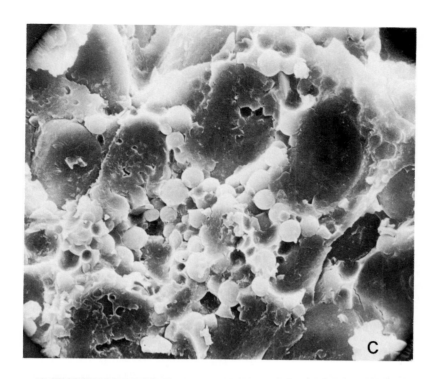

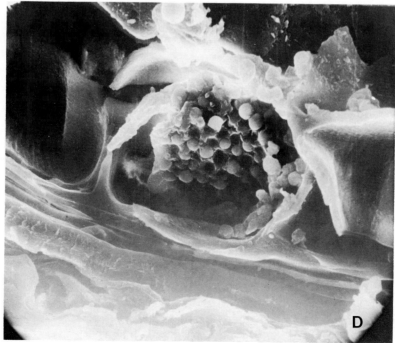

(*continued*)

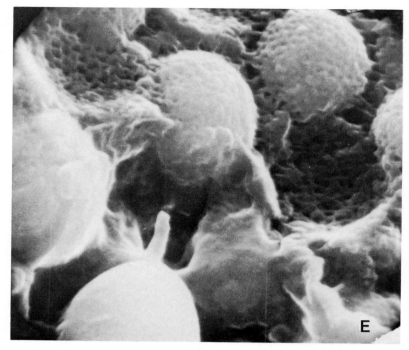

Fig. 10.3E (*Continued*)

of wheat also differ in density, it is possible to utilize air currents to remove small amounts of light bran particles from the heavier endosperm particles.

The differences in friability of the bran and endosperm are accentuated by tempering, or conditioning. In this process water is added to the wheat to toughen the bran and mellow or soften the endosperm. This is performed several hours before the wheat reaches the mill rolls. The differing responses of cellulosic bran and starchy endosperm to added water result from the formation of a moisture gradient in the kernel (most of the water is retained in the outside layers) and from the differences in physical modification of wheat components during interaction with water.

Maturing involves oxidative changes that take place during prolonged storage or after addition of trace amounts of oxidants, which generally improve the rheological (handling) properties of dough and enhance bread quality (LV, texture, and freshness retention). Malt supplements are added to flour to provide fermentable sugars and to modify the proteins and the starch; vitamins and minerals are added to enrich nutritional value. The added vitamins and minerals restore the nutrient contents of white flour to levels present in whole wheat flour.

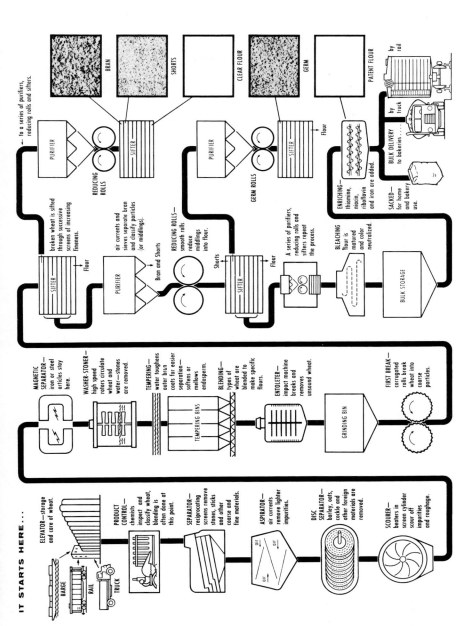

Fig. 10.4. How flour is milled. This chart is greatly simplified. The sequence, number, and complexity of different operations vary in different mills. (Courtesy of the Wheat Flour Institute, Chicago, Illinois.)

TABLE 10.6

Composition of Wheat Tissues

	Pericarp (%)	Aleurone layer (%)	Starchy endosperm (%)	Germ (%)
Kernel	9	8	80.0	3
Ash	3	16	0.5	5
Protein	5	18	10.0	26
Lipid	1	9	1.0	10
Crude fiber	21	7	0.5	3

B. From Flour to Bread

Bread making, by either the single batch process or the continuous process (in which ingredients are added to a mixer and the dough is processed into bread in an ongoing manner), includes several basic operations (Fig. 10.5). Flour, bakers' yeast, salt, and other optional ingredients such as sugar, shortening, and yeast food are mixed with water to develop a dough with the desired viscoelastic properties. The dough is fermented in bulk under controlled conditions, at about 30°C, and is then divided into uniform pieces of suitable size, which are rounded, shaped, and usually inserted into baking pans for a final fermentation called proofing. Proofing provides a fresh supply of carbon dioxide for leavening, modifies dough proteins, and imparts desirable rheological properties. The proofed dough is baked at about 230°C, and during the early stages of baking "oven spring"—rapid expansion—takes place. This is followed by gelatinization of the starch and coagulation of the proteins, accompanied by formation of a rigid bread structure. Chemical changes during baking produce the flavor and crust color—the browning is caused by the Maillard reaction, which involves the

TABLE 10.7

Composition of Wheat Flours

	Milling extraction (%)		
Flour	75.0	85.0	100.0
Percent of flour weight			
Ash	0.5	1.0	1.5
Protein	11.0	12.0	12.0
Lipid	1.0	1.5	2.0
Crude fiber	0.5	0.5	2.0

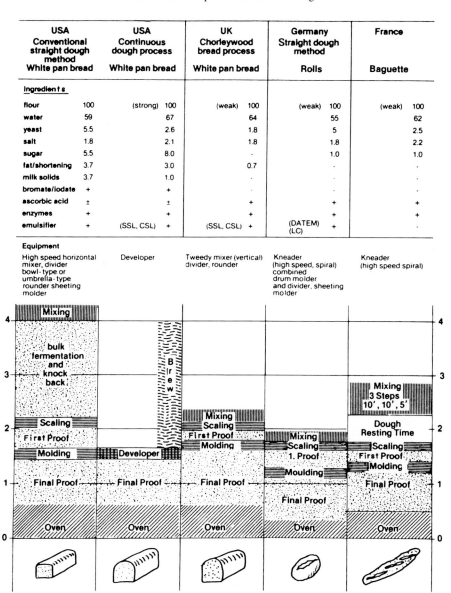

	USA Conventional straight dough method White pan bread		USA Continuous dough process White pan bread		UK Chorleywood bread process White pan bread		Germany Straight dough method Rolls		France Baguette	
Ingredients										
flour	100		(strong)	100	(weak)	100	(weak)	100	(weak)	100
water	59			67		64		55		62
yeast	5.5			2.6		1.8		5		2.5
salt	1.8			2.1		1.8		1.8		2.2
sugar	5.5			8.0		.		1.0		1.0
fat/shortening	3.7			3.0		0.7		.		.
milk solids	3.7			1.0		.		.		.
bromate/iodate	+			+		.		.		.
ascorbic acid	±			±		+		+		+
enzymes	+			+		+		+		+
emulsifier	+		(SSL, CSL)	+	(SSL, CSL)	+	(DATEM) (LC)	+		.

Equipment

High speed horizontal mixer, divider bowl-type or umbrella-type rounder sheeting molder	Developer	Tweedy mixer (vertical) divider, rounder	Kneader (high speed, spiral) combined drum molder and divider, sheeting molder	Kneader (high speed spiral)

Fig. 10.5. (A) Production of wheat-baked products in various countries. Strong, good bread making potential; weak, poor bread making potential. SSL, sodium stearoyl lactylate; CSL, calcium stearoyl lactylate; DATEM, diacetyl tartaric acid esters, monoglycerides and diglycerides; LC, lecithin. Numerals at top by weight, flour basis. 10′, 10′, 5′, 10, 10, and 5 min with intermediate rest. (From Schuster and Adams, 1984.) (B) Continuous bread-making process. (From Seiling, 1969.) *(continued)*

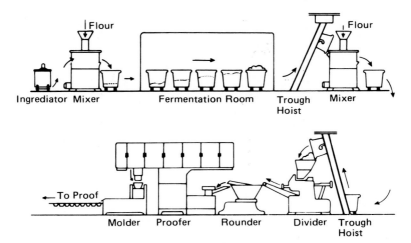

Fig. 10.5B *(Continued)*

interaction between a reducing sugar and an amine (the name comes from the French chemist L. C. Maillard, who did pioneering research on the basic factors of browning in foods). For changes in baking as affected by temperature see Fig. 10.6.

It has been known for many years that flours milled from certain wheat varieties grown in some locations consistently produce bold loaves with good internal crumb grain and color and with excellent shelf life. Other wheat flours produce compact bread that has a crumbly interior and stales rapidly. Researchers surmised that those differences result from qualitative and quantitative compositional variations. The problems of relating chemical composition and structure of wheat flour components to functional properties in bread making is complicated by several factors: large number of components; high molecular weight, limited solubility, and difficulty of separating or isolating pure components without altering them; and the interaction of the components during dough mixing, fermentation, and baking. Some classical fractionation methods make it possible to separate flour components; some modern biochemical techniques enable the researcher to characterize those components and to follow interactions among macromolecules. A combination of separation and characterization techniques can be used to elucidate the roles of wheat flour components in bread making.

C. Baking Tests

To establish a causative relationship between chemical composition and structure of wheat flour components and their functional properties in bread making, it is necessary to (a) extract, fractionate, and characterize flour components; (b) reconstitute the isolated moieties; and (c) ascertain that neither the isolation nor the reconstitution procedure impairs the functional properties of the components.

Historically and logically the latter requirement was met first. An optimized baking test, rather than test bakings under fixed and arbitrarily selected conditions, was developed to determine the characteristics inherent in wheat and flour components. In the optimized baking test, four factors—mixing time, water

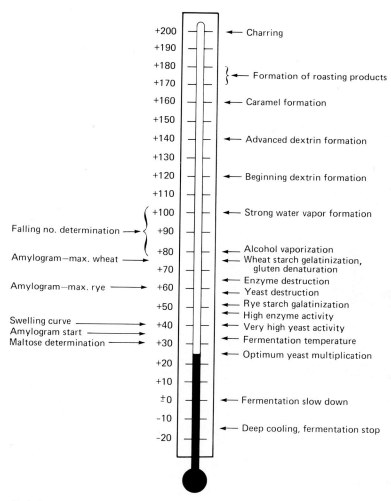

Fig. 10.6. Indirect and direct baking changes as affected by temperature (°C). (Courtesy of J. Brümmer.)

absorption, yeast activity, and fermentation time—are optimized and balanced. Oxidation requirements are met by use of the optimum amount of either potassium bromate ($KBrO_3$) or a mixture of $KBrO_3$ and ascorbic acid (Shogren and Finney, 1974). Because flours from commercial wheats show a relatively narrow range of oxidation requirements, 50 ppm ascorbic acid alone effectively replaces $KBrO_3$ as a dough oxidant and developer. The added ingredients—shortening, sugar, and malt—are supplied in excess because an insufficient quantity of any one would limit the performance of the wheat flour ingredients. Fig. 10.7 illustrates the effects and importance of dough ingredients on LV and crumb; Fig. 10.8 shows some differences between good and poor quality flours. When 100 gm of flour was mixed with water, as needed, the LV of 230 cm^3 equaled the dough volume after mixing. LV was increased to 483 cm^3 by the addition of 6% sugar and yeast; and to 732 cm^3 by the further addition of 1.5% salt, but the crumb grains were not satisfactory in any of the loaves. When 4% soy flour,

Fig. 10.7. Breads baked from (A) flour and water plus (B) yeast; (C) yeast and sugar; (D) yeast, sugar, and salt; (E) yeast, sugar, salt, malt, ascorbic acid, and shortening; (F) yeast, sugar, salt, malt, ascorbic acid, shortening, and soy flour. (From Finney, 1978.)

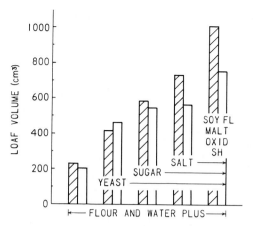

Fig. 10.8. Loaf volumes of breads baked from good and poor flours. FL, flour; OXID, ▨, good quality flour; ☐, poor quality flour. (From Finney, 1978.)

0.25% malt, 50 ppm ascorbic acid, and 3% shortening were added to the five basic ingredients (wheat flour, water, sugar, yeast, and salt), LV increased to 1006 cm³ and crumb grain was excellent. The optimized baking test can be performed with the precision of most biological assays, and it establishes several parameters (the most important being loaf volume) as standards of comparison.

Loaf volume is highly correlated, over a relatively wide range, with dough handling properties, consumer acceptance (as defined by crumb texture and freshness retention), and technological versatility. For a given type of commercial bread, taste and overall freshness retention depend on the types and amounts of wheat flour components as well as on the method used to produce the bread, optional ingredients, skill of the baker, and storage conditions.

1. Protein Content

In initial investigations, bread was baked from flours milled from wheat varieties grown under widely different climatic and soil conditions. Protein content was the major factor that accounted for variation in LV within a single variety. Indeed, the relation between LV (provided the bread was baked by the optimized test) and protein content is linear (Fig. 10.9). Because the protein content–loaf volume relation is linear *within a single variety,* the bread-making quality of new wheat varieties can be easily determined.

In practice, a plant breeder can determine the slope of the regression line from two or three samples of a new variety at different protein levels. As shown in Fig. 10.9 the regression of LV with *protein quantity* (content) differs widely among varieties. This variation results in differences in the slopes of regression lines for individual varieties: the greater the slope, the greater the increase in LV

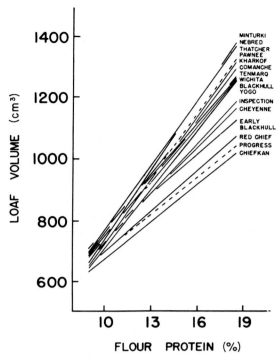

Fig. 10.9. Loaf volume–protein content regression lines for hard red winter wheats (—) and hard red spring wheats (- - - -). Each line represents many samples harvested throughout the Great Plains during several crop years. (From Finney and Barmore, 1948.)

per 1% increase of protein. Those differences in slope reflect differences in bread-making quality of the protein in the different varieties of wheat. The regression lines of LV with protein content for different wheat varieties are shown in Fig. 10.10 The regression curves form a family of lines for predicting LV associated with any desired protein content.

2. Fractionation

After the linear relation between protein content and loaf volume was established in the optimized baking test, wheat flours representing a wide range in quality (as indicated by LV response per 1% of flour protein) were fractionated into starch, gluten, and water-soluble components. Fractions from one variety were recombined in their original and in different proportions, and various fractions of different varieties were interchanged (Finney, 1943). Invariably, flours that were reconstituted in their original proportions gave bread that was identical to bread made with the original, unfractionated flour; this established that neither the fractionation nor reconstitution techniques impaired bread-making qualities.

When wheat gluten, starch, and water-soluble fractions from different varieties were interchanged and the flours so formed were baked, gluten proteins accounted for differences in bread quality of the varieties studied, as defined by LV increase per 1% protein. Although these results indicated that gluten proteins govern differences in bread-making potential of wheat varieties, bread of acceptable volume, texture, and freshness retention cannot be produced unless other flour components—namely, starch, water-solubles, and lipids—are also present in approximately the same quantities found in normal, unfractionated wheat flours.

3. Gluten Proteins

Once the key role of gluten proteins in governing varietal differences in bread-making potential was established, the next logical step was to identify the gluten component(s) responsible for those differences. When water is added to wheat flour during dough mixing, the water-insoluble proteins hydrate and form gluten—a complex, coherent mass in which starch, added yeast, and other dough components are embedded. Thus, the gluten is the framework of wheat flour dough and is responsible for gas retention, which makes possible the production of light, leavened products. Endosperm proteins of wheat possess the unique property of forming gluten, and it is this gluten formation, rather than any distinctive nutritive property, that gives wheat its prominence in the diet.

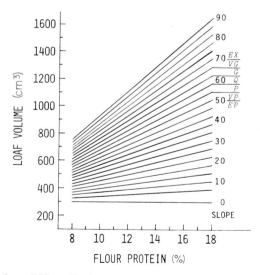

Fig. 10.10. Loaf volume (100 gm flour) versus protein content regression lines for correcting loaf volumes of wheat varieties to a constant protein basis. Slope is rate of change in loaf volume per 1% protein. EX, excellent; VG, very good; G, good; Q, questionable; P, poor; VP, very poor; EP, extremely poor. (From Finney, 1979.)

The first scientific report on flour components was made by Italian scientist Becari. In a lecture before the Academy of Bologna in 1745, he described experiments in the separation of gluten and starch. Our present knowledge of wheat proteins is based on the foundations of seed protein chemistry laid down by the German scientist H. Ritthausen at the end of the last century. Those foundations were expanded by T. B. Osborne in the United States at the beginning of this century. Osborne's work led to a useful classification of wheat proteins based on separation (alas incomplete, as shown by recent chromatographic and electrophoretic studies) according to differences in solubility. Gluten is formed of two proteins: gliadin and glutenin (Fig. 10.11). In its hydrated form, gluten is elastic and cohesive; the glutenin fraction, forming about one-half of the total, is tougher and less easily stretched, whereas the gliadin fraction has less cohesiveness and elasticity (Fig. 10.12).

Fractionation of gluten components on the basis of their differing solubility damages the bread making properties of those components. Separation by ultracentrifugation of the two main fractions of gluten into a supernatant rich in gliadins and into a centrifugate containing glutenin has been shown (by reconstitution studies) not to impair the bread-making properties of the gluten components. The gliadin fraction controls LV and varies in flours that differ in bread-making potential (Hoseney *et al.*, 1969b). The factor responsible for mixing time and dough development is in the glutenin fraction. During dough mixing, the protein mass is converted from granular protein bodies into a homogeneous network in which starch granules are embedded. Proper dough development is essential to good performance in bread making.

4. Amino Acids

Automated techniques to determine the amino acid composition of proteins facilitate the assay of amino acids in gluten proteins. Gluten proteins are characterized by high concentrations of glutamic acid (about 32%) and proline (about 10%) (Table 10.8A from Krull and Wall, 1969; also see Fig. 10.11). The acidic amino acids, glutamic and aspartic, occur mainly as amides. Thus far, studies on the amino acid composition of cereals have been disappointing because they fail to explain the differences in bread-making characteristics of wheat varieties. However, many studies have shown that both ionic and nonionic groups govern the unique viscoelastic properties of wheat gluten. The four types of groups/ bonds that are known to contribute to viscoelastic properties of dough are (a) amide groups, (b) sulfhydryl groups and disulfide bonds, (c) hydrogen bonds, and (d) hydrophobic interactions (see also Chapter 5).

When the acidic moieties of amides in gluten, glutenin, or gliadin are methylated, the properties of solubility, intrinsic viscosity, and cohesion are changed significantly. Dough handling properties can be modified, either beneficially or

A

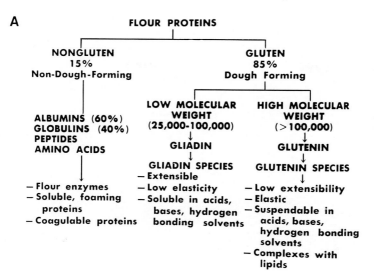

FLOUR PROTEINS

NONGLUTEN
15%
Non-Dough-Forming

GLUTEN
85%
Dough Forming

ALBUMINS (60%)
GLOBULINS (40%)
PEPTIDES
AMINO ACIDS

LOW MOLECULAR
WEIGHT
(25,000-100,000)

GLIADIN

GLIADIN SPECIES

HIGH MOLECULAR
WEIGHT
(>100,000)

GLUTENIN

GLUTENIN SPECIES

— Flour enzymes
— Soluble, foaming
   proteins
— Coagulable proteins

— Extensible
— Low elasticity
— Soluble in acids,
   bases, hydrogen
   bonding solvents

— Low extensibility
— Elastic
— Suspendable in
   acids, bases,
   hydrogen bonding
   solvents
— Complexes with
   lipids

B

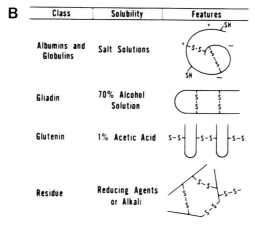

Class	Solubility	Features
Albumins and Globulins	Salt Solutions	
Gliadin	70% Alcohol Solution	
Glutenin	1% Acetic Acid	
Residue	Reducing Agents or Alkali	

Fig. 10.11. A, The main protein fractions of wheat flour. (From Holme, 1966.) B, Types of proteins in wheat flour as separated by solubility. (From Wall and Huebner, 1981.)

adversely, by the addition of minute amounts of reducing agents or sulfhydryl-blocking reagents. As stated previously, the performance of a flour in bread making can be improved significantly by addition of an appropriate amount of oxidizing agent, such as 20–50 ppm $KBrO_3$.

The oxidation requirement (the amount of oxidant needed to produce the best loaf of bread in terms of volume, crumb texture, and freshness retention) is related to total protein content and to protein sulfhydryl groups and disulfide

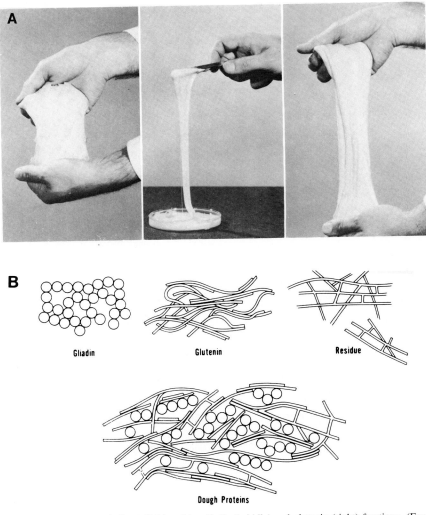

Fig. 10.12. A, Hydrated gluten (left) and its gliadin (middle) and glutenin (right) fractions. (From Dimler, 1965.) B, Effect of wheat protein structures on molecular associations and viscoelastic properties. (From Wall and Huebner, 1981.)

linkages. Continuous interchange reactions between sulfhydryl and disulfide groups and the reactivity of protein sulfhydryl groups affect oxidation requirements. Some studies indicate that protein quality may be affected significantly by the disulfide to sulfhydryl ratio. That hydrogen bonds are involved in stabilizing dough structure was demonstrated by use of deuterium oxide instead of water in dough mixing.

Table 10.8A

Structure of Amino Acid Residues and the Amino Acid Composition of Wheat Proteins

Structure	Name	Character of side chain	Composition of protein fractions (%)		
			Albumins–globulins	Gliadin	Glutenin
	Glutamic acid	Acidic	5.6	1.0	2.3
	Aspartic acid		6.5	1.0	1.2
	Lysine		3.4	0.7	1.2
	Histidine	Basic	4.3	1.8	1.7
	Arginine		7.1	2.5	2.8
	Tyrosine		3.5	2.9	4.1
	Serine	Hydroxyl	5.6	3.7	4.6
	Threonine		2.2	1.9	2.6
	Glutamine	Amides	10.0	37.0	34.0
	Asparagine		0.4	1.2	1.1
	Cysteine				
	Cystine	Sulfur	2.8	2.2	1.3
	Methionine		1.5	1.5	1.1
	Glycine		6.1	1.3	4.2
	Alanine		3.3	1.6	2.0
	Valine	Hydrocarbon	5.2	3.4	3.3
	Isoleucine		4.8	3.8	2.9
	Leucine		7.4	6.7	5.9
	Phenylalanine	Aromatic	3.6	6.3	4.3
	Tryptophan		2.0	0.8	1.7
	Proline	Cyclic	10.8	13.9	12.5

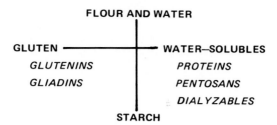

Fig. 10.13. Scheme for fractionating wheat flour into gluten, water-solubles, and starch. (Courtesy of K. F. Finney.)

5. Water-Solubles

The three fractions obtained when wheat flour dough is washed under a gentle stream of water are gluten, starch, and water-soluble components (Fig. 10.13). The water-soluble components of flour can be separated, in turn, by dialysis and centrifugation into four fractions: (a) membrane-permeable dialysate (a mixture of soluble carbohydrates, amino acids, peptides, minerals, and growth factors essential for yeast fermentation); (b) globulins, which precipitate in the dialysis bag in the absence of minerals and can be removed after centrifugation; (c) heat-coagulable albumins in the supernatant; and (d) residual supernatant (see Fig. 10.14).

The role of water-soluble components in bread making is twofold. The membrane-permeable dialysate is essential to panary fermentation because it contributes to gas formation of yeast; however, that contribution can be replaced by synthetic yeast food. Neither the globulins nor the albumins are essential to produce a normal loaf of bread. The residual supernatant is rich in water-soluble pentosans and glycoproteins, and contributes to gluten extensibility and to retention of gas in a fermented dough.

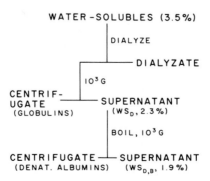

Fig. 10.14. Fractionation scheme for water-soluble fractions of wheat flour. WS, water soluble; D, dialyzed; B, boiled. (From Hoseney *et al.,* 1969a.)

6. Lipids

Wheat flour lipid composition is shown schematically in Fig. 10.15 (MacMurray and Morrison, 1970, see also Chapter 7). Total wheat flour lipids consist of about equal amounts of nonpolar and polar components. Triglycerides (TG) are a major component of nonpolar lipids, digalactosyl diglycerides (DGDG) of glycolipids; and lysophosphatidyl cholines (LPC) and phosphatidyl cholines (PC) are major components of phospholipids.

The free lipids (extracted with petroleum ether, PE) can be fractionated according to their elution from a silicic acid column. About 70% of the total free lipids can be eluted with chloroform; and they form what is arbitrarily called the nonpolar fraction, containing TG as a major component. The residual 30% free lipids can be eluted from the column with a more polar solvent, such as methanol, and constitute a mixture of free polar lipids. Among the free polar lipids, about two-thirds are glycolipids, containing DGDG as a major component, and one-third are phospholipids, with PC as a major component.

About 0.6–1.0% bound lipids can be extracted from flour with water-saturated butanol after PE extraction. Bound lipids contain about 30% nonpolar and 70% polar lipids. Bound polar lipids are rich in phospholipids, with LPC as a major

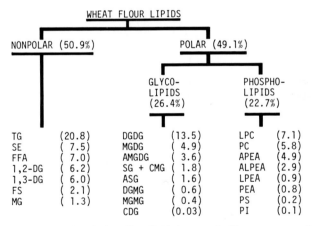

Fig. 10.15. Composition of total wheat flour lipids (extracted with water-saturated butanol). TG, triglycerides; SE, stearyl esters; FFA, free fatty acids; 1,2-DG, 1,2-diglycerides; 1,3-DG, 1,3-diglycerides; FS, free sterols; MG, monoglycerides; DGDG, digalactosyl diglycerides; MGDG, monogalactosyl diglycerides; AMGDG, O-acylmonogalactosyl diglycerides; SG, stearylglucoside; CMG, ceramide monoglycerides; ASG, 6-O-acylstearyl glucosides; DGMG, digalactosyl monoglycerides; MGMG, monogalactosyl monoglycerides; CDG, ceramide diglycosides; LPC, lysophosphatidylcholines; PC, phosphatidylcholines; APEA, N-acylphosphatidylethanolamines; ALPEA, N-acyllysophosphatidylethanolamines; LPEA, lysophosphatidylethanolamines; PEA, phosphatidylethanolamines; PS, phosphatidylserines; PI, phosphatidylinositols. (Adapted from MacMurray and Morrison, 1970.)

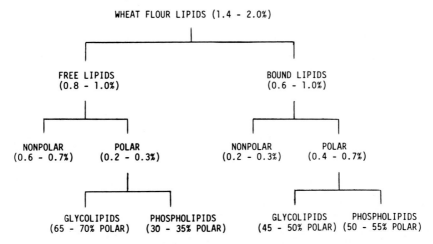

Fig. 10.16. Free and bound lipids in wheat flour. (From Pomeranz and Chung, 1978.)

component. Although in terms of percentage the free polar lipids are richer in glycolipids than the bound polar lipids are, the actual amounts of both glycolipids and phospholipids are higher in the bound polar than in free polar lipids (Fig. 10.16; see also Chapter 7).

7. Lipids in Bread Making

To demonstrate the role of lipids in bread making, gluten was isolated from both untreated and defatted flours and then remixed with starch and water-soluble components in different proportions. Loaves were then baked from these doughs. At each gluten level, LV was higher in bread baked from untreated (that is, lipid containing) than from defatted flour (Fig. 10.17). Studies with defatted flours have shown that nonpolar lipids are deleterious and polar lipids are beneficial in bread making (Daftary *et al.*, 1968). See Fig. 10.18. The deleterious effects of nonpolar lipids can be counteracted by polar lipids, and the effect on LV depends on the levels of nonpolar and polar lipids as well as on their ratio. The effects of lipids demonstrated that they are minor components that have major importance in bread making.

The significance of protein–lipid interactions was examined by Chung and Pomeranz (1978). Defatting decreased glutenin to gliadin ratios. The glutenin and gliadin fractions of strong and weak flours and of nondefatted and defatted flours differed in number and intensity of electrophoretic bands. In a subsequent study, Chung and Pomeranz (1979) examined the binding of acid-soluble proteins from untreated and defatted good and poor baking quality flours to phenyl-Sepharose CL-4B. Acid-soluble proteins from good and poor flours differed in their apparent hydrophobic properties.

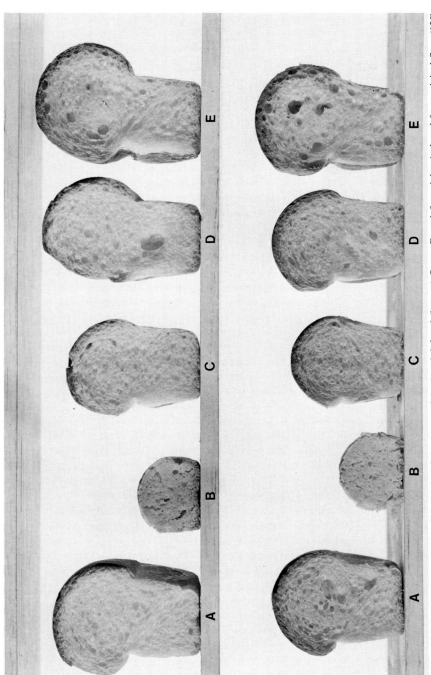

Fig. 10.17. Microloaves (10 gm flour) baked from untreated (top) and defatted (bottom) flours. From left to right: A, bread from original flour (13% protein); B, bread made from flour minus gluten (a mixture of starch and water-solubles); C, bread from a mixture of starch and water-solubles plus 10% gluten; D, bread from a mixture of starch and water-solubles plus 13% gluten; E, bread from a mixture of starch and water-solubles plus 16% gluten. (From Chiu *et al.*, 1968.)

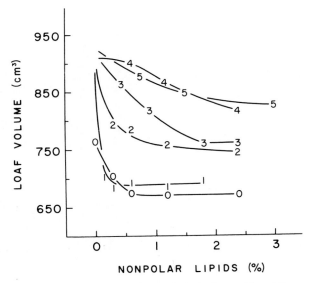

Fig. 10.18. Effects of nonpolar and polar lipids on loaf volume of bread from petroleum ether-extracted flour baked without shortening and with various combinations of nonpolar and polar lipids (1–5 denote 0.1–0.5 gm polar lipids/100 gm flour.) (From Daftary *et al.*, 1968.)

8. *Glycolipids*

Olcott and Mecham postulated in 1947 that a gluten–lipid complex is formed during dough mixing or gluten preparation. The structure of that complex and the manner in which it contributes to bread making have been elucidated as a result of intensive investigations of wheat flour glycolipids. The structural formulas of glycolipids were given in Fig. 7.1 (see also Chapter 7).

Certain types of glycolipids were first isolated from wheat flour by Carter and co-workers (1961). The interactions of glycolipids with wheat flour macromolecules and the complexes that glycolipids form with starch, gliadin, and glutenin were studied by infrared and nuclear magnetic resonance spectroscopy (Wehrli and Pomeranz, 1970a). Those spectroscopic studies and subsequent investigations (Hoseney *et al.*, 1970) on differences in solubility and migration of gluten proteins in an ultracentrifuge have shown that, in gluten, glycolipids are bound to glutenin proteins by hydrophobic bonds and to gliadin proteins by hydrogen bonds. In unfractionated gluten, the glycolipid is apparently bound to both proteins at the same time and forms a complex that can be pictured as units of gliadin and glutenin proteins bound together by glycolipids (Fig. 10.19).

Additional studies determined which interactions take place in dough and bread containing both starch and proteins. Tritium-labeled glycolipids were synthesized, and sections prepared from dough and bread containing the labeled glycolipid were studied by autoradiography (Wehrli and Pomeranz, 1970b). In the dough, most of the glycolipid was in the gluten; in the bread, most of it was in the oven-heated gelatinized starch granules. The findings on the macromolecular interaction between glycolipid and wheat flour components are summarized in Table 10.8B.

Basically, the important increase in LV during bread making is controlled by

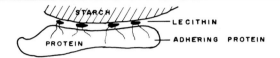

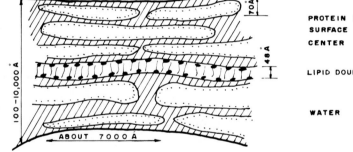

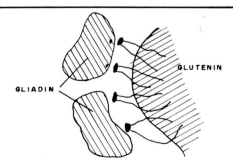

Fig. 10.19. Proposed models of the complex formed in bread making: (top) starch–lipid–adhesive protein complex in flour of Hess and Mahl (1954); (middle) lipoprotein model of Grosskreutz (1961); (bottom) gliadin–glycolipid–gluten complex of Hoseney *et al.* (1970).

TABLE 10.8B

Bonds in Glycolipids and Wheat Flour Macromolecule Complexes[a]

| Method of study | Type of bond between glycolipid and | | |
	Starch	Gliadin	Glutenin
Solvent extraction of gluten proteins	—	Hydrogen	Hydrophobic
Lipid binding in starch dough	Hydrogen	—	—
Infrared	Hydrogen	Van der Waals, hydrogen	Van der Waals, hydrogen
Nuclear magnetic resource	Hydrogen, some induced dipole interaction	—	Hydrophobic and hydrogen
Autoradiography	Strong interaction in bread	—	Interaction in dough
Baking test	Hydrophobic and hydrogen bonds are essential for improvement in bread making		

[a]From Wehrli and Pomeranz, 1970a.

the interaction between glycolipids and gluten proteins, principally the gliadin fraction. To form the complex, adequate amounts of glycolipids are required. The strength of interaction apparently depends on the composition of gliadin proteins. Gliadins from various flours differ consistently in their electrophoretic mobilities. The key to bread strength seems to be in the gliadin fraction and the mode of its interaction with other flour components.

The occurrence of a gliadin–glycolipid–glutenin complex explains, at least in part, the importance of hydrophobic and hydrogen bonding in gluten structure. The complex is of special significance in view of the high sensitivity of hydrogen-bonds to heat and the increased stability of hydrophobic bonds at elevated temperatures (gliadin is bound to glycolipids by hydrogen bonds). The complex also explains the relatively substantial effect of small amounts of glycolipids on the bread making performance of wheat flours (Wehrli and Pomeranz, 1969a).

D. Wheat Flour Lipids and the Shortening Response

Interaction between lipids and wheat flour macromolecules was observed during studies of the "shortening effect," that is, the increase in loaf volume and improvement of crumb grain by the addition of 1–3% shortening or hardened vegetable fat (see Bell *et al.*, 1977; Chapter 7).

Wheat flour lipids and their role in bread making have been the subject of several comprehensive reviews (for a detailed list, see Chung and Pomeranz,

1977; Chung *et al.*, 1978). The roles of polar and nonpolar lipids are modified by shortening and/or surfactants. Interactions among wheat flour components are important in the shortening response (Pomeranz and Chung, 1978). According to Wehrli (1969), since shortening in the absence of glycolipids is detrimental to LV, shortening probably interferes with the formation of a stable membrane between starch and proteins. Such a membrane, presumably, can form only if the starch surface is covered with glycolipids. Shortening probably is beneficial when the quantity and quality of gluten proteins and free lipids, especially polar glycolipids, in wheat flour are favorable.

E. Protein-Enriched Bread

Experiments with glycolipids synthesized in the laboratory (Wehrli and Pomeranz, 1969b) indicated that a certain hydrophilic–hydrophobic balance is required for best effects in bread making (Pomeranz and Wehrli, 1969; see also Chapter 9). The significance of those findings with synthetic compounds was threefold. First, they confirmed the role of natural flour glycolipids in bread making. Second, the findings suggested the possibility of synthesizing glycolipids superior to those found in flour. Finally, the studies provided the basis for the production of protein-enriched bread through the addition of small amounts of glycolipids (Pomeranz *et al.*, 1969; Chung *et al.*, 1976).

Wheat flour contains fairly large concentrations of proteins, but the proteins are not well balanced in amino acids and are particularly low in lysine. The quality of protein poses practically no problem if bread is part of a diet rich in protein foods of high biological value, such as milk, eggs, meat, fish, and legumes. In countries where high protein foods are not generally available, however, wheat flour enriched with plant and animal proteins, including defatted oilseed flours, edible yeasts, and fish flour, might be used to improve nutrition in inexpensive protein-enriched breads and other baked products.

Without the addition of extra glycolipids, even relatively low levels (about 5%) of lysine-rich protein supplements (i.e., soy flour) decrease LV, impair crumb texture, reduce shelf life. If small amounts of glycolipids are added, however, at least 10% soy flour containing about 50% protein can be included in the formula without impairing consumer acceptance of the bread (Fig. 10.20). Soy protein contains about 6% lysine (compared to only 2% in wheat flour), so the addition of 10% soy flour easily doubles the amount of nutritionally limiting lysine that would be supplied by flour. The use of glycolipids in the production of acceptable, nutritionally improved bread is particularly promising because little or no changes would be required in bread-making processes, formulations, schedules, or equipment (Pomeranz and Finney, 1972; Pomeranz and Wehrli, 1973).

An "ideal" nutritionally improved, high protein bread should be economical;

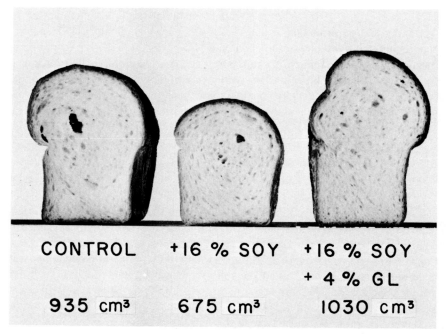

CONTROL   + 16 % SOY   + 16 % SOY
                                        + 4 % GL

935 cm³   675 cm³   1030 cm³

Fig. 10.20. Cut loaves of bread baked from 100 gm flour. Left, with 4 gm milk powder and 3 gm shortening; middle, with 16 gm soy flour; right, with 16 gm soy flour and 4 gm sucrose tallowate, a glycolipid. (From Pomeranz *et al.*, 1969.)

require no special formulations, schedules, and equipment to produce; and be acceptable to the consumer with regard to overall eating qualities, including freshness retention. Although today's high protein bread is superior to those produced earlier, it still is not possible to satisfy all the requirements for an ideal bread. The use of flour from germinated soybeans (in the presence of small amounts of surfactants such as sucrose esters or sodium stearoyl lactylate) overcomes some, but not all, of the difficulties (Pomeranz *et al.*, 1977).

F. Bread Structure

Microscopy can help us understand differences between flours that vary in bread-making potential. The technique also helps us learn how to improve nutritional value (such as by the addition of protein-rich supplements) without adversely affecting consumer acceptance of bread.

Bechtel *et al.* (1978) examined the structures of a water–flour dough, a fermented bread dough, and a wheat flour bread. They found that protein strands provide a network in a mixed dough; adequate mixing is required for matrix formation while excessive mixing destroys the matrix. The effects of overmixing and undermixing depend on the mixing requirements and tolerance of the flour. Fermentation of a complete dough produces gas vacuoles. After oven spring, the protein strands are thin with small vacuoles. Starch granules in the bottom center of a loaf vary widely in degree of gelatinization after oven spring; starch begins to gelatinize from the interior of the granule and appears fibrous. In the bread loaf, most of the starch is gelatinized into fibrous strands interwoven with thin protein strands (Fig. 10.21).

A well-mixed dough formed from a good bread-making composite flour results in even and continuous distribution of inclusions within the protein. The undermixed composite-flour dough does not have all these characteristics. In the overmixed dough, distribution of inclusions within the protein is even, but the protein is not continuous; moreover, protein strands are disrupted easily. These results agree with the findings of Baker and Mize (1946), who measured (by changes in dough density) vacuole formation in doughs and effects of vacuole formation on bread quality. Doughs that are inadvertently overmixed can be relaxed and then remixed to form an optimally mixed dough (Bloksma, 1971).

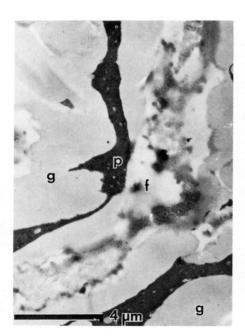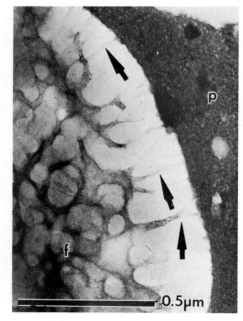

Fig. 10.21. Left, micrograph showing top center of loaf immediately after baking; note gelatinized starch (f) between thin protein strands; note lack of vacuoles in protein (p); gas vacuoles (g). Right, highly magnified protein (p)–gelatinized starch (f) interface. Note fine connections between starch and protein (arrows). (From Bechtel *et al.*, 1978.)

When an overmixed dough was allowed to relax, the protein vacuoles decreased in size and number. It is possible that relaxation and gentle remixing "mends" the broken protein strands so that an optimally mixed dough is restored.

Poor quality flour dough was seen to differ strikingly from the good quality flour dough. The protein strands of the poor flour dough broke easily, even before the protein inclusions were uniformly distributed. In good quality flour, the tensile strength of the hydrated proteins apparently governs the structure of the dough and the bread. This effect was quite apparent when a good quality flour was grossly overmixed; the protein strands remained continuous, with relatively few being broken. This "stability" in the face of overmixing (and even undermixing) is one of the most important and desirable characteristics of good quality flours.

These results pointed to the significance of starch and starch–protein interaction in the baked bread. They agree with the findings of Yasunaga et al. (1968) on factors that govern gelatinization of starch during baking, of Derby et al. (1975) on the significance of limiting amounts of water in gelatinization of starch in baked or cooked products, and of Wehrli and Pomeranz (1970a,b) on the significance of starch–protein interactions in dough and bread.

III. Malt Products

Based on a condensation and update of papers by Pomeranz (1972, 1973b, 1975a,b), this review of ingredients and processes used in the production of beer covers the malting process; barley and malts; adjuncts in brewing; hops; and various aspects of the brewing process (chemical changes, types of beer, by-products, and new developments). Beer is a fermented, hopped, malt beverage. The three most important ingredients are malt, hops, and yeast.

A. Malting

Malting involves controlled wetting (by steeping) and germination of seeds under conditions conducive to production of desired physical and chemical changes associated with the germinative process. During the malting process, weight losses due to respiration must be kept to a minimum. Malting develops the amylolytic enzymes that modify starch to fermentable carbohydrates, the source of alcohol in beer. Malt also contains proteases and gumases. The products of proteolytic degradation (along with other components) act as flavor precursors and as nutrients for the yeast during fermentation. The malted grain is dried to halt growth and stop enzymatic activity and to produce a storable product of desired color and flavor. Drying is followed by removal of malt sprouts. The malting process is depicted in the form of a flowsheet in Fig. 10.22.

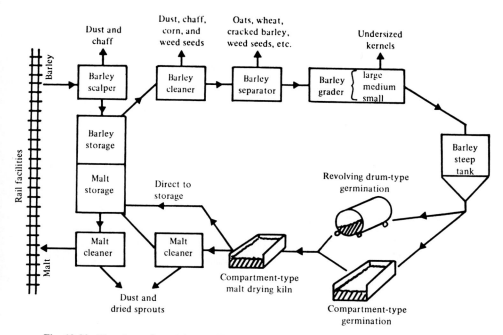

Fig. 10.22. Flowsheet of a malt house. (From U.S. Department of Agriculture, 1968. Courtesy of A. D. Dickson.)

In the old system of malting, steeped barley was germinated on concrete floors in cool, moist rooms and turned by hand. Floor malting was replaced by pneumatic-type malting, in which conditioned air was forced through the grain. Now malting is performed in drums or compartments. Use of compartments and greater mechanization reduces labor costs and allows a more uniform product. Pneumatic malting systems require large quantities of air adjusted to a desired temperature and saturated with water vapor.

B. Barley in Malting

Barley occupies a unique position in malting and brewing. During malting barley produces hydrolytic enzymes, including relatively large amounts of α- and β-amylases. The combination of the two amylolytic enzymes results in a more complete and rapid degradation of starch than in can be obtained using malts from most other cereal grains. This degradation of starch is accompanied by breakdown of other grain components (mainly proteins and nonstarchy polysaccharides) and yields an optimally modified malt.

In barley, the husks are "cemented" to the kernel and remain attached after threshing. The husks protect the kernel from mechanical injury during commercial malting, strengthen the texture of the steeped barley, and contribute to a more uniform germination of all kernels. The husks are important as a filtration aid in the separation of extract components during mashing (described later) and they contribute to the flavor of the malt and the beer.

Cultivated barley was used by Neolithic cultures in Egypt between 5000 and 6000 B.C. Major gene centers where cultivated types of barley may have developed include Ethiopia and the highlands of Sikkim and southern Tibet. Barley is a relatively winter-hardy and drought-resistant grain; it generally matures more rapidly than wheat, oats, or rye. It is grown worldwide.

Barley belongs to the family Gramineae, subfamily Festucoideae, tribe Hordeae, and genus *Hordeum*. Most cultivated barleys are covered with a hull. In some Far Eastern countries and mountainous areas of Africa and Asia, where barley is used for human food, naked (hull-less) barleys are found. The two main types of cultivated covered barleys are two-rowed and six-rowed, according to the arrangement of grains in the ear. The former predominates in Europe, parts of Australia, and western United States. The latter, more resistant to extremes of temperature, is grown in North America, India, and the Middle East (Wiebe, 1968). The axis of the barley ear has nodes throughout its length; the nodes alternate from side to side. In the six-rowed type of barleys, three kernels develop at each node, one central kernel and two lateral kernels (Fig. 10.23). In the two-rowed barleys, the two lateral kernels are sterile and only the central kernel develops.

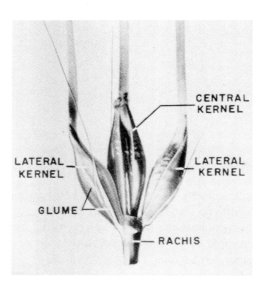

Fig. 10.23. Arrangement of kernels in six-rowed barley. (Courtesy of Malting Barley Improvement Association, Milwaukee, Wisconsin.)

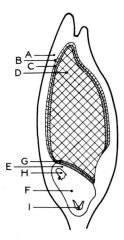

Fig. 10.24. Diagram of longitudinal section through a barley kernel. A, Husks; B, pericarp; C, aleurone layer; D, starchy endosperm; E, scutellum; F, embryo; G, epithelium; H, plumule; I, radicle.

1. The Kernel

The kernel of covered barleys consists of the caryopsis and the flowering glumes (or husks). A diagram of a longitudinal section through a barley kernel is shown in Fig. 10.24. The husks consist of two membranous sheaths that completely enclose the caryopsis (Fig. 10.25). One of the husks (the lemma) is drawn out into a long awn. The color of the grain in covered barleys depends on the color of the caryopsis and of the second husk (the palea). Color in the caryopsis is due to anthocyanin pigment or to a black melanin-like compound. Anthocyanin, when present, is red in the pericarp and blue in the aleurone layer. During the development of the growing barley, a cementing substance causing husk adherence is secreted by the caryopsis within the first two weeks after pollination.

The caryopsis is a one-seeded fruit in which the outer pericarp layers enclose the embryo and the two endosperm layers, aleurone and starchy endosperm. The starchy endosperm is the main storage tissue; the aleurone layer is at least two cells thick and forms the peripheral layer of the endosperm. The caryopsis has a furrow (crease) in the side opposite to the embryo and is covered by the palea.

The husk amounts to 7–25% of the grain (average 13%). The proportion varies according to type, variety, grain size, and latitude of cultivation. The proportion of husk increases as the latitude where the barley is grown approaches the equator. Large and heavy grains have less husk than small, lightweight grains. The husk content of two-rowed barleys (8.0–8.9%) is lower than in four- or six-rowed barleys (10.4–12.9%) (Kent, 1966).

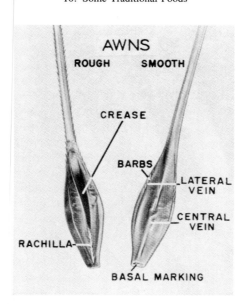

Fig. 10.25. Awns, hulls, and caryopsis of barley. (Courtesy of the Malting Barley Improvement Association, Milwaukee, Wisconsin.)

The barley kernel weight is 21–45 mg, the length is 6.0–12.0 mm, the width is 2.7–5.0 mm, and the thickness is 1.8–4.5 mm. In covered barleys in which the husks amount to about 13%, the pericarp is (plus testa) 2.9%, the aleurone 4.8%, the starchy endosperm 76.2%, the embryo 1.7%, and the scutellum 1.3% of the whole grain.

2. Composition

On a dry-matter basis, covered barley contains 63–65% starch, 1–2% sucrose, about 1% of other sugars, 1.0–1.5% soluble gums, 8–10% hemicellulose, 2–3% lipids, 8–13% protein ($N \times 6.25$), 2–2.5% ash, and 5–6% other components (MacLeod, 1969). In regular barley, the linear starch component (amylose) constitutes 24% of the total starch.

The proteins in barley (as in all cereal grains) are composed of four groups varying in solubility. The albumin fraction is less than 10% of the proteins, the globulins about 20%, the hordeins (soluble in 70% alcohol) 30%, and the remaining 40% of the proteins are glutelins. About one-half of the amino acid residues in hordeins are either glutamine or proline; the amounts of aspartic acid, glycine, and lysine are small. The amino acid composition of the glutelins resembles that of hordeins.

Barley lipids are concentrated in the embryo and the aleurone layer. Although the whole grain has only 2% petroleum ether-extractable material, isolated embryos contain 15% lipids. The predominant constituent fatty acids are linoleic, oleic, and palmitic, with the unsaturated components accounting for nearly 80% of the total.

Mature barley may contain more than 2% of fructosans. Unlike starch, which is restricted to the endosperm, the fructosans are distributed throughout the grain. Sucrose is virtually restricted to the embryo and aleurone; it represents 12–15% of the embryo but only 1–2% of the whole grain. Raffinose is also a major embryo constituent, about 5% of the dry weight. The husks contain more than two-thirds of the grain's cellulose; the cell walls of the central starchy endosperm lack true cellulose.

C. Malt Types

Two general types of malts are produced commercially—brewers' and distillers' malts. Brewers' malts are made from barleys of plumper, heavier kernels with a mellow or friable starch mass. They are steeped and germinated at moisture contents ranging from 43 to 46%; the final temperatures used in drying the malts range from 71 to 82°C. These malts are dried to about 4% moisture content. The high final drying temperature reduces the enzymatic activities of the malt, darkens the malt and the wort made from it, and increases malt flavor and aroma. Distillers' (or high-diastatic) malts are made from small kerneled barleys high in protein content and enzymatic potentialities. The barleys are steeped and malted at higher moisture contents (45–49%) and dried at lower temperatures (49–60°C) to higher finished moisture contents (5–7%) than the brewers' type of malt.

Comparative ranges in composition of barley and malt are given in Table 10.9. Cross-sections through an untreated barley kernel and kilned malt are compared in Fig. 10.26. In addition to the structure of the central endospern, Fig. 10.26(A) shows the aleurone layer with the aleurone grains and the pericarp. Whereas in the untreated barley kernel there is a protein matrix in which intact starch granules are embedded, in the kilned malt the protein matrix has basically disintegrated and is in part coagulated on the surface of starch granules. Several of the latter are degraded extensively. See Fig. 10.26(B).

D. Adjuncts in Brewing

Adjuncts are used in brewing to impart certain characteristics to wort and beer, but mainly for reasons of economy. More than 90% of brewers' extract consists of carbohydrates, of which 70% is fermented in the production of beer. The use of cereal adjuncts has increased in recent years, and brewers in many countries incorporate up to 30% of these materials in their grist. In the United States and

TABLE 10.9

Comparative Ranges in Composition of Barley and Malt[a]

Property	Barley	Malt (brewers' and distillers')
Kernel weight (mg)	32–36	29–33
Moisture (%)	10–14	4–6
Starch (%)	55–60	50–55
Sugars (%)	0.5–1.0	8–10
Total nitrogen (%)	1.8–2.3	1.8–2.3
Soluble nitrogen (% of total)	10–12	35–50
Diastatic power (°L)	50–60	100–250
α-amylase (20° units)[b]	Trace	30–60
Proteolytic activity[c]	Trace	15–30

[a] From U.S. Department of Agriculture, Courtesy of A. D. Dickson.
[b] 20°C dextrinizing units, a unit of α-amylase activity.
[c] Arbitrary units.

Canada, the proportion of adjuncts used may in some instances be as great as 60%.

The adjuncts most commonly used in the United States are corn or rice grits. Corn and rice tend to have a reciprocal usage relation due to market price fluctuations of the two adjuncts. Some brewers believe that the alteration of flavor resulting from use of corn is not great enough to justify using brewers' rice if there is any price differential. Others prefer brewers' rice to any other adjunct if the cost differential is not prohibitive. Materials used per barrel of beer (31 gal) total about 45 lb; they include malt, corn, rice, and syrup. The amount of hops used today is only half that used in the 1940s. Of the cereals used in breweries, malt and malt products comprise over 65%, corn and corn products over 25%, and rice and rice products about 8%. Amounts of other cereals are small. Barley and wheat products are used in countries where they are more economical sources of fermentable carbohydrates than other cereals.

In addition to their economic advantages, corn and rice adjuncts are used to obtain the paler, blander, and less-filling beers preferred by most consumers in the United States. They also improve certain functional properties in brewing and

Fig. 10.26. A, SEM transverse section through pericarp, seed coat, and endosperm of barley (×1000). PE, multilayered, cordlike pericarp; AW, aleurone cell wall; AG, aleurone grain; PM, protein matrix of subaleurone layer; S, starch granule; EW, cell wall of subaleurone layer. B, Central endosperm of malted and kilned barley (×1250). AP, adhering "proteinaceous" material; DS, highly damaged starch granules. (From Pomeranz, 1972.)

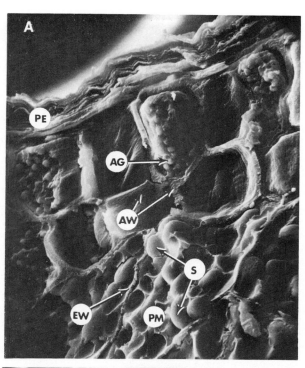

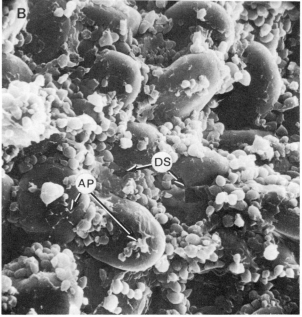

TABLE 10.10

Typical Analyses of Brewing Adjuncts[a]

Composition	Amount (%)
Cereal adjuncts	
Corn grits	
Moisture	10.9
Extract, dry-basis	91.4
Oil, dry-basis	0.76
Corn flakes	
Moisture	9.0
Extract, dry-basis	92.4
Oil, dry-basis	0.50
Refined Grits	
Moisture	9.6
Extract, dry-basis	103.3
Oil, dry-basis	0.03
Rice	
Moisture	12.0
Extract, dry-basis	93.0
Oil, dry-basis	0.86
Liquid adjuncts	
Corn syrup	
Extract, as is	82.0
Extract, dry-basis	100.0
Fermentable extract, as is	60.2
Reducing sugars (as dextrose), as is	50.5
Ash, as is	0.19
pH (10% solution)	4.95

[a]From Westermann and Huige, 1979.

in the final product. Typical analyses of brewing adjuncts are given in Table 10.10.

E. Hops

Hops were first used in brewing in Finland and were introduced by the Estonians to the Slavs and Germans. The famous Hallertau hop fields of Germany were established by 840 A.D. According to Hind (1950), hops have medicinal and dietetic properties, act as preservatives, ensure the soundness and stability of beer, and—probably most importantly—impart a unique and pleasant flavor and aroma to fermented malt liquors. The hops of commerce are dried inflorescences of the female plant of the diecious, perennial species *Humulus lupulus* L. The hop cone, male flower, and lupulin gland are shown in Fig. 10.27.

The axis of the hop cone is known as the "strig." It is bent at obtuse angles along its length and has at each angle four bracteoles enclosed by two bracts. The bracteoles are oval and slightly incurved at the base to carry the seed. The bracts are coarser, somewhat larger, and generally pointed at the tip. The rachis (strig) is covered with downy, linear, epidermal hairs. Specialized cupshaped glandular hairs (lupulin glands) develop on the surface of the lower parts of bracteoles and over the surface of the perianth; sometimes a few are found on the bracts. The lupulin glands contain the brewing principles, the bitter resins, and the essential oils. A membranous covering on the glands prevents escape of the lupulin.

The strig of a cone of hops is shown at low magnification in Fig. 10.28(A). The picture shows bracteoles attached to a pubescent axis, bract scars (which are important in typing of hops), and lupulin glands. The fine structure of a lupulin gland is shown in Fig. 10.28(B).

The initial process in brewing of beer is mashing (Fig. 10.29). The term means extraction of malt with water, though in practice the process is accompanied by additional changes. Prior to mashing, the dried malt is milled (Fig. 10.29). The milled product and water are put into a mash tub in which a series of controlled time–temperature treatments enhance dissolution of materials present in malt in a soluble form and enzymatic degradation of insoluble compounds.

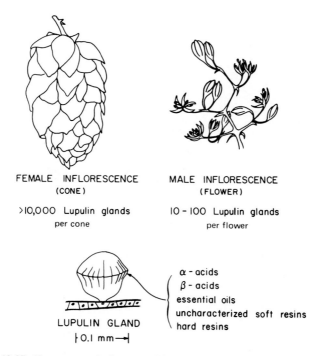

FEMALE INFLORESCENCE
(CONE)

>10,000 Lupulin glands
per cone

MALE INFLORESCENCE
(FLOWER)

10 - 100 Lupulin glands
per flower

LUPULIN GLAND
⊢ 0.1 mm ⟶

α - acids
β - acids
essential oils
uncharacterized soft resins
hard resins

Fig. 10.27. Hop cone, male flower, and lupulin gland. (Courtesy of J. S. Hough.)

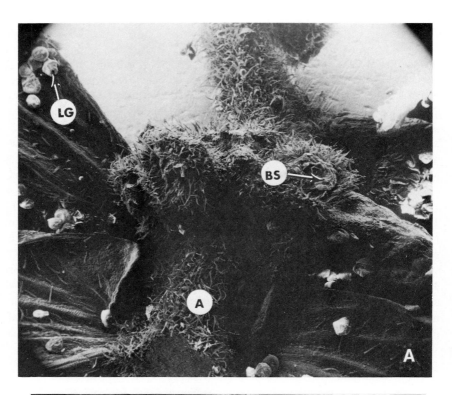

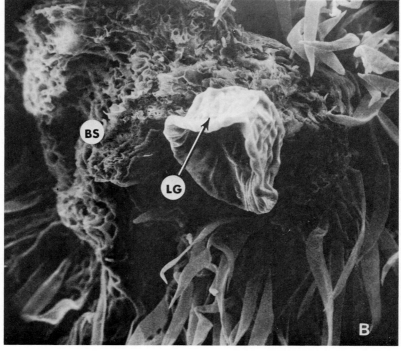

There are two basic methods of mashing. Infusion mashing is the traditional and simplest method, in which the mixed grist and liquor are allowed to stand in a mash tub and the temperature is gradually raised but kept below the boiling point. This process somewhat resembles the making of tea. Decoction mashing is the traditional method in the preparation of a malt extract for "bottom-fermented" beers and is the main mashing method used in Central Europe. In the decoction method, part of the mash is withdrawn, boiled, and returned to the mash tub to raise the temperature of the whole mash.

After extraction the mash passes to a lauter tub, which has a false bottom. The grain husks collect on this false bottom and form a filter bed, through which the sugar-rich extract (wort) is strained. The spent grains are separated from the wort. The clear extract (sweet wort) is boiled with hops. Then the hot wort is strained to remove hop leaves and stems. The hopped wort is cooled, yeast is mixed in, and the mixture is pumped to settlers, where it remains 10–12 hr. It is then transferred to fermentation tanks, where it remains until fermentation is completed and the fermented sugars are converted to carbon dioxide and ethanol. The fermented wort is allowed to age, prefiltered, chillproofed (by proteases to hydrolyze undegraded proteins and prevent their precipitation during chilling), filtered, carbonated, and poured into containers. Bottled or canned beer is either pasteurized or sterilized by ultrafiltration. Typical wort analysis is given in Table 10.11.

F. Chemical Changes during Brewing

The modifications during mashing involve simple dissolution of materials, already present in malt in a soluble form, and enzymatic changes. The most important quantitative reaction during mashing is the conversion of starch to dextrins and maltose.

The question of changes in nitrogenous compounds is more complex. According to deClerck (1957) a great qualitative difference obtains between the soluble nitrogenous materials formed during malting and during mashing. About 40% of the solubilized nitrogen formed at malting was in form of formol-nitrogen, but only 20% of formol-nitrogen was formed at mashing. Clapperton (cited by Anonymous, 1971) concluded that mashing favored liberation of larger peptides rather than degradation of smaller ones.

Boiling the wort stabilizes its composition by inactivating amylases and cytoclastic enzymes, by elimination (coagulation) of unstable colloidal protein (trub), and by extraction of wort components from hops. According to Hough *et al.* (1971), the carbohydrates account for 91–92% of the hopped wort extract, of

Fig. 10.28. A, Strig of a cone of hops (SEM, ×24): bracteoles attached to pubescent axis (A); lupulin glands (LG); and end of pedicel which supported a bracteole (BS). B, Lupulin gland (×250). (From Pomeranz, 1972.)

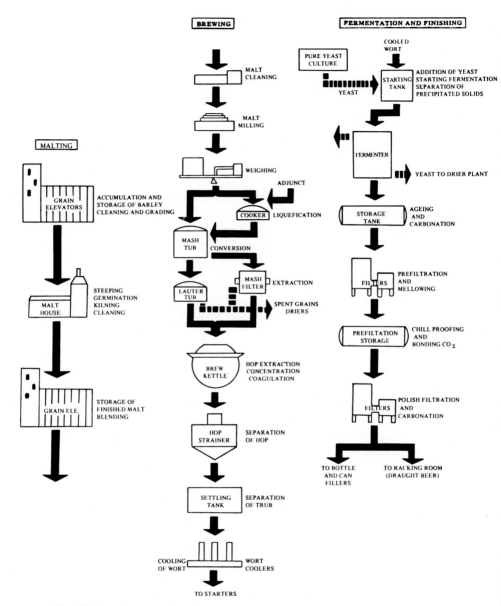

Fig. 10.29. Malting and brewing flowchart. (Courtesy of Jos. Schlitz Brewing Co., Milwaukee, Wisconsin.)

TABLE 10.11

Typical Wort Analysis[a]

Wort color (°Lov. 52–1 in. cell)	6.0
Specific gravity (20°C/20°C)	1.04545
Extract (°Plato)	11.3
Limit attenuation (°Plato)	2.5
Fermentable extract (%)	63.0
Wort protein (N × 6.25) (%)	0.41
Soluble nitrogen (gm/100 gm)	0.065
Wort pH	5.35
Titr. acidity; ml N/10 alkali/100 gm	9.0
Calcium (ppm)	72
Iodine reaction	Negative

[a]From Westermann and Huige, 1979.

which 68–75% is fermentable by yeast. Of the wort carbohydrates, maltotetraose, more complex polysaccharides, and dextrins account for 23–28% of the extract; are not normally fermented; and persist into the finished beer. The nitrogenous constituents of wort account for up to 5% of the extract. They include amino acids, peptides, polypeptides, proteins, nucleic acids, and their degradation products. Most of the amino acids in wort (except proline) are assimilated during fermentation. The majority of the proteins are not assimilated during fermentation, and those that persist into the beer slowly react with other constituents to form nonbiological haze.

Beer solids consist of about 80% carbohydrates (mainly dextrins with a small amount of unfermented mono- and disaccharides). Gum-like materials are 10–20 percent of the extract, but only about one-eighth of those are high-molecular material (see Chapter 3). The nitrogenous materials show a wide range of complexity. Comparison of beer with wort indicated that the high molecular weight nitrogenous compounds were substantially unaltered by fermentation and persisted in the beer. As molecular weight decreased, increasing amounts disappeared during fermentation, and the simplest compounds (i.e., most amino acids) were present in beer in trace amounts only. The extract usually contains 3–4% of mineral matter and small amounts of miscellaneous compounds from hops and fermentation.

Haze formation in beer has been the subject of numerous investigations. Three kinds can be distinguished: biological haze, due to microorganisms; chemical haze, from the action of various chemical agents (starch, metallic aggregates); and colloidal haze, due to aggregation of colloids. According to deClerck (1957), most beer components are not in a molecular form but rather are in the colloidal state. They are present as molecular aggregates and carry an electric charge. Colloidal stability of the beer depends to a large extent upon the size and complexity of those aggregates. Interadsorption of colloids also occurs; this accounts for the heterogeneous nature of the colloidal deposits from beer.

Both proteins and polyphenols are essential for haze formation; the haze also contains carbohydrates and metallic elements. Most, if not all, barley proteins are found in beer, though some predominate more than others. The mineral content and composition of hazes differ appreciably from the parent beer, indicating selective precipitation. The importance of the metallic constituents in haze formation is not clear. They may be active initiators of haze formation or they may be involved (as cations or chelates) indirectly through association with proteins and polyphenols that form the haze. Some reports indicate that electronegative proteins are associated with haze formation (cited in Anonymous, 1970a). According to Bateson (cited in Anonymous, 1970b), however, no significant correlation could be established between protein components of low isoelectric point and shelf life of beer.

G. Types of Beer

Ancient tablets found in Mesopotamia describe 16 types of beer made as early as 4000 B.C. An Assyrian tablet from 2000 B.C. indicates that beer was among the provisions aboard Noah's Ark. The Chinese had in 2300 B.C. a beerlike beverage called "kun." The ancient Incas had a brew called "sora," which was replaced by a brew called "chicha."

A beverage resembling English beer was brewed in America at least 33 years before the Pilgrims landed. The Mayflower was bound for Virginia but put in at Plymouth Rock because the ship drifted off course and, as the ship's log recorded, "we could not take time for further search or consideration, our victuals being much spent, especially our beer." Thomas Jefferson persuaded some Bohemian master brewers to emigrate to America and teach their advanced skills to American brewers.

Today the many types of beer differ according to the composition of the ingredients and the brewing methods. A major difference between British and Continental European practices of beer making arises from the behavior of the yeast in fermentation. In Britain, brewers still employ species of yeast (i.e., *Saccharomyces cerevisiae*) that rise to the surface during fermentation. "Top-fermentation" beers come in many varieties, from relatively pale ales to porter (a dark, heavy ale popular in Ireland) and stout (a dark, heavy beer with a relatively high alcohol content). Top-fermented beers (mainly ales) are still preferred in New England. Elsewhere in the United States, the brewing industry derives its traditions from the brewers who came from Germany in the nineteenth century to set up large commercial breweries. They brought with them the new technique of bottom-fermentation, which employs a yeast (*Saccharomyces carlsbergensis*) that settles to the bottom during fermentation. The German "lager" beers (the term indicates that the beer has been aged-lagered after fermentation) are usually paler than ales and are more mellow in flavor. There are three main types of lager beer—Münchener, Dortmunder, and Pilsener—named after the towns in which they were first brewed, and listed here in the order of their colors, from dark to pale. Pilsener beer has a high hop content and a stable head. Münchener beer has a more pronounced malt flavor; darker malt and less hops are used in its manufacture. Under the pressures of standardization arising from mass distribution, United States beers are becoming increasingly pale and now resemble Pilsener beer, except that they have less hop flavor and are more highly carbonated. An exception to the trend is the traditional bock beer: a dark, full-flavored beer that is brewed in winter and marketed in spring. Caramel and sometimes black malts are used in its production. The influence of raw materials and yeast on various beer types is summarized in Table 10.12.

H. By-Products

The main byproducts of the brewing process are spent grains, trub (break), spent hops, and yeast. Most byproducts are incorporated into animal feed compositions. The spent grains constitute about 22% (dry-weight) of the original malt.

For each 100 barrels (1630 hl) of processed wort 115–230 kg of wet spent hops and 35–45 kg wet weight of hot break (trub) are removed. According to Luers (1950), 100 hl of a 12% wort produce 2.4–4.0 kg trub; 100 kg malt yield 0.25 kg trub (dry-matter). The protein of the trub comprises 1.1–1.3% of the malt proteins. The hot break contains, on a dry-matter basis, 50–60% proteins, 16–20% hop resins, 20–30% polyphenols, and 2–3% ash (Hough *et al.*, 1971).

For every kg of yeast that the brewer adds to his wort, he separates at the end of fermentation 6 kg of yeast. The brewer can make available 1½ kg of yeast cake per barrel of beer.

I. New Developments in Brewing

During the past two decades numerous reports were published on the feasibility of beer brewing based on the addition of an enzyme complex to barley and other unmalted raw materials. Even though feasible, widespread use of unmalted barley plus enzymes as substitute for malt in United States brewing practices has not been adopted. The most attractive methods from the standpoint of economy and quality control involve the use of high levels of ungerminated barley in combination with special high enzyme malts (i.e., abraded, treated with gibberellic acid, or from triticale) and mixtures of amylases and proteases from

TABLE 10.12

Types of Beer[a]

Name	Original extract (%)	Bottom-fermented	Top-fermented	Alcohol (% by weight)	Characteristics
Vollbier	11–14			3.0–4.5	
Maltbeer with sugar added	12–13		X	0.5–1.5	Dark, palateful, malt flavor, sweet
Lagerbier	11.0–12.5	X		3.5–4.0	Pale, light hop bitterness
Wheat-beer	11–12		X	4	Pale, malt flavor, light hop bitterness, much CO_2
Dietetic beer	11.3	X		3.7–4.8	Pale, accentuated hop bitterness
Alt-beer	11.2–12.0		X	3.5–3.9	Mostly dark, aromatic, hop bitter
Kolsch	11.2–11.8		X	3.5–3.9	Pale, aromatic, hop bitter
Pils	11.5–12.0	X		3.8–4.0	Pale, bitter, prickling, accentuated hop bitterness
Export beer	12.5–14.0	X		4.0–4.2	Pale, palateful, less bitter than Pils
Export wheat-beer	12.5–14.0		X	4	Pale, malt flavor, light bitterness, much CO_2
Marzen	12.5–14.0	X		3.8–4.3	Dark golden color, palateful, malt flavor
Spezial	13–14	X		4.0–4.3	Pale, light hop bitterness
Rauchbier	13.5	X		4.5	Dark, bitter, aromatic, smoke flavor
Einfach bier	2.0–5.5			0.5–1.5	Pale or dark, thin, without marked character
Low-gravity beer	7–8			0.5–2.6	
Malt beer	7		X	0.5–1.5	Dark, palateful, malt flavor, sweet
Berliner-Weisse	7–8		X	2.6	Pale, light hop bitterness, much CO_2
Alcohol-free beer	7.5	X		0.5	Pale, without marked character
Low-alcohol beer	7.5	X		1.5	Pale, without marked character

[a]From Kieninger, 1983.

microbial sources. The use of liquid adjuncts (syrups) has seen a steady increase in the United States.

The mashing process may exert a great effect on wort composition. Many patents have been granted for continuous mashing processes but none has gained wide acceptance (Hudson, 1973). The continuous processes have high energy requirements and are very complicated mechanically.

The trend in the area of hops is toward the utilization of α-acids added after fermentation. The addition of hop extracts affords both economic advantages and improved control of bitterness. The tendency is, however, to add a small amount of hops at the boiling stage to impart the odor of hop oils to the brew (Pollock, 1971).

Proteins precipitated during the boiling of wort can be removed by centrifuges or filters. In a separator, the wort is injected tangentially into a vertical cylindrical vessel and the protein precipitates and is deposited in a central cone. The vertical vessels are used in conjunction with a hop strainer, which removes the coarse spent hops. The energy requirements of a separator are less than the requirements of conventional centrifuges or filters (Hudson, 1973).

Considerable advances have been made in the technology of beer fermentations, both from production and developmental points of view. Recent developments in fermentation technology help reduce costs by using large batches. The capital costs of the equipment can be further reduced by fermenting concentrated worts and then diluting the beer to the desired concentration. In other developments, the practice of separately fermenting malt components and adjunct carbohydrates is used to introduce still further economies while retaining the overall quality of consumer acceptable beers (Pollock, 1973).

Introduction of centrifugal yeast separators has reduced the total fermentation time and introduced much flexibility in selection of yeast strains and fermentation vessels. Cylindroconical fermentation vessels can be used for the production of top-fermentation and bottom-fermentation beers. According to Pollock (1973), when large fermenters are used the carbon dioxide concentration in the beer increases and it is necessary to take special precautions to preserve beer sterility. Considerable savings can be obtained by continuous fermentations. One continuous fermentation process employs a multivessel system, which allows the free escape of yeast, with the stream of beer, from the main fermentation zone. In a second type (called plug–flow fermentation), wort is forced through a yeast-Kieselguhr mixture in a plate filter. In this process, the concentration of yeast relative to wort is very high, the flow of the wort is rapid, and little or no growth of yeast cells takes place. Beer produced by plug–flow fermentation techniques may have an undesirable flavor, but it can be eliminated by a second fermentation. Since there is no growth of yeast, this rapid fermentation requires an extraneous supply of yeast.

Because pasteurization can alter the flavor of beer there is great interest in sterile filtration through membranes, mostly of synthetic polymers, and with chemical preservatives. Whereas membrane filtration is used quite extensively, chemical preservatives have yet to be accepted by brewers.

IV. Soybean Products

One acre of land used for beef production can provide only one-fifth of the yearly requirements of one adult male; the same acre produces enough soybean protein to feed six adult males for a whole year (Daniels, 1981). As long as this relationship continues, soybeans are likely to remain the cheapest and potentially most versatile source of proteins for human consumption. When adequate protein is supplied with adequate energy, soy protein is a satisfactory source of protein for human growth and maintenance. Including soy protein in foods from animal sources makes it possible to regulate the fat content of foods. The character of the fats can be modified and the protein content can be increased as well. All the advantages of soybean supplementation may be achieved at low cost without sacrificing consumer acceptance, food diversity, and nutritional needs.

According to Wolf (1967) and Rackis (1979), some soybean research problems are

1. Production—yield, worldwide adaptability
2. Regulatory—standards, nutritional labeling, detection methods for soy in foods
3. Nutrition—quality and digestibility of protein
4. Antinutritional—factors that affect growth and availability of vitamins and minerals
5. Organoleptic—flavor, odor, texture
6. Flatulence—intestinal fermentation of oligosaccharides
7. Functionality—dispersibility, fat–water absorption, hydration, gelation

Progress in utilization of soybeans as food must address three key issues: nutrition, flavor, and functional properties.

A. Nutrition

For human consumption, soybeans are cooked and fermented or processed by moist heating. Such processes improve texture, palatability, and nutritional quality. Soybean proteins are an excellent source of available lysine, the limiting amino acid of most cereals. Their nutritive value is restricted by the relatively low content of sulfur-containing amino acids (present in relatively high concentrations in storage proteins of cereal grains). In processing soybeans for use in blends with cereals, however, the most significant considerations are inactivation of antinutrients, reduction of losses in available lysine, and reduction of other losses in nutrients as a result of excessive heat treatment.

The nutritional quality of soy products is determined by the quantity and quality of amino acids and their balance and by the processing conditions employed in manufacturing soy foods (Liener, 1981). The most important factor is heat treatment (live steam, toasting, cooking, or baking) to inactivate antinutritional factors present in raw soybeans. These substances can elicit adverse physiological responses in humans and animals; unless they are destroyed they can decrease the nutritional value of soy products. The antinutritional factors are either heat labile (trypsin inhibitors, hemagglutinins-lectins, goitrogens, antivitamins, or phytates) or heat stable (saponins, estrogens, flatulence factors, lysinoalanine, and allergens). The effect of the heat-stable factors on nutritional value is questionable. Harsh processing conditions, as in excessive heat treatment or extraction under alkaline conditions, however, may result in destruction of essential amino acids or formation of lysinoalanine.

Flatulence is caused by the presence of the oligosaccharides raffinose and stachyose in soybean meals. Processing of soybean meals into concentrates and isolates removes the causes of flatulence. In addition, enzyme technology is available to hydrolyze the oligosaccharides (Rackis, 1979).

B. Flavor

Among the functional properties of soybeans, color and especially flavor are seldom considered positive attributes (Martinez, 1979). One of the main motivations in the production of soy concentrates and isolates is to reduce or eliminate undesirable soy flavors. One of the most limiting factors in utilization of soy protein products is their objectionable "beany" flavor (Rackis *et al.,* 1979). Numerous simple and azeotropic extractions, and chemical, physical (mainly heat), biological (including germination), and enzymatic treatments have been proposed to produce bland products from soybeans. While much progress has been made in reducing their objectionable flavor, the levels at which soy ingredients must be used to contribute significantly to the protein contents of some foods means the flavors still cause problems (Wolf, 1981). This problem is of particular significance in bland foods, such as baked or dairy products. Soy products also suffer from the absence of any desirable flavors. When textured soy ingredients are used as meat extenders, they actually dilute the natural meat flavor. When other flavors are added to extended meat products, they may be released too rapidly or interact with soy proteins to produce "off" flavors.

C. Processing Soybeans

On the average, soybeans contain (on a moisture-free basis) 40% crude protein, 22% crude fat, 32% nitrogen-free extract, and about 5% ash. A major part of the soybean crop in the United States is processed into oil and meal. The basic steps in production of soy grits or flours for human foods involve cleaning, cracking, dehulling, conditioning, flaking and extracting with hexane (Fig. 10.30).

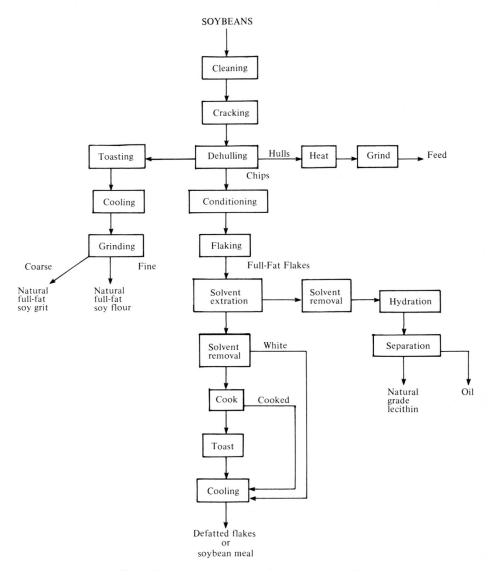

Fig. 10.30. Soybean processing. (From Rakosky, 1973.)

Five general types of edible-grade soy flours are manufactured in the United States to meet specific needs.

1. *Defatted soy flour* is produced by the nearly complete removal of the oil from soybeans through the use of hexane and usually contains 1% or less of fat (ether-extracted).

2. *Low fat soy flour* is produced either as a result of partial removal of the oil from soybeans or by the addition of soybean oil to defatted soy flour at a specified level, usually 5–6%.
3. *High fat soy flour* is produced by adding soybean oil to defatted soy flour at a specified level, usually about 15%.
4. *Full fat soy flour* contains all of the oil originally present in the raw soybeans, usually 18–22%.
5. *Lecithinated soy flour* is also being manufactured by some processors in limited amounts for specific food uses. This type of flour is produced by adding soybean phosphatides (lecithin) to defatted soy flour at a specified level, usually up to 15%.

Typical or average analyses for the three main types of soy flours and grits are summarized in Table 10.13. Depending on the relative humidity, the moisture content of soy flours and grits ranges from 5 to 10%.

The manner in which the residual solvent is removed from the defatted flakes determines the degree of denaturation of the proteins and their nutritional and functional properties. Denaturation is measured by determining the nitrogen solubility index (NSI) or the protein dispersibility index (PDI). Undenatured, dehulled, and defatted flakes containing about 50% protein are converted into concentrates by leaching out soluble materials (mainly sugars) or they may be converted into isolates by dissolving the proteins in the flakes with an alkaline solution and coagulating them into a curd by shifting the pH to 4.5 at the isoelectric point (Fig. 10.31). The concentrates and isolates may exist as water-insoluble products or as soluble proteinates. Typical analyses of soy products and nutrient contents are given in Table 10.14. The powdery soy flour or granular soy grits can be converted into products having fibrous textures by fiber spinning or thermoplastic extrusion. Such textures are described as "chewy" to enhance their appeal to consumers.

TABLE 10.13

Composition of Soy Flours and Grits[a]

	Defatted soy flour and grits	Low fat soy flour and grits	Full fat soy flour
Protein (N × 6.25) (%)	50.5	46.0	41.0
Fat (ether extracted) (%)	1.5	6.5	20.5
Fiber (%)	3.2	3.0	2.8
Ash (minerals) (%)	5.8	5.5	5.3
Carbohydrates (total) (%)	34.2	33.5	25.2

[a]From Hayward and Diser, 1961; Horan, 1974.

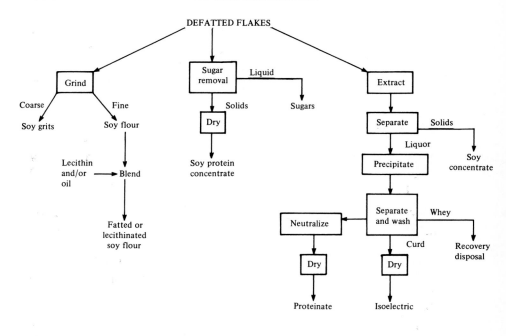

Fig. 10.31. Soy protein products from defatted flakes. (From Rakosky, 1973.)

D. Functional Properties

Soybean proteins have numerous functional properties useful in food systems (Bookwalter, 1978; Cowan and Wolf, 1974; Circle and Smith, 1978; Kinsella, 1979; Smith and Circle, 1978; Wolf, 1970, 1973, 1974, 1976). While most of the discussion in this section centers around the role of soybean protein, the contribution of other soybean components, such as oligosaccharides, is also noteworthy. These components may affect functional properties by themselves or through interaction with proteins.

As mentioned previously, heat treatment affects the functionality of soy proteins. The effects of heat on some physical properties of soy proteins are summarized in Table 10.15. The relation between PDI and soybean use in foods is given in Table 10.16. Functional properties of soybean proteins in food systems are listed in Table 10.17.

The proteins of oilseeds can be classified into two groups. The bulk (at least 70%) are storage proteins, that is, the proteins within the protein bodies. A smaller amount are nonstorage proteins, that is, the proteins that are part of the structural (membranes) and functional (enzymes, nucleoproteins, and cytoplasm)

TABLE 10.14

Typical Analysis of Soy Protein Products and Nutrient Content[a]

Analysis	Defatted soy flour	Soy protein concentrate	Isolated soy protein
Composition (%)[b]			
Moisture	6.5	4.9	4.7
Protein (N × 6.25)	53.0	67.6	91.8
Fat	1.0	0.3	—
Carbohydrates	31.0	18.8	—
Fiber, crude	2.5	2.6	0.1
Ash	6.0	4.8	3.4
Caloric content (per kg)	3670	3482	3672
Weight[b]	27.2	28.6	27.2
Composition[b]			
Moisture	1.8	1.4	1.3
Protein (N × 6.25)	14.4	19.3	25.0
Fat	0.3	0.1	—
Carbohydrates	8.4	5.4	—
Fiber, crude	0.7	0.7	—
Ash	1.6	1.4	0.9
Essential amino acids[b]			
Isoleucine	0.66	0.93	1.23
Leucine	1.09	1.51	1.93
Lysine	0.92	1.22	1.53
Methionine	0.20	0.27	0.28
Cystine (sparing action on methionine)	0.17	0.31	0.30
Phenylalanine	0.76	1.00	1.35
Threonine	0.60	0.81	0.93
Tryptophan	0.20	0.29	0.35
Valine	0.71	0.95	1.20
Vitamin content[b]			
Ascorbic acid	3.4 mg	1.6 mg	3.0 mg
Carotene	2.7 μg	2.9 μg	2.7 μg
Thiamin	92.5 μg	128.7 μg	22.8 μg
Riboflavin	72.9 μg	51.8 μg	40.0 μg
Niacin	574.0 μg	331.2 μg	163.2 μg
Folic Acid	71.3 μg	104.1 μg	2.5 μg
Mineral content[b]			
Calcium	46.2 mg	62.9 mg	81.6 mg
Phosphorus	199.1 mg	183.0 mg	209.4 mg
Potassium	734.4 mg	600.6 mg	54.4 mg
Sodium	3.0 mg	1.4 mg	299.2 mg
Iron	1.8 mg	2.9 mg	4.5 mg

[a]From Central Soya, Chicago, Illinois.
[b]Per 100 caloric portions (gm).

TABLE 10.15

Effect of Heat on Some Physical Properties of Soy Protein[a]

Property	80°C	100°C	120°C	140°C	160°C
Subunit structure	dissociation–unfolding—————————————degradation				
Solubility	decrease–precipitation—————————————increase in solubility				
Viscosity	increase–decrease——————————decrease				
Hydration	increase—————————————decrease				
Gelation (following heating)	regular————hard-fragile————————————soft elastic——solid				

[a]From Kinsella, 1979.

components of the cells (Martinez, 1979). The high molecular weight (above 100,000) storage proteins are relatively homogeneous and the low molecular weight nonstorage proteins are heterogeneous, high in cystine–cysteine, and capable of diverse interactions.

The undenatured soybean storage proteins hydrate and disperse readily in water. To separate them from the nonstorage proteins, Ca^{2+} at low ionic strength is added to prevent their hydration and dispersion. One of the most important characteristics of the storage proteins in terms of functionality is their propensity for aggregation–dissociation. Aggregation of 7 S and 11 S soy pro-

TABLE 10.16

Typical Uses of Soy Flours[a]

Product type[b]	Use
PDI 90–95	Bleaching agent in white bread
PDI 70–80	Bakery mix, doughnut mix, beverage, hydrolyzed vegetable protein, baby cereals
PDI 35–45	Pharmaceutical, baby cereals, meat processing, beverage, hydrolyzed vegetable protein, bakery, pet foods, animal milk replacers
PDI 8–20	Pharmaceutical, baby cereals, bakery, meat processing, pet foods, animal milk replacers

[a]From Horan, 1967.
[b]Protein dispersibility index.

TABLE 10.17

Functional Properties of Soy Protein Preparations[a]

Functional property	Mode of action	Food system	Preparation used[b]
Solubility	Protein solvation, pH dependent	Beverages	F,C,I,H
Water absorption and binding	Hydrogen-bonding of HOH, entrapment of HOH, no drip	Meats, sausages, breads, cakes	F,C
Viscosity	Thickening, HOH binding	Soups, gravies	F,C,I
Gelation	Protein matrix formation and setting	Meats, curds, cheese	C,I
Cohesion–adhesion	Protein acts as adhesive material	Meats, sausages, baked goods, pasta	F,C,I
Elasticity	Disulfide links in gels deformable	Meats, bakery	I
Emulsification	Formation and stabilization of fat emulsions	Sausages, bologna, soup, cakes	F,C,I
Fat adsorption	Binding of free fat	Meats, sausages, doughnuts	F,C,I
Flavor binding	Adsorption, entrapment, release	Simulated meats, bakery goods	C,I,H
Foaming	Forms stable films to entrap gas	Whipped toppings, chiffon desserts, angel cakes	I,W,H
Color control	Bleaching of lipoxygenase	Breads	F

[a]From Kinsella, 1979.
[b]F, soy flour; C, concentrate; I, isolate; H, hydrolysate; W, soy whey.

teins (discussed later) is affected by protein concentration and temperature; their dissociation is affected by ionic strength, pH, and presence of reducing agents that can cleave disulfide bonds. The storage 7 S and 11 S proteins are globulins, insoluble at pH 4.5. They can be precipitated, therefore, and recovered at this pH value, at which the other proteins are soluble and remain in the whey. As stated previously, although functional properties of soy products in the presence of Ca^{2+} are generally due to proteins, other soy components (i.e., cell wall constituents) can affect those properties. Additional contributing factors include particle size, heat treatment, and general storage and processing history.

Soy proteins consist of discrete groups of proteins covering a broad range of molecular weights. A typical ultracentrifuge pattern for water-extractable proteins shows four major fractions designated 2, 7, 11, and 15 S on the basis of their sedimentation rates (see Table 10.18). The storage proteins (7 S—con-

TABLE 10.18

Approximate Amounts and Components of Ultracentrifuge Fractions
of Water-Extractable Soybean Proteins[a]

Fraction	Percentage of total	Components	Molecular weight
2 S	8	Trypsin inhibitors	8,000, 21,500
		Cytochrome c	12,000
7 S	35	Hemagglutinin	110,000
		Lipoxygenase	102,000
		α-Amylase	61,700
		7 S Globulin	180,000–210,000
11 S	52	11 S Globulin	350,000
15 S	5	—	~600,000

[a]From Kinsella, 1979; Wolf, 1970.

glycinin and 11 S—glycinin) are the main components. According to research data, relative quantities of these proteins vary widely (Kinsella, 1979). This may be attributed to their association–dissociation properties.

Two types of foods have been selected to describe functional properties of soy proteins: meat-based products and oriental foods.

1. Meat-Based Products

Functions of soy proteins in meat-based products are listed below (Johnson, 1970):

1. Improves uniform emulsion formation and stabilization
2. Reduces cooking shrinkage and drip by entrapping and binding fats and water
3. Prevents fat separation
4. Enhances binding of meat particles without stickiness
5. Improves moisture-holding and mouth-feel
6. Gelation improves firmness, pliability, and texture
7. Facilitates cleaner, smoother slicing
8. May impart antioxidant effects
9. Improves nutritional value

Some general principles of the regulations that affect inclusion of processed plant protein products in meat and dairy food systems in the United States are described in Table 10.19 (Altschul, 1981; based on Food and Drug Administration, 1978). Table 10.20 provides examples of meat-type foods in which soy protein products can be used.

According to Rakosky (1973), isolated soy proteins have several important properties of great interest to the meat industry. Emulsification, fat-binding, water-absorption, adhesiveness, cohesiveness, film formation, thickening, suspending, stabilizing, limited foaming, and unique gelation are a few of the attributes of protein from soybeans. A 12–17% dispersion in water of certain isolates gels in 30 min at 68°C. This reflects typical sausage smokehouse conditions. The gel is irreversible below 115°C and has excellent binding properties.

TABLE 10.19

General Principles of Regulating Processed Plant Protein Products in Meats or Dairy Foods[a]

Category	Comment
Nutritional equivalence	
A. Protein	
1. Ingredient	
a. When not added as source of protein	Regulations do not apply
b. Short-term measures of protein quality	PER depending on intended use
c. Methionine content	Not specified
d. Methionine addition	Not permitted to levels below PER 2.5
e. Chemical definition	Protein content only
f. Limitations on amount added to animal product	<30%, PER 80% of casein >30%, PER 100% of casein
2. Final product	
a. Protein content	Specified depending on type of product
b. Protein quality	Not specified
B. Vitamins and minerals	
1. Ingredient	
a. Vitamin and mineral content	Not specified
2. Final product	
a. Vitamin and mineral content	Specified for a group of vitamins and minerals depending on product type
Labeling	
A. When nutritionally equivalent	
a. Name of ingredient	Includes source and product type (e.g., flour, concentrate, isolate)
b. Physical state	Optional (e.g., textured)
c. Special ingredients	Sodium and potassium levels intended
d. Product disclosure	to be full
B. When not nutritionally equivalent	Labeled "imitation"

[a]From Altschul, 1980.

TABLE 10.20

Meat-Type Foods with Soy Protein[a]

Manufactured product	Soy product permitted level	Comments
Cooked sausage[b]	Soy flour, 3.5% Soy protein concentrate 3.5% Isolated soy protein 2%	Individually or collectively with other approved extenders Where isolated soy protein is used, 2% is equivalent to 3.5% of others
Fresh sausage[b]	Same as cooked	Same as above
Chili con carne	Soy flour 8% Soy grits 8% Soy protein concentrate 8% Isolated soy protein 8%	Individually or collectively with other approved extenders
Spaghetti with meat balls[b] Salisbury steak[b]	Soy flour 12% Soy grits 12% Soy protein concentrate 12% Isolated soy protein 12%	Same as above
Imitation sausage, soups, stews, non specific loaves, scrapple, tamales, meat pies, pork with barbecue sauce, beef with barbecue sauce, patties	Sufficient for the purpose	Provided meat and moisture requirements are met where such requirements may exist

[a]From Rakosky, 1967.

[b]The use of soy protein products in both cooked and fresh sausage, meat balls, and salisbury steak must be shown on the label in a prominent manner, contiguous to the name of the product.

An "ideal" frankfurter should contain 12% protein, 58% water, 27% fat, and 3% mineral components (ash). The binding protein component (meat) is the most expensive component. To avoid failures in production, excess amounts of meat must be available to take into account variability in meat properties. A margin of safety can be provided by adding soy protein isolate. This can be accomplished by replacing 10% of the meat with 2% protein isolate. Isolated soy protein performs well because it has myosin-like functional properties. In addition to this "insurance" function, isolated soy protein makes it possible for the sausage manufacturer to use an emulsion temperature of 27°C. The frankfurters can be heated at elevated temperatures (up to 115°C). This allows much faster than normal processing of a stabilized product.

2. Oriental Foods

Production of soy beverages is predicated on (a) inactivation of lipoxygenases, (b) functionality with regard to dispersibility, viscosity, and mouth-feel, and elimination of flatulence effects (Rackis, 1979). Production of two beverage-base types from whole soybeans or full-fat soy flour are described in Figs. 10.32 and 10.33 (see Mustakas, 1974; Mustakas *et al.*, 1971). Although it has not been possible to eliminate the unique taste of soybeans in milk replacements, several fermented (acid milk and yoghurt) products of reasonable acceptance are available.

Fermented soy foods are listed in Table 10.21. Among them, shoyu (soy sauce) is the most important fermented food. It is used mainly as a flavoring agent (Ogawa and Fujita, 1980). Flowsheets for production of miso, shoyu, and tempeh (Hesseltine, 1961) are shown in Figs. 10.34, 10.35, and 10.36.

Tofu, as a bean curd, dry product, and fried, is consumed in larger amounts than any other soy food in the Orient. It is a highly hydrated, gelatinous product that resembles cottage cheese. It is made by curdling (with Ca^{2+} or Mg^{2+}) a heated extract of whole soybeans (soy milk). Yuba is the film of protein and oil that forms on the surface of soybean milk that is heated to near boiling. Fresh tofu is a bland, soft, smooth gel (Saio, 1979). Commercial tofus (Table 10.22) are classified as "momen," soft or "kinu," and packed. Two derivatives of tofu are called "aburage" and "kori."

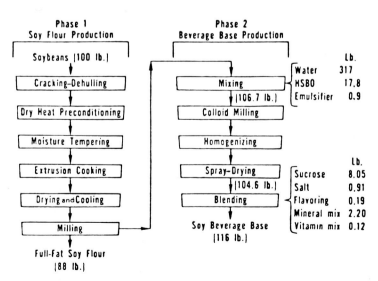

Fig. 10.32. Preparation of full-fat soy beverage base. HSBO, hydrogenated soybean oil. (From Mustakas *et al.*, 1971.)

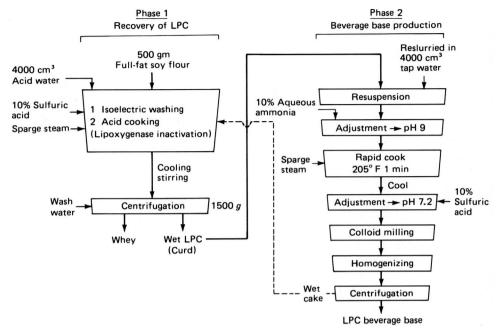

Fig. 10.33. Process of manufacturing a beverage base from soy lipid concentrate (LPC). (From Mustakas, 1974.)

TABLE 10.21

Fermented Soy Foods[a]

Product	Organisms used	Substrate	Form
Miso	*Aspergillus oryzae;* *Saccharomyces rouxii*	Cereals, soybeans	Paste
Natto	*Bacillus subtilis*	Soybeans	Solid
Shoyu	*Aspergillus oryzae;* *Lactobacillus; Hansenula;* *Saccharomyces*	Soybeans, wheat	Liquid
Sufu	*Actinomucor elegans;* *Mucor* sp.	Soybeans	Solid
Tempeh	*Rhizopus oligosporus*	Soybeans	Solid

[a]From Hesseltine, 1965.

In tofu making, soybean milk is heated for about 3 min after boiling. Limited heating results in a paste-like gel; excessive heating reduces gel cohesion. These changes involve sulfhydryl–disulfide-mediated aggregation–polymerization of soybean proteins. In making kinu and packed tofu, soybean milk is coagulated without elimination of whey. Increasing the solid content of soybean milk from 11 to 14% increases hardness of kinu tofu. Momen and kori tofu are prepared by heating dilute soy milk solutions and coagulation with elimination of whey. Raising the coagulation temperature increases the hardness of tofu. Accelerating the stirring after the addition of coagulant to soy milk decreases the volume and softness of tofu. Traditionally tofu was coagulated with a concentrate of seawater; today calcium salts are used. The concentration and nature of the salts

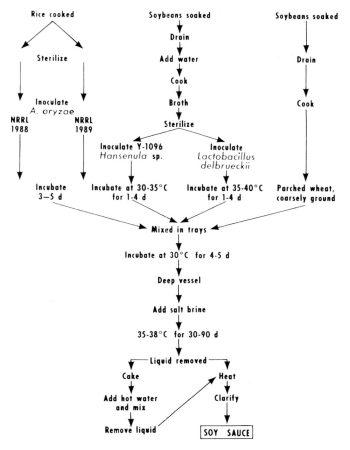

Fig. 10.34. Flowsheet for production of soy sauce (shoyu). (From Hesseltine, 1961.)

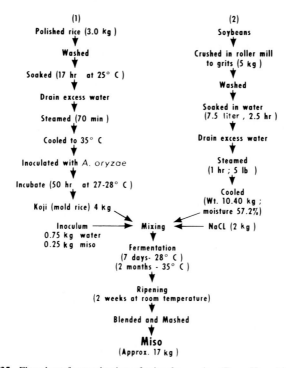

Fig. 10.35. Flowsheet for production of miso from grits. (From Hesseltine, 1961.)

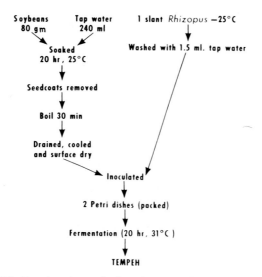

Fig. 10.36. Flowsheet for production of tempeh. (From Hesseltine, 1961.)

TABLE 10.22

Tofu and Its Derivatives[a]

	Fresh tofu				Tofu derivatives		
	Momen	Soft	Kinu	Packed	Aburage	Kori tofu	
Water added (no. of times)	10	7	5	5	10	15	
Coagulant	CaSO$_4$	CaSO$_4$	CaSO$_4$ and/or GDL	GDL and/or CaCl$_2$, CaSO$_4$	CaSO$_4$	CaCl$_2$	
Processing	Thorough elimination of whey	Elimination of whey	Coagulation of whole soybean milk without elimination of whey	Cooked soybean milk is packed immediately after addition of coagulant and reheated Whole soybean milk is coagulated	Soybean milk is moderately heated and coagulated with thorough elimination of whey Fresh tofu is fried in oil	Soybean milk is coagulated with continuous stirring and elimination of whey Fresh tofu if frozen at −20°C overnight, kept at 0–5°C for 2–3 weeks, thawed, and dried	
Chemical composition							
Moisture (%)	86.8	88.9	89.4	90.0	44.0	8.1	
Crude protein (%)	6.8	5.7	5.0	4.5	18.6	50.2	
Crude fat (%)	5.0	3.8	3.3	3.2	33.1	33.4	
Ash (%)	0.6	0.6	0.6	0.6	1.4	2.8	
Ca (mg %)	120.0	90.0	90.0	35.0	300.0	590.0	
Texture	Hard, rough	Intermediate between momen and kinu	Soft, smooth	Soft, smooth, and fragile	Before frying fresh tofu for aburage is harder than tofu for momen Rough, coarse aburage is texturized and chewy	Before freeze-drying, fresh tofu for kori tofu is harder than tofu for aburage Coarse, lumpy kori tofu is spongy, elastic, and chewy	

[a]From Saio, 1979.

(CaCl$_2$ or CaSO$_4$) are critical in determining the volume, hardness, and optical properties of tofu. Glucono-δ-lactone (GDL) causes soy milk to coagulate only after being reheated. This is useful in modern processing but introduces yet another variable in fine structure.

Disulfide bonds in soy proteins play an important role in GDL-tofu and calcium tofu but not in acid-precipitated tofu. The difference in solubility of GDL-tofu and calcium-tofu suggests a role for a calcium-bridge in tofu gel. Properties of the tofu gel are affected considerably by the ratio of 7 S:11 S proteins in soy milk. Calcium-tofu made from 11 S proteins is harder, more cohesive, and more springy than tofu made from 7 S proteins. The latter imparts slight adhesiveness properties (Saio, 1979).

In heat-denatured products, subunits of 11 S proteins form a soluble aggregate that is rapidly converted to an insoluble aggregate as a result of sulfhydryl–disulfide interchange. The aggregate increases as the amount of free sulfhydryl groups increases. The 11 S proteins precipitate faster and form larger aggregates at low calcium concentrations than do 7 S proteins. Increase in sulfhydryl groups increases hardness, cohesiveness, and springiness of 11 S tofu but not of 7 S tofu (Saio, 1979). Phytic acid is extracted into soybean milk and coprecipitates with proteins in tofu during coagulation. Phytic acid reacts both with Ca^{2+} and proteins and produces a colloidal precipitate that can hold more water in the gel. Relationships between fine structure and texture of tofu are summarized in Fig. 10.37. Network density is correlated with hardness; it varies with the coagulant, phytic acid, and incorporation of whey. Increase of 11 S protein aggregates increases tofu hardness and is related to sulfhydryl–disulfide interchange.

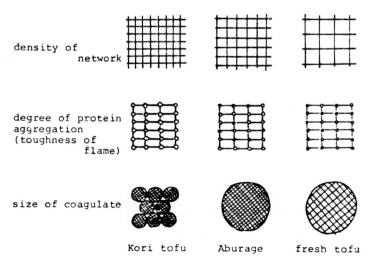

Fig. 10.37. Relationship between fine structure and textural properties of tofu. (From Saio, 1979.)

REFERENCES

Altschul, A. M. (1981). Plant proteins, the North American experience. *In* "Utilization of Protein Resources" (D. W. Stanley, E. D. Murray, and D. H. Lees, eds.), Chapter 18, pp. 349–361. Food and Nutrition Press, Inc., Westport, Connecticut.

Anonymous (1970a). *J. Inst. Brew.* **76**, 110.

Anonymous (1970b). *J. Inst. Brew.* **76**, 212.

Anonymous (1971). *J. Inst. Brew.* **77**, 231.

Baker, J. C., and Mize, M. D. (1946). Gas occlusion during dough mixing. *Cereal Chem.* **23**, 39–51.

Bechtel, D. B., Pomeranz, Y., and DeFrancisco, A. (1978). Breadmaking studied by light and transmission electron microscopy. *Cereal Chem.* **55**, 392–401.

Bell, B. M., Daniels, D. G. H., and Fisher, N. (1977). Physical aspects of the improvement of dough by fat. *Food Chem.* **2**, 57–70.

Bloksma, A. H. (1971). Rheology and chemistry in dough. *In* "Wheat Chemistry and Technology" (Y. Pomeranz, ed.), pp. 523–584. Am. Assoc. Cereal Chem., St. Paul, Minnesota.

Bookwalter, G. N. (1978). Soy protein utilization in food systems. *In* "Nutritional Improvement of Food and Feed Proteins" (M. Friedman, ed.), Chapter 36, pp. 749–763. Plenum, New York.

Carter, H. E., Hendry, R. A., and Stanacew, N. Z. (1961). Wheat flour lipids. II. Structure of the mono- and digalactosyl glyceride lipids. *J. Lipid. Res.* **2**, 223–228.

Chiu, C.-M., Pomeranz, Y., Shogren, M. D., and Finney, K. F. (1968). Lipid binding in wheat flours varying in bread making potential. *Food Technol.* **22**, 1157–1162.

Chung, K. H., and Pomeranz, Y. (1978). Acid soluble proteins of wheat flours. I. Effect of delipidation on protein extraction. *Cereal Chem.* **55**, 230–243.

Chung, K. H., and Pomeranz, Y. (1979). Acid soluble proteins of wheat flours. II. Binding to hydrophobic gels. *Cereal Chem.* **56**, 196–201.

Chung, O. K., and Pomeranz, Y. (1977). Wheat flour lipids and surfactants: A three way contribution to breadmaking. *Baker's Dig.* **51**(5), 32–44, 153.

Chung, O. K., Pomeranz, Y., Goforth, D. R., Shogren, M. D., and Finney, K. F. (1976). Improved sucrose esters in breadmaking. *Cereal Chem.* **53**, 615–626.

Chung, O. K., Pomeranz, Y., and Finney, K. F. (1978). Wheat flour lipids in breadmaking. *Cereal Chem.* **55**, 598–618.

Circle, S. J., and Smith, A. K. (1978). Processing soy flours, protein concentrates and protein isolates. *In* "Soybeans: Chemistry and Technology" (A. K. Smith and S. J. Circle, eds.), Rev. ed., Vol. 1, Chapter 9, pp. 294–338. Avi Publ., Co. Inc., Westport, Connecticut.

Cowan, J. C., and Wolf, W. J. (1974). Soybeans. *In* "Encyclopedia of Food Technology" (A. H. Johnson and M. S. Peterson, eds.), pp. 818–828. Avi Publ. Co., Westport, Connecticut.

Daftary, R. D., Pomeranz, Y., Shogren, M. D., and Finney, K. F. (1968). Functional bread-making properties of lipids. II. The role of flour lipid fractions in bread-making. *Food Technol.* **22**, 327–329.

Daniels, N. W. R. (1981). Utilization of oilseed proteins. *In* "Utilization of Protein Resources" (D. W. Stanley, E. D. Murray, and D. H. Lees, eds.), Sect. IV, Chapter 9, pp. 188–207. Food and Nutrition Press, Inc., Westport, Connecticut.

deClerck, J. (1957). "A Textbook of Brewing" (Engl. transl.), Vol. I. Chapman & Hall, London.

Derby, R. I., Miller, B. S., Miller, B. F., and Trimbo, H. B. (1975). Visual observation of wheat-starch gelatinization in limited water systems. *Cereal Chem.* **52**, 702–713.

Dimler, R. J. (1965). Exploring the structure of proteins in wheat gluten. *Baker's Dig.* **39**(5), 35–38, 40–42.

Finney, K. F. (1943). Fractionating and reconstituting techniques as tools in wheat flour research. *Cereal Chem.* **20**, 381–396.

Finney, K. F. (1978). Contribution of individual chemical constituents to the functional (breadmaking) properties of wheat. *In* "Cereals '78: Better Nutrition for the World's Millions" (Y. Pomeranz, ed.), Chapter 10, pp. 139–158. Am. Assoc. Cereal Chem., St. Paul, Minnesota.

Finney, K. F. (1979). Wheat proteins: What they do. *Proc. Wheat Protein Conf. 1978,* ARM-NC-9, pp. 50–61.

Finney, K. F., and Barmore, M. A. (1948). Loaf volume and protein content of hard winter and spring wheats. *Cereal Chem.* **25,** 291–312.

Food and Drug Administration (1978). Common or usual names for vegetable protein products or substitutes for meat, seafood, poultry, eggs or cheeses which contain vegetable protein products as sources of protein. *Fed. Regist.* **43**(136), 20473–20491.

Grosskreutz, J. C. (1961). A lipoprotein model of wheat gluten structure. *Cereal Chem.* **38,** 336–349.

Hayward, J. W., and Diser, G. M. (1961). Soy protein as soy flour and grits for improving dietary standards in many parts of the world. *Soybean Dig.* **21,** 14–26.

Hess, K., and Mahl, H. (1954). Electron microscopic study of wheat flour and wheat flour products. *Mikroskopie* **9,** 81–86.

Hesseltine, C. W. (1961). Research at Northern Regional Research Laboratory on fermented foods. *Proc. Conf. Soybean Prod. Protein Hum. Foods, 1961* pp. 67–74.

Hesseltine, C. W. (1965). A millenium of fungi, food and fermentation. *Mycologia* **57,** 149–197.

Hind, H. L. (1950). "Brewing Science and Practice." Chapman & Hall, London.

Holme, J. (1966). A review of wheat flour proteins and their functional properties. *Baker's Dig.* **40**(5), 38–42.

Horan, F. E. (1967). Defatted and full-fat soy flours by conventional processes. *Proc. Int. Conf. Soybean Protein Foods, 1966* ARS-USDA ARS 71–35, pp. 129–141.

Horan, F. E. (1974). Soy protein products and their production. *J. Am. Oil Chem. Soc.* **51,** 67A–73A.

Hoseney, R. C., Finney, K. F., Shogren, M. D., and Pomeranz, Y. (1969a). Functional (breadmaking) and biochemical properties of wheat flour components. II. Role of water-solubles. *Cereal Chem.* **46,** 117–125.

Hoseney, R. C., Finney, K. F., Shogren, M. D., and Pomeranz, Y. (1969b). Functional (breadmaking) and biochemical properties of wheat flour components. III. Characterization of gluten protein fractions obtained by ultracentrifugation. *Cereal Chem.* **46,** 126–135.

Hoseney, R. C., Finney, K. F., and Pomeranz, Y. (1970). Functional (breadmaking) and biochemical properties of wheat flour components. VI. Gliadin-lipid-glutenin interaction in wheat gluten. *Cereal Chem.* **47,** 135–140.

Hough, J. S., Briggs, D. E., and Stevens, R. (1971). "Malting and Brewing Science." Chapman & Hall, London.

Hudson, J. F. (1973). Advances in brewing technology. *Food Manuf.,* April 1973, pp. 23, 24, 27, 28.

Johnson, D. W. (1970). Functional properties of oilseed proteins. *J. Am. Oil Chem. Soc.* **47,** 402–407.

Kent, N. L. (1966). "Technology of Cereals with Special Reference to Wheat." Pergamon, Oxford.

Kieninger, H. (1983). The influence of raw materials and yeast on various beer types. *Brew. Dig.* **58**(10), 44–49.

Kinsella, J. E. (1979). Functional properties of soy proteins. *J. Am. Oil Chem. Soc.* **56,** 242–258.

Krull, L. H., and Wall, J. S. (1969). Relationship of amino acid composition and wheat protein properties. *Bakers Dig.* **43**(4), 30–39.

Leavell, G. (1942). Brewers' and distillers' by-products and yeast in livestock feeding. *U.S. Dep. Agric., Bur. Anim. Ind., Bull.* **58.**

Liener, I. E. (1981). Factors affecting the nutritional quality of soya products. *J. Am. Oil Chem. Soc.* **58,** 406–415.

Luers, H. (1950). "Die Wissenschaftlichen Grundlagen von Malzerei und Brauerei." Huber, Nuremberg, West Germany.

MacLeod, A. M. (1969). The utilization of cereal seed reserves. *Sci. Prog. (Oxford)* **57**, 99.

MacMurray, T. A., and Morrison, W. R. (1970). Composition of wheat flour lipids. *J. Sci. Food Agric.* **21**, 520–528.

Martinez, W. (1979). The importance of functionality of vegetable protein in foods. *In* "Soy Protein and Human Nutrition" (H. L. Wilcke, D. T. Hopkins, and D. H. Waggle, eds.), pp. 53–77. Academic Press, New York.

Mustakas, G. C. (1974). A new soy lipid-protein concentrate for beverages. *Cereal Sci. Today* **19**, 62–64, 69–73.

Mustakas, G. C., Albrecht, W. J., Bookwalter, G. N., Sohns, V. E., and Griffin, E. L., Jr. (1971). New process for low-cost high protein beverage base. *Food Technol.* **25**, 534–538, 540.

Ogawa, G., and Fujita, A. (1980). Recent progress in soy sauce production in Japan. *In* "Cereals for Food and Beverages" (G. E. Inglett and L. Munck, eds.), pp. 381–394. Academic Press, New York.

Olcott, H. S., and Mecham, D. K. (1947). Characterization of wheat gluten. I. Protein-lipid complex formation during doughing of flours. Lipoprotein nature of the glutenin fraction. *Cereal Chem.* **24**, 407–414.

Pollock, J. R. A. (1971). Trends in brewing technology. *Brauwissenschaft* **24**, 158–165.

Pollock, J. R. A. (1973). Recent developments in fermentation. *Brauwelt* **113**, 743–747.

Pomeranz, Y. (1964). Lactase (beta-D-galactosidase). I. Occurrence and properties. II. Possibilities in the food industries. *Food Technol.* **18**, 88–93, 96–103.

Pomeranz, Y. (1971). Composition and functionality of wheat-flour components. *Monogr. Ser.— Am. Assoc. Cereal Chem.* **3**(rev.), 585–674. St. Paul, Minnesota.

Pomeranz, Y. (1972). From barley to beer—as seen under the microscope. *Proc. Annu. Meet., Am. Soc. Brew. Chem.* pp. 24–29.

Pomeranz, Y. (1973a). From wheat to bread. A biochemical study. *Am. Sci.* **61**(6), 683–691.

Pomeranz, Y. (1973b). Industrial uses of barley. *In* "Industrial Uses of Cereals" (Y. Pomeranz, ed.), Chapter 11, pp. 371–392. Am. Assoc. Cereal Chem., St. Paul Minnesota.

Pomeranz, Y. (1975a). Nutritional properties of beer—a little known aspect of the beverage of moderation. *Brew. Dig.* **50**(10), 38, 40, 42.

Pomeranz, Y. (1975b). From barley to beer—a biochemical study. *Bull. Assoc. Oper. Millers,* pp. 3503–3512.

Pomeranz, Y. (1980a). Molecular approach to breadmaking—an update and new perspectives. *Baker's Dig.* **54**(1), 20–27, (2); 12, 14, 16, 18, 20, 24, 25.

Pomeranz, Y. (1980b). Wheat flour components in breadmaking. *In* "Cereals for Food and Beverages" (G. E. Inglett and L. Munck, eds.), pp. 201–232. Academic Press, New York.

Pomeranz, Y. (1980c). What? How much? Where? What function? *Cereal Foods World* **25**(10), 656–662.

Pomeranz, Y., and Chung, O. K. (1978). Interaction of lipids with proteins and carbohydrates in breadmaking. *J. Am. Oil Chem. Soc.* **55**(2), 285–289.

Pomeranz, Y., and Finney, K. F. (1972). Protein-enriched baked products and method of making same. U.S. Patent 3,679,433.

Pomeranz, Y., and Wehrli, H. P. (1969). Synthetic glycosylglycerides in breadmaking. *Food Technol.* **23**, 109–111.

Pomeranz, Y., and Wehrli, H. P. (1973). Synthesis of glycosyl glycerides. U.S. Patent 3,729,461.

Pomeranz, Y., Shogren, M. D., and Finney, K. F. (1969). Improving breadmaking properties with glycolipids. I. Improving soy products with sucroesters. II. Improving various protein-enriched products. *Cereal Chem.* **46**, 503–511, 512–518.

Pomeranz, Y., Shogren, M. D., and Finney, K. F. (1977). Flour from germinated soybeans in high-protein bread. *J. Food Sci.* **42**, 824–842.

Rackis, J. J. (1979). Soy protein foods. *In* "Tropical Foods: Chemistry and Nutrition" (G. E. Inglett and G. Charalambous, eds.), Vol. 2, pp. 485–510. Academic Press, New York.

Rackis, J. J., Sessa, D. J., and Honig, D. H. (1979). Flavor problems of vegetable food proteins. *J. Am. Oil Chem. Soc.* **56**, 262–270.

Rakosky, J. (1967). Soy proteins—their preparation and uses in comminuted meat products. "Meat Hygiene," Vol. 8, Part 6, pp. 1–13. Consumer Marketing Service-USDA, Washington, D.C.

Rakosky, J. (1973). "Soy Products as Functional Ingredients in Food and Industrial Applications." Central Soya, Chemurgy Div., Chicago, Illinois.

Saio, K. (1979). Tofu—relationship between texture and fine structure. *Cereal Foods World* **24**, 342–345, 350–354.

Schuster, G., and Adams, W. F. (1984). Emulsifiers as additives in bread and fine baked goods. *Adv. Cereal Sci. Technol.* **6**, 139–287.

Seiling, S. (1969). Equipment demands of changing production requirements. *Baker's Dig.* **43**(5), 54–56, 58–59.

Shogren, M. D., and Finney, K. F. (1974). A mixture of ascorbic acid and potassium bromate to optimize quickly loaf volume. *Cereal Foods World* **19**, 397.

Shogren, M. D., Pomeranz, Y., and Finney, K. F. (1979). Low lactose bread. *Cereal Chem.* **56**, 465–468.

Smith, A. K., and Circle, S. J., eds. (1978). "Soybeans: Chemistry and Technology," rev. ed., Vol. 1. Avi Publ. Co., Inc. Westport, Connecticut.

U.S. Department of Agriculture (1968). *U.S. Dep. Agric., Agric. Handb.* **338**.

Wall, J. S., and Huebner, F. R. (1981). Adhesion and cohesion. *In* "Protein Functionality in Foods," ACS Symposium Series 147, Chapter 6, pp. 111–130. Am. Chem. Soc., Washington, D.C.

Wehrli, H. P. (1969). The synthesis of glycolipids and their role in breadmaking. Ph.D. Thesis, Kansas State University, Manhattan.

Wehrli, H. P., and Pomeranz, Y. (1969a). Chemical bonds in dough. *Baker's Dig.* **43**(6), 22–26.

Wehrli, H. P., and Pomeranz, Y. (1969b). Synthesis of galactosylglycerides and related lipids. *Chem. Phys. Lipids* **3**, 357–370.

Wehrli, H. P., and Pomeranz, Y. (1970a). A note on the interaction between glycolipids and wheat flour macromolecules. *Cereal Chem.* **47**, 160–166.

Wehrli, H. P., and Pomeranz, Y. (1970b). A note on autoradiography of tritium-labeled galactolipids in dough and bread. *Cereal Chem.* **47**, 221–224.

Westermann, D. H., and Huige, N. J. (1979). Beer brewing. *In* "Microbial Technology" (H. J. Peppler and D. Perlman, eds.), 2nd ed., Vol. 2, Chapter 1, pp. 1–37. Academic Press, New York.

Wiebe, G. A. (1968). Barley: Origin, botany, culture, winterhardiness, genetics, utilization, pests. *U.S. Dep. Agric. Agric., Agric. Handb.* **338**.

Wolf, W. J. (1967). Trypsin inhibitors, hemagglutinin, saponins, and isoflavones of soybeans. *Proc. Int. Conf. Soybean Protein Foods 1966* ARS-USDA-71-35, pp. 112–128.

Wolf, W. J. (1970). Soybean proteins: Their functional, chemical, and physical properties. *J. Agric. Food Chem.* **18**, 969–976.

Wolf, W. J. (1973). Processing soybeans into protein products. *Bull. Assoc. Oper. Millers,* pp. 3403–3408.

Wolf, W. J. (1974). Soybean proteins: Their production, properties and food uses. A selected bibliography. *J. Am. Oil Chem. Soc.* **51**, 63A–66A.

Wolf, W. J. (1976). Chemistry and technology of soybeans. *Adv. Cereal Sci. Technol.* **1**, 325–377.

Wolf, W. J. (1981). Progress and future needs for research in soya protein utilization and nutrition. *J. Am. Oil Chem. Soc.* **58**, 467–473.

Yasunaga, T., Bushuk, W., and Irvine, G. N. (1968). Gelatinization of starch during breadbaking. *Cereal Chem.* **45**, 269–279.

11

Foods of the Future

Modern plant breeders rely on sexual or genetic recombination for the reassortment of genes within species that interbreed freely. Plant breeders' decide the direction and intensity of the selection to be applied. Genetic engineering is the intentional genetic manipulation of species to produce special forms for improving crop productivity. In the more exotic forms of genetic engineering, the major emphasis is on transfer of genes between species that normally do not interbreed. Modern research findings suggest new ways to expand genetic diversity and to select desired germ plasm. Wide sexual crosses of an interspecific, intergeneric, or interfamilial nature may be achieved by techniques of overcoming incompatibility. Other techniques, such as producing haploid plants from anther and pollen culture, organelle or DNA feeding, and somatic cell hybridization, may bring about an unprecedented expansion of genetic diversity in plants. Techniques involving culture of single cells permit rapid, large-scale selection.

Although the great expectations of these approaches have not yet been realized, the initial findings are still quite recent, and the goal is worthy of much additional effort in plant breeding. Plant cell culture offers great potential for use in genetic engineering. According to Morris (1981), genetic engineering (in particular, recombinant DNA) will be the next frontier in food manufacturing. Applications of this technology to food processing will be designed to reduce costs and improve yields in two major areas: first, the production of enzymes and miscellaneous food ingredients either in current fermentations or by shifts to fermentations to produce substances that presently are extracted from natural sources or synthesized chemically; and second, food processes in which enzymes are used, such as fermentation systems and immobilized enzyme systems. Potential specific applications include production of the following items:

1. Amino acids for use in nutritional enrichment or as flavoring agents (e.g., monosodium glutamate)
2. Vitamins for use in nutritional enrichment
3. Food preservatives from petrochemicals
4. Enzymes (e.g., papain, glucose isomerase)
5. Short peptides (e.g., aspartame as artificial sweeteners)
6. Nucleotides as flavoring agents
7. Sugars and alcohols from cellulose and various waste products
8. Polysaccharides for new gums
9. Alternate hydrocolloids to replace scarce and expensive natural gums of widely fluctuating composition and functionality
10. Colorants (e.g., carotene from algal cultures)
11. Organic acids through fermentation

Dally *et al.* (1981) listed among the potential new products and processes made possible by recombinant DNA two new proteins (monellin and thaumatin) as sweeteners and high lysine protein for SCP. Dally and colleagues visualize several developments of more efficient processes, including (a) the use of cheaper substrates (utilization of cellulose biomass for fermentations by introducing cellulase genes); (b) the use of less energy (in microorganisms and plants); and (c) the introduction of more efficient enzymes in processing (higher thermostability and reaction rates, and reactivity not subject to feedback inhibition).

I. Synthesis versus Modification

Photosynthesis is still the most effective and cheapest method to produce food. Consequently, synthesis is not likely to play an important role in the production of basic food raw materials (Pomeranz, 1980). Chemical modification of basic materials for structural and nutritional reasons is likely to become more widespread, however. We can expect that flexibility in choice of raw materials from which to fabricate foods will increase and that textured engineered foods will be on the rise. Production of SCP will continue to occupy the attention of food chemists and technologists.

A new food based on *Fusarium* fungus recently was described as delicious.

Moreover, it has the advantages of year-round consistency of production and quality, with little or no waste (Anonymous, 1981). The fungus is produced on glucose and ammonia. It is a multicellular protein with structural properties; it has an NPU of its protein of 70–75%, and when supplemented with 0.2% methionine of 100% (equal to egg proteins). It is low in ribonucleic acid, is simple to produce, and has high protein efficiency (it takes 2–3 kg carbohydrate to produce 1 kg mucoprotein). Still, its economic feasibility is questionable.

During the past decade, much attention has been devoted to four classes of nonconventional foods: (a) microorganisms—including algae, bacteria, yeasts, molds, and higher fungi; (b) plant products—including textured plant protein isolates and concentrates and plant cells grown in tissue culture; (c) animal products—including fish and meat protein concentrates and animal cells from tissue culture; and (d) completely synthetic foods prepared by chemical synthetic methods. A few of these are at the drawing-board stage, some are theoretically feasible, some are technologically possible, and some offer real promise. While the production of these foods will continue to be of great interest to food scientists and technologists, the extent to which they actually reach the market will be determined by nutritional, economic, social, legal, and even political considerations.

Before discussing novel foods, a few words about complete synthetic foods are necessary. The term "synthetic" encompasses foods containing all the nutrients that are known, or suspected, to be necessary for human growth, maintenance, and repair, which are created from well-defined chemical components (Beigler, 1976). The development of a single complete food presumes a knowledge of all nutrients needed by man, and some doubt remains about whether such complete knowledge is available. Still, sufficient information on individual nutrients exists to design formulations that are adequate to sustain human life when eaten over extended periods. Such diets have been fed as the sole source of nutrition to humans for more than 4 years with no observed deleterious effects and with the apparent ability to support life. Such an elemental diet is described in Table 11.1.

Synthetic diets may be developed that are economically competitive with balanced diets of conventional foods. Commercial elemental formulations are produced in the form of dry powder in sealed packets, the contents of which may be reconstituted into a complete liquid synthetic food. Semisolid preparations for use in conventional dietetics are based on starch and alignate thickeners and fillers. Oligosaccharides and starches provide a pudding-like base. With gelatin or various polymers as models, one can construct a gel-like elemental food in which the required amino acids are attached to a polymer "backbone." The simplest method of making synthetic foods is to mix all the powders and press them into a tablet, wafer, or cookie to yield a webbed-type material, which serves as base for extrusion or spinning into solid forms.

TABLE 11.1

Typical Composition of a Synthetic or Elemental Diet[a,b]

Amino acids (gm)			
L-Lysine·HCl	3.50	Sodium L-aspartate	6.40
L-Leucine	3.80	L-Threonine	2.40
L-Isoleucine	2.40	L-Proline	10.30
L-Valine	2.60	Glycine	1.60
L-Phenylalanine	1.70	L-Serine	5.30
L-Arginine·HCl	2.50	L-Tyrosine ethyl ester·HCl	6.80
L-Histidine·HCl·H₂O	1.50	L-Tryptophan	0.70
L-Methionine	1.70	L-Glutamine	9.00
L-Alanine	2.50	L-Cysteine ethyl ester·HCl	0.90

Vitamins (mg)			
Thiamin·HCl	1.00	Ascorbic acid	62.50
Riboflavin	1.50	Vitamin B₁₂	0.00167
Pyridoxine·HCl	1.67	Choline bitartrate	231.25
Niacinamide	10.00	Vitamin A acetate	3.60
Inositol	0.83	Vitamin D	0.05
D-Calcium pantothenate	8.33	α-Tocopherol acetate	57.00
D-Biotin	0.83	Menadione	4.00
Folic acid	1.67		

Salts (mg)			
Potassium iodide	0.20	Magnesium oxide	300
Manganous acetate·4H₂O	18.30	Sodium chloride	4700
Zinc benzoate	2.80	Ferrous gluconate	800
Cupric acetate·H₂O	2.50	Calcium chloride·2H₂O	2400
Sodium glycerophosphate·5½ H₂O	5200	Potassium hydroxide	4000

Carbohydrates (gm)		Fat (gm)	
Glucose	555.0	Ethyl linoleate	2.0

[a]From Beigler, 1976.
[b]The daily ration for males was diluted with water to 2700 ml containing 1 kcal/ml.

Most elemental diets have "pungent" or "musty" odors with excessively sweet and unpleasant background aftertastes characteristic of amino acid mixtures. Some improvements have been made with masking agents and through pH control (Beigler, 1976). Another problem is stability against microorganisms in liquid foods, oxidative rancidity, and excessive browning during prolonged storage. Complete synthetic foods provide the dietician with a wide choice of applications for patients with inborn errors of metabolism (such as phenylketonuria), patients with nonfunctional pancreases who should avoid the need for digestion, or patients who suffer from gluten-sensitive enteropathies.

II. The Nutritional Dimension

A. Introduction

Nutrition is the link between diet and health. A few decades ago we were concerned about minimum daily requirements to combat dietary diseases; today we are concerned about establishing maximum levels of nutrients to prevent diseases linked with dietary affluence or overabundance (Chou, 1979b; Dubos, 1979).

In January 1977, a Senate Select Committee on Nutrition and Human Needs called on Americans to use more fruits, vegetables, whole grains, poultry, fish, skim milk, and vegetable oils in their diets to reduce their consumption of whole milk, meat, eggs, butterfat, and foods high in sugar, salt, and fat. Specifically, the committee asked consumers to:

1. Increase carbohydrate consumption to account for 55–60% of energy (caloric) intake
2. Reduce overall fat consumption from approximately 40 to 30% of energy intake
3. Reduce saturated fat consumption to account for about 10% of total energy intake, and balance that reduction by consuming polyunsaturated and monounsaturated fats, which should each account for about 10% of energy intake
4. Reduce cholesterol consumption to about 300 mg a day
5. Reduce sugar consumption by about 40% to account for about 15% of total energy intake
6. Reduce salt consumption by about 50–85% to approximately 3 gm a day

In December 1977, a second edition of the report on dietary goals appeared. Instead of suggesting that consumers eat less meat, the new report recommended that people reduce their intake of animal fat "and choose meats, poultry and fish which will reduce saturated fat intake." The new edition did not contain advice about reducing the whole milk and egg consumption by young children and, for adults, raised the suggested limit for salt to 5 gm. Changes regarding dietary allowances continue to be made (Anonymous, 1979). The United States Department of Agriculture and the Department of Health, Education, and Welfare joint report entitled "Dietary Guidelines for Americans" largely follows the dietary recommendations published by the Senate Select Committee on Nutrition and Human Needs (Anonymous, 1980).

B. Restoration, Fortification, and Enrichment

In the United States, cereal enrichment is a major factor contributing to the virtual elimination of vitamin deficiency diseases that killed thousands of Americans annually as recently as 50 years ago. If future diets are to emphasize less meat and more vegetables and cereals, it seems essential that cereals continue to be enriched with the nutrients that are present in meat (Chou, 1979b). Mass use of synthesized nutrients, as well as improved technology, has driven food costs down dramatically. A good example is the shrinking cost of enrichment. For example the cost to enrich 100 lb of flour was $0.17 in 1941, $0.02 in 1967, and $0.0004 in 1978. It may be estimated that 70–85% of all flour produced in United States mills is enriched. The interest in foods fortified with proteins, amino acids, and vitamins may be replaced by concern over the content of fat, sugar, salt, and minerals such as calcium, zinc, and iron (Chou, 1979a). In developing countries where there is little choice between a meat or vegetarian diet, cereal enrichment and fortification offers cheap effective solutions to health and nutrition problems. These answers may have more positive effect on human health than comparable investments in medical care.

Development of fortification procedures is a major consideration in planning for the future. With increased ability to engineer foods and better understanding of the role of trace elements to destroy or enhance the effects of other elements beneficial or harmful to human health, food processors will confront decisions about the use as fortifiers just as in the past they faced decisions about the addition of vitamins and minerals. Responsible policies will favor fortifying certain staples and basic grain products and restoring nutrients in conventional and formulated foods (Chou, 1979a).

Flexibility in ingredient composition and the trend toward the production of complete foods will be made possible through formulations selected on the basis of computer-stored compositional tables. Low cost diets to meet any nutritional requirement will be engineered and tailored to combine enjoyment with regular and special nutritional needs, including metabolic constraints (Altschul and Hornstein, 1972).

III. Modern Foods and Food Additives

The constantly changing world affects foods and food manufacture. For example, raw materials are no longer produced in proximity to where foods are consumed (Kent-Jones, 1971). Storage, transportation, packaging, shelf life, and sanitary practices are important elements in the food marketing chain, which make preservatives and technologically useful additives necessary. The use of food additives is essential if the public is to receive attractive, tasty, nutritious foods. We use preservatives and antioxidants to check rancidity, humectants to offset effects of variations in humidity and to prevent desiccation, sequestrants to destroy deleterious effects of metals, glazing and anticaking agents, and firming and crisping agents. We consume instant coffee and tea, cake mixtures that require only the addition of liquids and baking, prepared dinners, deep-frozen and dehydrated foods, and dehydrated ready-packed meals—and expect them to be of high quality (that is, attractive, tasty, and convenient). None of these foods can be prepared by conventional methods or without food additives.

IV. Engineered Foods

A. Introduction

Engineering complete foods involves "building-in" six functional properties: nutrition, palatability, safety, shelf life, economy, and convenience. The objectives of engineered foods vary from imparting novel, pleasing organoleptic sensations (in beverages, snacks, and desserts), to coping with raw material shortages and high ingredient costs (in extended meat products and dairy analogs), to improved nutrition, diet control, and convenience (Trauberman, 1975).

Engineered foods are functional systems which have as building blocks specially processed proteins, carbohydrates, and fats. These building blocks are supplemented with a variety of other engineered ingredients designed to ensure that the final product has the required functional properties. Many "alternative" food proteins suffer from the disadvantage of having no specific, desirable structural properties. In the development of alternative foods, much progress has been made in imparting to those proteins more useful properties. Chemical restructuring of fats is already a major industry. It is probable that techniques already developed will become more sophisticated as we learn more about the importance of the nutritional effects of various kinds of fats. As restructuring of foods becomes more common, attention is turning toward the structural properties of carbohydrates. Thus far, most modified carbohydrates are used mainly in nonfood industries.

Proteins from plant sources contribute unique nutritional and some textural properties. Carbohydrates are used primarily to impart sweetness and such attributes as mouth-feel and viscosity; gums and modified starches are used for thickening, gelling, binding, stabilizing, and film forming (Trauberman, 1975). Vegetable fats are used to replace butterfat in various foods because of lower cost and cholesterol content; they are used in dairy analogs and chocolate confections. Production of acceptable engineered foods has been greatly facilitated by advances in fabrication techniques, for example, high-temperature short-time (HTST) extrusion cooking, fiber spinning, microwave–vacuum processing, ultrafiltration, freeze concentration, reverse osmosis, enzyme immobilization, aseptic packaging, retorting in flexible pouches, and automation. The process

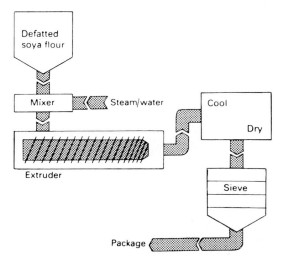

Fig. 11.1. Extruded soy protein production sequence. (From Seal, 1977.)

that has contributed most to the growing acceptance of plant proteins in new foods is texturization by spinning and extrusion. The processes are described diagrammatically in Fig. 11.1 and 11.2. See also Gutcho, 1977; El-Dash, 1981; Linko *et al.,* 1981; Smith, 1975; Smith and Crocco, 1975. The physical properties and functional advantages of textured soy protein are listed in Table 11.2 (Horan, 1974).

Engineering of new foods is predicated on and makes possible better use of traditional raw materials. Some processes that were patented in the 1970s may be

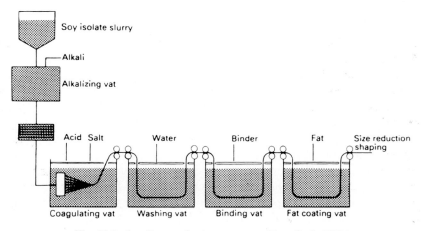

Fig. 11.2. Soy fiber production sequence. (From Seal, 1977.)

TABLE 11.2

Physical Properties and Functional Advantages
of Textured Soy Protein[a]

Properties	Advantages
Texture	Crunchy when dry
	Fibrous and chewy when hydrated
	Particle retention of integrity during cooking and retorting
Color	Light tan, or maybe colored to meet specific need
Flavor	Slight toasted; flavoring may be incorporated to specific needs; compatible with most food products
Hydration	Hydration to three times its dry weight; increases retention of natural flavors, juices, and moisture

[a]From Horan, 1974.

particularly useful. If the wheat kernel (rather than wheat flour) is separated into its components in a wet-process plant located adjacent to a bakery, the desirable fractions could be used to produce a variety of foods. The wet separation could be substantially more efficient than present dry-milling, and the products of wet separation could be used without the great expense of drying. Alternatively, the fractions could be used in production of gluten, starch, or dextrose (Fig. 11.3).

According to Skatrud and Sfat (1981), critical fermentation elements that control flavor, aroma, structure, and nutritional value can be manipulated to produce a variety of new foods and beverages. The principles involve optimizing critical and/or functional components in the raw materials. Examples include cheese flavor concentrates, coffee substitutes, fermented bread flavorings, and cheese ripening agents. In addition, new products can be constructed by mixing elements from different fermentations to achieve unique qualities.

In conventional brewing, for example, barley is converted to malt, corn to grits or syrup, and hops to an extract (See Chapter 10). These fractions are entered into the brewing process and extracted to form a single wort and a by-product (brewers' grain). The wort is fermented. In a new versatile process it is technically feasible to make a product comparable to regular beer—but without using malt. In this process, barley proteins are converted enzymatically to a 40% protein extract and carbohydrates are converted enzymatically to fermentable sugars (see Fig. 11.4). The two main streams (hydrolyzed proteins and hydrolyzed carbohydrates) can be combined in various proportions and toasted to yield a variety of malt-flavored beverages with a great range of caloric value, and

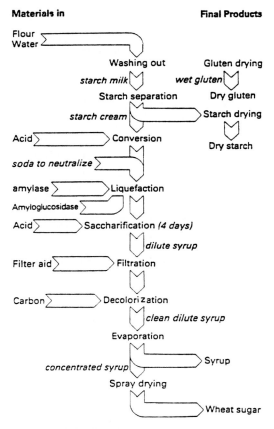

Fig. 11.3. Flow diagram of wheat flour separation. (From Selby, 1977.)

carbohydrate, alcohol, and protein content. The proteins and carbohydrates obtained may serve as sources of flavor concentrates for use in many other foods too (Brenner, 1980).

B. Biotechnology

To the consumer, food usually is characterized by a unique combination of texture and flavor (Kirsop, 1981). Food processing is the conversion of agricultural products to substances that have specific textural and organoleptic characteristics using technologically and economically feasible methods. Some raw materials determine a food's physical attributes during manufacture and in finished products. Food ingredients may be part of a formulation to impart particular physical properties (such as emulsification, gas retention, foam formation, solution, or adhesion) or to modify textural properties to alter the mouth-feel of a food, appearance, and the extent to which flavored substances are perceived by the consumer. Some developments relevant to engineered foods are discussed in this section.

1. Enzymes

Milk caseins consist of α-, β-, and κ-caseins (see Chapters 6 and 10). They are almost all in the form of micelles, which are highly hydrated particles of about

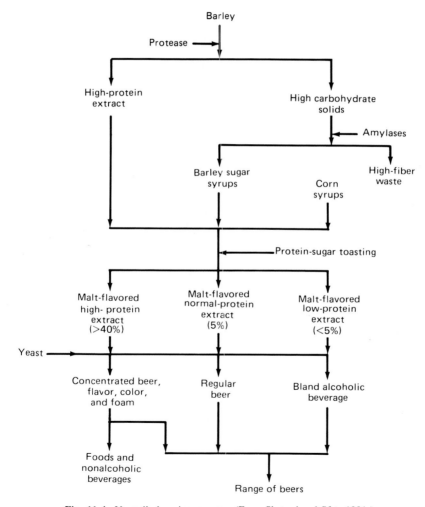

Fig. 11.4. Versatile brewing process. (From Skatrud and Sfat, 1981.)

120 nm in diameter and contain calcium–phosphate–citrate. At the calcium concentration of milk, the α- and β-caseins precipitate and κ-casein stabilizes the micelles. Rennet induces the precipitation of caseins to form curd by hydrolyzing κ-casein at a specific peptide link between phenylalanine and methionine and yields an acidic large peptide and a basic hydrophobic para-κ-casein. When about 60% of the κ-casein is hydrolyzed, coagulation takes place. Continuous processing systems are being developed in which peptide bond hydrolysis is separated from the more gradual aggregation of proteins. These two phenomena have different temperature relationships. Consequently, it is possible to allow hydrolysis at temperatures that retard curd formation and to accelerate curd production by raising the temperature. There are good prospects for selective modification of β-casein by milk's proteolytic enzyme plasmin into γ-casein and smaller fragments which have useful functional properties in foods.

The plastein reaction to improve the amino acid composition of food proteins (with regard to their nutritional and functional properties) has been described by many researchers (Arai et al., 1978; von Hofsten and Lalasides, 1976; Yamashita et al., 1971, 1972; see also Chapter 5). The process involves incorporation of amino acid esters into concentrated protein hydrolysates during the plastein reaction with papain. Prior to this synthesis–incorporation, it is necessary to prepare a protein hydrolysate as substrate. The entire process requires two independent steps: enzymatic protein hydrolysis and resynthesis. It is also possible to incorporate amino acid esters directly into proteins by one step under certain reaction conditions (Aso et al., 1977). The plastein reaction is highly complex because the substrate is an ill-defined protein hydrolysate, a complex mixture of oligosaccharides (Arai and Fujimaki, 1978). Both condensation and transpeptidation reactions are involved; as a result the substrate comprises a low molecular weight fraction and a high molecular weight (plastein) fraction. It has been postulated that peptidyl-enzyme intermediates interact with a peptide acting as a nucleophile with formation of a higher molecular weight peptide (Arai and Fujimaki, 1978). The plastein generally has hydrophobic properties and can be separated by precipitation. During incorporation of amino acids in the ester form, each acts as a nucleophile, while interacting with a peptidyl-enzyme intermediate that is incorporated into the C-terminal. In the incorporation reaction catalyzed by papain, there is a specificity that depends on the amino acid chain structure. The application of the plastein reaction to some casein components and to unfractionated milk powder gives rise to thixotropic gels with unique functional properties (Kirsop, 1981).

Protein modification is another possibility. Acylation of sunflower proteins with succinic anhydride leads to desirable changes in emulsifying and foaming characteristics. In general, however, acylation is relatively nonspecific and of less potential value than modification by enzymes. Proteolytic enzymes have been used to modify the functional properties of fish, soybean, rapeseed, and cottonseed proteins (Kirsop, 1981). More basic specifically targeted modifications are being investigated and may contribute to incorporation of such proteins into novel engineered foods.

2. Pectins

An excessive degree of acetylation of galacturonic acid residues in pectin impairs gelling properties. A specific deacetylase might be of substantial value in improving viscosity. Gel formation is also affected by the degree of methylation of galacturonic acid. Chemical methods of demethylation, but not of increasing the extent of methylation, are available. Prospects of introducing new methyl groups in pectins by the use of enzymes are good.

3. Flavors

The production of specific flavoring materials by plant cells in tissue culture, rather than in intact plants in the field, has several potential advantages: (a) better control and higher yields; (b) elimination of contamination and uncertainties of supply; and (c) production of desirable and selected mixtures of compounds. Much progress has been made in the production of the ''hot'' principle of the fruit *Capsicum frutescens* (African chillies)—capsaicin—which can be formed in relatively high yield from its precursors. Progress is also being made in synthesis of essential oils (Kirsop, 1981).

4. Sugars

Blends of sucrose and corn syrups for practically all fluid systems and uses have been engineered (see Chapter 4; Godzicki, 1975). They include proportionate combinations of the following:

1. Sucrose or invert sugar and 62 DE corn syrup for canned products (80:20), condiments (50:50), and frozen foods (67:33)
2. Sucrose or invert and high fructose corn syrup (HFCS) for carbonated beverages (75:25 or 50:50)
3. HFCS and 42 DE corn syrup for pie fillings (50:50), or HFCS and 95 DE corn syrup for snack cakes and cream fillings (50:50), or 95 DE corn syrup and 42 DE corn syrup for frozen desserts (70:30)
4. Sucrose and 43 DE corn syrup for confections, dairy products, preserves, and syrup blending (80:20), confections and syrup blending (50:20), and sucrose–dextrose–43 DE corn syrup for preserves (50:20:30)
5. Invert and dextrose for beverages and canned products (90:10), condiments and frozen foods (80:20), and pickles (70:30)
6. Sucrose and dextrose for baking, beverages, confections and dairy prod-

ucts (90:10), preserves (80:20), baking, beverages, confections, and preserves (70:30), baking and beverages (60:40), and condiments (50:50)

5. Meat and Fish

The reports of experts at a conference on vegetable food proteins have been
evaluated in terms of the prospects of such proteins in engineered foods by
Attiyate (1979). (See also Chapters 6 and 10.)

The development of special soy protein isolates and refined meat curing technology permits use of isolated proteins in whole meat cuts. Isolates with low-
flavor profiles can be incorporated through injection or massaging–tumbling.
Typical applications include cooked hams and smoked pork shoulders, boneless
pork loins, roast and corned beef, and cooked poultry breasts. Soy isolate–salt–
phosphate brines are useful in retorted products such as meat balls, frankfurters,
luncheon meats, and meat loaves. The soy protein isolates are used to enhance
emulsification and fat binding and gelation properties at elevated temperatures.
The brines can be injected or massaged into whole cuts or mixed into coarsely
comminuted canned meats.

Much of the research on spun vegetable proteins is aimed at creating products
of excellent nutritional quality and high consumer acceptance, rather than simply
imitating meat. A combination of soy isolates, wheat gluten, whey proteins, and
egg albumin may be texturized by spinning into such a product. Texturized
proteins can be hydrated, mixed with ground or minced fish and a matrix-
forming material composed of cereal flours, gums, and spices. This mixture is
then extruded, and battered or breaded, fried, and frozen. Textured protein from
soybeans can be used as an extender in canned tuna or salmon, fish cakes or
patties, or batter systems.

6. Dairy Products

Soy isolates with improved flavor characteristics find use in dairy products
because of their excellent functional properties, such as whipping capability,
viscosity control, emulsification, and emulsion stability. Engineered dairy-type
products include whipped toppings (for example, frozen prewhipped toppings
with excellent stability in the face of several freeze–thaw cycles), yoghurt with
improved viscosity and gel strength, beverages for infants who are allergic to
bovine milk, and cheap nutritious soy–whey drink mixes.

7. Fortified Snacks and Bakery Items

A variety of soy protein isolates are used in pie crusts, cakes, Danish and puff
pastry, pasta, biscuits, pizza mixes, meat pies, doughnuts, cakes, breads, and
other snack foods. The purposes range from increasing water absorption, brown-

ing, tenderness, shelf life, and nutritional value, to reducing oil saturation and improving machinability, color, and texture.

8. Protein Whipping Agents

Enzyme-modified vegetable proteins can achieve whipping and foaming properties comparable to egg white. An ideal whipping agent should (a) be soluble in water and sometimes in concentrated sugar solutions over a wide pH range; (b) have excellent foam formation and foam stability properties; (c) be active over a wide temperature range; (d) be stable against lipids; (e) be compatible with uses in foods; and (f) have a reasonable price–performance ratio. Enzyme-modified whipping agents from plant sources come close to meeting all these requirements. These spray-dried powders are bland, light-colored, soluble in hot and cold water, and functional over a wide pH range.

V. Creating Foods

A. Systems Approach

As stated previously, engineered foods are functional systems whose building blocks are proteins, carbohydrates, and fats, supplemented with other engineered ingredients required for functional properties. (Anonymous, 1975). This systems approach is illustrated in Figs. 11.5, 11.6, and 11.7, which show the creation of foods based on proteins, carbohydrates, and lipids, supplemented with surfactants and polyols.

The surfactants used as supplements in engineered foods are:

1. *Mono- and diglycerides.* Mono- and diglycerides are the most commonly used surfactants in the food industry. They are usually made by glycerolysis, in which glycerol is reacted with natural fats or oils. They may also be made by direct esterification of glycerol with selected fatty acids. These surfactants are available in a wide range of physical forms—liquids, soft plastics, hard waxy solids, flaked, or powdered.

2. *Sorbitan fatty acid esters.* Sorbitan monostearate is a sorbitol-derived analog of the traditional monostearate. It is slightly more hydrophilic than the glycerol fatty acid esters and offers versatility in formulation. By combining sorbitan fatty acid esters and polysorbates, a wide range of HLB values and surfactant effects can be achieved.

3. *Polysorbates.* Polysorbates are made by the reaction of ethylene oxide with lipophilic sorbitan esters, thereby creating a hydrophilic and efficient surfactant in the stabilization of oil-in-water emulsions. Polysorbates are used because of their ability to solubilize essential and vitamin oils; they can be blended with

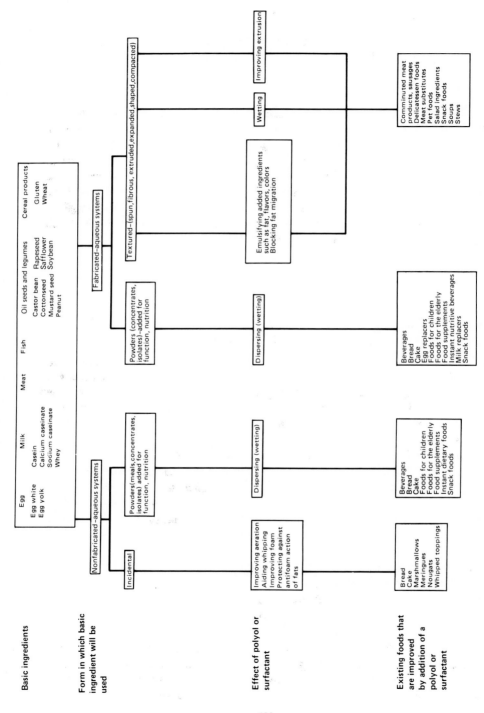

Fig. 11.5. Engineered food based on proteins. (Courtesy of ICI United States Inc., Wilmington, Delaware.)

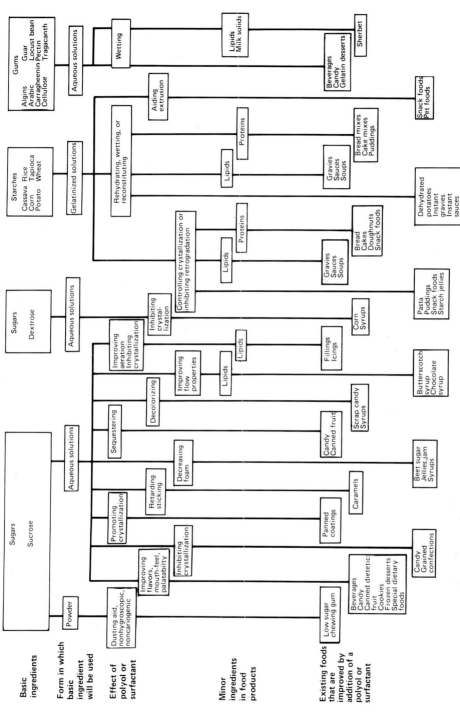

Fig. 11.6. Engineered food based on carbohydrates. (Courtesy of ICI United States Inc., Wilmington, Delaware.)

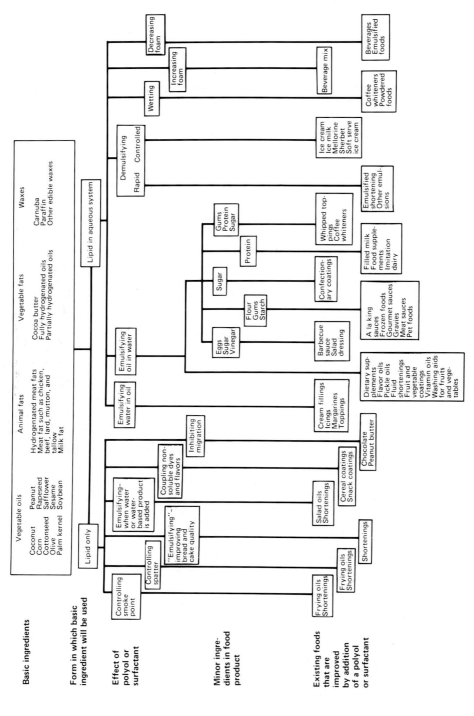

Fig. 11.7. Engineered food based on lipids. (Courtesy of ICI United States Inc., Wilmington, Delaware.)

lipophilic surfactants of either the glycerol or sorbitan ester type to produce a variety of effects over a wide HLB range.

4. *Polyethylene stearates.* These stearates are excellent antifoaming agents for processed foods, fruit jellies, preserves, and sauces.

The polyols shown in Figs. 11.5, 11.6, and 11.7 include

1. *Sorbitol.* The six-carbon, straight-chain polyhydric alcohol, sorbitol, is used as a bodying agent, humectant, flavor, and crystal modifier.
2. *Mannitol.* The other six-carbon, straight chain polyhydric alcohol, mannitol, is used as a flavor modifier and as an inert base for tablet form foods.

The following is a list of the functional properties of surfactants and polyols in various foods (from LG-130-2, Create a Food Based on Lipids; LG-130-3, Create a Food Based on Carbohydrates; LG-130-4, Create a Food Based on Proteins; 238–7, 10M 11/76, Atlas Products for Foods, ICI United States, Inc., Wilmington, Delaware.

1. *Bread.* Mono- and diglycerides retard crumb firming. A mixture of mono- and glycerides and polysorbates, besides slowing crumb firming, strengthens bread doughs, reduces mixing time, and increases water absorption. In white canned bread, crystalline sorbitol inhibits excessive browning.

2. *Beverages.* Polysorbates act as foaming agents in nonalcoholic beverage mixes; sorbitol improves the body and mouth-feel of special dietary beverages.

3. *Cakes.* Mono- and diglycerides improve volume and texture. Combinations of mono- and diglycerides, sorbitan fatty acid esters, and polysorbates improve volume, texture, and shelf life.

4. *Caramels, candy, and grained confections.* Mono- and diglycerides enhance chewing quality and reduce stickiness in caramels. Sorbitol controls sugar crystallization, retards moisture loss, improves palatability, and extends shelf life in candies.

5. *Chewing gums.* Sorbitol, generally in combination with mannitol, improves texture and palatability of special dietary gums.

6. *Coconut.* Sorbitol acts as a humectant, retards moisture loss, and prevents yellowing and ''off'' flavors in shredded coconut.

7. *Chocolate.* In dark sweet chocolate products, polysorbates improve palatability. Combinations of sorbitan fatty acid esters and polysorbates provide initial gloss and retard gloss loss and fat bloom. In milk chocolate, sorbitan fatty acid esters provide initial gloss and retard gloss loss and fat bloom.

8. *Coatings.* In nonstandard and confection coatings, sorbitan fatty acid esters are used to provide initial gloss and retard gloss loss and fat bloom. Polysorbates improve coating palatability. In panned coatings, polysorbates reduce panning time and by modifying sugar crystals produce coatings with increased opacity.

9. *Coffee Whiteners.* In solid powder whitener types, mono- and di-glycerides provide good dispersibility without "oiling-off"; the effectiveness is increased in mixtures with sorbitan fatty acid esters or/and polysorbates. In liquid types, mono- and diglycerides aid in preparation of stable emulsions and improve dispersibility.

10. *Doughnuts.* In yeast-leavened doughnut mixes, mono- and diglycerides retard crumb firming and provide optimum fat absorption. Mixtures of mono- and diglycerides and polysorbates, in addition to retarding crumb firming, act as dough conditioners.

11. *Ice creams, ice milk, mellorine, and soft-serve.* Mono- and diglycerides in ice creams promote relative dryness and overrun. Some polysorbates provide exceptional dryness and can be added at the freezer.

12. *Icings and fillings, including mixes.* Mono- and diglycerides and/or poly-sorbates provide superior volume, texture, and stability in fillings and icings. Sorbitol controls sugar crystallization and retards moisture loss; it also improves palatability and extends shelf life.

13. *Margarine.* Mono- and diglycerides improve emulsion stability and pal-atability during cutting and packaging of margarine.

14. *Peanut butter.* Mono- and diglycerides inhibit oil separation and improve palatability in peanut butter.

15. *Pickles and pickle products.* Polysorbates disperse flavors and colors in pickles.

16. *Potatoes, dehydrated.* Mono- and diglycerides make rehydration easier and improve palatability of dehydrated potatoes.

17. *Process foods, defoaming.* Mono- and diglycerides are effective antifoam agents in sugar–protein syrup systems; polyoxyethylene stearates, sorbitan fatty acid esters, and polysorbates are effective in certain foods. Some polysorbates are particularly effective as antifoaming agent in beet sugar products.

18. *Salad dressings.* Polysorbates provide emulsion stability in dressings.

19. *Salt.* Polysorbates control salt crystal size.

20. *Sausages, cooked.* Sorbitol improves removal of sausage casing and re-duces caramelization and charring flavors.

21. *Sherbet.* Polysorbates vary widely in function in sherbets. Some increase whip, provide smoother texture, and improve overall palatability; others impart dryness and texture; still others enhance aeration and eating qualities.

22. *Shortenings.* In bakers' multipurpose products mono- and diglycerides bestow good cake baking and icing properties; sorbitan fatty acid esters (usually in combination with polysorbates and mono- and diglycerides) ensure good baking and icing performance. In bread and sweet dough products mono- and diglycerides retard crumb firming; combinations of mono- and diglycerides with polysorbates retard crumb firming, act as dough conditioners, reduce mixing time, and increase water absorption. In cake mix shortenings, mono- and di-

glycerides improve cake volume and texture, in combinations of sorbitan fatty acid esters and polysorbates, they provide rapid batter aeration. In retail shortenings, mono- and diglycerides supply good cake baking and icing properties with minimum lowering of smoke point. Polysorbates, alone or in combination with mono- and diglycerides and sorbitan fatty acid esters, improve frying properties and quality of cakes baked with consumer shortenings.

23. *Starch jellies.* Mono- and diglycerides retard starch crystallization, resulting in improved shelf life.

24. *Sucrose-free gums, mints, tablets.* The addition of mannitol and/or sorbitol provides body and mouth-feel in "sugar-free" gums.

25. *Sweet doughs.* Mono- and diglycerides retard crumb firming. In combination with polysorbates, they also strengthen sweet doughs, reduce mixing time, and increase water absorption.

26. *Whipped toppings, vegetable oil.* Mono- and diglycerides, and/or sorbitan fatty acid esters, and/or polysorbates improve aeration, dryness, body, and texture in toppings and vegetable oils.

27. *Yeast.* Polysorbates act as dispersing agents for defoamer yeast compositions.

The systems approach is illustrated in the following sections by discussions of the preparation of two engineered foods: whipped toppings and coffee whiteners (238-13 and 238-16, respectively.) (Courtesy of ICI United States, Inc., Wilmington, Delaware.)

B. Whipped Toppings

Whipped toppings are marketed in a variety of forms: liquid, aerosol, and powdered. Variations of these forms are listed in the following paragraphs.

1. *Liquid whipped topping:* An o/w emulsion containing fat, protein, sugar, stabilizer, and emulsifier is mechanically whipped to produce a basic liquid topping with the desired overrun and dryness features.

2. *Liquid whipped topping concentrate* is an o/w emulsion similar to liquid whipped topping but containing less water. Prior to use, skim or whole milk, cream, juice, or water is added and the mixture is agitated to produce the desired overrun and dryness.

3. *Aerosol topping:* An o/w emulsion similar to a liquid topping is prepared, but it is packaged in a pressurized container. The topping is automatically whipped as it passes through the aerosol spray nozzle.

4. *Powdered topping:* An o/w emulsion containing a minimum of water is spray-dried to create one of the most difficult toppings to manufacture. When reconstituted with whole or skim milk or water, the topping is mechanically

whipped to attain the desired stiffness and overrun. This topping offers long shelf life and ranks high in consumer acceptance.

5. *Proteinless topping* is an o/w emulsion similar to a whipped topping concentrate, except that it contains no protein. This topping is highly compatible with acidic foods, such as lemon pie.

The typical ranges of basic ingredient percentages of final whipped toppings are given in Table 11.3.

TABLE 11.3

Composition of Final Whipped Topping Product[a]

Component	% of total weight
Fat	25.00–35.00
Protein[b]	1.00–6.00
Sucrose	6.00–12.00
Corn syrup solids	2.00–5.00
Stabilizers	0.10–0.80
Emulsifiers	0.40–1.00
Salts	0.025–0.15
Water	46.00–64.00

[a]Courtesy of ICI United States, Inc., Wilmington, Delaware.
[b]Omitted in proteinless formula.

1. Individual Ingredients

1. *Emulsifiers.* Selection of the best emulsifier system for a whipped topping is critical since overrun, dryness, mix stability, topping stability, body, and texture depend on it. Emulsifier concentration ranges from 0.4 to 1.0% of the total weight of topping, depending on the emulsifier. Some commonly used emulsifiers are (a) mono- and diglycerides, free-flowing, granular, and white emulsifiers, which at a level of 0.6–1.0% produce high overrun, soft peak, good body, and above-average stability; (b) a combination of sorbitan monostearate and polysorbate (60:40), which at a level of 0.4% produces an exceptionally dry, stiff-peaked topping (this blend is recommended when overrun is not as important as dryness and stiffness); and (c) a combination of mono- and diglycerides and polysorbates (80:20), which at a level of 0.5–0.7% produces a topping with excellent overrun and dryness qualities.

2. *Fat.* Toppings should contain 31–35% vegetable fat to achieve whipping characteristics and body equivalent to natural cream. Lower fat content may be used since additional body in low fat toppings can be achieved by raising the levels of sugar, stabilizer, and emulsifier. Selection depends on the economics of

the formula. A topping containing less than 25% fat generally exhibits slack body and poor mouth-feel, stability, and texture. For best results, the fat should have a narrow plastic range and low melting point (between 35.5 and 39°C). Such fat reduces the tendency toward a waxy, greasy aftertaste.

3. *Protein.* "Bite," chewiness, and proper cell structure are obtained with protein, which also controls air entrapment and distribution. The selection of the proper caseinate is important to peaking and body properties and general quality of topping. The protein content in the formula ranges from 1 to 6%, depending on the other ingredients, mainly fat. Excessive protein levels decrease the quality of body and texture; inadequate levels produce a topping of poor body, texture, aeration, and "stand-up" properties. Nonfat dry milk solids offer high functionality. They are rich in protein (containing about 38%), easy to store, and relatively inexpensive. On a comparable weight basis, nonfat dry milk solids' functional advantages are better than those of caseinates, in terms of better body, dryness, and aeration. The normal use level of low fat or nonfat dry milk solids is 5–7%. Protein-containing toppings often curdle when used in acid foods (e.g., fruit pies). For such uses, the protein is often replaced with other film-forming ingredients as carboxymethylcellulose (CMC) or related compounds. These film-formers, when used at levels higher than the proteins they replace, perform the same entrapment function as do proteins. When protein is eliminated from a formula, the emulsifier level must be increased 0.10–0.15% to provide for emulsification previously available from the protein. When corn syrup solids are used, it may be necessary to increase the sugar level 2–3%.

4. *Sugars.* Sucrose is used widely in toppings; various corn syrups and sugar–corn syrup blends are also common. Mixtures of sucrose and corn syrup often improve the body and texture of the topping, making it stiffer and chewier. In the typical liquid whipped topping, 6–10% sucrose is sufficient. Replacing part of the sucrose with too much corn syrup produces unnatural sweetness and a "gummy" body, and tends to make the topping too viscous and difficult to aerate. In toppings with less than 30% fat, a blend of sucrose and corn syrup (ratio of 60:40 to 80:20, depending on the DE of the corn syrup) is more functional than sucrose alone.

5. *Stabilizers.* Carboxymethyl cellulose (CMC) is widely used. High viscosity CMC at levels of 0.1–0.5% provides excellent stability. CMC is often used in combination with carrageenan to reduce tendency to separate or "whey-off."

6. *Stabilizing salts.* Phosphate and citrate salts stabilize proteins and improve colloidal stability. Sodium citrate, tetrasodium pyrophosphate, dipotassium phosphate, and disodium phosphate may also be used. Salt levels higher than 0.025–0.15% may cause undesirable flavors.

7. *Flavors and colors.* In many cases, processing and storing conditions volatilize and/or bleach out certain flavors and colors in foods. Stability of flavor and color additives in the formulation is an important consideration.

TABLE 11.4

Starting Formulation for Whipped Toppings[a]

Component	%
Fat (35.5°C melting point)	30.00
Sodium caseinate	6.00
Sucrose	6.00
Corn syrup solids (42 DE)	2.00
Stabilizer	0.50
Emulsifiers	0.4–0.9
Disodium phosphate	0.05
Color	as needed
Flavor	as needed
Water	as needed to 100%

[a]Courtesy of ICI United States, Inc., Wilmington, Delaware.

The compositions of starting formulations, liquid whipped toppings, and proteinless whipped toppings are described in Tables 11.4, 11.5, and 11.6.

C. Coffee Whiteners

Coffee whiteners are marketed in three forms: liquid, frozen liquid, and spray-dried. Spray-dried powders exhibit the most stability and widest consumer acceptance. The liquid and frozen liquid forms must be refrigerated.

TABLE 11.5

Liquid Whipped Topping[a]

Component	%
Fat (35.5°C melting point)	30.0
Sodium caseinate	3.0
Sucrose	6.0
Corn syrup solids (42 DE)	2.0
Stabilizer	0.5
Emulsifier	0.4–0.9
Disodium phosphate	0.05%
Color	As much as suffices
Flavor	As much as suffices
Water	As much as suffices to 100

[a]Courtesy of ICI United States, Inc., Wilmington, Delaware.

TABLE 11.6

Proteinless Whipped Topping[a]

Component	%
Phase I	
Stabilizer	0.05
Hot water	19.50
Phase II	
Water (cold)	As much as suffices to 100
Sucrose	10.00
Salt	0.80
Lecithin	0.05
Fat (38°C melting point)	25.00
Emulsifier	0.80

[a]Courtesy of ICI United States, Inc., Wilmington, Delaware.

In the preparation of liquid coffee whiteners, stability, viscosity, color, and flavor are important considerations. A high degree of stability is required. The product must display no "oiling-off" or "feathering" when added to hot coffee and must withstand freeze–thaw cycles without separating. The viscosity must simulate that of natural dairy milk and cream. A heavy-bodied whitening product does not disperse well in coffee. The product must have a satisfactory whitening effect; this is governed by the concentration of total solids and fineness of the dispersed phase. Finally, a whitener must be bland-flavored and odor-free. Typical formulation of a liquid coffee whitener is given in Table 11.7.

TABLE 11.7

Liquid Coffee Whitener[a]

Component	Typical %[a]	Suggested %[b]
Fat	3.0–18.0	10.0
Protein (sodium caseinate)	1.0–3.0	2.0
Corn syrup solids	1.5–3.0	2.5
Sucrose	1.5–3.0	2.5
Monoglyceride	0.3–0.5	0.4
Carrageenan	0.1–0.2	0.2
Stabilizer salts (sodium citrate)	0.1–0.3	0.15
Flavor	As much as suffices	As much as suffices
Color	As much as suffices	As much as suffices
Water	As much as suffices to 100%	As much as suffices to 100%

[a]Courtesy of ICI United States, Inc., Wilmington, Delaware.
[b]Based on total weight.

The characteristics and manufacturing techniques for frozen liquid whiteners are similar to those of the regular liquid whitener. Because it is maintained in the frozen state, freeze–thaw stability is important.

The spray-dried, free-flowing whitening powders require no refrigeration. While the main use is as replacement for cream in coffee, the powder also may be reconstituted for various uses, including whipped toppings. In addition to the requirements for liquid whiteners, the dry product must have good flow properties, resist clumping or caking, disperse easily in hot liquids, be relatively non-hygroscopic, and packaged properly for convenient use and protection against moisture "pick-up."

A typical formulation for a spray-dried whitener is given in Table 11.8. The final product should have a moisture content of no more than 1%, a particle size of 125–150 μm in diameter, and entrapped fat globules no larger than 1–3 μm in diameter. Residual heat of fat crystallization should be allowed to dissipate from the powder prior to packaging to avoid clumping. In some cases it may be necessary to "instantize" the whitener by agglomerating the powder to make it more dispersible. In preparing coffee whiteners, a surfactant or surfactant system of HLB 5–6 is required to prepare a stable emulsion.

1. Individual Ingredients

1. *Fat.* The concentration of fat ranges from 5 to 18% in liquid products (average 10%); in powdered whiteners the average fat content is 35–40%. A fat with a low melting point and narrow plastic range (e.g., hydrogenated coconut or palm kernel oil, or their fractions) is satisfactory. High melting point fats (above

TABLE 11.8

Spray-Dried Coffee Whitener[a]

Component	% (dry-basis)
Vegetable fat	35–40
Corn syrup solids (42 DE)	55–60
Sodium caseinate	4.5–5.5
Dipotassium phosphate	1.2–1.8
Emulsifier (mixture of monoglycerides, sorbitan monostearate, and polysorbate)	0.3–0.5
Color	As much as suffices
Flavor	As much as suffices
Anticaking agent	As much as suffices

[a]Courtesy of ICI United States, Inc., Wilmington, Delaware.

46°C) produce heavy-bodied whiteners of poor palatability. Low melting point or liquid fats interfere with dispersibility because of their absorption by proteins and because the emulsified fats or oils tend to coalesce.

2. *Protein.* The concentration of protein ranges from 1.5 to 3.0% in liquid products and from 4.0 to 5.5% in powdered whiteners. Raising the protein content increases viscosity and whitening power; it also may impart too much body, however. Calcium caseinate is less rapidly dissolved in the aqueous phase and is less stable to separation than sodium caseinate. Soybean protein isolate may be used to create an all-vegetable product.

3. *Carbohydrates.* Liquid coffee whiteners contain about 6% carbohydrates and powdered whiteners 50–60%. The function of carbohydrates is mainly as a bodying agent and for some sweetening effect. In powdered whiteners, the main function is as carrier of fat. Corn syrup solids and sucrose are usually blended for optimum body and flavor characteristics in liquid products. Corn syrup and lactose are used in powdered types.

4. *Stabilizer.* A stabilizer's function is twofold: to inhibit syneresis or "wheying-off" and to support the body or viscosity of the liquid product. If the formula calls for less than 10% fat or less than 1.5% protein, viscosity can be achieved by adding CMC, locust bean gum, alginates, or guar gums. In most liquid formulations, 0.1–0.2% carrageenan provides the required stability and body.

5. *Stabilizing salts.* Sodium citrate, tetrasodium pyrophosphate, and dipotassium phosphate are recommended at a 0.1% level. Excessive amounts of salts may impair flavor. The stability of liquid whiteners is modified drastically by slight changes in surfactant level and/or system or by minor changes in stabilizer level.

6. *Anti-caking agents.* In powdered coffee whiteners, especially when high levels of corn syrup are used, caking may be a problem. Under such circumstances, it may be necessary to add an anticaking agent. Adding mannitol may reduce or eliminate caking and allow a more free-flowing product.

References

Altschul, A. M., and Hornstein, I. (1972). Foods of the future. *Agric. Food Chem.* **20,** 532–536.

Anonymous (1975). Engineering ingredients: Building blocks for food design. *Food Eng.* **47**(7), 46, 47.

Anonymous (1979). The ever-shifting dietary goals. *Science* **204,** 1177.

Anonymous (1980). Government issues dietary guidelines. *Milling & Baking News* **58**(54), 1, 12.

Anonymous (1981). Food from a fermenter, cooks and tastes like meat. *Food Eng.* **53** (5), 117–118.

Arai, S., and Fujimaki, M. (1978). The plastein reaction. Theoretical basis. *Ann. Nutr. Aliment.* **32**(2/3), 701–707.

Arai, S., Yamashita, M., and Fujimaki, M. (1978). Nutritional improvement of food proteins by means of the plastein reaction and its novel modification. *In* "Nutritional Improvement of Food and Feed Proteins" (M. Friedman, ed.), Chapter 32, pp. 663–680. Plenum, New York.

Aso, K., Yamashita, M., Arai, S., Suzuki, J., and Fujimaki, M. (1977). Specificity for incorporation of α-amino acid esters during the plastein reaction by papain. *Agric. Food Chem.* **25,** 1138–1141.

Attiyate, Y. (1979). Special report: World conference on vegetable proteins. *Food Eng.* **51**(1), EF 3–6, 10–13, 16–20, 22, 24, 28–31.

Beigler, M. A. (1976). Complete synthetic foods. *In* "New Protein Foods" (A. M. Altschul, ed.), Vol. 2, Part B, Chapter III, pp. 62–85. Academic Press, New York.

Brenner, M. W. (1980). Beers of the future. *Master Brewers' Assoc. America Tech. Q.* **17**(4), 185–195.

Chou, M. (1979a). Changing food policies. *In* "Critical Food Issues of the Eighties" (M. Chou and D. P. Harmon, eds.), Chapter 7, pp. 103–118. Pergamon, Oxford.

Chou, M. (1979b). Changing attitudes and lifestyles—shaping food technology in 1980's. *In* "Critical Food Issues of the Eighties" (M. Chou and D. P. Harmon, eds.), Chapter 10, pp. 149–190. Pergamon, Oxford.

Dally, E. L., Eveligh, D. E., Montecourt, B. S., and Stokes, H. W. (1981). Recombinant DNA technology: Food for thought. *Food Technol.* **35**(7), 26–32.

Dubos, R. (1979). The intellectual basis of nutritional science and practice, *In* "Critical Food Issues of the Eighties" (M. Chou and D. P. Harmon, eds.), Chapter 6, pp. 95–102. Pergamon, Oxford.

El-Dash, A. A. (1981). Application and control of thermoplastic extrusion of cereals for food and industrial uses. *In* "Cereals: A Renewable Resource; Theory and Practice" (Y. Pomeranz, ed.), Chapter 10, pp. 165–216. Am. Assoc. Cereal Chem. St. Paul, Minnesota.

Godzicki, M. M. (1975). Engineering "sugar." *Food Eng.* **47**(10), E14–E17.

Gutcho, M. C. (1977). "Texturized Protein Products," Food Technol. Rev. No. 44. Noyes Data Corp., Park Ridge, New Jersey.

Horan, F. E. (1974). Soy protein products and their production. *J. Am. Oil Chem. Soc.* **51,** 67A–73A.

Kent-Jones, D. W. (1971). Modern food and food additives. *Chem. Ind. (London)* 1275–1283.

Kirsop, H. H. (1981). Biotechnology in the food processing industry. *Chem. Ind.,* pp. 218–222.

Linko, P., Colonna, P., and Mercier, C. (1981). High-temperature, short-time extrusion cooking. *Adv. Cereal Sci. Technol.* **4,** 145–235.

Morris, C. E. (1981). Genetic engineering: Its impact on the food industry. *Food Eng.* **53**(5), 57–69.

Pomeranz, Y. (1980). Cereal science and technology at the turn of the decade. *Adv. Cereal Sci. Technol.* **3,** 1–40.

Seal, R. (1977). Soya products: A food processor's guide. *Chem. Ind. (London)* **11,** 441–446.

Selby, K. (1977). The role of cereal-based products. *Chem. Ind. (London)* **11,** 494–498.

Skatrud, T. J., and Sfat, M. R. (1981). Designing new fermented foods. *Food Eng.* **53**(7), 80–82.

Smith, O. B. (1975). Extrusion and forming: Creating new foods. *Food Eng.* **47**(7), 48–50.

Smith, O. B., and Crocco, S. C. (1975). Engineering meat. *Food Eng.* **47**(10), EF 25, 26, 28, 30.

Trauberman, L. (1975). Industry's goal: The "compleat" food. *Food Eng.* **47**(7), 43–45.

von Hofsten, B. and Lalasides, G. (1976). Protease-catalyzed formation of plastein products and some of their properties. *Agric. Food Chem.* **24,** 460–465.

Yamashita, M., Arai, S., Tsai, S.-J., and Fujimaki, M. (1971). Plastein reaction as method for enhancing the sulfur containing amino acid level of soybean protein. *Agric. Food Chem.* **19,** 1151–1154.

Yamashita, M., Arai, S., Aso, K., and Fujimaki, M. (1972). Location and state of methionine residues in a papain-synthesized plastein from a mixture of soybean protein hydrolyzate and L-methionine ethyl ester. *Agric. Biol. Chem.* **36,** 1353–1360.

III

Information and Documentation

12

Information and Documentation

I. Introduction

Availability, uses, and limitations of tables on composition of foods were described and discussed by Murphy *et al.* (1973). Information on audiovisual resources in foods and nutrition and on food and nutrition bibliography are available from the Oryx Press, Phoenix, Arizona (for the National Agricultural Library, United States Department of Agriculture).

The subject of information and documentation was reviewed in six papers presented during a symposium held during the thirty-fifth annual meeting of the Institute of Food Technologists (June 8–11, 1975, Chicago, Illinois) (Burton, 1976; Cuadra and Boyle, 1976; Fisher, 1976; Hopper, 1976; Mann, 1976; Stadelman, 1976). Mann described the scope and contents of the monthly journal *Food Science and Technology Abstracts*. Cuadra and Boyle (with the SDC Search Service, System Development Corporation, Santa Monica, California) described the availability and success in using on-line retrieval systems for bibliographical and other textual data. Fisher (with BioSciences Information Services, Philadelphia, Pennsylvania) reviewed services provided by his company. Information support services available from Agricultural Research Service, United States Department of Agriculture, were summarized by Burton. Hopper discussed the three basic kinds of information needed by the food industry: technical information, safety data, and worldwide government regulations.

He then compared primary sources of information for the food industry and some major on-line information services available to the food industry. Stadelman reviewed the main objectives of users of information at universities for training students at all levels and for conducting research.

II. Data Bases

The availability of a data base on magnetic tape was described by Van Dyke and Ayer (1972) and Caponio and Moran (1975). Mermelstein (1977) described the available on-line retrieval systems as an alternative to manual searching. The meeting of information needs of food scientists through computerized literature searching was described by Sze (1981). Dialog Information Services, Inc., Palo Alto, California, has a broad information retrieval system that covers over 180 data bases. Data bases of greatest interest to food scientists and technologists include (as of January 1984) the following:

AGRICOLA, 1970 to present, 1,834,000 records, monthly updates (National Agricultural Library, Beltsville, Maryland). AGRICOLA (formerly CAIN) is the cataloging and indexing data base of the National Agricultural Library (NAL). This file provides coverage of worldwide journal and monographic literature on agriculture and related subjects. Since AGRICOLA represents the actual holdings of the National Agricultural Library, there is substantial coverage of subject matter contained in a large library (see also Gilreath, 1979 and Turner, 1983).

BIOSIS Previews, 1969 to the present, 4,054,000 records, biweekly updates (BioSciences Information Service, Philadelphia, Pennsylvania). BIOSIS Previews contains citations from *Biological Abstracts* and *Biological Abstracts/ RRM* (formerly entitled *Bioresearch Index*), the major publications of Bio-Sciences Information Service of *Biological Abstracts*. Together, these publications constitute the major English language service providing worldwide coverage of research in the life sciences. Over 9000 primary journals and monographs as well as symposia, reviews, preliminary reports, semipopular journals, selected institutional and government reports, research communications, and other secondary sources provide citations on all aspects of the biosciences and medical research. Searchable abstracts are available from *Biological Abstracts*'s records from July 1976 to the present.

CA Search, 1967 to the present, 6,174,000 records, biweekly updates (Chemical Abstracts Service, Columbus, Ohio). The CA Search data base contains bibliographic data, keyword phrases, and index entries for documents covered by Chemical Abstracts Service. CA Search is an expanded data base that contains the basic bibliographic information in *Chemical Abstracts*. Index entries containing CA General Subject Headings from a controlled vocabulary, and the CAS

Registry Numbers, a unique number assigned to each specific chemical compound, appear complete with modifying phrases. Additional keyword index phrases and cross-referenced CA General Subject Headings are also included. This merged data base provides the user with many points to each citation and increases the ease of searching for broadly defined topics and specific compounds.

CAB Abstracts, 1972 to the present, 1,559,000 records, monthly updates (Commonwealth Agricultural Bureaux, Farnham Royal, Slough, England). CAB Abstracts is a comprehensive file of agricultural and biological information containing records in the 26 principal abstract journals published by Commonwealth Agricultural Bureaux. Over 8500 journals in 37 languages are scanned, as well as books, reports, and other publications. In some instances, less accessible literature is abstracted by scientists working in other countries. About 130,000 items are selected for publication yearly; significant papers are abstracted, while less important works are reported with bibliographic details only.

CRIS/USDA, 1981 to present, 30,000 records, monthly updates (U.S. Department of Agriculture, Washington, D.C.). CRIS (Current Research Information System) is a current-awareness data base for agriculturally related research projects. The projects cover current research in agriculture and related sciences, sponsored or conducted by USDA research agencies, state agricultural experiment stations, state forestry schools, and other cooperating state institutions. Currently active and recently completed projects within the last 2 years are included.

Dissertation Abstracts Online, 1861 to the present, 842,500 records, monthly updates (University Microfilms International, Ann Arbor, Michigan). Dissertation Abstracts Online is a subject, title, and author guide to American dissertations accepted at accredited institutions since 1861, when academic doctoral degrees were first granted in the United States. In addition, Dissertation Abstracts Online serves to disseminate citations for thousands of Canadian dissertations and an increasing number of papers accepted in institutions abroad.

Food Science and Technology Abstracts, 1969 to the present, 251,000 records, monthly updates (International Food Information Service, Reading, Berkshire, England). Food Science and Technology Abstracts provides access to research and new development literature in areas related to food science and technology. Allied disciplines, such as agriculture, chemistry, biochemistry, and physics, are also covered. Related disciplines, such as engineering and home economics, are included when relevant to food science. FSTA provides indexing to over 1200 journals from over 50 countries, patents from 20 countries, and books in any language. Information in this data base supports research by scientists, technologists, marketing personnel, teachers, and scholars working in areas related to food science and technology.

Foods Adlibra, 1974 to the present, 74,400 records, monthly updates (Komp

Information Services, Louisville, Kentucky). Foods Adlibra contains information on developments in food technology and packaging. New food products introduced since 1974 are covered, and nutritional and toxicology information is also included. Foods Adlibra provides information on retailers, processors, brokers, equipment suppliers, gourmet food importers, and general company and food association news. Major significant research, such as technological advances in processing methods and packaging is reported. Brief abstracts from over 250 trade periodicals constitute the bulk of the Foods Adlibra data base; over 500 highly technical research journals are scanned for additional information. United States patents and some British patents are included. Foods Adlibra is a useful source of information on government guidelines and regulations on the processing and packaging of foods. Marketing and Management news and statistics as well as information on world food economics can be found.

Life Sciences Collection, 1978 to the present, 480,000 records, monthly updates, (Cambridge Scientific Abstracts, Bethesda, Maryland). Life Sciences Collection contains abstracts of information in the fields of animal behavior, biochemistry, ecology, entomology, genetics, immunology, microbiology, toxicology, and virology. The worldwide coverage is of journal articles, books, conference proceedings, and report literature.

NTIS, 1964 to the present, 1,025,000 records, biweekly updated [National Technical Information Service (NTIS), U.S. Department of Commerce, Springfield, Virginia]. The NTIS data base consists of government-sponsored research, development, and engineering plus analyses prepared by federal agencies and their contractors or grantees. It is the means through which unclassified, publicly available, unlimited distribution reports are made available for sale. State and local government agencies are now beginning to contribute their reports to the file.

The NTIS data base includes material from both the hard and soft sciences, including substantial material on technological applications, business procedures, and regulatory matters. Many topics of immediate broad interest are included, such as environmental pollution and control, energy conversion, technology transfer, behavioral/societal problems, and urban and regional planning.

REFERENCES

Burton, H. D. (1976). Computer-based literature searching at the USDA/ARS. *Food Technol.* **30**(5), 70, 72.

Caponio, J. F., and Moran, L. (1975). CAIN: A computerized literature system for the agricultural sciences. *J. Chem. Inf. Comput. Sci.* **15**(3), 168–161.

Cuadra, C. A., and Boyle, H. (1976). On-line information services for food science and technology. *Food Technol.* **30**(5), 60, 62, 63.

Fisher, D. A. (1976). Keeping current through information services. *Food Technol.* **30**(5), 66, 68.

Gilreath, C. L. (1979). "AGRICOLA Users' Guide. Agricultural Reviews and Manuals," ARM-H-7. Science Ed. Admin., U.S. Department of Agriculture, Beltsville, Maryland.

Hopper, P. F. (1976). Information systems in industry. *Food Technol.* **30**(5), 74, 75, 76.

Mann, E. J. (1976). The international food information service: Past, present, and future. *Food Technol.* **30**(5), 54, 56, 58.

Mermelstein, N. H. (1977). Retrieving information from the food science literature. *Food Technol.* **31**(9), 46–48, 52–55.

Murphy, E. W., Watt, B. K., and Rizek, R. L. (1973). Tables of food composition: Availability, uses and limitations. *Food Technol.* **27**(1), 40–51.

Stadelman, W. J. (1976). Information systems in the university. *Food Technol.* **30**(5), 78, 84.

Sze, M. C. (1981). Meeting the information needs of food scientists through computerized literature searching. *Food Technol.* **35**(10), 92–97.

Turner, P. A. (1983). The use of mini and macrocomputers at the National Agricultural Library. *Agr. Libr. Inf. Notes* **9**(9), 1–5.

Van Dyke, V. J., and Ayer, N. L. (1972). Multipurpose cataloging and indexing system (CAIN) of the National Agricultural Library. *J. Libr. Autom.* **5**(1), 21–29.

III. Bibliography

A. Food Processing—Technology

Brennan, J. F., Butters, J. R., Cowell, N. D., and Lilly, A. E. V. (1969). "Food Engineering Operations." Am. Elsevier, New York.

Charm, S. E. (1963). "Food Engineering." Avi Publ. Co., Westport, Connecticut.

Clarke, R. J. (1957). "Process Engineering in the Food Industries." Heywood, London.

Desrosier, N. W., ed. (1977). "Elements of Food Technology." Avi Publ. Co., Westport, Connecticut.

Desrosier, N. W., and Desrosier, J. N. (1977). "The Technology of Food Preservation," 4th ed. Avi Publ. Co., Westport, Connecticut.

Duffy, J. I., ed. (1971). "Snack Food Technology. Recent Developments." Noyes Data Corp., Park Ridge, New Jersey.

Earle, R. L. (1966). "Unit Operations in Food Processing." Pergamon, Oxford.

Farrall, A. W. (1963). "Engineering for Dairy and Food Products." Wiley, New York.

Farrall, A. W. (1976). "Food Engineering Systems," Vol. 1. Avi Publ. Co., Westport, Connecticut.

Farrall, A. W. (1979). "Food Engineering Systems," Vol. 2. Avi Publ. Co., Westport, Connecticut.

Farrall, A. W., and Basselman, J. A., eds. (1979). "Dictionary of Agricultural and Food Engineering." Interstate Printers and Publishers, Danville, Illinois.

Fiechter, A., ed. (1980). "Advances in Biochemical Engineering." Springer-Verlag, Berlin and New York.

Friberg, S., ed. (1976). "Food Emulsions." Dekker, New York.

Ghose, T. K., and Fiechter, A. (1971). "Advances in Biochemical Engineering," Vol. 1. Springer-Verlag, Berlin and New York.

Ghose, T. K., Fiechter, A., and Blackbrough, N., eds. (1972). "Advances in Biochemical Engineering," Vol. 2. Springer-Verlag, Berlin and New York.

Gillies, M. T. (1974). "Compressed Food Bars." Noyes Data Corp., Park Ridge, New Jersey.

Glicksman, M. (1971). Fabricated foods. *CRC Crit. Rev. Food Technol.* **2**, 21–43.

Goldblith, S. A. (1963). "Exploration in Future Food-Processing Techniques." MIT Press, Cambridge, Massachusetts.

Goldblith, S. A., Rev, L., and Rothmayr, W. N., eds. (1974). "Freeze Drying and Advanced Food Technology." Academic Press, New York.

Graham, H. D., ed. (1977). "Food Colloids." Avi Publ. Co., Westport, Connecticut.

Gutcho, M. (1973). "Prepared Snack Foods," Food Technol. Rev. No. 2. Noyes Data Corp., Park Ridge, New Jersey.

Gutterson, M. (1972). "Food Canning Technology." Noyes Data Corp., Park Ridge, New Jersey.

Hall, C. W. (1979). "Dictionary of Drying." Dekker, New York.

Hall, C. W., Farrall, A. W., and Rippen, A. L. (1971). "Encyclopedia of Food Engineering." Avi Publ. Co., Westport, Connecticut.

Harper, J. C. (1976). "Elements of Food Engineering." Avi Publ. Co., Westport, Connecticut.

Harper, J. M. (1981). "Extrusion of Foods," 2 vols. CRC Press, Boca Raton, Florida.

Harper, W. J., and Hall, C. W. (1976). "Dairy Technology and Engineering." Avi Publ. Co., Westport, Connecticut.

Heid, J. L., and Joslyn, M. A. (1967). "Fundamentals of Food Processing Operations." Avi Publ. Co., Westport, Connecticut.

Heldman, D. R. (1975). "Food Process Engineering." Avi Publ. Co., Westport, Connecticutt.

Hendderson, S. M., and Perry, R. L. (1976). "Agricultural Process Engineering," 3rd ed. Avi Publ. Co., Westport, Connecticut.

Honig, P., ed. (1953). "Principles of Sugar Technology." Elsevier, Amsterdam.

Honig, P. (1963). "Principles of Sugar Technology," Vol. 3. Am. Elsevier, New York.

Inglett, G. E., ed. (1975). "Fabricated Foods." Avi Publ. Co., Westport, Connecticut.

Irani, R. R., and Callis, C. F. (1963). "Particle Size: Measurement, Interpretation, and Application." Wiley, New York.

Johnson, A. H., and Peterson, M. S. (1974). "Encyclopedia of Food Technology." Avi Publ. Co., Westport, Connecticut.

Joslyn, M. A., and Heid, J. L., eds. (1963–1964). "Food Processing Operations," 3 vols. Avi Publ. Co., Westport, Connecticut.

Judge, E. E. (1983). "The Almanac of the Canning, Freezing, Preserving Industries." E. E. Judge, Westminster, Maryland.

Judson King, C. (1970). Freezedrying of Foodstuffs. *CRC Crit. Rev. Food Technol.* **1**, 379–452.

Karel, M., Fennema, O. R., and Lund, D. B. (1975). "Principles of Food Science," Part 2. Dekker, New York.

Leniger, H. A., and Beverloo, W. A. (1975). "Food Process Engineering." Reidel Publ., Dordrecht, Netherlands.

Linko, P. K., Malkki, Y., Olkku, J., and Larinkari, J. (1980). "Food Process Engineering," Vols. 1 and 2. Appl. Sci. Publ., Barking, Essex, England.

Loncin, M. (1969). "The Basis of Processing Technology in the Food Industry." Verlag Sauerlander, Aarau and Frankfurt am Main (in German).

Loncin, M., and Merson, R. L. (1979). "Food Engineering: Principles and Selected Applications," Academic Press, New York.

McCrone, W. C., and Delly, J. G. (1975). "The Particle Atlas," 2nd ed., Vols. 1–4. Ann Arbor Sci. Publ., Ann Arbor, Michigan.

Matz, S. A. (1976). "Snack Food Technology." Avi Publ. Co., Westport, Connecticut.

Parker, M. E., Harvey, E. H., and Stateler, E. S. (1952). "Elements of Food Engineering," 3 vols. Van Nostrand-Reinhold, Princeton, New Jersey.

Peterson, M. S., and Tressler, D. K., eds. (1963). "Food Technology the World Over," Vol. 1. Avi Publ. Co., Westport, Connecticut.

Peterson, M. S., and Tressler, D. K., eds. (1965). "Food Technology the World Over," Vol. 2. Avi Publ. Co., Westport, Connecticut.

Petrowski, G. E. (1976). Emulsion stability and its relation to foods. *Adv. Food Res.* **22,** 310–365.

Pitcher, W. H. (1980). "Immobilized Enzymes for Food Processing." CRC Press, Boca Raton, Florida.

Priestley, R. J., ed. (1979). "Effects of Heating on Foodstuffs." Applied Sci. Publ., Barking, Essex, England.

Robbins, P. M. (1976). "Convenience Foods; Recent Technology." Noyes Data Corp., Park Ridge, New Jersey.

Smith, A. L., ed. (1976). "Theory and Practice of Emulsion Technology." Academic Press, New York.

Steel, R., ed. (1958). "Biochemical Engineering." Macmillan, New York.

Toledo, R. T. (1980). "Fundamentals of Food Process Engineering." Avi Publ. Co., Westport, Connecticut.

Tressler, D. K., Van Arsdel, W. B., and Copley, M. J. (1968). "The Freezing Preservation of Foods," Vols. 1–4. Avi Publ. Co., Westport, Connecticut.

Van Arsdel, W. B., Copley, M. J., and Morgan, A. I. (1973). "Food Dehydration," Vols. 1 and 2. Avi Publ. Co., Westport, Connecticut.

Wiseman, A., ed. (1980). "Topics in Enzyme and Fermentation Biotechnology," Vol. 4. Wiley, New York.

B. Foods—General

Allaby, M. (1977). "World Food Resources—Actual and Potential." Appl. Sci. Publ., Barking, Essex, England.

Anonymous (1969). "Food for Us All." Yearbook of Agriculture, U.S. Dept. of Agriculture, Washington, D.C.

Anonymous (1974). "Shopper's Guide." Yearbook of Agriculture, U.S. Dept. of Agriculture, Washington, D.C.

Anonymous (1975). "Agriculture: Food and Man." Brigham Young Univ. Press, Provo, Utah.

Anonymous (1975). "That We May Eat." Yearbook of Agriculture, U.S. Dept. of Agriculture, Washington, D.C.

Anonymous (1977). "Proceedings of the World Food Conference of 1976." Iowa State Univ. Press, Ames.

Anonymous (1979). "What's to Eat." Yearbook of Agriculture, U.S. Dept. of Agriculture, Washington, D.C.

Anonymous (1981). "Will There Be Enough Food?" Yearbook of Agriculture, U.S. Dept. of Agriculture, Washington, D.C.

Anonymous (1982). "Food—From Farm to Table." Yearbook of Agriculture, U.S. Dept. of Agriculture, Washington, D.C.

Aurand, L. W., and Woods, A. E. (1973). "Food Chemistry." Avi Publ. Co., Westport, Connecticut.

Barrons, K. C. (1975). "The Food in Your Future—Steps to Abundance." Van Nostrand-Reinhold, Princeton, New Jersey.

Bender, F. E., Kramer, A., and Kahan, G. (1977). "Systems for the Food Industry." Avi Publ. Co., Westport, Connecticut.

Bennion, M. (1980). "The Science of Food." Harper & Row, New York.

Bennion, M., and Hughes, O., eds. (1975). "Introductory Foods," 6th ed. Macmillan, New York.

Berk, Z. (1976). "Braverman's Introduction to the Biochemistry of Foods." Elsevier, Amsterdam.

Birch, G. G., Cameron, A. G., and Spencer, M. (1972). "Food Science." Pergamon, Oxford.

Birch, G. G., Parker, K. J., and Worgan, J. T., eds. (1976). "Food from Waste." Appl. Sci. Publ., Barking, Essex, England.

Black, F. S., ed. (1955). "Handbook of Food and Agriculture." Van Nostrand-Reinhold, Princeton, New Jersey.

Borgström, G. A. (1976). "Principles of Food Science," Vol. 1. Food & Nutrition Press, Westport, Connecticut (2nd Printing).

Brouk, B. (1975). "Plants Consumed by Man." Academic Press, New York.

Considine, D. M., and Considine, G. D., eds. (1982). "Foods and Food Production Encyclopedia." Van Nostrand-Reinhold, Princeton, New Jersey.

Darby, W. J., Ghalioungui, P., and Grivetti, L. (1977). "Food: The Gift of Osiris," Vols. 1 and 2. Academic Press, New York.

Deatherage, F. E. (1975). "Food for Life." Plenum, New York.

Deman, J. M. (1976). "Principles of Food Chemistry." Avi Publ. Co., Westport, Connecticut.

Duckham, A. N., Jones, J. G. W., and Roberts, E. H., eds. (1976). "Food Production and Consumption." Elsevier/North-Holland, New York.

Economic Research Service, USDA (1974). "The World Food Situation and Prospects to 1985," Foreign Agric. Econ. Rep. No. 98. ERS, USDA, Washington, D.C.

Eskin, N. A. M., Henderson, H. M., and Townsend, R. J. (1971). "Biochemistry of Foods." Academic Press, New York.

Fennema, O. R., ed. (1976). "Principles of Food Science," Part I. Dekker, New York.

Finney, E. E., ed. (1981). "CRC Handbook of Transportation and Marketing in Agriculture," Vols. I and II. CRC Press, Boca Raton, Florida.

Gaman, P. M., and Sherrington, K. B. (1977). "The Science of Food: An Introduction to Food Science, Nutrition, and Microbiology." Pergamon, Oxford.

Garard, I. D. (1974). "The Story of Food." Avi Publ. Co., Westport, Connecticut.

Hale, N. C. (1975). "Impact of Technology on the Food Supply: Alternative Food Ingredients." Arthur D. Little, Cambridge, Massachusetts.

Hawthorn, J. W. (1981). "Foundations of Food Science." Freeman, San Francisco, California.

Heimann, W. (1980). "Fundamentals of Food Chemistry." Avi Publ. Co., Westport, Connecticut.

Heiser, C. B., Jr. (1973). "Seed to Civilization: The Story of Man's Food." Freeman, San Francisco, California.

Hoff, J. E., and Janick, J., eds. (1973). "Food-Readings from Scientific American." Freeman, San Francisco, California.

Hollingworth, D., and Morse, E., eds. (1976). "People and Food Tomorrow." Appl. Sci. Publ., London.

Inglett, M. J., and Inglett, G. E. (1982). "Food Products Formulary Series," Vol. 4. Avi Publ. Co., Westport, Connecticut.

Jackson, H., and Wolfe, F. H. (1978). "Food in Perspective." Univ. of Alberta Press, Edmonton, Alberta.

Jensen, L. B. (1953). "Man's Foods." Garrard Publ. Co., Champaign, Illinois.

Kramer, A. (1977). "Food and the Consumer," 2nd ed. Avi Publ. Co., Westport, Connecticut.

Kretovich, V. L., and Pijanowski, E. (1963). "Biochemical Principles of the Food Industry" (transl. from Russian). Pergamon, Oxford.

Lee, F. A. (1975). "Basic Food Chemistry." Avi Publ. Co., Westport, Connecticut.

Leitch, J. M., and Rhodes, D. N., eds. (1963). "Recent Advances in Food Science," Vol. 3. Butterworth, London.

Lowenberg, M. E., Todhunter, E. N., Wilson, E. D., Savage, J. R., and Lubawski, J. L. (1974). "Food and Man." Wiley, New York.

McWilliams, M. (1966). "Food Fundamentals." Wiley, New York.

Mallette, M. F., Althouse, P. M., and Claggett, C. O. (1960). "Biochemistry of Plants and Animals." Wiley, New York.

Meyer, L. H. (1960). "Food Chemistry." Van Nostrand-Reinhold, Princeton, New Jersey.

Morr, M. L., and Irmiter, T. F. (1970). "Introductory Foods." Macmillan, New York.

Morton, E. N., and Rhodes, D. N., eds. (1974). "The Contribution of Chemistry to Food Supplies." Butterworth, London.

National Academy of Sciences—National Research Council (1974). "Food Science in Developing Countries: A Selection of Unsolved Problems." Natl. Acad. Sci., Washington, D.C.

Nickerson, J. T. R., and Ronsivalli, L. J. (1980). "Elementary Food Science," 2nd ed. Avi Publ. Co., Westport, Connecticut.

Paul, P. C., and Palmer, H. H., eds. (1972). "Food Theory and Applications." Wiley, New York.

Peddersen, R. B. (1977). "Specs.—The Comprehensive Food Service Purchasing & Specifications Manual." Cahner Books, Boston, Massachusetts.

Pirie, N. W. (1969). "Food Resources, Conventional and Novel." Penguin Books, London.

Potter, N. N. (1973). "Food Science." Avi Publ. Co., Westport, Connecticut.

Pyke, M. (1968). "Food and Society." John Murray, London.

Pyke, M. (1970). "Synthetic Food." John Murray, London.

Raw, I., Bromley, A., Pariser, E. R., and Vournakis, J. (1975). "What People Eat: An Introduction to Chemistry and Food Sciences," Rev. Exp. Ed. William Kaufmann, Los Altos, California.

Robinson, T. (1975). "The Organic Constitutents of Higher Plants," 3rd ed. Cordus Press, North Amherst, Massachusetts.

Steele, F., and Bourne, A., eds. (1975). "The Man/Food Equation." Academic Press, New York.

Stewart, G. F., and Amerine, M. A. (1982). "Introduction to Food Science and Technology," 2nd ed. Academic Press, New York.

Tarrant, J. R. (1980). "The Food Problem: National and International Policies." Wiley, New York.

Teranishi, R., ed. (1978). "Agricultural and Food Chemistry: Past, Present, Future." Avi Publ. Co., Westport, Connecticut.

Thomas, G. W., Curl, S. E., and Bennett, W. F., Sr. (1976). "Food and Fiber for a Changing World." Interstate Printers and Publishers, Danville, Illinois.

Trager, J. (1970). "The Foodbook." Grossman Publ., New York.

Walcher, D. N., Kretchmer, N., and Barnett, H. L. (1976). "Food, Man, and Society." Plenum, New York.

Wolff, I. A., ed. (1982). "CRC Handbook of Processing and Utilization in Agriculture," Vol. I. CRC Press, Boca Raton, Florida.

Wolff, I. A., ed. (1982). "CRC Handbook of Processing and Utilization in Agriculture," Vol. II, Part 1. CRC Press, Boca Raton, Florida.

Woollen, A. H., ed. (1956). "Food Industries Manual," 20th ed. Leonard Hill, London.

C. Foods—Specific

1. Fruits and Vegetables

Arthey, V. D. (1975). "Quality of Horticultural Products." Wiley, New York.

Beets, M. G. J. (1977). "Edible Nuts of the World." Horticultural Books, Stuart, Florida.

Biale, J. B. (1960). The postharvest biochemistry of tropical and subtropical fruits. *Adv. Food Res.* **10,** 293–354.

Braverman, J. D. S. (1949). "Citrus Products—Chemical Composition and Chemical Technology." Wiley (Interscience), New York.

Charley, V. L. S. (1977). "Black Currant Juice Processing Technology." Verlag Gunter Hempel, Braunschweig, Germany (in German).

Cruess, W. V. (1958). "Commercial Fruit and Vegetable Products," 3rd ed. McGraw-Hill, New York.

Dennis, C., ed. (1983). "Post-Harvest Pathology of Fruits and Vegetables." Academic Press, New York.

Goose, P. G., and Binsted, R. (1973). "Tomato Paste and Other Tomato Products." Food Trade Press, London.

Gould, W. A. (1974). "Tomato Production, Processing and Quality Evaluation." Avi Publ. Co., Westport, Connecticut.

Gutterson, M. (1971). "Fruit Processing." Noyes Data Corp., Park Ridge, New Jersey.

Gutterson, M. (1971). "Vegetable Processing." Noyes Data Corp., Park Ridge, New Jersey.

Hanson, L. P. (1976). "Commercial Processing of Fruits." Noyes Data Corp., Park Ridge, New Jersey.

Hulme, A. C. (1958). Some aspects of the biochemistry of apple and pear fruits. *Adv. Food Res.* **8,** 297–312.

Hulme, A. C., ed. (1970–1971). "The Biochemistry of Fruits and Their Products," 2 Vols. Academic Press, New York.

Joslyn, M. A., and Goldstein, J. L. (1964). Astringency of fruit and fruit products in relation to phenolic content. *Adv. Food Res.* **13,** 179–217.

Joslyn, M. A., and Ponting, M. D. (1951). Enzyme-catalyzed oxidative browning of fruit products. *Adv. Food Res.* **3,** 1–44.

Kefford, J. F. (1959). The chemical constituents of citrus fruits. *Adv. Food Res.* **2,** 259–296.

Luh, B. S., and Woodroof, J. G. (1975). "Commercial Vegetable Processing." Avi Publ. Co., Westport, Connecticut.

Paul, J. K. (1975). "Fruit and Vegetable Juice Processing." Noyes Data Corp., Park Ridge, New Jersey.

Ryall, A. L., and Lipton, W. J. (1972). "Handling, Transportation and Storage of Fruits and Vegetables." Avi Publ. Co., Westport, Connecticut.

Salunkhe, D. K. (1974). "Storage, Processing and Nutritional Quality of Fruits and Vegetables." CRC Press, Cleveland, Ohio.

Salunkhe, D. K., and Wu, M. T. (1974). Developments in technology of storage of fresh fruits and vegetables. *CRC Crit. Rev. Food Technol.* **5,** 15–54.

Smith, O. (1977). "Potatoes: Production, Storing and Processing." Avi Publ. Co., Westport, Connecticut.

Swiss Fruit Union (1982). "IXth International Congress of Fruit Juices, Munich, 1982." Swiss Fruit Union, Zurich, Switzerland.

Talburt, W. F., and Smith, O. (1975). "Potato Processing," 3rd ed. Avi Publ. Co., Westport, Connecticut.

Tressler, D. K., and Joslyn, M. A. (1971). "Fruit and Vegetable Juice Processing Technology," 2nd ed. Avi Publ. Co., Westport, Connecticut.

Tressler, D. K., and Woodroof, J. G. (1976). "Food Products Formulary," Vol. 3. Avi Publ. Co., Westport, Connecticut.

Van Buren, J. (1970). Current concepts on the texture of fruits and vegetables. *CRC Crit. Rev. Food Technol.* **1,** 5–24.

Walker, S. E., ed. (1972). "American Potato Yearbook." Am. Potato Soc., Scotch Plains, New Jersey.

Walker, S. E., ed. (1976). "American Potato Yearbook." Am. Potato Soc., Scotch Plains, New Jersey.

White, P. L., and Selvey, N., eds. (1974). "Nutritional Qualities of Fresh Fruits and Vegetables." Futura Publ. Co., Mt. Kisco, New York.

Woodroof, J. G. (1979). "Tree Nuts," 2nd ed. Avi Publ. Co., Westport, Connecticut.

Woodroof, J. G. (1980). "Coconuts: Production, Processing, Products." Avi Publ. Co., Westport, Connecticut.

Woodroof, J. G., and Luh, B. S. (1974). "Commercial Fruit Processing." Avi Publ. Co., Westport, Connecticut.

2. Meat, Fish, and Eggs

American Meat Institute Foundation (1960). "The Science of Meat and Meat Products." Freeman, San Francisco, California.

Bailey, A. J. (1972). The basis of meat texture. *J. Sci. Food Agric.* **23**, 995–1007.

Bate-Smith, E. C. (1948). The physiology and chemistry of rigor mortis, with special reference to the aging of beef. *Adv. Food Res.* **1**, 1–38,

Bendall, J. R. (1962). The structure and composition of muscle. *Recent Adv. Food Sci.* **1**, 58–67.

Bendall, J. R. (1969). "Muscles, Molecules, and Movement." Heinemann, London.

Borgstrom, G., ed. (1961–1965). "Fish as Food," 4 vols. Academic Press, New York.

Briskey, E. J. (1964). Etiological status and associated studies of pale, soft, exudative porcine musculature. *Adv. Food Res.* **13**, 90–178.

Briskey, E. J., and Fukazama, T. (1971). Myofibrillar proteins of skeletal muscle. *Adv. Food Res.* **19**, 279–360.

Briskey, E. J., Cassens, R. G., and Trauman, J. C. (1966). "The Physiology and Biochemistry of Muscle as a Food." Univ. of Wisconsin Press, Madison.

Briskey, E. J., Cassens, R. G., and Marsh, B. B., eds. (1970). "Physiology and Biochemistry of Muscle as Food," Vol. 2. Univ. of Wisconsin Press, Madison.

Brody, J. (1965). "Fishery By-Products Technology." Avi Publ. Co., Westport, Connecticut.

Carter, T. C., ed. (1968). "Egg Quality: A Study of the Hen's Egg." Oliver & Boyd, Edinburgh.

Cassens, R. G. (1972). "Muscle Biology," Vol. 1. Dekker, New York.

Cassens, R. G., and Cooper, C. C. (1971). Red and white muscle. *Adv. Food Res.* **19**, 1–74.

Cole, D. J. A., and Lawrie, R. A. (1975). "Meat." Avi Publ. Co., Westport, Connecticut.

Connell, J. J., ed. (1980). "Advances in Fish Science and Technology." Fishing News Books, Farnham, Surrey, England.

Connell, J. J., and Hardy, R. (1982). "Trends in Fish Utilization." Fishing News Books, Farnham, Surrey, England.

Donnely, T. H., Rongey, E. H., and Barsuko, V. J. (1966). Protein composition and functional properties of meat. *J. Agric. Food Chem.* **14**, 196–200.

Feeney, R. E., and Allison, R. G. (1969). "Evolutionary Biochemistry of Proteins. Homologous and Analogous Proteins from Avian Egg Whites, Blood Sera, Milk, and Other Substances." Wiley, New York.

Finch, R. (1970). Fish protein for human foods. *CRC Crit. Rev. Food Technol.* **1**, 519–580.

Forrest, J. C., Aberle, E. D., Heddrick, H. B., Judge, M. D., and Merkel, R. A. (1975). "Principles of Meat Science." Freeman, San Francisco, California.

Gillies, M. (1971). "Seafood Processing." Noyes Data Corp., Park Ridge, New Jersey.

Goll, D. E., Robson, R. M., and Stromer, M. H. (1977). Muscle proteins. *In* "Food Proteins" (J. R. Whitaker and S. R. Tannenbaum, eds.), pp. 121–174. Avi Publ. Co., Westport, Connecticut.

Gutcho, M. (1978). "Dairy Products and Eggs. Recent Developments." Noyes Data Corp., Park Ridge, New Jersey.

Hamm, R. (1960). Biochemistry of meat hydration. *Adv. Food Res.* **10**, 355–463.

Karmas, E. (1972). "Sausage Processing," Food Process. Rev. No. 24. Noyes Data Corp., Park Ridge, New Jersey.

Karmas, E. (1982). "Meat, Poultry and Seafood Technology. Recent Developments." Noyes Data Corp., Park Ridge, New Jersey.

Komarik, S. L., Tressler, D. K., and Long, L. (1974). "Food Products Formulary," Vol. 1. Avi Publ. Co., Westport, Connecticut.

Kramlich, W. E., Pearson, A. M., and Tauber, F. W. (1973). "Processed Meats." Avi Publ. Co., Westport, Connecticut.

Kreuzer, R. (1975). "Fishery Products." Fishing News (Books), West Byfleet, Surrey, England.

Lawrie, R. A. (1962). The conversion of muscle to meat. *Recent Adv. Food Sci.* **1,** 68–82.

Lawrie, R. A. (1966). "Meat Science." Pergamon, Oxford.

Levie, A. (1967). "The Meat Handbook," 2nd ed. Avi Publ. Co., Westport, Connecticut.

Long, L., Komarik, S. L., and Tressler, D. K. (1982). "Food Products Formulary Series," 2nd ed., Vol. I. Avi Publ. Co., Westport, Connecticut.

McLoughlin, J. V. (1969). Relationship between muscle biochemistry and properties of fresh and processed meats. *Food Manuf.* **44**(1), 36–40.

Martin, R. E., Flick, G. J., Hebard, C. E., and Ward, D. R. (1982). "Chemistry and Biochemistry of Marine Food Products." Avi Publ. Co., Westport, Connecticut.

Meade, T. L. (1971). Recent advances in the fish by-products industry. *CRC Crit. Rev. Food Technol.* **2,** 1–19.

Mountney, G. J. (1976). "Poultry Products Technology," 2nd ed. Avi Publ. Co., Westport, Connecticut.

Price, J. F., and Schweigert, B. S., eds. (1971). "The Science of Meat and Meat Products," 2nd ed. Freeman, San Francisco, California.

Romanoff, A. L., and Romanoff, A. J. (1949). "The Avian Egg." Wiley, New York.

Rust, R. E. (1976). "Sausage and Processed Meats Manufacturing." Am. Meat Inst., Washington, D.C.

Saffle, R. L. (1968). Meat emulsions. *Adv. Food Res.* **16,** 105–160.

Stadelman, W. J., and Cotterill, O. J., eds. (1973). "Egg Science and Technology." Avi Publ. Co., Westport, Connecticut.

Stansby, M. E. (1963). "Industrial Fisheries Technology." Van Nostrand-Reinhold, Princeton, New Jersey.

Stansby, M. E., ed. (1967). "Fish Oils, Their Chemistry, Technology, Stability, Nutritional Properties and Uses." Avi Publ. Co., Westport, Connecticut.

Vadehra, D. V., and Nath, K. R. (1973). Eggs as a source of protein. *CRC Crit. Rev. Food Technol.* **4,** 193–309.

Weiss, G. H. (1976). "Commercial Processing of Poultry." Noyes Data Corp., Park Ridge, New Jersey.

Whitaker, J. R. (1959). Chemical changes associated with aging of meat with emphasis on proteins. *Adv. Food Res.* **9,** 1–60.

Windsor, M., and Barlow, S. (1981). "Introduction to Fishery By-Products." Fishing News Books, Farnham, Surrey, England.

3. Dairy Products

Arbuckle, W. S. (1972). "Ice Cream," 2nd ed. Avi Publ. Co., Westport, Connecticut.

Brink, M. F., and Kritchevsky, D., eds. (1968). "Dairy Lipids and Lipid Metabolism." Avi Publ. Co., Westport, Connecticut.

Coulter, S. T., Jenness, R., and Geddes, W. F. (1951). Physical and chemical aspects of the production, storage and utility of dry milk products. *Adv. Food Res.* **3,** 45–106.

Davis, J. G. (1955). "A Dictionary of Dairying," 2nd ed. Leonard Hill, London.

Davis, J. G. (1965). "Cheese." Am. Elsevier, New York.

Davis, J. G. (1974). "Cheese," Vol. 4. Am. Elsevier, New York.

Fox, P. F. (1982). "Developments in Dairy Chemistry," Vol. 1. Appl. Sci. Publ., Barking, Essex, England.

Gillies, M. T. (1974). "Dehydration of Natural and Simulated Dairy Products," Food Technol. Rev. No. 15. Noyes Data Corp., Park Ridge, New Jersey.

Gillies, M. T. (1974). "Whey Processing and Utilization," Food Technol. Rev. No. 19. Noyes Data Corp., Park Ridge, New Jersey.

Gutcho, M. (1978). "Dairy Products and Eggs. Recent Developments." Noyes Data Corp., Park Ridge, New Jersey.

Hall, C. W., and Hedrick, T. I. (1971). "Drying of Milk and Milk Products," 2nd ed. Avi Publ. Co., Westport, Connecticut.

Harper, W. J., and Hall, C. W. (1976). "Dairy Technology and Engineering." Avi Publ. Co., Westport, Connecticut.

Henderson, J. L. (1971). "The Fluid Milk Industry," 3rd ed. Avi Publ. Co., Westport, Connecticut.

Jenness, R., and Patton, S. (1959). "Principles of Dairy Chemistry." Wiley, New York.

Kon, S. K. (1959). "Milk and Milk Products in Human Nutrition," FAO Nutr. Stud. No. 17. Food Agric. Organ., Rome.

Kosikowski, F. V. (1977). "Cheese and Fermented Milk Foods," 2nd ed. Edwards, Ann Arbor, Michigan.

Lampert, L. M. (1970). "Modern Dairy Products." Chem. Publ. Co., New York.

Lyster, R. L. J. (1972). Reviews of the progress of dairy science. Section C. Chemistry of milk proteins. *J. Dairy Res.* **39,** 279–318.

National Dairy Council (1968). "Newer Knowledge of Milk," 3rd ed. Natl. Dairy Council, Chicago, Illinois.

Packard, V. S. (1982). "Human Milk and Infant Formula." Academic Press, New York.

Reibold, G. W. (1972). "Swiss Cheese Varieties." Pfizer Milwaukee Operations, Milwaukee, Wisconsin.

Schormuller, J. (1968). The chemistry and biochemistry of cheese ripening. *Adv. Food Res.* **16,** 335–456.

Schwartz, M. E. (1973). "Cheese-Making Technology," Food Technol. Rev. No. 7. Noyes Data Corp., Park Ridge, New Jersey.

Scott, R. (1981). "Cheesemaking Practice." Appl. Sci. Publ., Barking, Essex, New England.

Simon, A. L. (1956). "Cheeses of the World." Faber & Faber, London.

Webb, B. H., Johnson, A. H., and Alford, J. A. (1974). "Fundamentals of Dairy Chemistry," 2nd ed. Avi Publ. Co., Westport, Connecticut.

Wilcox, G. (1971). "Milk, Cream, and Butter Technology," Food Process. Rev. No. 18. Noyes Data Corp., Park Ridge, New Jersey.

Wilcox, G. (1971). "Eggs, Cheese, and Yogurt Processing." Noyes Data Corp., Park Ridge, New Jersey.

4. Cereals, Oilseeds, and Legumes

Adrian, W. (1959). "And Thus Bread was Made from Stock and Blood." Ceres-Verlag, Bielefeld, Germany (in German).

Anonymous (1965). "From Wheat to Flour." Wheat Flour Institute, Chicago, Illinois.

Anonymous (1966). "Cottonseed and Its Products," 7th ed. National Cottonseed Products Association, Memphis, Tennessee.

Anonymous (1971). Grain sorghum research in Texas—1970. *Tex. Agric. Exp. Stn. Consolidated Prog. Rep.* pp. 2938–2949.

Araullo, E. V., De Padua, D. B., and Graham, M., eds. (1976). "Rice—Post Harvest Technology," IDRC-053e. International Development Centre, Ottawa, Ontario, Canada.

Aykroyd, W. R., and Doughty, J. (1970). "Wheat in Human Nutrition," FAO Nutr. Stud. No. 23. Food Agric. Organ., Rome.

Bailey, A. E. (1948). "Cottonseed and Cottonseed Products." Wiley (Interscience), New York.

Bailey, C. H. (1944). "The Constituents of Wheat and Wheat Products." Van Nostrand-Reinhold, Princeton, New Jersey.

Barber, S., Mitsuda, H., and Desikachar, H. S. R., eds. (1975). "Rice Report." Instituto de Agroquimica y Tecnologia de Alimentos, Valencia, Spain.

Bennion, E. B. (1967). "Breadmaking: Its Principles and Practice," 4th ed. Oxford Univ. Press, London and New York.

Bohn, R. M. (1957). "Biscuit and Cracker Production." American Trade Publishing Co., New York.

Brown, L. R., and Eckholm, E. P. (1975). "By Bread Alone." Pergamon, Oxford.

Bushuk, W., ed. (1976). "Rye: Production, Chemistry, and Technology." Am. Assoc. Cereal Chem., St. Paul, Minnesota.

Caldwell, B. E., Howell, R. W., Judd, R. W., and Johnson, H. W., eds. (1973). "Soybeans: Improvement, Production, and Uses." Am. Soc. Agron., Madison, Wisconsin.

Canadian International Grains Institute (1975). "Grains and Oilseeds, Handling, Marketing, Processing," 2nd ed. Can. Intern. Grains, Inst., Winnipeg, Manitoba.

Christensen, C. M., ed. (1974). "Storage of Cereal Grains and Their Products." Am. Assoc. Cereal Chem., St. Paul, Minnesota.

Christensen, C. M., and Kaufmann, H. H. (1969). "Grain Storage—The Role of Storage in Quality Loss." Univ. of Minnesota Press, Minneapolis.

Cook, A. H. (1962). "Barley and Malt: Biology, Biochemistry, and Technology." Academic Press, New York.

Daniels, R. (1970). "Modern Breakfast Cereal Processes." Noyes Data Corp., Park Ridge, New Jersey.

Daniels, R. (1974). "Breakfast Cereal Technology." Noyes Data Corp., Park Ridge, New Jersey.

De Clerk, J. (1957–1958). "A Textbook of Brewing" (transl. by K. Barton-Wright), Vols. I and II. Chapman & Hall, London.

De Renzo, D. J. (1975). "Bakery Products—Yeast Leavened," Food Technol. Rev. No. 200. Noyes Data Corp., Park Ridge, New Jersey.

De Renzo, D. J. (1975). "Doughs and Baked Goods; Chemical, Air and Nonleavened." Noyes Data Corp., Park Ridge, New Jersey.

Doggett, H. (1970). "Sorghum." Longmans, Green, New York.

Dunlap, F. L. (1945). "White versus Brown Flour." Wallace & Tiernan Co., Newark, New Jersey.

Fance, W. J. (1969). "Breadmaking and Flour Confectionary," 3rd ed. Avi Publ. Co., Westport, Connecticut.

Feistritzer, W. P., ed. (1975). "Cereal Seed Technology." FAO, Rome.

Findley, W. P. K., ed. (1971). "Modern Brewing Technology." Macmillan, New York.

Hehn, E. R., and Barmore, M. A. (1965). Breeding wheat for quality. *Adv. Agron.* **17,** 85–114.

Hind, H. L. (1950). "Brewing Science and Practice,' Vols. I and II. Chapman & Hall, London.

Hopkins, R. H., and Krause, B. (1947). "Biochemistry Applied to Malting and Brewing." Van Nostrand-Reinhold, Princeton, New Jersey.

Horder, T. J., Dodds, C., and Moran, T. (1954). "Bread: The Chemistry and Nutrition of Flour and Bread." Constable, London.

Hough, J. S., Briggs, D. E., and Stevens, R. (1971). "Malting and Brewing Science." Chapman & Hall, London.

Hough, J. S., Briggs, D. E., Stevens, R., and Young, T. W. (1982). "Malting and Brewing Science," 2nd ed., Vol. 2. Methuen, London.

Houston, D. F., ed. (1972). "Rice: Chemistry and Technology." Am. Assoc. Cereal Chem., St. Paul, Minnesota.

Houston, D. F., and Kohler, G. O. (1970). "Nutritional Properties of Rice." Food Nutr. Board, Natl. Res. Counc., Washington, D.C.

Huelsen, W. A. (1954). "Sweet Corn." Wiley (Interscience), New York.

Hummel, C. (1966). "Macaroni Products—Manufacture, Processing, and Packaging," 2nd ed. Food Trade Press, London.

Inglett, G. E., ed. (1970). "Corn: Culture, Processing, Products." Avi Publ. Co., Westport, Connecticut.

Inglett, G. E., ed. (1974). "Wheat: Production and Utilization." Avi Publ. Co., Westport, Connecticut.

Ingram, J. S. (1975). "Standards, Specifications and Quality Requirements for Processed Cassava Products." Tropical Products Institute, London.

Jacob, H. E. (1944). "Six Thousand Years of Bread, Its Holy and Unholy History." Doubleday, Garden City, New York.

Jugenheimer, R. W. (1976). "Corn: Improvement, Seed Production, and Uses." Wiley, New York.

Kay, D. E. (1979). "Food Legumes." Tropical Products Inst., London.

Kent, N. L. (1966). "Technology of Cereals with Special Reference to Wheat." Pergamon, Oxford.

Kent-Jones, D. W., and Amos, A. J. (1967). "Modern Cereal Chemistry," 6th ed. Food Trade Press, London.

Kent-Jones, D. W., and Mitchell, S. (1962). "The Practice and Science of Breadmaking," 3rd ed. Northern Publ. Co., Liverpool.

Kramer, J. K. G., Sauer, F. D., and Pidgen, W. J., eds. (1983). "High and Low Erucic Acid Rapeseed Oils: Production, Usage, Chemistry and Toxicological Evaluation." Academic Press, New York.

Kretovich, V. L., ed. (1965). "Biochemistry of Grain and of Breadmaking" (transl. from Russian). Israel Program for Scientific Translations, Jerusalem.

Kuprits, Ya. N. (1965). "Technology of Grain Processing and Provender Milling (transl. from Russian, 1967). U.S. Department of Commerce, Springfield, Virginia (TT 67-51273).

Lockwood, J. F. (1960). "Flour Milling," 4th ed. Northern Publ. Co., Liverpool.

Luh, B. S., ed. (1980). "Rice: Production & Utilization." Avi Publ. Co., Westport, Connecticut.

McCance, R. A., and Widdowson, E. M. (1956). "Breads, White and Brown: Their Place in Thought and Social History." Lippincott, Philadelphia, Pennsylvania.

McFarlane, W. D. (1970). "Industry-Sponsored Research on Brewing. Needs and Priorities." Brewing Industries Research Institute, Chicago, Illinois.

Mac Intyre, R., and Campbell, M. (1974). "Triticale." International Development Research Center, Ottawa, Ontario.

Manley, D. J. R. (1983). "Technology of Biscuits, Crackers, and Cookies." Ellis Horwood, Chichester, England.

Matz, S. A., ed. (1959). "The Chemistry and Technology of Cereals as Food and Feed." Avi Publ. Co., Westport, Connecticut (out of print).

Matz, S. A. (1968). "Cookie and Cracker Technology." Avi Publ. Co., Westport, Connecticut.

Matz, S. A. (1969). "Cereal Science." Avi Publ. Co., Westport, Connecticut.

Matz, S. A. (1970). "Cereal Technology." Avi Publ. Co., Westport, Connecticut.

Matz, S. A. (1972). "Bakery Technology and Engineering," 2nd ed. Avi Publ. Co., Westport, Connecticut.

Matz, S. A. (1976). "Snack Food Technology." Avi Publ. Co., Westport, Connecticut.

Mertz, E. T., ed. (1975). "High-Quality Protein Maize." Dowden, Hutchinson & Ross, Stroudsburg, Pennsylvania.

Neuman, M. P., and Pelshenke, P. F. (1954). "Bread Grain and Bread." Parey, Berlin (in German).

Pomeranz, Y. (1968). Relationship between chemical composition and breadmaking potentialities of wheat flour. Adv. Food Res. 16, 335–455.

Pomeranz, Y. (1970). Protein-enriched bread. CRC Crit. Rev. Food Technol. 1, 453–478.

Pomeranz, Y. (1971). Biochemical and functional changes in stored cereal grains. CRC Crit. Rev. Food Technol. 2, 45–80.

Pomeranz, Y., ed. (1971). "Wheat Chemistry and Technology," 2nd ed., Monogr. 3. Am. Assoc. Cereal Chem. St. Paul, Minnesota.

Pomeranz, Y. (1973). Food uses of barley. *CRC Crit. Rev. Food Technol.* **4**, 377–395.

Pomeranz, Y., ed. (1973). "Industrial Uses of Cereals." Am. Assoc. Cereal Chem., St. Paul, Minnesota.

Pomeranz, Y., ed. (1976–1984). "Advances in Cereal Science and Technology," Vols. I–VI. Am. Assoc. Cereal Chem., St. Paul, Minnesota.

Pomeranz, Y., ed. (1978). "Cereals '78: Better Nutrition for the World's Millions." Am. Assoc. Cereal Chem., St. Paul, Minnesota.

Pomeranz, Y., and MacMasters, M. M. (1970). Wheat and other cereal grains. *Kirk-Othmer's Encycl. Chem. Technol.,* 2nd ed., **22**, 253–308.

Pomeranz, Y., and Munck, L., eds. (1981). "Cereals: A Renewable Resource, Theory and Practice." Am. Assoc. Cereal Chem., St. Paul, Minnesota.

Pomeranz, Y., and Shellenberger, J. A. (1971). "Bread Science and Technology." Avi Publ. Co., Westport, Connecticut.

Preece, I. A. (1954). "Biochemistry of Brewing." Oliver & Boyd, Edinburgh.

Purseglove, J. W. (1972). "Tropical Crops: Monocotyledons," Vol. I. Wiley, New York.

Pyler, E. J. (1976). "Baking Science and Technology," 2 vols. Siebel Publ. Co., Chicago, Illinois.

Quisenberry, K. S., and Reitz, L. P., eds. (1967). "Wheat and Wheat Improvement." Am. Soc. Agron., Madison, Wisconsin.

Rohrlich, M., and Bruckner, G. (1966). "Cereals," Vol. I. Parey, Berlin (in German).

Scott, J. H. (1951). "Flour Milling Processes," 2nd ed. Chapman & Hall, London.

Sherman, H. C., and Pearson, C. S. (1942). "Modern Bread from the Viewpoint of Nutrition." Macmillan, New York.

Smith, A. K., and Circle, S. J. eds. (1978). "Soybeans: Chemistry and Technology. Proteins," 2nd ed., Vol. 1. Avi Publ. Co., Westport, Connecticut.

Smith, L. (1944). "Flour Milling Technology," 3rd ed. Northern Publ. Co., Liverpool, England.

Smith, W. H. (1972). "Biscuits, Crackers, & Cookies," 2 vols. Appl. Sci. Publ., Ltd., Barking, Essex, England.

Spicer, A., ed. (1975). "Bread: Social, Nutritional and Agriculture Aspects of Wheat Bread." Appl. Sci. Publ., London.

Storck, J., and Teague, W. D. (1952). "A History of Milling. Flour for Man's Bread." Univ. of Minnesota Press, Minneapolis.

Sultan, W. J. (1965). "Practical Baking." Avi Publ. Co., Westport, Connecticut.

Swanson, C. O. (1938). "Wheat and Flour Quality." Burgess, Minneapolis, Minnesota.

Tsen, C. C., ed. (1974). "Triticale: First Man-Made Cereal." Am. Assoc. Cereal Chem., St. Paul, Minnesota.

Wall, J. S., and Ross, W. M., eds. (1970). "Sorghum Production and Utilization." Avi Publ. Co., Westport, Connecticut.

Weiss, T. J. (1983). "Food Oils and Their Uses," 2nd ed. Avi Publ. Co., Westport, Connecticut.

Wiebe, G. A., ed. (1968). "Barley: Origin, Botany, Culture, Winterhardiness, Genetics, Utilization, Pests," Agric. Handb. No. 338. ARS-USDA, Washington, D.C.

Yamazaki, W. T., and Greenwood, C. T., eds. (1981). "Soft Wheat: Production, Breeding, Milling, and Uses." Am. Assoc. Cereal Chem., St. Paul, Minnesota.

5. *Miscellaneous*

Alikonis, J. J. (1979). "Candy Technology." Avi Publ. Co., Westport, Connecticut.

Amerine, M. A. (1971). Recent advances in enology. *CRC Crit. Rev. Food Technol.* **2**, 407–515.

Amerine, M. A., and Joslyn, M. A. (1970). "Table Wines—The Technology of their Production," 2nd ed. Univ. of California Press, Berkeley and Los Angeles.

Amerine, M. A., Berg, H. W., and Cruess, W. V. (1972). "Technology of Wine Making," 3rd ed. Avi Publ. Co., Westport, Connecticut.

Andersen, A. J. C., and Williams, P. N. (1965). "Margarine," 2nd ed. Pergamon, Oxford.

Anonymous (1972). "Gelatin and Gelling Agents," Proc. Symp. No. 13. British Food Manufacturing Research Association, Leatherhead, Surrey, England.

Association Scientifique Internationale du Cafe (1976). "7th International Colloquium on the Chemistry of Coffee." Paris (in English, French, German, and Spanish).

Bacila, M., Horecker, B. L., and Stoppani, A. O. M., eds. (1978). "Biochemistry and Genetics of Yeasts: Pure and Applied Aspects." Academic Press, New York.

Binsted, R., and Dewey, J. D. (1970). "Soup Manufacture," 3rd ed. Food Trade Press, London.

Binsted, R., and Dewey, J. D. (1971). "Pickle and Sauce Making," 3rd ed. Food Trade Press, London.

Bokuchava, M. A., and Skobelova, N. I. (1969). The chemistry and biochemistry of tea and tea manufacture. *Adv. Food Res.* **17**, 215–280.

Chatt, E. M. (1953). "Cocoa," Vol. III. Wiley (Interscience), New York.

Cook, A. H., ed. (1958). "The Chemistry and Biology of Yeasts." Academic Press, New York.

Cook, L. R. (1963). "Chocolate Production and Use." Magazines for Industry, New York.

Crane, E., ed. (1975). "Honey, A Comprehensive Survey." Crane, Russak & Co., New York.

Cruess, W. V. (1943). The role of microorganisms and enzymes in wine making. *Adv. Enzymol.* **3**, 349–386.

Eden, T. (1965). "Tea," 2nd ed. Longmans, Green, New York.

Formo, M. S., Jungermann, E., Norris, F. A., and Sonntag, N. V. (1979). "Bailey's Industrial Oil and Fat Products." Wiley, New York.

Forsyth, W. G. C., and Quesnel, V. S. (1963). The mechanism of cocoa curing. *Adv. Enzymol.* **25**, 457–492.

Garattini, S., Pagliadunga, S., and Scrimshaw, N. S. (1979). "Single-Cell Protein Safety for Animal and Human Feeding." Pergamon, Oxford.

Gillies, M. T. (1974). "Shortenings, Margarines, and Food Oils 1974," Food Technol. Rev. No. 10. Noyes Data Corp., Park Ridge, New Jersey.

Gillies, M. T. (1979). "Candies and Other Confections." Noyes Data Corp., Park Ridge, New Jersey.

Gray, W. D. (1970). The use of fungi as food and in food processing. *CRC Crit. Rev. Food Technol.* **1**, 225–329.

Guenther, E. (1948). "The Essential Oils," Vols. 1–6. Van Nostrand-Reinhold, Princeton, New Jersey (reprinted 1955).

Gutcho, M. H. (1973). "Textured Foods and Allied Products," Food Technol. Rev. No. 1. Noyes Data Corp., Park Ridge, New Jersey.

Gutcho, M. H. (1976). "Edible Oils and Fats. Recent Developments." Noyes Data Corp., Park Ridge, New Jersey.

Gutcho, M. H. (1977). "Fortified and Soft Drinks." Noyes Data Corp., Park Ridge, New Jersey.

Gutcho, M. H. (1979). "Alcoholic Beverage Processes." Noyes Data Corp., Park Ridge, New Jersey.

Hamilton, R. J., and Bhatti, A., eds. (1980). "Fats and Oils: Chemistry and Technology." Appl. Sci. Publ., Ltd., Barking, Essex, England.

Hendrickson, R. (1976). "The Great American Chewing Gum Food." Chilton Book Co., Radnor, Pennsylvania.

Johnson, J. C. (1977). "Yeast for Food and Other Purposes." Noyes Data Corp., Park Ridge, New Jersey.

Lees, R., and Jackson, B. (1973). "Sugar Confectionery and Chocolate Manufacture." Leonard Hill Books, Leonard Hill, London.

Lopez, A. (1974). "Manufacturing Cooking Oil, Salad Oil, and Margarine," Res. Bull. No. 96. Virginia Polytech. Inst., State University, Blacksburg.

Minifie, B. W. (1980). "Chocolate, Cocoa and Confectionery: Science and Technology," 2nd ed. Avi Publ. Co., Westport, Connecticut.

Parry, J. W. (1962). "Spices, Their Morphology, Histology and Chemistry." Chem. Publ. Co., New York.

Pederson, C. S. (1971). "Microbiology of Food Fermentations." Avi Publ. Co., Westport, Connecticut.

Phaff, H. J., Miller, M. W., and Mrak, E. M. (1978). "The Life of Yeasts," 2nd ed. Harvard Univ. Press, Cambridge, Massachusetts.

Pintauro, N. D. (1977). "Tea and Soluble Tea Products Manufacture." Noyes Data Corp., Park Ridge, New Jersey.

Pollock, J. R. A., ed. (1979). "Brewing Science," Vol. 1. Academic Press, New York.

Pollock, J. R. A., ed. (1981). "Brewing Science," Vol. 2. Academic Press, New York.

Prescott, S. C., and Dunn, C. G. (1959). "Industrial Microbiology." McGraw-Hill, New York.

Reed, G., and Peppler, H. J. (1973). "Yeast Technology." Avi Publ. Co., Westport, Connecticut.

Roberts, E. A. H. (1942). The chemistry of tea fermentation. *Adv. Enzymol.* **2,** 113–133.

Roelofsen, P. A. (1958). Fermentation, drying and storage of cacao beans. *Adv. Food Res.* **8,** 225–296.

Rose, A. H. (1961). "Industrial Microbiology." Butterworth, London.

Rose, A. H., ed. (1982). "Fermented Foods." Academic Press, New York.

Sanderson, G. W. (1972). The chemistry of tea and tea manufacturing. *Recent Adv. Phytochem.* **5,** 247–316.

Schwartz, M. E. (1974). "Confections and Candy Technology," Food Technol. Rev. No. 12. Noyes Data Corp., Park Ridge, New Jersey.

Shalleck, J. (1972). "Tea." Viking Press, New York.

Shurtleff, W., and Aoyagi, A. (1979). "The Book of Tempeh; Tempeh Production: The Book of Tempeh," Vol. II. New Age Foods Study Center, Lafayette, California.

Sivetz, M., and Desrosier, N. W. (1979). "Coffee Technology." Avi Publ. Co., Westport, Connecticut.

Sivetz, M., and Foote, H. E. (1963). "Coffee Processing Technology," Vols. 1 and 2. Avi Publ. Co. Westport, Connecticut.

Skinner, F. A., and Carr, J. G., eds. (1976). "Microbiology in Agriculture, Fisheries and Food." Academic Press, New York.

Solms, J., ed. (1973). "Functional Properties of Fats in Foods." Forster Verlag, Zurich, Switzerland.

Stahl, W. H. (1962). The chemistry of tea and tea manufacturing. *Adv. Food Res.* **11,** 201–262.

Steinkraus, K. H., ed. (1983). "Handbook of Indigenous Fermented Foods." Dekker, New York.

Tannenbaum, S. R., and Wang, D. I. C., eds. (1975). "Single Cell Protein," Vol. II. MIT Press, Cambridge, Massachusetts.

Ukers, W. H. (1976). "All about Coffee," 2nd ed. Gale Research Co., Book Tower, Detroit, Michigan.

Ward, A. G., and Courts, A. (1977). "Science and Technology of Gelatin." Academic Press, New York.

Webb, A. D., ed. (1974). "Chemistry of Winemaking," Adv. Chem. Ser. No. 137. Am. Chem. Soc., Washington, D.C.

Weiser, H. H. (1962). "Practical Food Microbiology and Technology." Avi Publ. Co., Westport, Connecticut.

Weiss, T. J. (1983). "Fatty Oils and Their Uses," 2nd ed. Avi Publ. Co., Westport, Connecticutt.
Woodroof, J. G., and Phillips, G. F. (1974). "Beverages: Carbonated and Non-Carbonated." Avi Publ. Co., Westport, Connecticut.

D. Food Components—Functional Properties

Aspinall, G. O., ed. (1983). "The Polysaccharides," Vol. 1. Academic Press, New York.
Aspinall, G. O., ed. (1983). "The Polysaccharides," Vol. 2. Academic Press, New York.
Banks, W., and Greenwood, C. T. (1976). "Starch and Its Constituents." Halsted Press, New York.
Becher, P. (1983). "Encyclopedia of Emulsion Technology," Vol. 1. Dekker, New York.
Birch, G. G., and Green, L. F., eds. (1973). "Molecular Structure and Function of Food Carbohydrate." Wiley (Halsted), New York.
Birch, G. G., and Parker, K. J., eds. (1979). "Sugar: Science and Technology." Appl. Sci. Publ., Barking, Essex, England.
Birch, G. G., Blakebrough, N., and Parker, K. J., eds. (1981). "Enzymes and Food Processing." Appl. Sci. Publ., Barking, Essex, England.
Cherry, J. P., ed. (1981). "Protein Functionality in Foods," ACS Symp. Ser. 147. Am. Chem. Soc., Washington, D.C.
Daussant, J., Vaughan, J., and Mosse, J., eds. (1983). "Seed Proteins." Academic Press, New York.
Davies, R., Birch, G. G., and Parker, K. J., eds. (1976). "Intermediate Moisture Foods." Appl. Sci. Publ., Barking, Essex, England.
Dickinson, E., and Stavinsky, G. (1982). "Colloids in Foods." Appl. Sci. Publ., Barking, Essex, England.
Drawert, F., ed. (1975). "Geruch und Geschmackstoffe." Verlag H. Carl, Nuremberg, West Germany (in German).
Dupuy, P., ed. (1982). "The Use of Enzymes in Food Technology." Librairies Lavoisier, Techniques et Documentation. Paris.
Fenema, O., ed. (1979). "Proteins at Low Temperatures." Am. Chem. Soc., Washington, D.C.
Fogarty, W. M., ed. (1983). "Microbial Enzymes and Biotechnology." Appl. Sci. Publ., Barking, Essex, England.
Fondu, M. (1982). "Food Additives Tables," Classes I–IV. Elsevier, Amsterdam.
Fox, P. F., and Condon, J. J., eds. (1982). "Food Proteins." Appl. Sci. Publ., Barking, Essex, England.
Furia, T. E., ed. (1980). "The Regulatory Status of Direct Food Additives." CRC Press, Boca Raton, Florida.
Glicksman, M. (1969). "Gum Technology in the Food Industry." Academic Press, New York.
Glicksman, M. (1982). "Food Hydrocolloids," Vol. I. CRC Press, Boca Raton, Florida.
Grant, R. A., ed. (1980). "Applied Protein Chemistry." Appl. Sci. Publ., Barking, Essex, England.
Gunstone, F. D., and Norris, F. A. (1983). "Lipids in Foods; Chemistry, Biochemistry and Technology." Pergamon, Oxford.
Gutcho, M. H. (1977). "Textured Protein Products." Noyes Data Corp., Park Ridge, New Jersey.
Hanson, L. P. (1974). "Vegetable Protein Processing," Food Technol. No. 16. Noyes Data Corp., Park Ridge, New Jersey.
Hough, C. A. M., Parker, K. J., and Vlitos, A. J. (1979). "Development in Sweeteners," Vol. 1. Appl. Sci. Publ., Barking, Essex, England.
Hudson, B. J. F., ed. (1982). "Developments in Food Proteins," Vol. 1. Appl. Sci. Publ., Barking, Essex, England.

Hudson, B. J. F., ed. (1983). "Developments in Food Proteins." Appl. Sci. Publ., Barking, Essex, England.

Iglesias, H. A., and Chirife, J. (1982). "Handbook of Food Isotherms: Water Sorption Parameters for Food and Food Components." Academic Press, New York.

Inglett, G. E., ed. (1972). "Symposium: Seed Proteins." Avi Publ. Co., Westport, Connecticut.

Johnson, J. C. (1979). "Emulsifiers and Emulsifying Technology." Noyes Data Corp., Park Ridge, New Jersey.

Johnson, J. C., ed. (1983). "Food Additives. Recent Developments." Noyes Data Corp., Park Ridge, New Jersey.

Koivistoinen, P., and Hyvönen, L., eds. (1980). "Carbohydrate Sweeteners in Foods and Nutrition." Academic Press, New York.

Lee, C. K., and Lindley, M. G., eds. (1982). "Developments in Food Carbohydrate," Vol. 3. Applied Sci. Publ., Barking, Essex, England.

Lineback, D. R., and Inglett, G. E., eds. (1982). "Food Carbohydrates." Avi Publ. Co., Westport, Connecticut.

Maltz, M. A. (1981). "Food Protein Supplements. Recent Advances." Noyes Data Corp., Park Ridge, New Jersey.

Markakis, P., ed. (1982). "Anthocyanins as Food Colors." Academic Press, New York.

Milner, M., Scrimshaw, N. S., and Wang, D. I. C. (1978). "Protein Resources and Technology. Status and Research Needs." Avi Publ. Co., Westport, Connecticut.

Morton, I. D., and MacLeod, A. J., eds. (1982). "Food Flavors," Part A. Elsevier, Amsterdam.

Neukom, H., and Pilnik, W., eds. (1980). "Gelling and Thickening Agents in Foods." Forster Verlag, Zurich, Switzerland (in German).

Norton, G., ed. (1977). "Plant Proteins." Butterworth, London.

Noyes, R. (1969). "Protein Food Supplements," Food Process. Rev. No. 3. Noyes Data Corp., Park Ridge, New Jersey.

Ory, R. L., and Angelo, A. J., eds. (1977). "Enzymes in Food and Beverage Processing." Am. Cheese Soc., Washington, D.C.

Pancoast, H. M., and Junk, W. R. (1980). "Handbook of Sugars," 2nd ed. Avi Publ. Co., Westport, Connecticut.

Parry, D. A. D., and Creamer, L. K., eds. (1980). "Fibrous Proteins: Scientific, Industrial and Medical Aspects," Vol. 2. Academic Press, New York.

Pintauro, N. D. (1979). "Protein Food Supplements," Food Process. Rev. No. 3. Noyes Data Corp., Park Ridge, New Jersey.

Pour-El, A., ed. (1979). "Functionality and Protein Structure," ACS Symp. 92. Am. Chem. Soc., Washington, D.C.

Pyke, M. (1981). "Food Science and Technology," 4th ed. Crane, Russak & Co., New York.

Robinson, T. (1980). "Organic Constituents of Higher Plants," 4th ed. Cordus Press, North Amherst, Massachusetts.

Rockland, L. B., and Stewart, G. F., eds. (1981). "Water Activity: Influences on Food Quality." Academic Press, New York.

Sanford, P. A., and Laskin, A., eds. (1977). "Extracellular Microbial Polysaccharides." Am. Chem. Soc., Washington, D.C.

Sanford, P. A., and Matsuda, K., eds. (1979). "Fungal Polysaccharides." Am. Chem. Soc., Washington, D.C.

Schwimmer, S. (1981). "Source Book of Food Enzymology." Avi Publ. Co., Westport, Connecticut.

Shallenberger, R. S. (1982). "Advanced Sugar Chemistry: Principles of Sugar Stereochemistry." Avi Publ. Co., Westport, Connecticut.

Shallenberger, R. S., and Birch, G. G. (1975). "Sugar Chemistry." Avi Publ. Co., Westport, Connecticut.

Supran, M. K., ed. (1978). "Lipids as a Source of Flavor." Am. Chem. Soc., Washington, D.C.

Taylor, R. J. (1980). "Food Additives." Wiley, New York.

Troller, J. A., and Christian, J. H. B. (1978). "Water Activity and Food." Academic Press, New York.

VanStraten, S., ed. (1983). "Volatile Compounds in Foods: Quantitative Data," Vol. 1. CIVO Inst., TNO, Zeist, The Netherlands.

Whistler, R. L., and Hymowitz, T. (1979). "Guar: Agronomy, Production, Industrial Use, and Nutrition." Purdue Univ. Press, Lafayette, Indiana.

Whistler, R. L., Be Miller, J. N., and Paschall, E. F. (1983). "Starch: Chemistry and Technology," 2nd ed. Academic Press, New York.

Whitaker, J. R., and Fujimaki, M., eds. (1980). "Chemical Deterioration of Proteins." Am. Chem. Soc., Washington, D.C.

Whitaker, J. R., and Tannenbaum, S. R., eds. (1977). "Food Proteins." Avi Publ. Co., Westport, Connecticut.

Wieland, H. (1972). "Enzymes in Food Processing and Products," Food Process. Rev. No. 23. Noyes Data Corp., Park Ridge, New Jersey.

Winter, R. (1972). "A Consumer's Dictionary of Food Additives." Crown Publ., New York.

Wiseman, A., ed. (1983). "Topics in Enzyme and Fermentation Biotechnology," Vol. 7. Wiley, New York.

E. Physical, Sensory, and Quality Parameters

Amerine, M. A., Pangborn, R. M., and Roessler, E. B. (1965). "Principles of Sensory Evaluation of Food." Academic Press, New York.

Ammore, J. E. (1970). "Molecular Basis of Odor." Thomas, Springfield, Illinois.

Anonymous (1973). "Sensory Evaluation of Appearance of Materials," ASTM Spec. Tech. Publ. 454. Am. Soc. Test. Mater., Philadelphia, Pennsylvania.

Anonymous (1974). "The Role of Rheology in Foods." Dechema Monogr., Frankfurt, West Germany (in German).

APT, C. M., ed. (1978). "Flavor: Its Chemical, Behavioral, and Commercial Aspects." Westview Press, Boulder, Colorado.

Birch, G. G., Brennan, J. G., and Parker, K. J., eds. (1977). "Sensory Properties of Foods." Appl. Sci. Publ., Barking, Essex, England.

Borenstein, B., and Bunnell, R. H. (1966). Carotenoids: Properties, occurrence, and utilization in foods. Adv. Food Res. 15, 195–276.

Bourne, M. C. (1983). "Food Texture and Viscosity: Concepts and Measurements." Academic Press, New York.

Caul, J. F. (1957). The profile method of flavor analysis. Adv. Food Res. 7, 1–37.

Charalambous, G., and Inglett, G. E., eds. (1978). "Flavor of Foods and Beverages: Chemistry and Technology." Academic Press, New York.

Charalambous, G., and Inglett, G., eds. (1983–1984). "Instrumental Analysis of Foods," Vols. 1 and 2. Academic Press, New York.

Charalambous, G., and Katz, I., eds. (1976). "Phenolic, Sulfur, and Nitrogen Compounds in Food Flavors," ACS Symp. Ser. 26. Am. Chem. Soc., Washington, D.C.

Charm, S. E. (1962). The nature and role of fluid consistency in food engineering applications. Adv. Food Res. 11, 356–435.

deMan, J. M., Voisey, P. W., Rasper, V., and Stanley, D., eds. (1976). "Rheology and Texture in Food Quality." Avi Publ. Co., Westport, Connecticut.

Drake, B., and Johannson, B. (1969). "Sensory Evaluation of Food," Vols. 1 and 2. Academic Press, New York.

Ellis, B. H. (1961). "A Guide Book for Sensory Testing." Continental Can Co., Chicago, Illinois.

Ellis, G. P. (1959). The Maillard reaction. *Adv. Carbohydr. Chem.* **14**, 63–81.

Eskin, N. A. M. (1979). "Plant Pigments, Flavors and Textures: The Chemistry and Biochemistry of Selected Compounds." Academic Press, New York.

Farkas, L., Gabor, M., and Kallay, F., eds. (1975). "Topics in Flavonoid Chemistry and Biochemistry." Am. Elsevier, New York.

Fazzalari, F. A., ed. (1978). "Compilation of Odor and Taste Threshold Values Data." Am. Soc. Test. Mater., Philadelphia, Pennsylvania.

Finney, E. E. (1973). "Measurement Techniques for Quality Control of Agricultural Products." Am. Soc. Agric. Eng., St. Joseph, Missouri.

Furia, T. E., and Bellanca, N., eds. (1971). "CRC Fernaroli's Handbook of Flavor Ingredients, International Edition." Chem. Rubber Publ. Co., Cleveland, Ohio.

Gaffney, J. J. (1976). "Quality Detection in Foods." Am. Soc. Agric. Eng., St. Joseph, Missouri.

Geissman, T. A., ed. (1962). "The Chemistry of Flavonoid Compounds." Pergamon, Oxford.

Goodwin, T. W., ed. (1965). "Chemistry and Biochemistry of Plant Pigments." Academic Press, New York.

Gould, R. F., ed. (1966). "Flavor Chemistry," Adv. Chem. Ser. 56. Am. Chem. Soc., Washington, D.C.

Gould, W. A. (1977). "Food Quality Evaluation." Avi Publ. Co., Westport, Connecticut.

Harborne, J. B. (1964). "Biochemistry of Phenolic Compounds." Academic Press, New York.

Harborne, J. B. (1967). "Comparative Biochemistry of Flavonoids." Academic Press, New York.

Herschdoerfer, S. M. (1967). "Quality Control in the Food Industry," Vol. 1, Academic Press, New York.

Herschdoefer, S. M. (1968). "Quality Control in the Food Industry," Vol. 2. Academic Press, New York.

Herschdoerfer, S. M. (1972). "Quality Control in the Food Industry," Vol. 3. Academic Press, New York.

Hornstein, I., ed. (1966). "Flavor Chemistry," Adv. Chem. Ser. No. 67. Am. Chem. Soc., Washington, D.C.

Hunter, R. S., ed. (1973). "The Measurement of Appearance." Hunter Associates Lab., Fairfax, Virginia.

Institut International de Froid (1975). "Current Studies on the Thermophysical Properties of Foodstuffs." The IIF Institute, Paris.

Irving, G. W., and Hoover, S. R., eds. (1965). "Food Quality—Effects of Production Practices and Processing," Publ. No. 77. Am. Assoc. Adv. Sci., Washington, D.C.

Kramer, A., and Szczesniak, A. S., eds. (1973). "Texture Measurements of Foods." Reidel Publ., Dordrecht, The Netherlands.

Kramer, A., and Twigg, B. A. (1966). "Fundamentals of Quality Control for the Food Industry," 2nd ed. Avi Publ. Co., Westport, Connecticut.

Kramer, A., and Twigg, B. A., eds. (1970). "Quality Control for the Food Industry," 3rd ed., Vol. 1. Avi Publ. Co., Westport, Connecticut.

Kramer, A., and Twigg, B. A., eds. (1973). "Quality Control for the Food Industry," 3rd ed., Vol. 2. Avi Publ. Co., Westport, Connecticut.

Larmond, E. (1970). Methods for sensory evaluation of foods. *Can., Dept. Agric. Publ.* **1284.**

MacKinney, G., and Little, A. C. (1962). "Color of Foods." Avi Publ. Co., Westport, Connecticut.

Macrae, R., ed. (1982). "HPLC in Food Analysis." Academic Press, New York.

Maga, J. A. (1974). Bread Flavor. *CRC Crit. Rev. Food Technol.* **5**, 55–142.

Mathew, A. G., and Parpia, H. A. B. (1971). Food browning as a polyphenol reaction. *Adv. Food Res.* **19**, 75–145.

Matz, S. A. (1962). "Food Texture." Avi Publ. Co., Westport, Connecticut.

Meiselman, H. L. (1972). Human taste perception. *CRC Crit. Rev. Food Technol.* **3**, 89–119.
Mohsenin, N. N. (1970). "Physical Properties of Plant and Animal Materials," Vol. 1. Gordon & Breach, New York.
Mohsenin, N. N. (1981). "Thermal Properties of Foods and Agricultural Products." Gordon & Breach, New York.
Moskowitz, H. R., Scharf, B., and Stevens, J. C. (1974). "Sensation and Measurement." Reidel Publ., Dordrecht, The Netherlands.
Muller, H. G. (1973). "An Introduction to Food Rheology." Crane, Russack & Co., New York.
Ohloff, G., and Thomas, A. F. (1971). "Gustation and Olfaction." Academic Press, New York.
Peleg, M., and Bagley, E. B., eds. (1983). "Physical Properties of Food." Avi Publ. Co., Westport, Connecticut.
Pfaffman, C., ed. (1969). "Olfaction and Taste," Vol. III. Rockefeller Univ. Press, New York.
Pomeranz, Y., and Meloan, C. E. (1978). "Food Analysis: Theory and Practice." Avi Publ. Co., Westport, Connecticut.
Powers, J. J., and Moskowitz, H. R., eds. (1976). "Correlating Sensory Objective Measurements—New Methods for Answering Old Problems." Am. Soc. Test. Mater., Philadelphia, Pennsylvania.
Prentice, J. H. (1972). Rheology and texture of dairy products. *J. Text. Stud.* **3**, 415–458.
Reynolds, T. M. (1963). 1965. Chemistry of non enzymic browning. I. *Adv. Food Res.* **12**, 1–52.
Rha, C., ed. (1974). "Theory, Determination, and Control of Physical Properties of Food Materials." Reidel Publ., Dordrecht, The Netherlands.
Roberts, T. A., and Skinner, F. A., eds. (1983). "Food Microbiology; Advances and Prospects." Academic Press, New York.
Rose, A. H., ed. (1977). "Alcoholic Beverages and Potable Spirits." Academic Press, New York.
Rose, A. H., ed. (1978). "Primary Products of Metabolism." Academic Press, New York.
Rose, A. H., ed. (1979). "Microbial Biomass." Academic Press, New York.
Rose, A. H., ed. (1979). "Secondary Products of Metabolism." Academic Press, New York.
Rose, A. H., ed. (1981). "Microbial Enzymes and Bioconversions." Academic Press, New York.
Rose, A. H., ed. (1982). "Fermented Foods." Academic Press, New York.
Rose, A. H., ed. (1982). "Microbial Biodeterioration." Academic Press, New York.
Rose, A. H., ed. (1983). "Food Microbiology." Academic Press, New York.
Scanlan, R. A., ed. (1977). "Flavor Quality: Objective Measurement," ACS Symp. Ser. 51. Am. Chem. Soc., Washington, D.C.
Scott-Blair, G. W. (1945). "A Survey of General and Applied Rheology." Pitman, London.
Scott-Blair, G. W., ed. (1953). "Foodstuffs: Their Plasticity, Fluidity, and Consistency." Wiley (Interscience), New York.
Scott-Blair, G. W. (1958). Rheology in food research. *Adv. Food Res.* **8**, 1–61.
Scott-Blair, G. W., and Reiner, M. (1957). "Agricultural Rheology." Routledge & Kegan Paul, London.
Shankaranarayana, M. L., Raghvan, B., Abraham, K. O., and Natarajan, C. P. (1973). Volatile sulfur compounds in food flavors. *CRC Crit. Rev. Food Technol.* **4**, 395–435.
Sherman, P. (1970). "Industrial Rheology." Academic Press, New York.
Sherman, P., ed. (1979). "Food Texture and Rheology." Academic Press, London.
Sjöström, L. B. (1972). "The Flavor Profile." A. D. Little, Cambridge, Massachusetts.
Society of Chemical Industry (1960). "Texture in Foods," Monogr. No. 7. S.C.I., London.
Sone, T. (1972). "Consistency of Foodstuffs." Reidel Publ., Dordrecht, The Netherlands.
Stahl, W. H. (1973). "Compilation of Odor and Taste Threshold Values Data." Am. Soc. Test. Mater., Philadelphia, Pennsylvania.
Teranishi, R., Hornstein, I., Issenberg, P., and Wick, E. L. (1971). "Flavor Research—Principles and Techniques." Dekker, New York.

Torrey, S., ed. (1980). "Fragrances and Flavors: Recent Developments." Noyes Data Corp., Park Ridge, New Jersey.

Van Nazer, J. R., Lyons, J. W., Kim, K. Y., and Colwell, R. E. (1963). "Viscosity and Flow Measurement—A Laboratory Handbook of Rheology." Wiley (Interscience), New York.

Whorlow, R. W. (1980). "Rheological Techniques." Wiley, New York.

Zotterman, Y., ed. (1965). "Olfaction and Taste." Pergamon, Oxford.

F. Composition, Nutrition, Enrichment, and Labeling

Abelson, P. H., ed. (1975). "Food: Policies, Economics, Nutrition, and Research." Am. Assoc. Adv. Sci., Washington, D.C.

Adams, C. F. (1975). "Nutritive Value of American Foods in Common Units," Agric. Handb. No. 456. ARS–USDA, Washington, D.C.

Adrian, J. (1974). Nutritional and physiological consequences of the Maillard reaction. *World Rev. Nutr. Diet.* **19,** 71–122.

Agricultural Research Council (1975). "Food and Nutrition Research—Report of the ARC/MRC Committee." Am. Elsevier, New York.

American Medical Association (1974). "Nutrients in Processed Foods." Publ. Sci. Group, Acton, Massachusetts.

American University of Beirut (1963). "Food Composition: Tables for Use in the Middle East," Publ. No. 20. Div. Food Technol. Nutr., American University, Beirut, Lebanon.

Anonymous (1970). "White House Conference on Food Nutrition and Health," Final Rep. Supt. of Documents, U.S. Govt. Printing Office, Washington, D.C.

Anonymous (1970). "A Select List of Books on Food Science and Technology." National College of Food Technology, University of Reading, Weybridge, Surrey, England.

Anonymous (1974). "Guidelines for a National Nutrition Policy." Supt. of Documents, U.S. Govt. Printing Office, Washington, D.C.

Arlin, M. T. (1972). "The Science of Nutrition." Macmillan, New York.

Aylward, F., and Jul, M. (1975). "Protein and Nutrition Policy in Low-Income Countries." Wiley, New York.

Baker, E. A., and Foskett, D. J. (1958). "Bibliography of Food." Butterworth, London.

Barlow, S. M., and Stansby, M. E., eds. (1982). "Nutritional Evaluation of Long-Chain Fatty Acids in Fish Oil." Academic Press, New York.

Bender, A. E. (1967). "Food Processing and Nutrition." Academic Press, New York.

Bender, A. E. (1968). "Dietetic Foods." Chem. Publ. Co., New York.

Bender, A. E. (1976). "Dictionary of Nutrition and Food Technology." Chem. Publ. Co., New York.

Berg, A., Scrimshaw, N., and Call, D. (1973). "Nutrition National Development and Planning." MIT Press, Cambridge, Massachusetts.

Birch, G. G., Green, L. F., and Plaskett, L. G., eds. (1972). "Health and Food." Wiley, New York.

Block, R. J., and Mitchell, H. H. (1946–1967). The correlation of the amino acid composition of proteins with their nutritive value. *Nutr. Abstr. Rev.* **16,** 249–278.

Bodwell, C. E., ed. (1977). "Evaluation of Proteins for Humans." Avi Publ. Co., Westport, Connecticut.

Borenstein, B. (1971). Rational and technology of food fortification with vitamins, minerals, and amino acids. *CRC Crit. Rev. Food Technol.* **2,** 171–186.

Borgström, G. A. (1973). "Man, Food and Disease. Food and Population—A Critical Appraisal." Stanford Medical School, Palo Alto, California.

Boyd, E. M. (1973). "Toxicity of Pure Foods." CRC Press, Cleveland, Ohio.

Burton, B. T. (1976). "The Heinz Handbook of Nutrition." McGraw-Hill, New York.

Carter, R. (1964). "Your Food and Your Health." Harper & Row, New York.

Catsimpoolas, N., ed. (1977). "Immunological Aspects of Foods." Avi Publ. Co., Westport, Connecticut.

Chaney, M. S., and Ros, M. L. (1966). "Nutrition," 7th ed. Houghton, Boston, Massachusetts.

Chicago Nutrition Association (1975). "Nutrition References and Book Reviews." Chicago Nutrition Association, Chicago, Illinois.

Clydesdale, F. M., ed. (1979). "Food Science and Nutrition: Current Issues and Answers." Prentice-Hall, Englewood Cliffs, New Jersey.

Clydesdale, F. M., and Francis, F. J. (1977). "Food, Nutrition & You." Prentice-Hall, Englewood Cliffs, New Jersey.

Crampton, E., and Lloyd, L. (1959). "Fundamentals of Nutrition." Freeman, San Francisco, California.

Davidson, S., and Passmore, R. (1969). "Human Nutrition and Dietetics." Williams & Wilkins, Baltimore, Maryland.

Davidson, S., Meiklejohn, A. P., and Passmore, R. (1961). "Human Nutrition and Dietetics." Williams & Wilkins, Baltimore, Maryland.

Fleck, H. (1976). "Introduction to Nutrition," 3rd ed. Macmillan, New York.

Fleck, H., and Munves, E. (1965). "Introduction to Nutrition." Macmillan, New York.

Food and Agriculture Organization (1957). "Protein Requirements," Nutr. Stud. No. 16. FAO, Rome.

Food and Agriculture Organization (1970). "Amino-Acid Content of Foods and Biological Data on Protein." FAO, United Nations, New York.

Food and Agriculture Organization (1974). "Food and Nutrition Terminology," Terminol. Bull. No. 28. FAO, Rome.

Food and Agriculture Organization of the United Nations (1976). Amino-Acid Content of Foods and Biological Data on Proteins," FAO Nutr. Stud. No. 24. FAO, Rome (3rd printing).

Food and Agriculture Organization/World Health Organization (1965). "Protein Requirements," WHO Tech. Rep. Ser. No. 301 or FAO Nutr. Meet. Rep. 37. FAO/WHO, Geneva, Switzerland.

Food and Agriculture Organization/World Health Organization (1973). "Energy and Protein Requirements," WHO Tech. Rep. Ser. 522. FAO/WHO, Geneva, Switzerland.

Fox, B. A., and Cameron, A. G. (1961). "A Chemical Approach to Food and Nutrition." Univ. of London Press, London.

Fox, F. W. (1966). "Studies on the Chemical Composition of Food Commonly Used in Southern Africa." S.A. Inst. Med. Res., Johannesburg.

Friedman, M., ed. (1975). "Nutritional Quality of Foods and Feeds," Part 1. Dekker, New York.

Friedman, M., ed. (1975). "Protein Nutritional Quality of Foods and Feeds," Part 2. Dekker, New York.

Furda, I., ed. (1983). "Unconventional Sources of Dietary Fiber: Physiological and in Vitro Functional Properties." Am. Chem. Soc., Washington, D.C.

Goodhart, R. S., and Shils, M. E. (1975). "Modern Nutrition in Health and Disease," 5th ed. Lea & Febiger, Philadelphia, Pennsylvania.

Gunderson, F. L., Gunderson, H. W., and Ferguson, E. R., Jr. (1963). "Food Standards and Definitions in the United States." Academic Press, New York.

Hansen, R. G., Wyse, B. W., and Sorenson, A. W. (1981). "Nutrition Quality Index of Foods." Avi Publ. Co., Westport, Connecticut.

Harris, R. S., and Karmas, E., eds. (1975). "Nutritional Evaluation of Food Processing," 2nd ed. Avi Publ. Co., Westport, Connecticut.

Heaton, K. W., ed. (1982). "Dietary Fibre: Current Developments and Importance to Health." Food & Nutrition Press, Westport, Connecticut.

Hollingsworth, D., and Russell, M., eds. (1973). "Nutritional Problems in a Changing World." Wiley, New York.

Hui, Y. H. (1979). "United States Food Laws, Regulations and Standards." Wiley, New York.

Hulse, J. H., and Laing, E. M. (1974). "Nutritive Value of Triticale Protein." International Development Research Centre, Ottawa, Ontario, Canada.

Inglett, G. E., ed. (1983). "Nutritional Bioavailability of Zinc." Am. Chem. Soc., Washington, D.C.

Inglett, G. E., and Falkehag, S. I., eds. (1979). "Dietary Fibers: Chemistry and Nutrition." Academic Press, New York.

International Food Information Service (1975). "Food Science and Technology Books," Food Annot. Bibliogr. No. 47. Commonwealth Bureau of Dairy Science and Technology, Shinfield, Reading, England.

Jeannes, A., and Hodge, J., eds. (1975). "Physiological Aspects of Food Carbohydrates." Am. Chem. Soc., Washington, D.C.

Johnston, B. F., and Greaves, J. F., eds. (1969). "Manual on Food and Nutrition Policy," FAO Nutr. Stud. No. 22. FAO, Rome.

Kies, C., ed. (1982). "Nutritional Bioavailability of Iron." Am. Chem. Soc., Washington, D.C.

Kreutler, P. A. (1980). "Nutrition in Perspective." Prentice-Hall, Englewood Cliffs, New Jersey.

Lang, K. (1970). "Biochemistry of Nutrition," 2nd rev. ed. Steinkopff Verlag, Darmstadt, Germany (in German).

Leung, A. Y. (1980). "Encyclopedia of Common Natural Ingredients Used in Foods, Drugs, and Cosmetics." Wiley, New York.

Leung, W. T. W., Busson, F., and Jardin, C. (1968). "Food Composition Tables for Use in Africa." Food Agric. Organ., Rome.

Liener, I. E. (1969). "Toxic Constituents of Plant Foodstuffs." Academic Press, New York.

Liener, I. E., ed. (1980). "Toxic Constituents of Plant Foodstuffs," 2nd ed. Academic Press, New York.

Livingston, G. E., Moshy, R. J., and Chang, C. M., eds. (1982). "The Role of Food Product Development in Implementing Dietary Guidelines." Food & Nutrition Press, Westport, Connecticut.

McCance, R. A., and Widdowson, E. M. (1960). "The Composition of Foods," Med. Res. Counc. Spec. Rep. Ser. No. 297. H. M. Stationery Office, London.

Mayer, J. (1972). "Human Nutrition—Its Physiological, Medical, and Social Aspects." Thomas, Springfield, Illinois.

Mayer, J., ed. (1973). "U.S. Nutrition Policies in the Seventies." Freeman, San Francisco, California.

Miller, D. F., ed. (1958). "Composition of Cereal Grains and Forages," Publ. No. 585. Natl. Acad. Sci.—Natl. Res. Counc., Washington, D.C.

Milner, M., ed. (1969). "Protein Enriched Cereal Foods for World Needs." Am. Assoc. Cereal Chem., St. Paul, Minnesota.

Milner, M., ed. (1975). "Nutritional Improvement of Food Legumes by Breeding." Wiley, New York.

Minor, L. J. (1983). "Nutritional Standards." Avi Publ. Co., Westport, Connecticut.

Muller, H. G., and Tobin, G. (1980). "Nutrition and Food Processing." Avi Publ. Co., Westport, Connecticut.

Munck, L. (1972). Improvements of nutritional value in cereals. *Hereditas* **72**, 1–128.

National Academy of Sciences (1975). "World Food and Nutrition Study. Enhancement of Food Production for the United States," Report of the Board on Agriculture and Renewable Resources. Natl. Acad. Sci., Washington, D.C.

National Academy of Sciences (1975). "Technology of Fortification of Foods." Natl. Acad. Sci., Washington, D.C.

National Academy of Sciences (1976). "Genetic Improvement of Seed Proteins." Natl. Acad. Sci., Washington, D.C.

National Academy of Sciences (1977). "World Food and Nutrition Study. The Potential Contributions of Research." Natl. Acad. Sci., Washington, D.C.

National Academy of Sciences (1980). "Recommended Dietary Allowances," 9th ed. NAS, Washington, D.C.

National Academy of Sciences (1980). "Towards Healthful Diets. Food and Nutrition Board." NAS, Washington, D.C.

National Academy of Sciences—National Research Council (1959). "Evaluation of Protein Nutrition," Publ. No. 711. NAS—NRC, Washington, D.C.

National Academy of Sciences—National Research Council (1974). "Recommended Dietary Allowances," 8th rev. ed. NAS/NRC, Washington, D.C.

National Academy of Sciences—National Research Council (1974). "Improvement of Protein Nutriture." Committee on Amino Acids, Food, and Nutrition Board, NAS/NRC, Washington, D.C.

Nelson, O. E. (1969). Genetic modification of protein quality in plants. *Adv. Agron.* **21,** 171–194.

Nutrition Foundation (1976). "Present Knowledge in Nutrition," 4th ed. Nutr. Found., Office of Education, Washington, D.C.

Orr, M. L., and Watt, B. K. (1957). Amino acid content of foods. *U.S., Dep. Agric., Home Econ. Res. Rept.* **4.**

Ory, R. L., ed. (1981). "Antinutrients and Natural Toxicants in Foods." Food & Nutrition Press, Westport, Connecticut.

Packard, V. S., Jr. (1976). "Processed Foods and the Consumer: Additives, Labeling, Standards, and Nutrition." Univ. of Minnesota Press, Minneapolis.

Paul, A. A., and Southgate, D. A. T. (1978). "The Composition of Foods." Elsevier/North-Holland Biomedical Press, Amsterdam.

Pearson, P. B., and Greenwell, J. R., eds. (1980). "Nutrition, Food & Man. An Interdisciplinary Perspective." Univ. of Arizona Press, Tucson.

Pennington, J. A. (1976). "Dietary Nutrient Guide." Avi Publ. Co., Westport, Connecticut.

Perkins, E. G., and Visek, W. J., eds. (1983). "Dietary Fats and Health," Am. Oil Chem. Soc., Champaign, Illinois.

Pike, R. L., and Brown, M. L. (1967). "Nutrition: An Integrated Approach." Wiley, New York.

Pintauro, N. D. (1975). "Nutrition Technology of Processed Foods." Noyes Data Corp., Park Ridge, New Jersey.

Platt, B. S. (1962). "Tables of Representative Values of Foods Commonly Used in Tropical Countries." H. M. Stationery Office, London.

Rathman, D. M., Stockton, J. R., and Melnick, D. (1970). Dynamic utilization of recent nutritional findings. *CRC Crit. Rev. Food Technol.* **1,** 331–378.

Rechcigl, M., Jr. (1973). "Man, Food and Nutrition." Chem. Rubber Publ. Co., Cleveland, Ohio.

Rechcigl, M., Jr. (1975). "World Food Problem: A Selective Bibliography of Reviews." CRC Press, Cleveland, Ohio.

Rechcigl, M., Jr., ed. (1982). "Handbook of Nutritive Value of Processed Food," Vol. 1. CRC Press, Boca Raton, Florida.

Robinson, C. H. (1968). "Fundamentals of Normal Nutrition." Macmillan, New York.

Rutgers University (1971). "Food Stability and Open Dating." Rutgers University, New Brunswick, New Jersey.

Scrimshaw, N. S., and Altschul, A. M., eds. (1971). "Amino Acid Fortification of Protein Foods." MIT Press, Cambridge, Massachusetts.

Scrimshaw, N. S., and Behar, M., eds. (1976). "Nutrition and Agricultural Development: Significance and Potential for the Tropics." Plenum, New York.

Silverstone, T. ed. (1976). "Appetite and Food Intake." Heyden, London.

Souci, S. W., Fachmann, W., and Kraut, H. (1974). "Composition of Foods: Nutritional Tables." Wiss. Verlagsges., Stuttgart, Germany (in German).

Spiller, G. A., ed. (1978). "Topics in Dietary Fiber Research." Plenum, New York.

Spiller, G. A., and Amen, F. J., eds. (1976). "Fiber in Human Nutrition." Plenum, New York.

Stare, F. J., and McWilliams, M. (1977). "Living Nutrition," 2nd ed. Wiley, New York.

Underwood, B. A., ed. (1983). "Nutrition Intervention Strategies in National Development." Academic Press, New York.

U.S. Department of Agriculture (1977). "Nutritive Value of Foods," Home and Garden Bull. No. 72. Supt. of Documents, USDA-ARS, U.S. Govt. Printing Office, Washington, D.C.

Waller, G. R., and Feather, M. S., eds. (1983). "The Maillard Reaction in Foods and Nutrition." Am. Chem. Soc., Washington, D.C.

Waterlow, J. C. (1970). Human protein requirements and malnutrition. In "Evaluation of Novel Protein Products" (A. E. Bender, R. Kihlberg, B. Lofquist, and L. Munck, eds.). Pergamon, Oxford.

Watt, B. K., and Merrill, A. L. (1963). Composition of foods. *U.S., Dep. Agric., Agric. Handb.* **8.**

Wayler, T. J., and Klein, R. S. (1965). "Applied Nutrition." Macmillan, New York.

Wilson, E. D., Fisher, K. H., and Fuqua, M. E. (1965). "Principles of Nutrition," 2nd ed. Wiley, New York.

Winkoff, B., ed. (1978). "Nutrition and National Policy." MIT Press, Cambridge, Massachusetts.

Wittwer, S. H. (1975). "World Food and Nutrition Study: Enhancement of Food Production for the United States." Natl. Acad. Sci., Washington, D.C.

Wu Leung, W. T., Butrum, R. R., and Chang, F. H. (1972). "Food Composition Table for Use in East Asia and Selected Bibliography on East Asian Food Composition Table." National Institutes of Health, Nutrition Program, Atlanta, Georgia.

Index

Y

Z

FOOD SCIENCE AND TECHNOLOGY

A SERIES OF MONOGRAPHS

Maynard A. Amerine, Rose Marie Pangborn, and Edward B. Roessler, PRINCIPLES OF SENSORY EVALUATION OF FOOD. 1965.

S. M. Herschdoerfer, QUALITY CONTROL IN THE FOOD INDUSTRY. Volume I — 1967. Volume II — 1968. Volume III — 1972.

Hans Riemann, FOOD-BORNE INFECTIONS AND INTOXICATIONS. 1969.

Irvin E. Leiner, TOXIC CONSTITUENTS OF PLANT FOODSTUFFS. 1969.

Martin Glicksman, GUM TECHNOLOGY IN THE FOOD INDUSTRY. 1970.

L. A. Goldblatt, AFLATOXIN. 1970.

Maynard A. Joslyn, METHODS IN FOOD ANALYSIS, second edition. 1970.

A. C. Hulme (ed.), THE BIOCHEMISTRY OF FRUITS AND THEIR PRODUCTS. Volume 1 — 1970. Volume 2 — 1971.

G. Ohloff and A. F. Thomas, GUSTATION AND OLFACTION. 1971.

C. R. Stumbo, THERMOBACTERIOLOGY IN FOOD PROCESSING, second edition. 1973.

Irvin E. Liener (ed.), TOXIC CONSTITUENTS OF ANIMAL FOODSTUFFS. 1974.

Aaron M. Altschul (ed.), NEW PROTEIN FOODS: Volume 1, TECHNOLOGY, PART A — 1974. Volume 2, TECHNOLOGY, PART B — 1976. Volume 3, ANIMAL PROTEIN SUPPLIES, PART A — 1978. Volume 4, ANIMAL PROTEIN SUPPLIES, PART B — 1981.

S. A. Goldblith, L. Rey, and W. W. Rothmayr, FREEZE DRYING AND ADVANCED FOOD TECHNOLOGY. 1975.

R. B. Duckworth (ed.), WATER RELATIONS OF FOOD. 1975.

Gerald Reed (ed.), ENZYMES IN FOOD PROCESSING, second edition. 1975.

A. G. Ward and A. Courts (eds.), THE SCIENCE AND TECHNOLOGY OF GELATIN. 1976.

John A. Troller and J. H. B. Christian, WATER ACTIVITY AND FOOD. 1978.

A. E. Bender, FOOD PROCESSING AND NUTRITION. 1978.

D. R. Osborne and P. Voogt, THE ANALYSIS OF NUTRIENTS IN FOODS. 1978.

Marcel Loncin and R. L. Merson, FOOD ENGINEERING: PRINCIPLES AND SELECTED APPLICATIONS. 1979.

Hans Riemann and Frank L. Bryan (eds.), FOOD-BORNE INFECTIONS AND INTOXICATIONS, second edition. 1979.

N. A. Michael Eskin, PLANT PIGMENTS, FLAVORS AND TEXTURES: THE CHEMISTRY AND BIOCHEMISTRY OF SELECTED COMPOUNDS. 1979.

J. G. Vaughan (ed.), FOOD MICROSCOPY. 1979.

J. R. A. Pollock (ed.), BREWING SCIENCE, Volume 1 — 1979. Volume 2 — 1980.

Irvin E. Liener (ed.), TOXIC CONSTITUENTS OF PLANT FOODSTUFFS, second edition. 1980.

J. Christopher Bauernfeind (ed.), CAROTENOIDS AS COLORANTS AND VITAMIN A PRECURSORS:
TECHNOLOGICAL AND NUTRITIONAL APPLICATIONS. 1981.

Pericles Markakis (ed.), ANTHOCYANINS AS FOOD COLORS. 1982.

Vernal S. Packard, HUMAN MILK AND INFANT FORMULA. 1982.

George F. Stewart and Maynard A. Amerine, INTRODUCTION TO FOOD SCIENCE AND TECHNOLOGY, second edition. 1982.

Malcolm C. Bourne, FOOD TEXTURE AND VISCOSITY: CONCEPT AND MEASUREMENT. 1982.

R. Macrae (ed.), HPLC IN FOOD ANALYSIS. 1982.

Héctor A. Iglesias and Jorge Chirife, HANDBOOK OF FOOD ISOTHERMS: WATER SORPTION PARAMETERS FOR FOOD AND FOOD COMPONENTS. 1982.

John A. Troller, SANITATION IN FOOD PROCESSING. 1983.

Colin Dennis (ed.), POST-HARVEST PATHOLOGY OF FRUITS AND VEGETABLES. 1983.

P. J. Barnes (ed.), LIPIDS IN CEREAL TECHNOLOGY. 1983.

George Charalambous (ed.), ANALYSIS OF FOODS AND BEVERAGES: MODERN TECHNIQUES. 1984.

David Pimentel and Carl W. Hall, FOOD AND ENERGY RESOURCES. 1984.

Joe M. Regenstein and Carrie E. Regenstein, FOOD PROTEIN CHEMISTRY: AN INTRODUCTION FOR FOOD SCIENTISTS. 1984.

R. Paul Singh and Dennis R. Heldman, INTRODUCTION TO FOOD ENGINEERING. 1984.

Maximo C. Gacula, Jr., and Jagbir Singh, STATISTICAL METHODS IN FOOD AND CONSUMER RESEARCH. 1984.

S. M. Herschdoerfer (ed.), QUALITY CONTROL IN THE FOOD INDUSTRY, second edition. Volume 1—1984.

Y. Pomeranz, FUNCTIONAL PROPERTIES OF FOOD COMPONENTS. 1985.

In preparation

S. M. Herschdoerfer (ed.), QUALITY CONTROL IN THE FOOD INDUSTRY, second edition. Volume 2—1985. Volume 3—1986. Volume 4—1987.

Herbert Stone and Joel L. Sidel, SENSORY EVALUATION PRACTICES. 1985.

Aaron M. Altschul and Harold L. Wilcke (eds.), NEW PROTEIN FOODS: VOLUME 5, SEED STORAGE PROTEINS. 1985.